Urban Sustainability

The Urban Sustainability Book Series is a valuable resource for sustainability and urban-related education and research. It offers an inter-disciplinary platform covering all four areas of practice, policy, education, research, and their nexus. The publications in this series are related to critical areas of sustainability, urban studies, planning, and urban geography.

This book series aims to put together cutting-edge research findings linked to the overarching field of urban sustainability. The scope and nature of the topic are broad and interdisciplinary and bring together various associated disciplines from sustainable development, environmental sciences, urbanism, etc. With many advanced research findings in the field, there is a need to put together various discussions and contributions on specific sustainability fields, covering a good range of topics on sustainable development, sustainable urbanism, and urban sustainability. Despite the broad range of issues, we note the importance of practical and policy-oriented directions, extending the literature and directions and pathways towards achieving urban sustainability.

The series will appeal to urbanists, geographers, planners, engineers, architects, governmental authorities, policymakers, researchers of all levels, and to all of those interested in a wide-ranging overview of urban sustainability and its associated fields. The series includes monographs and edited volumes, covering a range of topics under the urban sustainability topic, which can also be used for teaching materials.

Lorenzo Diana · Francesco Sommese ·
Simona Colajanni · Gaetano Sciuto
Editors

Greening and Resilience

Strategies, Approaches and Best Practices for the Built Environment

Springer

Editors
Lorenzo Diana
Department of Civil, Building
and Environmental Engineering
University of Naples Federico II
Naples, Italy

Francesco Sommese
Department of Civil, Building
and Environmental Engineering
University of Naples Federico II
Naples, Italy

Simona Colajanni
Department of Architecture
University of Palermo
Palermo, Italy

Gaetano Sciuto
Department of Civil Engineering
and Architecture
University of Catania
Catania, Italy

ISSN 2731-6483 ISSN 2731-6491 (electronic)
Urban Sustainability
ISBN 978-981-92-0418-2 ISBN 978-981-92-0419-9 (eBook)
https://doi.org/10.1007/978-981-92-0419-9

This Springer imprint is published by the registered company Springer Nature Singapore Pte Ltd.
The registered company address is: 152 Beach Road, #21-01/04 Gateway East, Singapore 189721, Singapore

Preface

The volume *Greening and Resilience*. Approaches and Best Practices for the Built Environment arises from the growing need to promote resilience in order to improve the health of the urban environment and to address the challenges associated with rapid urbanization and climate change. In this context, greening strategies play a central role in redefining the built environment, contributing to the mitigation of heat waves, the management of stormwater, the reduction of environmental stress, and the overall enhancement of urban liveability. Nature-Based Solutions (NbS), in particular, emerge as integrated tools capable of simultaneously addressing environmental, social, and energy-related challenges, while strengthening the resilience of cities and territories.

In recent years, governments, institutions, and territorial policies have increasingly promoted incentives, guidelines, and operational measures aimed at greening the built environment, acting at both the urban and building scales. Within this framework, the volume offers an updated overview of current measures, methodological tools, and innovative approaches related to the presence and role of green strategies in the built environment, fostering a dialogue between scientific research, design experimentation, and practical applications. The guiding aim is to offer an interdisciplinary contribution to the understanding of how green infrastructures and NbS can enhance climate resilience, urban health, life quality, and citizens' well-being, with implications also for the social sphere.

The book is structured into three parts, organized according to a coherent progression in terms of scale of intervention and level of investigation, ranging from climate risk management strategies to urban and building-scale applications.

Greening Strategies for Climate Risk Management

The first part addresses greening as a key strategy for climate risk management, with particular emphasis on the role of Nature-Based Solutions in adaptation and mitigation processes. The contributions explore collaborative governance frameworks, co-creation processes, and participatory validation of NbS, highlighting the socio-cultural and inclusive dimensions of resilience-oriented strategies. Alongside these perspectives, the section examines territorial and planning approaches focused on green infrastructure and the resilience of urban and peri-urban contexts, with specific attention to Mediterranean environments and inner areas in transition. A further set of chapters is devoted to the development of indicators, monitoring tools, remote sensing techniques, and environmental performance assessments, culminating in applied studies on green roofs, green walls, and urban green spaces as technical and ecological devices for reducing climate-related impacts.

Green Interventions at the Urban Scale: Approaches and Methods

The second part focuses on green transformations at the urban scale, investigating design approaches, analytical methodologies, and co-creation processes in contemporary cities. The contributions address strategies for sustainable urban transitions, the role of participation, and the integration of Nature-Based Solutions in complex contexts, including city-port systems and climate-transitioning urban environments. Particular attention is given to the experiential and biophilic dimensions of urban spaces, outdoor environmental comfort, and the perceptual quality of public environments. From a methodological perspective, this part gathers studies on thermal comfort assessment, environmental simulation, decision-support tools, and microclimatic analysis, applied to green and blue infrastructures, pocket green spaces, and open spaces in Mediterranean and historic urban contexts. Overall, the chapters demonstrate how urban greening interventions can significantly improve microclimatic conditions, environmental performance, and urban well-being.

Green Interventions at the Building Scale

The third part narrows the focus to the building scale, examining the role of greening interventions in enhancing the environmental and energy performance of buildings and their associated open spaces. The contributions adopt multi-scale perspectives that connect the building, neighborhood, and urban contexts, with particular reference to the cooling potential and climate mitigation effects of green roofs and green facades. The part also explores site-specific design methodologies, bioclimatic retrofitting strategies for existing architecture, and integrated interventions in educational and built environments aimed at increasing climate resilience. Specific attention is devoted to performance-based assessments grounded in field monitoring, energy simulations, and microclimatic analyses, as well as to the management and maintenance phases of vertical greenery systems and technical green solutions.

Taken together, the three parts provide a comprehensive and coherent framework that links strategies, methods, and best practices across different scales of the built environment, highlighting how greening represents a crucial design and policy lever for enhancing resilience, sustainability, and the health of contemporary urban environments.

Naples, Italy — Lorenzo Diana
Naples, Italy — Francesco Sommese
Palermo, Italy — Simona Colajanni
Catania, Italy — Gaetano Sciuto

Contents

About the Editors

Lorenzo Diana Assistant Professor (RTD-B) in Architectural Engineering (CEAR-08/A) at the Department of Civil, Building and Environmental Engineering, University of Naples Federico II (UNINA).

Ph.D. in "Architecture, Engineering and Urban Planning," his expertise is mainly referred to the assessment of qualities in existing buildings and urban contexts. His main research activities are related to the regeneration and reuse of existing buildings (public housing and cultural heritage buildings), the parametric reading of buildings, the definition of analysis indicators on the concept of transformability, and the relationship between typology, construction, and environmental aspects. His current research activities at UNINA focus on analyzing disused monumental hospitals in the city of Naples where he proposes regeneration strategies based both on the integrated sustainable retrofitting of buildings and the accessibility of internal courtyards and cloisters. He has been from November 2023 till February 2026 the Principal Investigator of the National Research Program "Greenwork: an interdisciplinary framework for urban health and urban resilience enhancement based on greening strategies on buildings and open spaces." The project aims to define a supporting framework for developing effective greening strategies in existing built environments, particularly addressing large-scale buildings and open spaces, with a specific emphasis on greening techniques applicable at both the building and neighborhood scales. The project examines the applicability of greening strategies, typically designed for new constructions, to existing urban contexts together with the definition of protocols for the assessment of the transformabilty and the technical feasibility for the application of green roofs and facades. This approach aligns with a broader vision aimed at increasing green in densely built-up areas, highlighting its benefit for environmental quality, public health, and urban resilience.

Author of approximately 80 scientific publications, he is currently responsible for several academic administrative activities such as the Coordination of the Thesis Workshop in Building Engineering-Architecture and the Organization of some modules within the orientation of high school students' program, and research projects such as the Collaboration Agreement with ASL NAPOLI 1 and ASL

NAPOLI 3 on the analysis, classification, evaluation, and definition of intervention guidelines for the regeneration of buildings for collective use like abandoned monumental hospitals in central areas.

Francesco Sommese Building Systems Engineer and Ph.D. in Civil Systems Engineering, cum Laude, for a thesis entitled "Biomimetic strategies for the design of climate adaptive building envelopes: a problem-based methodological approach."

Post-doctoral researcher in Architectural Engineering (CEAR-08/A) at the Department of Civil, Building and Environmental Engineering (DICEA), University of Naples Federico II.

His research focuses on biomimetic and climate-adaptive strategies for building envelopes, greening strategies for the built environment as nature-based solutions, as well as regenerative design and advanced building technologies.

He is a speaker at national and international conferences and co-supervisor of undergraduate theses. He has received awards for his research activity. He is the author of numerous articles published in international scientific journals, as well as author and editor of books, and reviewer for international journals.

He is an adjunct lecturer in Architectural Engineering at the Polytechnic and Basic Sciences School of the University of Naples Federico II.

He is assistant to the expert panel for the evaluation of research quality at ANVUR (National Agency for the Evaluation of Universities and Research Institutes), the agency of the Italian Ministry of Universities.

Simona Colajanni Associate Professor in Architectural Engineering (CEAR-08/A) at the Department of Architecture, University of Palermo, since 2010.

Graduated in Architecture and earned a Ph.D. in Building Engineering: Techniques of Planning, Production, and Building Recovery. A member of the International Ph.D. Program in Architecture, Arts, and Planning at the University of Palermo, she has supervised numerous Ph.D. theses and over 150 degree dissertations.

Deputy Coordinator of the Second-Level Master's Program in Innovative Materials and Techniques for Sustainable Construction (2nd, 3rd, and 4th editions) and an experienced lecturer in building construction and technology disciplines. Co-founder of PADesign—Passive Approach Design s.r.l., an academic spin-off specializing in energy-efficient building solutions for both new constructions and retrofitting. Co-inventor of multiple patented technological innovations arising from research projects.

Currently serves as the Delegate for Scientific Laboratories at the Department of Architecture and as Curator of the Design Cabinet Collection of the Royal Application School for Engineers and Architects. Previously, she managed departmental spaces and the D'ARCH Library.

Member of the Research Quality Assurance (AQ) Committee of the Department of Architecture.

Member of the Group of Evaluation Experts (GEV) for the Architecture Disciplinary Area, conducting assessment activities for the 2020-2024 Research Quality Evaluation (VQR) campaign regarding scientific outputs submitted by evaluated institutions on behalf of the National Agency for the Evaluation of Universities and Research Institutes (ANVUR).

Her research focuses on Sustainable Design and Innovative Materials, the application of Nature-Based Solutions (NBS) to existing architecture, including the development of advanced technological systems, natural passive systems applications, and bio-based materials derived from Sicilian agri-food waste.

An active participant in national and international research projects, she also serves on various scientific committees. Author of approximately 120 publications in national and international journals and conference proceedings.

Gaetano Sciuto Associate Professor of Architectural Engineering (CEAR-08/A), Department of Civil Engineering and Architecture, University of Catania, since 2007.

He received the Master's Degree in Civil Engineering, building section, at the University of Catania.

He received the Ph.D. Degree in Restoration of the building and environmental heritage.

He is member of the Ph.D. teaching staff in Catania.

He was vice-head of the Department of Civil Engineering and Architecture, University of Catania.

He was Coordinator of the Degree Program in Building Engineering-Architecture, University of Catania.

At the University of Catania, he has taught many subjects in the disciplinary sector of Buildings construction and techniques.

He has supervised numerous Ph.D. theses and over 200 degree dissertations.

He has coordinated research on the following topics: use procedures of the non-destructive instruments; application of non-destructive-test for stony materials of Eastern Sicily; historical buildings refurbishment; building typologies of Eastern Sicily towns; materials and traditional constructive techniques; relationship between technique and architecture; design of transferable housing.

Currently, his research interests focus on bioclimatic and echo-sustainable design; building envelopes and energy saving in existing buildings, including the integration of greening systems; characterization of performance and innovation in use of natural materials.

The publishing activity resulting from the research activity consisted of the publication of about 150 scientific publications: monographs and papers in national and international journals and conferences.

Greening and Resilience: Perspectives for the Built Environment

Francesco Sommese and Lorenzo Diana

Abstract Climate change calls for a profound revision of design and planning paradigms for the built environment, requiring approaches capable of integrating mitigation, adaptation, and the strengthening of urban resilience. This contribution examines the role of greening strategies at both urban and building scales as systemic levers to enhance environmental performance and support the sustainable transformation of cities. In particular, this study frames the concept of green transformability of the built environment as the capacity of the built environment to evolve through the progressive integration of green and blue infrastructure, Nature-Based Solutions, and vegetation-based architectural systems. The chapter highlights how greening strategies contribute to microclimatic regulation, stormwater management, the mitigation of urban heat island effects, and the enhancement of ecosystem services, generating measurable environmental, social, and energy-related benefits. It also discusses the main limitations and implementation challenges, including governance issues, maintenance requirements, and performance assessment complexities. Within the framework of European policies, the contribution interprets Nature-Based Solutions as mechanisms for integrating mitigation and adaptation and as key components in the transition from sustainable design to regenerative design. The study argues that the adoption of systemic, multiscale, and performance-based approaches represents a necessary condition for strengthening urban resilience and guiding the built environment towards long-term regenerative outcomes.

Keywords Nature-based solutions · Greening strategies · Built environment · Urban resilience · Regenerative design

F. Sommese (✉) · L. Diana
Department of Civil, Building and Environmental Engineering, University of Naples Federico II, Naples, Italy
e-mail: francesco.sommese@unina.it

L. Diana et al. (eds.), *Greening and Resilience*, Urban Sustainability,
https://doi.org/10.1007/978-981-92-0419-9_1

1 Introduction

Climate change represents a systemic and multidimensional challenge for the built environment, requiring a deep rethinking of conventional urban and architectural design paradigms. Scientific evidence indicates rising average global temperatures, an intensification of urban heat island effects, increasing frequency and severity of extreme climatic events, as well as disruptions to hydrological cycles and a progressive loss of urban biodiversity (Benaomar & Outzourhit, 2024; Lazzarini & Luttenberger, 2025). These phenomena directly affect the functionality, safety, and environmental quality of cities, placing growing pressure on buildings, urban spaces, and existing infrastructure (Diana et al., 2024).

In this context, the built environment assumes a central and dual role. On the one hand, the construction and building sector is responsible for a substantial share of greenhouse gas emissions, considering both operational and embodied carbon (IEA, 2019); on the other hand, it is highly exposed to the impacts of climate change.

However, the built environment offers the setting where consistent and effective actions can be implemented. Its complexity highlights the need for systemic and integrated actions capable of addressing both the causes and the consequences of climate change simultaneously. Sectoral or purely technology-driven approaches have proven insufficient when confronted with the interconnected nature of urban systems. By contrast, the integration of mitigation and adaptation strategies within a coherent framework strengthens urban space resilience, understood as the ability of urban systems to absorb climatic disturbances, adapt to evolving conditions and maintain essential functions over time. This integrated system of actions should act at various scales, from urban to building, and be composed of tangible and intangible greening interventions.

Within this perspective, the effectiveness of holistic approaches emerges clearly, as they conceptualise the built environment as a complex socio-ecological system characterised by dynamic interactions between buildings, open spaces, infrastructure, natural components, and social actors. In such approaches, greening strategies play a structural rather than an ancillary role. Urban greenery, that can be implemented through green and blue infrastructure, Nature-based solutions (NBS), and the integration of vegetation at the building scale, has been shown to contribute measurably to microclimatic regulation, stormwater management, the mitigation of urban heat island effects, and the overall environmental and social quality of urban spaces (Brears, 2023; European Commission, 2013; Veerkamp et al., 2021).

Within this conceptual framework, the notion of green transformability of the built environment refers to the capacity of urban and architectural systems to undergo progressive transformation through the systemic integration of greening strategies and ecological solutions. This transformability implies an evolution of design and planning processes aimed at making buildings and cities adaptive, resilient, and capable of actively contributing to the restoration of ecosystem services.

The objective of this chapter is to critically examine how greening strategies, applied at both urban and building scales, can enhance urban resilience while supporting the transition from conventional sustainability-oriented approaches towards regenerative design. By framing greening as a multiscale and systemic design driver, the chapter provides a theoretical basis for understanding its role within contemporary urban sustainability agendas and European policy frameworks.

2 Urban Greening Strategies and Resilience

At the urban scale, greening strategies constitute a core operational dimension through which resilience objectives can be translated into spatial and planning practices. Rather than referring to isolated green elements, urban greening encompasses a set of coordinated and complementary strategies that integrate vegetation, water systems, and ecological processes into the urban fabric, contributing to the adaptive capacity of cities under climate stress. Within contemporary urban sustainability research, these strategies are increasingly framed as green and blue infrastructure systems and NBS, designed to function as climate-responsive urban infrastructure (Andersson et al., 2019).

From a strategic perspective, urban greening can be articulated into several main typologies. Large-scale green and blue infrastructures, such as urban parks, metropolitan green belts, urban forests, restored river corridors, and floodable landscapes, play a fundamental role in regulating urban microclimates, managing stormwater, and providing spatial continuity for ecological processes (Sommese, 2024). These elements operate at the city or metropolitan scale and are particularly effective in addressing systemic climate risks, including heatwaves and pluvial flooding.

Complementing these large-scale interventions, distributed and fine-grained greening strategies—including street trees, green squares, permeable public spaces, and linear green corridors—act at the neighbourhood and district scale, directly influencing local thermal comfort and environmental quality. Their spatial diffusion allows for incremental implementation within consolidated urban areas, contributing to resilience through redundancy and spatial distribution. A further strategic layer is represented by urban ecological networks, which aim to connect fragmented green spaces into coherent systems, enhancing biodiversity, ecosystem functionality and long-term adaptability of the urban environment.

Across these typologies, the primary targets of urban greening strategies include the reduction of urban heat island effects, the regulation of surface runoff and stormwater, the enhancement of ecosystem services, and the improvement of environmental quality in public spaces. These targets are closely aligned with European climate adaptation and biodiversity policies, which promote greening as a multifunctional and cost-effective approach to urban resilience. Importantly, greening strategies are increasingly expected to deliver simultaneous environmental, social and spatial benefits, reinforcing the systemic nature of resilience-building processes. As growing awareness of climate change advances, greening measures increasingly align with adaptation needs, complementing mitigation-oriented strategies.

The advantages of urban greening strategies are widely documented in scientific literature (Gherri, 2024; Ma et al., 2025; Ng et al., 2012; Smalls-Mantey & Montalto, 2021). Empirical evidence demonstrates their effectiveness in reducing air and surface temperatures, increasing evapotranspiration and shading, improving water retention and infiltration, and supporting urban biodiversity (Lazzarini & Luttenberger, 2025). Beyond environmental performance, urban greening contributes to social resilience by enhancing thermal comfort, public health, and the liveability of urban spaces (Cuce et al., 2025; Meerow & Stults, 2016). When implemented as integrated systems rather than isolated interventions, these strategies show a high multifunctional degree and long-term efficiency compared to conventional grey infrastructure.

However, the deployment of urban greening strategies also involves significant challenges. Spatial constraints in dense urban contexts, competing land-use priorities, long-term maintenance requirements, and governance fragmentation often limit their effectiveness (Sommese & Diana, 2025). In addition, the lack of standardised performance indicators and long-term monitoring frameworks complicates the assessment of resilience outcomes (Sommese & Diana, 2025). Without a systemic planning approach, there is a risk that greening interventions remain symbolic or fragmented, failing to contribute meaningfully to urban resilience.

For these reasons, the effectiveness of urban greening in enhancing resilience depends on its integration within a multiscale and coordinated planning framework, in which different greening strategies interact with the built environment, infrastructure systems, and social dynamics. When conceived as interconnected components of urban systems, greening strategies play a crucial role in strengthening resilience and advancing the green transformability of cities.

3 Building-Scale Greening Strategies and Environmental Performance

At the building scale, greening strategies represent a critical interface between architectural design and urban resilience, acting as mediators between individual buildings and broader urban ecological systems. While often implemented at a smaller spatial scale, building-integrated greening strategies can generate cumulative and systemic effects when adopted across the urban fabric, thereby contributing significantly to climate adaptation and resilience objectives.

Building-scale greening strategies encompass a range of typologies, including green roofs, green façades and living walls, vegetated courtyards, and building-integrated vegetation systems (Diana et al., 2024; Sommese, 2024). These interventions introduce vegetation directly into the architectural envelope, enabling buildings to actively participate in microclimatic regulation, stormwater management, and ecosystem service provision (Dehghan Lotfabad et al., 2025). Unlike conventional passive design measures, greening strategies operate through biophysical processes such as evapotranspiration, shading, and insulation, making them particularly effective under increasing thermal stress conditions. From a performance perspective, these processes translate into quantifiable environmental effects that can be assessed through energy, thermal, and hydrological indicators.

With respect to environmental performance targets, building greening strategies primarily address the reduction of cooling energy demand, the attenuation of peak surface and air temperatures, the improvement of stormwater retention, and the enhancement of indoor and outdoor comfort conditions. Several studies report measurable reductions in summer cooling loads and surface temperatures when green roofs and façades are integrated into building envelopes, particularly under heatwave conditions (Dehghan Lotfabad et al., 2025; Jahangir et al., 2024; Zheng et al., 2021). In addition, greening strategies contribute to stormwater performance by delaying runoff peaks and reducing pressure on urban drainage systems, supporting adaptation to increasingly intense precipitation events (Kõiv-Vainik et al., 2022; Sun et al., 2024). In case of dense urban environments, where interventions in public or open spaces are not feasible due to the scarcity of available areas, building-scale greening strategies can compensate by adopting a systemic approach that integrates a large amount of buildings to amplify and generalise the effectiveness of the intervention.

The advantages of building greening strategies are therefore multifaceted. In addition to energy and microclimatic benefits, they support improved building envelope durability, acoustic performance, and occupant well-being. When evaluated through a life-cycle perspective, greening strategies can contribute to reductions in operational impacts, although their embodied impacts and maintenance requirements must also be considered. Moreover, in dense urban contexts, vertical and rooftop greening offers a land-efficient means of increasing vegetated surfaces, thereby enhancing environmental performance without additional land consumption.

However, the adoption of building greening strategies also entails significant performance-related challenges and trade-offs. Structural constraints, irrigation demand, maintenance intensity, and uncertainties regarding long-term performance can limit their effectiveness if not properly addressed at the design stage. Furthermore, the absence of standardised assessment frameworks and performance benchmarks complicates the comparison of different greening solutions, often leading to design choices driven by qualitative considerations rather than quantified performance outcomes.

For these reasons, the effectiveness of building greening strategies depends on their integration within performance-based design and assessment frameworks, where environmental indicators, simulation tools, and post-occupancy evaluation support informed decision-making. When embedded within multiscale urban greening networks, building-level interventions contribute cumulatively to urban environmental performance and resilience, reinforcing the green transformability of the built environment through incremental yet measurable improvements.

4 From Sustainability to Regenerative Design: European Perspective and Objective

The analysis of urban- and building-scale greening strategies presented in the previous sections highlights how greening has evolved from a complementary design feature into a core component of climate-responsive urban and architectural systems. At both scales, greening strategies have been shown to contribute to environmental performance, microclimatic regulation, and resilience by interacting with energy, water, and ecological processes. This multiscale perspective reveals that greening operates not only through isolated interventions, but through systemic configurations capable of producing cumulative and long-term effects across the built environment.

Within this framework, the discussion naturally extends beyond conventional sustainability-oriented approaches towards broader conceptual models capable of addressing the interlinked challenges of climate change, biodiversity loss, and urban vulnerability. Traditional sustainability paradigms, largely focused on impact reduction and efficiency, are increasingly recognised as insufficient to cope with the magnitude and persistence of these challenges. As a result, attention is shifting towards regenerative design, which seeks to enable the built environment to actively support ecological processes, enhance adaptive capacity and contribute to environmental restoration (Sommese et al., 2025).

In the European context, this shift is strongly reflected in policy frameworks such as the European Green Deal, the EU Climate Adaptation Strategy, and the EU Biodiversity Strategy 2030, which promote integrated and nature-based approaches to urban transformation (European Commission, 2020, 2021a, 2021b). Within these frameworks, NBS represent a key operational mechanism for linking mitigation and adaptation objectives. Interventions such as urban forests, parks, green roofs, and green façades exemplify how NBS simultaneously capture carbon, mitigate urban heat island effects, regulate microclimates, manage stormwater, and improve environmental quality and liveability. By integrating mitigation and adaptation, NBS foster a systemic understanding of the built environment as part of a complex and interdependent urban ecosystem, supporting coherent greening and resilience strategies aligned with European emissions reduction, climate adaptation, and biodiversity targets.

However, the transformative potential of NBS and greening strategies depends on their integration within broader planning, design, and governance frameworks. When deployed in a fragmented or symbolic manner, greening risks remaining an isolated sustainability measure rather than a driver of structural change. This limitation underscores the need for conceptual frameworks capable of embedding greening within long-term transformation processes of the built environment.

From this perspective, regenerative design offers a theoretical and operational framework that reframes buildings and cities as active agents within ecological and climate systems. This approach emphasises the importance of moving beyond static sustainability metrics towards adaptive and regenerative outcomes, where greening strategies support biodiversity, carbon sequestration, and human well-being through measurable environmental performance (Naboni & Havinga, 2019; Sommese et al., 2025).

Within this conceptual trajectory, the objectives of this chapter are addressed by demonstrating how greening strategies—implemented across urban and building scales and framed as NBS—contribute to enhancing resilience while enabling a transition from sustainability-oriented practices to regenerative design. By positioning greening as a multiscale, performance-based, and ecosystem-oriented design driver, the chapter articulates the concept of green transformability of the built environment as a necessary condition for long-term urban resilience and regeneration.

In conclusion, the progression from sustainability to regenerative design, supported by European policy agendas and informed by contemporary research on NBS and performance-based design, represents a critical evolution for urban development under climate uncertainty. Rather than merely limiting environmental impacts, regenerative approaches enable the built environment to actively participate in ecological restoration and resilience-building processes, offering a coherent pathway for aligning greening strategies, environmental performance, and urban resilience within a European sustainability framework.

Acknowledgements This research was founded by the European Union EU Next Generation Italian PRIN 2022 PNRR Programme under the "Greenwork—An interdisciplinary framework for urban health and urban resilience enhancement based on greening strategies on buildings and open spaces" project, CUP: E53D23019110001.

References

Andersson, E., Langemeyer, J., Borgström, S., McPhearson, T., Haase, D., Kronenberg, J., Barton, D. N., Davis, M., Naumann, S., Röschel, L., & Baró, F. (2019). Enabling green and blue infrastructure to improve contributions to human well-being and equity in urban systems. *BioScience, 69*(7), 566–574. https://doi.org/10.1093/biosci/biz058

Benaomar, K., & Outzourhit, A. (2024). Exploring the complexities of urban forms and urban heat islands: Insights from the literature, methodologies, and current status in Morocco. *Atmosphere, 15*(7), 822. https://doi.org/10.3390/atmos15070822

Brears, R. C. (2023). Blue-green infrastructure in managing urban water resources. In R. C. Brears (Ed.), *Blue and green cities: The role of blue-green infrastructure in managing urban water resources* (pp. 43–59). Springer International Publishing. https://doi.org/10.1007/978-3-031-41393-3_3

Cuce, P. M., Cuce, E., & Santamouris, M. (2025). Towards sustainable and climate-resilient cities: Mitigating urban heat islands through green infrastructure. *Sustainability, 17*(3), 1303. https://doi.org/10.3390/su17031303

Dehghan Lotfabad, A., Hosseini, S. M., Dabove, P., Heiranipour, M., & Sommese, F. (2025). Impacts of vertical greenery on outdoor thermal comfort and carbon emission reduction at the urban scale in Turin, Italy. *Buildings, 15*(3), 450. https://doi.org/10.3390/buildings15030450

Diana, L., Sommese, F., Ausiello, G., & Polverino, F. (2024). New green spaces for urban areas: A resilient opportunity for urban health. In A. Cheshmehzangi, M. Sedrez, H. Zhao, T. Li, T. Heath, & A. Dawodu (Eds.), *Resilience vs pandemics: Innovations in public places and buildings* (pp. 37–53). Springer Nature. https://doi.org/10.1007/978-981-99-8672-9_3

European Commission. (2013). *Green infrastructure*. https://environment.ec.europa.eu/topics/nature-and-biodiversity/green-infrastructure_en

European Commission. (2020). *Biodiversity strategy for 2030*. https://environment.ec.europa.eu/strategy/biodiversity-strategy-2030_en

European Commission. (2021a). *EU adaptation strategy*. https://climate.ec.europa.eu/eu-action/adaptation-climate-change/eu-adaptation-strategy_en

European Commission. (2021b). *The European green deal*. https://commission.europa.eu/strategy-and-policy/priorities-2019-2024/european-green-deal_en

Gherri, B. (2024). Urban green spaces and their role in responding to the heat Island effect in historical urban context. In J. R. Littlewood, L. Jain, & R. J. Howlett (Eds), *Sustainability in energy and buildings 2023* (pp. 457–470). Springer Nature. https://doi.org/10.1007/978-981-99-8501-2_40

IEA. (2019). *The critical role of buildings. Perspectives for the clean energy transition*. IEA. https://www.iea.org/reports/the-critical-role-of-buildings

Jahangir, M. H., Zarfeshani, A., & Arast, M. (2024). Investigation of green roofs effects on reducing of the urban heat islands formation (The case of a municipal district of Tehran City, Iran). *Nature-Based Solutions, 5*, 100100. https://doi.org/10.1016/j.nbsj.2023.100100

Kõiv-Vainik, M., Kill, K., Espenberg, M., Uuemaa, E., Teemusk, A., Maddison, M., Palta, M. M., Török, L., Mander, Ü., Scholz, M., & Kasak, K. (2022). Urban stormwater retention capacity of nature-based solutions at different climatic conditions. *Nature-Based Solutions, 2*, 100038. https://doi.org/10.1016/j.nbsj.2022.100038

Lazzarini, L., & Luttenberger, L. R. (2025). Spatial planning for urban biodiversity: Clashing values and transformative middle grounds in Rijeka (Croatia). *International Planning Studies*, 1–22. https://doi.org/10.1080/13563475.2025.2590519

Ma, X., Luo, Q., Yan, F., Lei, Y., Lu, Y., Chen, H., Yang, Y., Feng, H., Zhou, M., Ding, H., & Zhao, J. (2025). Evaluating the microclimatic performance of elevated open spaces for outdoor thermal comfort in cold climate zones. *Buildings, 15*(15), 2777. https://doi.org/10.3390/buildings15152777

Meerow, S., & Stults, M. (2016). Comparing conceptualizations of urban climate resilience in theory and practice. *Sustainability, 8*(7), 701. https://doi.org/10.3390/su8070701

Naboni, E., & Havinga, L. C. (2019). Regenerative design. In *Digital practice: A handbook for the built environment*. Eurac ISBN 978-3-9504607-2-8. https://build-up.ec.europa.eu/sites/default/files/content/regenerative_design_in_digital_practice_small.pdf

Ng, E., Chen, L., Wang, Y., & Yuan, C. (2012). A study on the cooling effects of greening in a high-density city: An experience from Hong Kong. *Building and Environment, International Workshop on Ventilation, Comfort, and Health in Transport Vehicles, 47*, 256–271. https://doi.org/10.1016/j.buildenv.2011.07.014

Smalls-Mantey, L., & Montalto, F. (2021). The seasonal microclimate trends of a large scale extensive green roof. *Building and Environment, 197*, 107792. https://doi.org/10.1016/j.buildenv.2021.107792

Sommese, F. (2024). Nature-based solutions to enhance urban resilience in the climate change and post-pandemic era: A taxonomy for the built environment. *Buildings, 14*(7), Article 7. https://doi.org/10.3390/buildings14072190

Sommese, F., & Diana, L. (2025). A holistic framework for the implementation of green roofs on existing buildings: A case study in the Mediterranean climate of Naples. *Building and Environment, 274*, 112811. https://doi.org/10.1016/j.buildenv.2025.112811

Sommese, F., Diana, L., Colajanni, S., Bellomo, M., Sciuto, G., & Lombardo, G. (2025). Towards a regenerative and climate-resilient built environment: Greening lessons from European cities. *Buildings, 15*, 1878. https://doi.org/10.3390/buildings15111878

Sun, C., Rao, Q., Chen, B., Liu, X., Adnan Ikram, R. M., Li, J., Wang, M., & Zhang, D. (2024). Mechanisms and applications of nature-based solutions for stormwater control in the context of climate change: A review. *Atmosphere, 15*(4), 403. https://doi.org/10.3390/atmos15040403

Veerkamp, C. J., Schipper, A. M., Hedlund, K., Lazarova, T., Nordin, A., & Hanson, H. I. (2021). A review of studies assessing ecosystem services provided by urban green and blue infrastructure. *Ecosystem Services, 52*, 101367. https://doi.org/10.1016/j.ecoser.2021.101367

Zheng, X., Kong, F., Yin, H., Middel, A., Liu, H., Wang, D., Sun, T., & Lensky, I. (2021). Outdoor thermal performance of green roofs across multiple time scales: A case study in subtropical China. *Sustainable Cities and Society, 70*, 102909. https://doi.org/10.1016/j.scs.2021.102909

Greening Strategies for Climate Risk Management

Governing Urban Nature Together: Principles and Factors for Effective Co-creation of Nature-Based Solutions

Diana Dushkova and Olga Ivlieva

Abstract Nature-based Solutions (NBS) are widely promoted as integrated strategies to address the interconnected crises of climate change, biodiversity loss, and pollution in urban areas. However, their effectiveness and legitimacy hinge on meaningful stakeholder engagement and collaboration. This chapter explores the principles and conditions for effective co-creation in governing and realizing urban NBS. Drawing on empirical insights from four major EU-funded projects (CONNECTING Nature, RECONECT, EmpowerUs, and BioAgora), the analysis examines how collaborative governance and participatory approaches can be operationalized. The study identifies key principles of co-creation (e.g. inclusivity, transparency, legitimacy, linking knowledge and action) and outlines its multi-stage process ranging from identifying local challenges to monitoring and evaluating outcomes. While co-creation enhances local ownership, knowledge integration, and social acceptance, significant barriers persist (e.g. capacity gaps, institutional silos, resource limitations, and challenges in sustaining stakeholder engagement). The chapter concludes that for NBS to achieve transformative change, urban governance must institutionalize long-term support for participatory processes, foster cross-sectoral collaboration, and empower communities to become active agents of change. Ultimately, co-creation serves as a vital mechanism to bridge science, policy, and practice with communities, ensuring NBS are ecologically effective, locally adapted, socially grounded and equitable, and firmly embedded in the sustainable urban development agenda.

Keywords Nature-based solutions · Co-creation · Sustainability · Resilience · Urban environment · Participatory governance

---

D. Dushkova (✉) · O. Ivlieva
Department of Conservation Biology and Social-Ecological Systems, Helmholtz Centre for Environmental Research—UFZ, Leipzig, Germany
e-mail: diana.dushkova@ufz.de

L. Diana et al. (eds.), *Greening and Resilience*, Urban Sustainability,
https://doi.org/10.1007/978-981-92-0419-9_2

1 Introduction

The global community currently faces a triple planetary crisis of climate change, biodiversity loss, and pollution, which reinforce each other, posing significant threats to human health, ecosystems, and economies. The crisis is largely driven by systemic issues related to current economic models, the imperatives of economic growth, unsustainable resource use, environmentally harmful production and consumption, fossil fuel dependence, and inadequate governance and policy silos. The resulting impacts include intensified droughts and extreme weather, habitat degradation, biodiversity decline, ecosystem collapse, and widespread environmental contamination, emphasizing the urgent need for systemic, integrated, and transformative responses (Adams et al., 2023; Biodiversa+, 2024; IPBES, 2024).

Against this background, nature-based solutions (NBS) have emerged as a promising strategy that bridges science, policy, and practice to deliver holistic and context-specific interventions. By taking into consideration all three dimensions of sustainability (e.g. environmental, social, and economic), NBS are particularly relevant for urban areas facing multiple challenges (Biodiversa+, 2024; EC, 2021). NBS are defined as actions that protect, manage, or restore ecosystems while simultaneously addressing societal challenges such as climate change, biodiversity loss, and pollution, thus offering benefits for both nature and people (IUCN, 2020). These benefits range from ecological restoration, climate mitigation and adaptation, to socio-economic gains, such as employment generation, improvement of human health and well-being, and fostering social cohesion (Adams et al., 2023; Dushkova & Haase, 2023).

As both a concept and a practical approach, NBS synthesise ideas from ecosystem-based adaptation, ecological restoration, disaster risk reduction, and blue-green infrastructure planning, providing a holistic pathway to urban sustainability (Dushkova & Haase, 2023; EC, 2021). More recently, NBS have been framed as instruments for transformative change, empowering communities, enabling participatory governance, and fostering systemic sustainability transitions (Adams et al., 2023; Biodiversa+, 2024; IPBES, 2024; Raymond et al., 2023).

Effective implementation of NBS requires close integration with local ecological and social conditions and active participation of diverse stakeholders to ensure both ecological effectiveness and social acceptance and equity. To achieve this, a range of participatory methods and tools is needed to engage communities throughout all stages of the NBS lifecycle—from problem assessment and intervention planning, through NBS design and implementation, to monitoring and evaluation of impacts (Dushkova & Kuhlicke, 2024). Over the past decade, many EU-funded projects have embraced co-creation as a central approach in developing their own frameworks, guidelines, and practical toolkits for stakeholder engagement (e.g. De Los Ríos-White et al., 2020; EC, 2023; Fohlmeister et al., 2019; Morello et al., 2018; UNaLab, 2020; van der Have et al., 2022).

While standard participatory processes often involve stakeholders in providing input or feedback at specific points, co-creation embeds them as equal partners throughout the entire NBS lifecycle, characterized by iterative collaboration rather than sporadic consultation. It emphasizes shared responsibility, mutual learning, and knowledge co-production, integrating scientific, professional, and community-based knowledge on an equal basis (Dushkova & Haase, 2023; Raymond et al., 2023). By doing so, co-creation enhances transparency and inclusiveness while supporting capacity building in public institutions to enable effective, collaborative, and shared governance arrangements (De Los Ríos-White et al., 2020; Dushkova & Kuhlicke, 2024).

Despite these advances, stakeholders often lack the resources, knowledge, and capacities needed to participate meaningfully in local sustainability and resilience initiatives (Dushkova & Kuhlicke, 2024; Dushkova et al., 2025; van der Have et al., 2022). Consequently, while participatory governance is widely recognized as essential for the success of urban sustainability interventions, including NBS, significant barriers still constrain effective stakeholder collaboration. These barriers often include limited financial and human resources, insufficient technical or ecological knowledge, institutional silos, short-term political cycles, and bureaucratic complexity (Adams et al., 2023; Dushkova et al., 2025; van der Have et al., 2022). Additionally, low trust in public authorities, lack of inclusive decision-making structures, and unequal representation of vulnerable and marginalized groups, further hinder meaningful stakeholder engagement and reduce the potential for locally adapted and socially equitable solutions (Gionfra et al., 2023; Han et al., 2025).

To better understand and reveal key mechanisms through which these barriers can be addressed in practice, this chapter draws on insights from four EU projects (CONNECTING Nature, RECONECT, EmpowerUs, and BioAgora), to explore the following research questions:

1. What co-creation principles and frameworks support the effective NBS realization in urban contexts?
2. What factors facilitate or hinder NBS co-creation processes and outcomes?

By examining these questions, the chapter discusses how meaningful stakeholder engagement and co-creation approaches can be operationalized to maximize the ecological, social, and governance benefits of NBS in cities.

2 NBS Concept Evolution and the Added Value of Co-creation

The concept of NBS evolved from early scientific discussions (1990s) on the critical role of ecosystems in climate mitigation, adaptation, and broader sustainability goals, and the need to integrate nature into climate and sustainability policy (Fig. 1). Over the subsequent decades, NBS emerged as an umbrella concept for ecosystem-based approaches, disaster risk reduction, blue-green infrastructure, gaining traction within academic and policy debates.

Fig. 1 NBS evolution timeline: from early conceptualization and recognition to transformative NBS. *Source* Visualization made by authors

The term NBS entered political discourse in the early 2000s: by 2009, it appeared in IUCN position papers, and between 2013 and 2015, major international and EU bodies formally adopted the concept, shifting from theory to policy. This was reinforced by Horizon 2020 funding and practical workshops that fostered shared understanding of the concept. In 2016, IUCN Resolution 069 formally defined NBS and acknowledged their relevance (IUCN, 2016). At the same time, EU-funded projects and networks, such as BiodivERsA+, began actively implementing NBS concepts on the ground (Biodiversa+, 2024). NBS have gained significant European and global recognition as essential, cost-effective tools for simultaneously tackling climate change, biodiversity loss, and societal challenges. They are increasingly recognized as a key delivery mechanism to achieve SDGs as well as emembedded in international agreements (e.g. the Kunming-Montreal Global Biodiversity Framework, the Paris Agreement), the European Green Deal and related EU strategies (e.g. EU Biodiversity Strategy 2030, Nature Restoration Law, EU Adaptation Strategy).

In 2017–2020, various robust operational frameworks, evidence-based planning methodologies, and knowledge-exchange platforms, such as Oppla and ThinkNature, were developed. Global initiatives and international frameworks, including the CBD, the EU Biodiversity Strategy, and IPBES assessments, highlighted the need to connect science, policy, and practice to ensure biodiversity protection and sustainable development (IPBES, 2024). Various science-policy interfaces in biodiversity and sustainability, including IPBES (2019, 2024), IPCC (2023) and recent projects such as BioAgora, also recognize NBS as a key transformative approach to achieve the ambitious goals of the EU's Biodiversity Strategy for 2030 and actively support the corresponding NBS knowledge-exchange network (KEN). Nevertheless, scientific knowledge alone was insufficient to drive policy change, as biodiversity loss and environmental degradation persisted despite substantial research efforts (Adams et al., 2023; Dushkova & Haase, 2023). This highlighted the need for a co-creation approach that actively involved stakeholders and local communities in the design, planning, implementation, and evaluation of NBS. Therefore, co-creation emerged as a key mechanism for integrating diverse knowledge, promoting stakeholder engagement, and supporting participatory governance (Debele et al., 2023) to achieve transformative change rather than merely deliver green interventions (Raymond et al., 2023; Sowińska-Świerkosz & García, 2022). Within the evolution of the NBS concept, the emergence of co-creation has driven a significant shift from "building in nature" to "building with nature and people".

To support the NBS realization and integration of co-creation into the NBS process, several practice-oriented handbooks, technical standards, and practitioner guidelines were produced in 2020–2025 to ensure inclusive and effective NBS realization (e.g. IUCN Global Standards for NBS [2020, 2025] and the EC Handbooks for NBS practitioners [EC, 2021] and stakeholder engagement [EC, 2023]). They advocate for this integrated approach, promoting the combination of scientific expertise with local and Indigenous knowledge to ensure NBS are socially acceptable, ecologically effective, and contextually grounded (Adams et al., 2023; IUCN, 2020).

3 Materials and Methods

This study builds on empirical insights from four European collaborative research and innovation projects funded by the EU Horizon 2020 and Horizon Europe programmes: CONNECTING Nature (https://connectingnature.eu/), RECONECT (http://www.reconect.eu/), EmpowerUs (https://empowerus-project.eu/), and BioAgora (https://bioagora.eu/). These projects aimed to support the co-creation and scaling of NBS and other sustainability solutions to address challenges related to climate change, biodiversity loss, and the blue-green economy, and facilitate local sustainability transitions (Table 1). Across all four projects, strong emphasis was placed on stakeholder engagement and collaboration, recognizing co-creation as a key mechanism for enhancing the relevance, legitimacy, and long-term impact of NBS. In addition, the projects sought to connect research-based knowledge with the needs of decision-making, fostering targeted and structured dialogue among scientists, other knowledge holders, and policy actors to support evidence-informed and context-sensitive policy processes for biodiversity conservation and restoration, climate resilience, and sustainable development.

The analysis draws on a qualitative review of project-related materials, including reports, policy documents, meeting and workshop minutes, and communication outputs. To capture process dynamics and learning outcomes, the study further employed co-creation diaries and reflective monitoring and evaluation frameworks, which supported the systematic documentation and assessment of changes emerging from co-creation activities over time.

This combined analytical approach allowed the critical reflection on co-creation approaches and principles developed and applied within the projects (Sect. 4.1), the added value of co-creation (Sect. 4.2), identification and thorough examination of key barriers and enabling factors influencing NBS realization and community engagement (Sect. 4.3), and the transformative potential of NBS (Sect. 4.4). The findings presented in the next sections are therefore grounded in comparative, practice-based evidence, reflecting lessons learned from the application of the established co-creation frameworks in diverse real-world settings.

Table 1 Overview and key characteristics of the analysed projects

Project name and duration	Focus/scope	Demonstration sites/key outputs	Type of NBS	Key benefits targeted
CONNECTING Nature, 2017–2022	Scaling various urban NBS; develop governance, finance, planning tools, and frameworks	Front-Runner Cities and Fast-Follower Cities, as well as Multiplayers, guidebooks, assessment toolkits, and Connecting Nature Framework	Urban green infrastructure: parks, street trees, green roofs, multifunctional public spaces, etc.	Climate adaptation (urban heat, flood management), biodiversity enhancement, social cohesion, and recreation
RECONECT, 2018–2024	Regenerate ecosystems and reduce hydrometeorological risks via NBS; integrate social innovation	Network of Demonstrators and Collaborators in multiple geographic contexts; monitoring ICT platforms; co-creation with living NBS labs	Riverine, wetland, urban-peri-urban and natural ecosystems; floodplains and multifunctional landscapes	Flood and drought risk reduction, biodiversity restoration, ecosystem service provision, and community resilience
EmpowerUs, 2022–2025	Empower coastal communities via tailored empowerment programmes, NBS, and adaptive transformation; capacity building	Six Transition Coastal Labs (TCLs) across EU coastal regions; collaborative planning, piloting, and evaluation of community-driven solutions, tailored empowerment programme framework	Coastal and marine ecosystem restoration; wetlands; blue-green infrastructure and multifunctional public spaces	Climate adaptation and resilience, biodiversity protection, socio-economic empowerment, sustainable livelihoods
BioAgora, 2022–2027	Develop Science Service for Biodiversity; strengthen science-policy interface and evidence-informed policy	Platforms for biodiversity knowledge exchange and KENs; demonstration cases in freshwater, marine, pollination, and NBS	Multi-ecosystem focus: freshwater, terrestrial, marine, and urban biodiversity-related NBS	Policy-informed decision-making, biodiversity conservation, climate adaptation, ecosystem service enhancement

4 Building with Nature and People: The Role of Co-creation in NBS Realization

4.1 Co-creation Approaches Developed and Applied

In all the projects mentioned above, co-creation was understood as a collaborative and iterative process encompassing the assessment, planning, design, implementation, monitoring, and evaluation of NBS with a broad range of local stakeholders, including community members, policymakers, practitioners, and experts. The projects developed and applied a variety of methods and frameworks to support co-creation, including targeted capacity-building activities, co-creation tools, and mechanisms for monitoring and evaluating the co-creation process and outcomes. In addition, they critically reflected on the barriers to NBS co-creation (Table 2) and developed a set of enablers to overcome them (Sect. 4.3). Together, these efforts allowed to advance knowledge co-production frameworks that integrated scientific expertise with local community knowledge, thereby strengthening meaningful stakeholder involvement and collaboration (Dushkova & Haase, 2023; Dushkova & Kuhlicke, 2024; Dushkova et al., 2025; van der Have et al., 2022).

The projects adopted a Living Lab approach (Table 2), which embeds research and innovation within real-world settings and promotes continuous interaction among multiple stakeholders (Dushkova et al., 2025). Through collaborative experimentation, learning, and reflection, this approach enabled the co-design, testing, and assessment of NBS in diverse local contexts. While the specific number and naming of co-creation phases varied across the projects, several common characteristics were observed. In general, NBS co-creation process followed a dynamic, multi-stage process, typically including six stages represented in Fig. 2. These begin with (1) collaboratively identifying local challenges and analysing the socio-ecological and institutional context, which informs (2) the joint development of shared future visions and scenarios. Subsequently, (3) potential sustainability impacts and trade-offs are assessed, leading to (4) the co-design of the NBS. This is followed by (5) the co-implementation, operationalization, and maintenance of the solutions, with the entire process being underpinned by (6) continuous co-monitoring and evaluation of both the process and its impacts.

To move beyond the simple consultation towards genuine collaboration, partnership, and empowerment and ensure meaningful and transformative stakeholder engagement, co-creation should consider the key principles developed in the mentioned projects (Fig. 3). The principles have been formulated within the four EU projects through an iterative, practice-based process, drawing on the relevant literature (e.g. Brandsen et al., 2018; Durham et al., 2014; EC, 2023; Morello et al., 2018; Sanders & Stappers, 2008; UNaLab, 2020; Voorberg et al., 2015) and on insights emerging from case study experiences (Living labs). Although the projects applied different context-specific co-creation approaches to govern the planning and implementation of NBS, they all built on the design principles for effective co-creation

Table 2 Co-creation approaches, participatory monitoring, and key barriers/challenges faced in the analysed projects

Project name and duration	Co-creation & knowledge co-production approaches	Co-monitoring & co-evaluation approaches	Key co-creation barriers/challenges
CONNECTING Nature, 2017–2022	Living Lab approach; peer-to-peer learning between front-runner and follower cities; co-production workshops with citizens, local authorities, SMEs; iterative monitoring and evaluation	Participatory monitoring (reflexive monitoring in action); stakeholder feedback workshops; ecological assessments; use of the Oppla platform for publishing results (cases) and tracking feedback and outcomes	Time-consuming stakeholder engagement; uneven local capacity; conflicting priorities between municipal departments; limited funding for long-term participation
RECONECT, 2018–2024	Social Innovation & co-creation framework; participatory stakeholder mapping; iterative design of interventions; local workshops for knowledge exchange	Multi-criteria assessment of ecosystem services; GIS mapping; local stakeholder surveys; iterative evaluation workshops	Complex governance contexts; difficulty aligning ecological and social objectives; challenges in sustaining community engagement over time
EmpowerUs, 2022–2025	Transition Coastal Labs as co-creation platforms; participatory scenario building; citizen engagement and capacity building; integration of local experiential knowledge and scientific expertise; co-monitoring and adaptive management of pilot interventions	Co-monitoring (incl. photo-voice, citizen science, etc.) with local communities; participatory evaluation; adaptive management logs; pilot impact indicators (ecological, social, economic)	Varied community capacities; limited resources for consistent participation; potential conflicts between local priorities and project goals
BioAgora, 2022–2027	Science–Policy Interface co-production; structured dialogue between scientists, policymakers, and knowledge holders; co-development of governance structures; participatory workshops; integration of policy needs with biodiversity research outputs	Knowledge tracking dashboards; co-production workshops; policy uptake metrics; integration of stakeholder feedback	Bridging science and policy requires trust-building; varying policy timelines; difficulties in translating scientific outputs into actionable policy

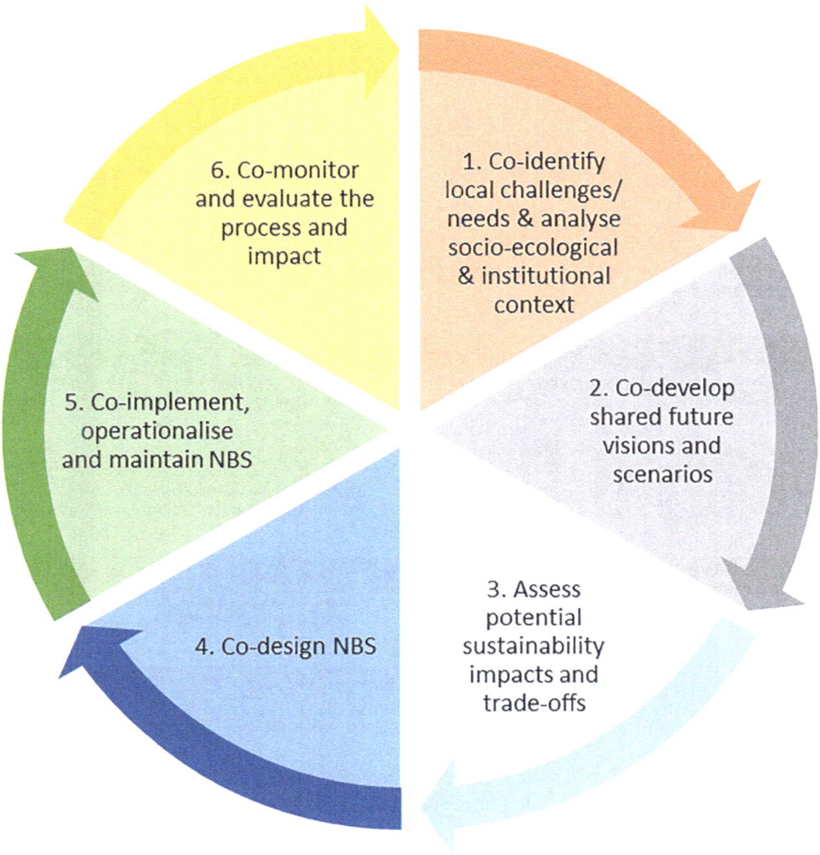

Fig. 2 Stages of NBS co-creation process. *Source* Authors

presented in Fig. 3 and explained below. These principles synthesise the key characteristics and enabling conditions observed across the projects and capture shared lessons learned on what leads to successful co-creation of NBS (more details in Dushkova & Haase, 2023; Dushkova & Kuhlicke, 2024; van der Have et al., 2022).

First, **inclusivity** is fundamental, requiring the engagement of all relevant stakeholder groups from the outset to foster greater buy-in and ensure NBS designs are better suited to their local context, and maximizes co-benefits. Such engagement must evolve beyond mere consultation into collective action and shared governance—the principle of "**doing together**". This shift is crucial for empowering communities as active partners rather than passive beneficiaries of intervention, ensuring the process is genuinely participatory.

Underpinning this collaboration are the principles of **openness, clarity, and transparency**. Providing clarity about the NBS process and benefits, stakeholder

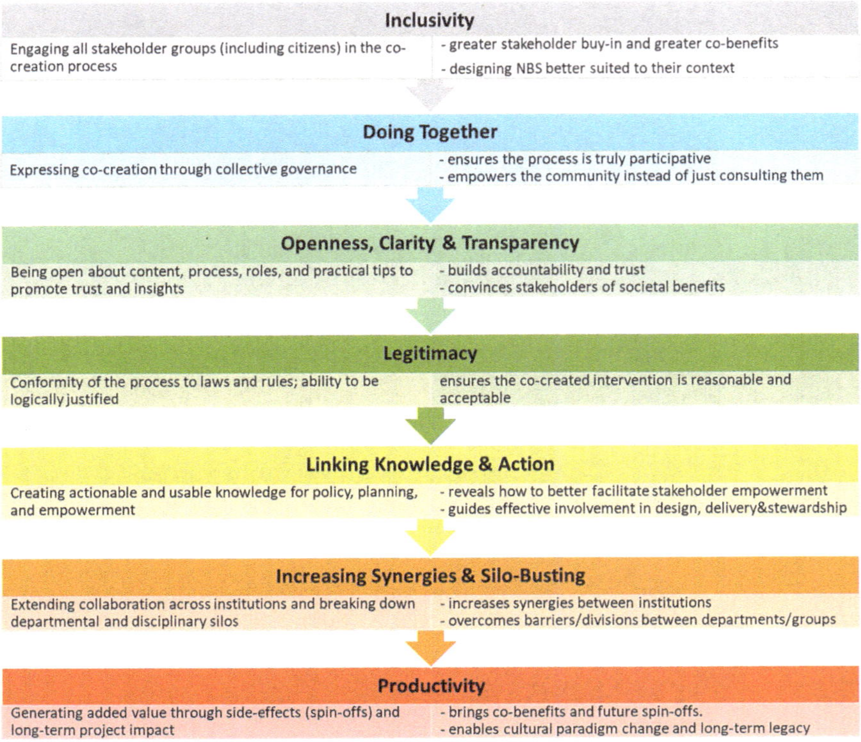

Fig. 3 Principles of NBS co-creation (interconnected factors of success) that guide the process from stakeholder engagement to long-term sustainability impact. *Source* Authors

roles, and practical information builds essential accountability and trust among stakeholders. This transparency helps convince stakeholders of the various NBS benefits in a fair and equitable way. Furthermore, the entire process must be built on **legitimacy** (conformity to laws and logical justification) to ensure the co-created interventions are reasonable, acceptable, and defensible.

A critical bridge must be built between **knowledge and action**, translating insights into actionable strategies for policy, planning, and community empowerment. This refers to better facilitating stakeholder involvement in the design, delivery, and stewardship of projects. To enable this, **increasing synergies and "silo-busting"** is necessary, extending collaboration across institutions and breaking down barriers between departments, disciplines, and decision-making procedures.

Ultimately, the process should aim for high **productivity**, defined not just by the primary goals but by the added value through experiencing side effects and spin-offs. This focus on generating co-benefits and enabling a cultural paradigm shift is what secures the long-term impact and legacy of co-created NBS projects.

4.2 Benefits of Co-creation

Co-creation provides multiple advantages for NBS implementation, with benefits evident across the projects.

First, co-creation enhances local ownership and social acceptance, which is critical for the sustained long-term impact of NBS. For instance, in CONNECTING Nature, the collaborative design of urban parks and green corridors with residents led to a stronger sense of community stewardship, meanwhile reducing vandalism and ensuring maintenance. Similarly, EmpowerUs enables coastal communities to co-design various sustainability projects (including NBS), directly increasing their commitment to protecting and sustaining these interventions over time.

Second, the process fosters knowledge co-production, creating more robust and context-specific solutions by merging scientific expertise with local and Indigenous knowledge. The BioAgora project institutionalizes this through its Science-Policy Interface, structuring dialogues where scientific models are refined with practitioner insights to produce actionable biodiversity governance tools. In RECONECT, the knowledge of local farmers about historical water flows was integrated with hydrological modelling to design more effective natural water retention measures. CONNECTING Nature advanced this through its peer-to-peer learning networks, where cities co-developed the Connecting Nature Framework for practical NBS realization by exchanging scientific benchmarks with on-the-ground experiences.

Third, co-creation builds capacity and facilitates empowerment, enabling communities to change their role as passive beneficiaries of solutions to active agents in sustainability transitions. For example, EmpowerUs is built on this principle, using its Transition Coastal Labs to provide communities with tailored skills and engage them in participatory planning and adaptive management (Dushkova et al., 2025). Likewise, CONNECTING Nature documented how fast-follower cities built internal municipal capacity by learning from front-runner cities, enabling them to independently develop and finance future NBS.

Finally, these processes encourage essential cross-sectoral collaboration, breaking down silos to link urban planners, ecologists, policymakers, and civil society. RECONECT exemplified this approach by creating consortia with multiple stakeholders (e.g. water authorities, environmental NGOs, spatial planners, and private companies) who jointly designed flood mitigation landscapes. BioAgora further facilitates this by developing various structured Knowledge Exchange Networks (KENs), ensuring that diverse expertise informs policy from the outset.

4.3 Barriers and Enablers in NBS Co-creation

Despite its benefits, co-creation faces various barriers, including structural, institutional, and social constraints (Table 2). To address these, five key factors (enablers) that facilitate NBS co-creation processes and maximizing the ecological, social, and governance benefits of NBS have been identified across all analysed projects (Fig. 4). First, **a shared understanding of NBS concepts** and long-term sustainability goals among stakeholders is essential to support coordinated actions. In this regard, **capacity building** is crucial, as the training of facilitators and local actors enhances the ability to implement NBS meaningfully and sustainably.

In all the above mentioned projects, **lack of political will, short-term policy cycles, and sectoral silos** have been identified as major barriers that hinder NBS uptake (Dushkova & Haase, 2023; Dushkova & Kuhlicke, 2024; Dushkova et al., 2025; Han et al., 2025; van der Have et al., 2022), which is aligned with academic literature (Adams et al., 2023; Davis et al., 2025; Gionfra et al., 2023). Decision-making structures often remain centralized, top-down, and technocratic, with limited inclusive mechanisms for incorporating local stakeholders in meaningful ways. Sometimes, policymakers use NBS as a buzzword/greenwashing to secure related funding or to demonstrate progress towards achieving specific SDGs. Moreover, in certain cases, participatory processes are rather tokenistic, lacking principles of genuine co-creation and not particularly influencing the outcomes (tick-the-box exercise). Bureaucratic complexity further discourages public involvement, especially when processes are perceived as unresponsive.

As we learned, **limited financial and human resources** reduce the ability of stakeholders to participate in planning processes, attend consultations, monitoring and evaluation, or mobilize around local sustainability initiatives (Han et al., 2025). This especially affects vulnerable and marginalized groups, leading to a lack of inclusivity and unequal representation in NBS decision-making. Additionally, as traditional grey infrastructure solutions continue to dominate planning approaches, NBS are often limited in their budgets. Moreover, land-use conflicts and delayed return on investment often hinder NBS implementation (spatial and temporal limitations). In some cases, participatory processes are **time-consuming** and require significant effort from participants, which might lead to stakeholder fatigue and resistance (Dushkova et al., 2025). To address this, various co-creation strategies and tools have been developed, which are based on principles of inclusivity, transparency, legitimacy, and respecting participants' time and capacities. This includes setting clear expectations, providing tangible benefits or incentives for involvement, ensuring transparent communication, and adapting participation methods to the local context (Dushkova & Kuhlicke, 2024; van der Have et al., 2022).

Effective participation in NBS processes often requires a baseline understanding of ecological principles, planning processes, and relevant policy frameworks (**NBS-related skills and expertise**) (Davis et al., 2025; EC, 2023; Gionfra et al., 2023). In certain cases, we observed that local residents, public authorities, decision-makers, and practitioners simply lacked access to the technical knowledge or educational opportunities needed to improve their expertise in the NBS realization process (Dushkova et al., 2025; Han et al., 2025).

Finally, the success of NBS co-creation depends greatly on **trust among stakeholders and the existence of social capital/networks** that facilitate collaboration, which is also emphasized by other studies (e.g. Gionfra et al., 2023; Morello et al.,

Fig. 4 Key factors of successful NBS co-creation. *Source* Authors with the icons from www.freepik.com

2018). Sometimes, communities were less willing or able to engage effectively if they had lower trust in authorities or where social fragmentation existed. Thus, establishing so-called think tank groups consisting of local core stakeholder groups knowledgeable in NBS can help transfer scientific knowledge, align efforts, and avoid fragmented actions (Dushkova et al., 2025; van der Have et al., 2022).

Strengthening these enabling conditions requires broader stakeholder involvement and long-term engagement strategies, capacity-building programmes, social capital, inclusive NBS design, political, and institutional support, and suitable financing mechanisms that help to embed co-creation into formal governance structures (Dushkova & Haase, 2023; EC, 2021; Han et al., 2025). As illustrated (Fig. 4), successful NBS co-creation depends on a cohesive set of enablers, including shared learning, institutional embedding, and dedicated funding. This highlights that systemic progress relies not on isolated efforts, but on the integrated reinforcement of complementary strategic, social, and procedural factors.

4.4 *Towards Transformative NBS*

The insights from the analysed EU-funded projects confirm that NBS provide an effective framework for tackling societal challenges through restoration, conservation, and sustainable management of ecosystems, generating benefits for biodiversity, climate resilience, and human well-being. However, their transformative potential as an imperative for addressing the triple planetary crisis is not automatic. When poorly implemented, NBS can even reinforce existing inequalities or deliver short-term benefits without addressing underlying drivers. In contrast, when grounded in approaches that promote participation, inclusivity, and justice, NBS can catalyse systemic change by reshaping governance, shifting practices, fostering innovation, and strengthening local capacities. For NBS to achieve this transformative change, they should move beyond pilot projects towards systemic integration. Building on project outcomes, below we provide evidence-based insights into how NBS can drive transformative change by reshaping enabling frameworks to deliver long-term ecological integrity, social equity, and resilience.

- **Embedding NBS in Policy and Regulatory Frameworks**, ensuring long-term continuity and institutional support (Adams et al., 2023; EC, 2021, 2023). This integration is crucial for transitioning from isolated pilots to mainstream practice. For example, the BioAgora directly contributes to this by advancing the science-policy interface. A key output, developed with the EC's Knowledge Centre for Biodiversity, is the Handbook of Tools for Informing and Monitoring Urban Greening Strategies and Nature Plans (Cortinovis et al., 2025). Developed in response to a policy request from the EC's Directorate-General for Environment (DG ENV), the Handbook maps and assesses a wide range of planning and decision-support tools, including those relevant to NBS (e.g. Nature-Based Solutions Opportunity Scan, NATURVATION Atlas), to directly support the implementation of the EU Biodiversity Strategy and the Nature Restoration Regulation. Similarly, CONNECTING Nature co-developed a Nature-based enterprise platform (https://naturebasedenterprise.com/) along with NBS business and financing models with city governments, creating replicable policy blueprints that were integrated into municipal strategic plans to secure long-term investment. These examples illustrate how well-designed science-policy and science-practice interfaces translate scientific knowledge into practical, policy-relevant tools that support the NBS uptake across governance levels and policy contexts, such as the EU Biodiversity Strategy for 2030 and the Nature Restoration Regulation.
- **Scaling Participatory Governance**, including co-creation and knowledge co-production as standard practice. Transformative NBS require shifting governance models to be genuinely inclusive. The EmpowerUs project operationalizes this through its Transition Coastal Labs, which serve as collaborative platforms for co-governance where communities, scientists, and local authorities jointly plan and steward coastal NBS (Dushkova et al., 2025). This institutionalizes participatory decision-making beyond project timelines. RECONECT further demonstrates this by establishing living NBS labs (network of cases) where iterative, community-driven design and monitoring become embedded in local risk management protocols (Dushkova & Kuhlicke, 2024).
- **Fostering Collaborative Networks** across scientists, practitioners, policymakers, and civil society to sustain learning and innovation. Systemic change relies on breaking down silos and creating continuing collaboration. CONNECTING Nature built a network of front-runner and fast-follower cities that continues to exchange knowledge and co-innovate (Dushkova & Haase, 2023). BioAgora is establishing structured Knowledge KENs to ensure dialogue between biodiversity science and policy actors, fostering a community of practice dedicated to evidence-informed NBS governance.

- **Empowering Communities, Building Trust, Ownership, and Access** to resources to enable local actors to become agents of change. True transformation redistributes power and capacity as demonstrated by all the projects. EmpowerUs is fundamentally designed for this, co-developing tailored empowerment programmes with coastal communities that build skills in advocacy, project management, and ecological stewardship, turning residents into leaders of local sustainability transitions (Dushkova et al., 2025). In RECONECT, communities co-manage restored floodplains, gaining not only increased resilience but also new governance rights and responsibilities over the landscape, ensuring they are active stewards rather than passive beneficiaries (Dushkova & Kuhlicke, 2024).

5 Concluding Remarks

This chapter examined the critical role of co-creation in governing and realizing NBS. Drawing on comparative evidence from four projects, the analysis reaffirms that NBS represent a powerful, integrated approach for addressing the triple planetary crisis. The transformative potential of NBS, however, is not inherent but unlocked through deliberate participatory governance.

The identified set of interconnected co-creation principles (e.g. inclusivity, doing together, openness & transparency, legitimacy, linking knowledge & action, silo-busting, and productivity) forms a framework for effective stakeholder collaboration. These principles are operationalized through a multi-stage co-creation process, from co-assessing challenges to co-monitoring outcomes, fostering local ownership, knowledge co-production, capacity building, and cross-sectoral collaboration. Empirical examples from the analysed projects clearly demonstrate these benefits, showing how co-creation leads to more socially accepted, ecologically robust, and sustainably managed interventions.

Nevertheless, the path to transformative NBS is marked by persistent barriers (e.g. capacity gaps, institutional silos, resource limitations, stakeholder fatigue, and deficits in trust and social capital). These barriers reveal a disconnect between participatory governance ambitions and implementation realities. Overcoming them requires systemic change, including long-term institutional support, dedicated funding for participatory processes, embedded capacity-building programmes, and governance reforms that legitimize community input in urban planning.

To catalyse systemic change and address the triple planetary crisis, NBS must evolve from isolated pilots into mainstream, policy-embedded practice. This requires embedding NBS in regulatory frameworks, scaling participatory governance models, fostering lasting collaborative networks, and empowering communities to become agents of change. Therefore, co-creation functions as both the essential mechanism and the guiding pathway for this transition, bridging science, policy, and practice, to ensure NBS are ecologically effective, socially equitable, locally adapted, and democratically legitimate. The future of sustainable and resilient cities depends on our collective commitment to this institutionalized collaborative approach, transforming urban governance to build with nature and people for shared benefits.

Acknowledgements The research work was supported by the project "Bio Knowledge Agora: Developing the Science Service for European Research and Biodiversity Policymaking" funded by the EU Horizon Europe programme (Grant agreement 101059438). The authors especially thank Victoria Novikova-Dickey for her great support in proofreading the manuscript.

References

Adams, C., Frantzeskaki, N., & Moglia, M. (2023). Mainstreaming nature-based solutions in cities: A systematic literature review and a proposal for facilitating urban transitions. *Land Use Policy, 130*, 106661. https://doi.org/10.1016/j.landusepol.2023.106661

Biodiversa+. (2024). *How nature-based solutions promote transformative change*. Biodiversa+, Paris. Available online: https://www.biodiversa.eu/wp-content/uploads/2024/07/D4.8-Desk-Study-How-NBS-promote-Transformative-change-1.pdf. Accessed 12 October 2025.

Brandsen, T., Steen, T., & Verschuere, B. (Eds). (2018). *Co-production and co-creation: Engaging citizens in public services*. Taylor & Francis. https://doi.org/10.4324/9781315204956

Cortinovis, C., Orta-Ortiz, M. S., Vasilakopoulos, P., & Velasco Gomez, M. (Eds.) (2025). *Handbook of tools for informing and monitoring urban greening strategies and nature plans*. Publications Office of the European Union, Luxembourg. https://doi.org/10.2760/8493055

Davis, M., Burgos, N., De Vreese, R., Lupp, G., Maestre-Andres, S., Xidous, D., & Zingraff-Hamed, A. (2025). *Co-creating nature-based solutions with commonly excluded stakeholders: Insights from practice and research*. NetworkNature+ Task Force 6.

Debele, S. E., Leo, L. S., Kumar, P., Sahani, J., Ommer, J., Bucchignani, E., Vranić, S., Kalas, M., Amirzada, Z., Pavlova, I., Shah, M. A. R., Gonzalez-Ollauri, A., & Di Sabatino, S. (2023). Nature-based solutions can help reduce the impact of natural hazards: A global analysis of NBS case studies. *Science of the Total Environment, 902*, 165824. https://doi.org/10.1016/j.scitotenv.2023.165824

De Los Ríos-White, M. I., Roebeling, P., Valente, S., & Vaittinen, I. (2020). Mapping the life cycle co-creation process of nature-based solutions for urban climate change adaptation. *Resources, 9*(4), 39. https://doi.org/10.3390/resources9040039

Durham, E., Baker, H., Smith, M., Moore, E., & Morgan, V. (2014) *The BiodivERsA stakeholder engagement handbook*. Paris. Available online: https://www.biodiversa.org/706/download. Accessed 12 October 2025.

Dushkova, D., & Haase, D. (2023). Resilient cities, healthy communities, and sustainable future: How do nature-based solutions contribute? In P. Liamputtong (Ed.), *Handbook of social sciences and global public health* (pp. 1–24). Springer. https://doi.org/10.1007/978-3-030-96778-9_133-1

Dushkova, D., & Kuhlicke, C. (2024). Making co-creation operational: A RECONECT seven-steps pathway and practical guide for co-creating nature-based solutions. *MethodsX, 12*, 102495. https://doi.org/10.1016/j.mex.2023.102495

Dushkova, D., Ivlieva, O., Pouget, C., & Vandewalle, M. (2025). How to support communities in the long-term sustainability transition: The tailored empowerment program. *Blue-Green Systems, 7*(1), 210–237. https://doi.org/10.2166/bgs.2025.049

EC—European Commission, Directorate-General for Research and Innovation. (2021). *Evaluating the impact of nature-based solutions: A handbook for practitioners*. Publications Office of the European Union, Brussels. Available online: https://doi.org/10.2777/244577. Accessed 12 October 2025.

EC—European Commission, Directorate-General for Research and Innovation. Ferreira, I., Lupp, G., & Mahmoud, I. (Eds.). (2023). *Guidelines for co-creation and co-governance of nature-based solutions: Insights from EU-funded projects*. Publications Office of the European Union. Available online: https://doi.org/10.2777/157060. Accessed 24 November 2025.

Fohlmeister S., Augenstein I., Jones C., Ramirez D., & Lupp G. (2019). *Starter toolbox for stakeholder knowledge mapping to co-design nature-based solutions at case study sites*. Deliverable D3.2 of PHUSICOS project.

Gionfra, S., Utkarsh, S., & Moncorps, L. (2023). *Leaving no one behind: Towards inclusive nature-based solutions*. Knowledge Brief drafted by IUCN and ICLEI for NetworkNature (H2020 project No. 887396). https://networknature.eu/sites/default/files/uploads/networknature-nbsknowledgebrief03.pdf

Han, S., Plavsic, J., Dushkova, D., Gebhardt, O., Izydorczyk, K., Kovačević-Majkić, J., Krauze, K., Kuzior, M., Włodarczyk-Marciniak, R., & Kuhlicke, C. (2025). Beyond individual barriers: A systems approach to understanding and addressing nature-based solutions implementation challenges. *Climate Risk Management, 50*, 100739. https://doi.org/10.1016/j.crm.2025.100739

IPBES. (2019). Global assessment report on biodiversity and ecosystem services of the Intergovernmental Science-Policy Platform on Biodiversity and Ecosystem Services. J. S. Brondizio, S. Díaz, H. T. Ngo (Eds.), IPBES, Bonn.

IPBES. (2024). *Summary for policymakers of the thematic assessment report on the underlying causes of biodiversity loss and the determinants of transformative change and options for achieving the 2050 vision for biodiversity*. In K. O'Brien, L. Garibaldi, A. Agrawal et al. (Eds.), IPBES Secretariat, Bonn, Germany.

IPCC (2023). AR6 Synthesis Report: Climate Change 2023. Available from: https://www.ipcc.ch/report/sixth-assessment-report-cycle/. Accessed November 30, 2026.

IUCN. (2016). *Defining nature-based solutions*. Resolution 069, World Conservation Congress, Honolulu, Hawai'i. International Union for Conservation of Nature (IUCN).

IUCN. (2020). *Global standard for nature-based solutions: A user-friendly framework for the verification, design, and scaling up of NbS*. IUCN.

IUCN. (2025). *Standards, methods, and guidelines for cross-referencing ecosystem classifications and maps to the IUCN global ecosystem typology*. Version 1.0. In D. A. Keith, A. Dayaram, A. Driver, J. R. Ferrer-Paris, R. Francis, S. Hay, N. J. Murray, E. Nicholson, J. Santos, A. L. Skowno, & M. G. Tozer (Eds.). IUCN, Gland, Switzerland.

Morello, E., Mahmoud, I., Gulyurtlu, S., Boelman, V., & Davis, H. (2018). *CLEVER cities guidance on co-creating nature-based solutions: PART I—Defining the co-creation framework and stakeholder engagement*. Deliverable 1.1.5, CLEVER Cities, H2020 grant no. 776604. Available online: https://clevercities.eu/files/fileadmin/user_upload/resources/d1.1_theme_5_co-creation_framework_fpm_12.2018.pdf. Accessed on 17 November 2025.

Raymond, C. M., Anderson, C. B., Athayde, S., Vatn, A., Amin, A. M., Arias-Arévalo, P., Christie, M., Cantú-Fernández, M., Gould, R. K., Himes, A., Kenter, J. O., Lenzi, D., Muraca, B., Murali, R., O'Connor, S., Pascual, U., Sachdeva, S., Samakov, A., & Zent, E. (2023). An inclusive typology of values for navigating transformations towards a just and sustainable future. *Current Opinion in Environmental Sustainability, 64*, 101301. https://doi.org/10.1016/j.cosust.2023.101301

Sanders, E. B. N., & Stappers, P. J. (2008). Co-creation and the new landscapes of design. *CoDesign, 4*(1), 5–18. https://doi.org/10.1080/15710880701875068

Sowińska-Świerkosz, B., & García, J. (2022). What are nature-based solutions (NBS)? Setting core ideas for concept clarification. *Nature-Based Solutions, 2*, 100009. https://doi.org/10.1016/j.nbsj.2022.100009

UNaLab. (2020). *UNaLab co-creation toolkit*. Available online: https://unalab.enoll.org/. Accessed on 28 November 2025.

van der Have, C., Hölscher, K., & Lodder, M. (2022). *A practical guide to using co-production for nature-based solutions*. Connecting Nature Project. Available online: https://connectingnature.eu/sites/default/files/downloads/CN-Co-production_for_NBS-Guidebook-MidRes.pdf. Accessed 6 May 2025.

Voorberg, W. H., Bekkers, V. J. J. M., & Tummers, L. G. (2015). A systematic review of co-creation and co-production: Embarking on the social innovation journey. *Public Management Review, 17*, 1333–1357.

Participatory Validation of Nature-based Solutions (NbS) for Enhanced Resilience

Awais Naeem Sarwar, Cristina Caramiello, Francesco Pugliese, Daniele Secci, Rafael Magnabosco de A. Gomes, Elif Bal, Chiara Ceseracciu, Angelo Avino, Irem Daloglu Cetinkaya, Nadim K. Copty, Seifeddine Jomaa, and Salvatore Manfreda

Abstract The rapid intensification of environmental, social, and economic challenges in urban areas, particularly under the stress of climate change, necessitates a shift toward more resilient planning models. This chapter aims to investigate the critical relevance of Nature-based Solutions (NbS) as a transformative strategy for enhancing urban resilience and sustainability, directly addressing climate adaptation. Conventionally, top-down approaches often fail to account for site-specific constraints and community acceptance, leading to implementation shortfalls. Therefore, this research aims to assess the effectiveness of participatory validation methods in the selection, design, and implementation of NbS within highly vulnerable urban ecosystems (cities) in the Mediterranean basin, a global biodiversity and climate change hotspot. The research objectives included evaluating the performance and social acceptance of the proposed NbS, identifying the roles of key stakeholders and local experts, and assessing the long-term viability of interventions across eight local demonstration sites within the OurMED Project. The methodology employed a rigorous, three-stage, spatially explicit approach: Catchment Classification (geospatial analysis), Evidence-Based Screening (a literature-driven decision matrix), and,

A. N. Sarwar (✉) · C. Caramiello · F. Pugliese · A. Avino · S. Manfreda
Department of Civil, Architectural and Environmental Engineering, University of Naples Federico II (UNINA), Naples, Italy
e-mail: awaisnaeem.sarwar@unina.it

D. Secci
Department of Engineering and Architecture, Università di Parma, Parma, Italy

R. M. de A. Gomes
Research Institute of Water and Environmental Engineering, Universitat Politècnica de València, Valencia, Spain

E. Bal · I. D. Cetinkaya · N. K. Copty
Institute of Environmental Sciences, Boğaziçi University, Istanbul, Turkey

C. Ceseracciu
Desertification Research Center (NRD), University of Sassari (UNISS), Sassari, Italy

S. Jomaa
Department of Aquatic Ecosystem Analysis and Management, Helmholtz Centre for Environmental Research UFZ, Magdeburg, Germany

L. Diana et al. (eds.), *Greening and Resilience*, Urban Sustainability,
https://doi.org/10.1007/978-981-92-0419-9_3

critically, local expert validation through a workshop. The results confirmed distinct NbS clusters based on hydro-climatic gradients, such as the prioritization of water harvesting in arid regions. Crucially, the validation phase revealed that one-third of the technically generated preliminary NbS proposals required significant adjustments, underscoring *"the need to integrate local knowledge."* This methodology is a transferable framework that successfully integrates technical rigor with community needs, advancing urban resilience theory and providing practical, policy-relevant tools for municipalities aiming for enhanced social acceptance and effective climate adaptation.

Keywords Nature-based solutions · Urban resilience · Participatory planning · Mediterranean cities · Climate adaptation · Sustainable urban development · Local knowledge · Stakeholders

1 Introduction

Urban ecosystems play a pivotal role in addressing global sustainability challenges, serving as both significant contributors to environmental stress and vital sites for innovative solutions. With over half of the global population currently living in cities, a figure expected to reach approximately 70% by 2050, the demand for sustainable urban development is unprecedented (UN DESA, 2018).

Cities are currently grappling with a multitude of stressors, including rapid expansion, habitat fragmentation, pollution, and the intensifying effects of climate change, such as heat waves, flooding, and water scarcity (Bush & Doyon, 2019; Gobster, 2010). These challenges are particularly pronounced in the Mediterranean basin, which is identified as a climate change hotspot and characterized by biodiversity loss, water stress, and increased vulnerability to extreme weather events (Ali et al., 2022).

In response, NbS are emerging as a transformative approach that leverages ecological processes to address societal challenges while delivering multiple co-benefits for biodiversity, human well-being, and economic resilience (EC, 2015; IUCN, 2020). NbS include various interventions, such as green roofs, urban wetlands, permeable pavements, and restored riparian corridors, all inspired by, supported by, or mimicking nature. When incorporated into urban planning, they can enhance resilience by mitigating urban heat island effects, managing stormwater, improving air quality, and creating recreational spaces (Lafortezza & Sanesi, 2018; Lehmann, 2021).

However, the successful implementation of NbS faces numerous challenges, including techno-centric, top-down planning that often overlooks local social and ecological contexts, resulting in solutions that may be technically sound but are socially contested or financially unfeasible (Nesshöver et al., 2017). This gap underscores the need for participatory validation designed to engage local stakeholders and experts in the co-design and evaluation of NbS to ensure they are contextually appropriate, socially acceptable, and institutionally feasible.

This chapter presents a case study of participatory validation within the OurMED Project, spanning over eight demonstration sites across the Mediterranean basin. Results reveal not only the technical adjustments required for NbS proposals but also the systemic barriers and enablers for implementation. By bridging the gap between scientific expertise and local knowledge, this chapter contributes to both the theory and practice of resilient urban planning.

2 Conceptual Framework: Integrating Technical and Social Dimensions of NbS

The development and execution of NbS operate at the intersection of intricate biophysical systems and evolving socio-institutional frameworks. To effectively address this complexity, our research is structured around an integrated conceptual framework that explicitly integrates technical assessment with social (participatory) engagement to create a robust and locally relevant NbS options map. The three-stage framework shown in Fig. 1 starts with catchment classification that integrates ecosystem types, hydro-climatic regimes, soil permeability, and geomorphology; then, in the 2nd stage, evidence-informed screening using a literature-derived NbS decision matrix is conducted, which was used to produce the preliminary NbS map to support participatory discussion. Then, in the 3rd stage, a spatial prioritization of NbS is coupled with stakeholder and local expert validation to ensure feasibility and social acceptance.

2.1 Participatory Action Research and Co-production

Our approach is based on local experts, stakeholders, and the co-production of knowledge (Norström et al., 2020). This framework emphasizes the importance of developing effective and actionable science for sustainability through collaboration with stakeholders, rather than solely for their benefit. Consequently, participatory validation is not merely an add-on; it is an integral part of the research process, ensuring that NbS are embedded in local contexts and align with community values and governance structures.

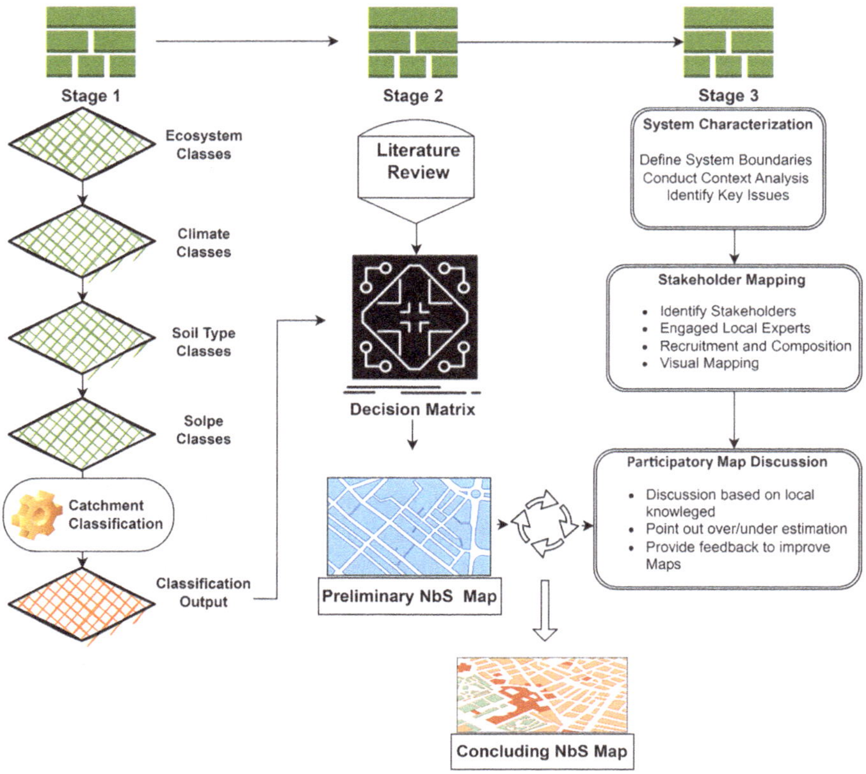

Fig. 1 Integrated conceptual framework for Nature-based Solutions (NbS) mapping and participatory approach

2.2 *Urban Resilience as an Emergent Property*

Urban resilience should not be viewed as a static outcome but rather as an emergent property arising from the interactions among ecological systems, built infrastructure, and social networks (Meerow et al., 2016). NbS enhances resilience by improving adaptive capacity, reducing vulnerability, and fostering learning and self-organization. Furthermore, participatory validation strengthens these features by building social capital, fostering trust, and ensuring shared ownership of solutions.

The strategic relevance of advancing the study of NbS in urban ecosystems stems from the significant shortcomings of traditional urban planning models, which have often prioritized economic growth and infrastructural expansion over ecological preservation and social inclusion. Conventional urban development approaches have contributed to increased greenhouse gas emissions, habitat fragmentation, reduced biodiversity, pronounced urban heat island effects, compromised air and water quality, and heightened vulnerability to extreme weather events (Bai et al., 2017).

A participatory validation process ensures that NbS are ecologically effective, socially acceptable, and economically viable (Jetol et al., 2024). Integrating community-based initiatives with strategic urban planning can reconnect urban populations with nature, promote local resilience, and foster inclusive development (Yiu, 2025).

The framework's third stage identifies the socio-ecological validation workshop as a pivotal mechanism for bridging empirical data and experiential knowledge. This process facilitates the transformation of broad solutions into contextually valid and widely accepted interventions.

3 Study Context: Mediterranean Cities

The Mediterranean basin presents a cogent yet challenging context for urban resilience. Its distinctive climatic gradients, rich biodiversity hotspots, and millennia of human landscape modification have created a complex socio-ecological system (Amico & Currà, 2014). Cities in the region face pronounced environmental pressures, including (i) intensifying heatwaves and urban heat island effects, (ii) water scarcity, exacerbated by seasonal variability and competing demands, (iii) increased flood risks due to intense rainfall events and impervious surfaces, and (iv) biodiversity loss driven by habitat fragmentation and land-use change.

In the Mediterranean basin, the concentration of urban development patterns makes the region vulnerable to significant environmental pressures stemming from escalating temperatures, including urban heat island effects, impacts on human health, and increased water demand, especially during the summer season (Jetol et al., 2024).

Incorporating NbS enhances urban resilience and slows biodiversity decline by developing new green neighborhoods and fostering the regeneration of neglected urban areas (Lehmann, 2021).

This study is being conducted for the OurMED Project's demonstration sites. The project encompasses eight demonstration sites, distributed across the Mediterranean Basin, representing the basin's diversity, as illustrated in Fig. 2.

4 Methodology: A Three-Stage Participatory Framework

Our methodology employs a spatially explicit, three-stage framework for identifying, prioritizing, and validating NbS. This mixed-methods approach strikes a balance between technical rigor and socio-ecological relevance, as shown in Fig. 1.

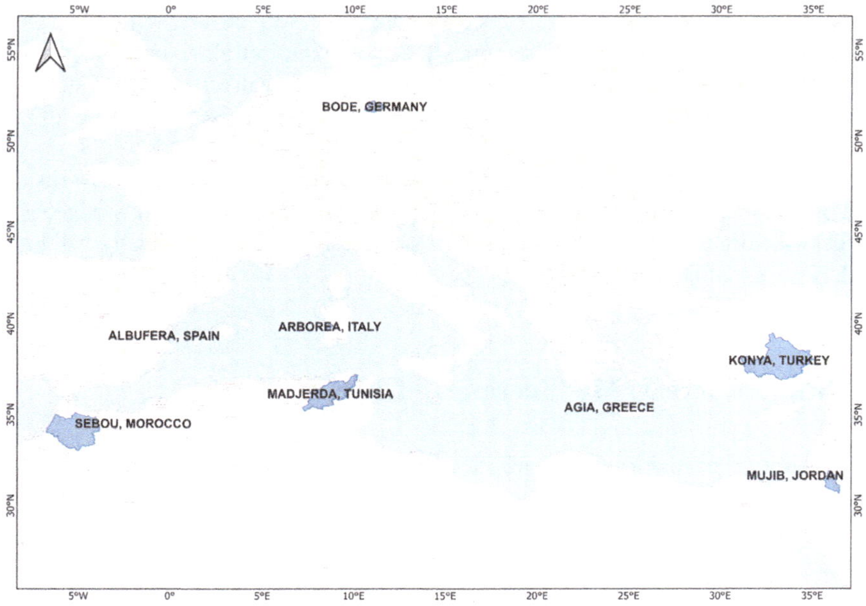

Fig. 2 DEMO sites available across the Mediterranean basin within the OurMED Project

4.1 *Stage 1: Catchment Classification and Biophysical Profiling*

The first stage involved geospatial analysis to assess the biophysical context of each demonstration site. Using GIS tools, we integrated data layers for the following:

- **Ecosystem types** (land cover, vegetation indices);
- **Climate zones** (precipitation, temperature, aridity indices);
- **Soil characteristics** (type, permeability);
- **Topography** (slope, drainage patterns).

This classification generated a baseline ecological profile, providing a foundational layer for assessing the suitability of NbS.

4.2 *Stage 2: Evidence-Based Screening and Decision Matrix*

In the second stage, we conducted a systematic literature review to identify NbS interventions with documented effectiveness under specific biophysical conditions. A decision matrix was developed to cross-reference site characteristics with potential NbS types, such as:

- **Green infrastructure**: green roofs, permeable pavements, urban forests;
- **Blue infrastructure**: constructed wetlands, rain gardens, retention ponds;
- **Hybrid solutions**: bioswales, infiltration trenches, water harvesting systems.

Each NbS was scored based on criteria including hydrological performance, biodiversity enhancement, cost, and maintenance requirements. This process yielded preliminary NbS suitability maps for each site (Sarwar et al., 2025).

4.3 Stage 3: Expert Validation Through a Workshop

The central component of the participatory approach was a structured workshop that engaged local experts from all eight demonstration sites. This workshop was designed as both a knowledge-sharing arena and a decision-support mechanism, with the aim of fostering interdisciplinary dialogue, integrating diverse forms of expertise, and systematically incorporating local perspectives into NbS mapping and planning. The workshop design involved establishing participant recruitment and composition, as well as developing a workshop protocol, as detailed in the following sections.

Participant Recruitment and Composition

Participants (3–8 per demo site) were selected from the OurMED Project Consortium for their expertise in water management, urban planning, ecology, and community engagement, ensuring that the workshop reflected a broad range of perspectives and knowledge relevant to the demonstration sites. All had more than ten years of experience in their respective fields.

Workshop Protocol

A standardized protocol was developed to guide the session, combining plenary discussions with breakout groups to foster both collective reflection and focused problem-solving. The protocol emphasized transparency, inclusivity, and iterative feedback, ensuring meaningful contributions from all participants. Structured facilitation techniques were employed to systematically document inputs and translate local knowledge into actionable insights for the project.

1. **Knowledge Presentation (30 min)**: Participants reviewed preliminary NbS maps and technical documentation.
2. **Structured Deliberation (60 min)**: Facilitated breakout sessions focused on:
   - Technical feasibility: "Are soil permeability assumptions accurate?"
   - Social acceptability: "Will this intervention align with community values?"

- Institutional viability: "What regulatory or funding barriers exist?"

3. **Iterative Mapping**: Participants annotated maps, suggesting modifications, vetoes, or new priorities.
4. **Synthesis and Feedback (30 min)**: Revised maps were presented, and consensus was documented.

Figure 3 illustrates the third stage of the participatory workshop, involving different stakeholders, focusing on the participatory map discussion of the conceptual framework. Participants engage in discussions based on local knowledge, provide feedback on preliminary NbS maps, and contribute to the iterative refinement of the proposed solutions. This process is crucial for validating the technical assessment and ensuring that the final NbS map is locally appropriate and implementable.

Fig. 3 Local experts participatory approach for NbS development. *Source* OurMED Project Website & Social Media

5 Results: Thematic Outcomes and Spatial Adjustments

The validation workshop generated critical insight, which proved instrumental in refining and significantly updating the NbS proposals. A summary of the key problems, relevant NbS interventions, and their co-benefits and rationale is provided in Table 1. Beyond confirming the relevance of previously identified measures, the discussions highlighted context-specific priorities, spatial adjustments, and thematic refinements that strengthened the overall coherence of the strategies. The collective input of local experts not only validated technical assumptions but also introduced new dimensions of social acceptability, governance feasibility, and ecological integration. As a result, the proposals evolved into more robust, place-sensitive interventions that were better aligned with the diverse realities of the demonstration sites. Key findings are organized below.

5.1 Thematic Analysis of Workshop Discussions

Qualitative analysis of the workshop discussions revealed recurring patterns and points of convergence across the diverse demonstration sites. By systematically coding and categorizing the participants' contributions, the analysis distilled complex dialogues into coherent thematic clusters. These themes capture both the shared challenges and site-specific distinctions that emerged during deliberations, offering a structured lens for interpreting the collective knowledge generated. Three central themes emerged from the qualitative analysis: contextualization overrides generalization; governance and maintenance as critical barriers; and uncovering hidden social landscapes.

Theme 1: Contextualization Overrides Generalization

Local experts consistently emphasized that generic NbS templates must be adapted to site-specific conditions. For example, in Konya, a centralized water-harvesting basin was replaced with distributed small-scale rain gardens due to the high infiltration potential and semiarid conditions. Also, the lake renaturalization was not considered a viable solution by locals, as the inflows required for restoring existing water bodies are largely diverted for agricultural use.

Theme 2: Governance and Maintenance as Critical Barriers

Discussions consistently shifted from technical design to long-term governance. Key concerns that emerged in all of the demo sites included the lack of dedicated municipal budgets for NbS maintenance; institutional fragmentation between water, planning,

Table 1 OurMED project demo sites: key water management problems, challenges, and potential for NbS

Demo site (Pop.)	Key problems	NbS interventions	Co-benefits & rationale
Bode, Germany (0.37)	Water quantity, quality deterioration	River restoration and renaturalization, Riparian green corridors, Urban retention basins, and wetlands	Reforestation and healthy soils stabilize banks, reduce erosion, filter pollutants (improving quality), and enhance infiltration, wetlands act as natural filters, removing nutrients
Jucar, Spain (1.0)	River pollution and eutrophication, flood risk in built-up areas, and soil sealing	Wetland Creation/ Restoration, Constructed Wetlands and Riparian Buffer Zones. Floodplain reconnection, permeable surfaces	Wetlands act as natural filters, removing nutrients (such as nitrogen and phosphorus) and suspended solids, which significantly improves water quality and provides habitat
Agia, Crete, Greece (0.07)	Water scarcity, seawater intrusion into municipal aquifers	Managed Aquifer Recharge (MAR), Restoration of Coastal Ecosystems, Protecting and restoring coastal wetlands and mangroves	Building up the freshwater head in the aquifer pushes the saltwater interface back toward the coast. Coastal ecosystems act as physical barriers, enhancing coastal resilience and potentially influencing local hydraulic gradients
Arborea, Italy (0.07)	Urban and agricultural water pollution, Inefficient water distribution, and high energy use in water services	Nature-based drainage systems, Vegetated treatment systems, and green buffers integrated with water energy management	Capture and infiltrate rainwater locally to reduce runoff and augment groundwater. Filter stormwater runoff contaminants. Green buffers reduce building cooling needs (lowering energy demand for power generation)
Mujib, Jordan (0.56)	Water scarcity, Unequal municipal water distribution, and constraints on wastewater reuse	Managed Aquifer Recharge (MAR), re-meandering rivers, restoring floodplains, and removing physical barriers	Restored floodplains and meanders allow natural filtering and purification processes to occur. Enhances natural flood/ drought regulation, as well as water flow dynamics
Sebou, Morocco (6.2)	Urban flooding, untreated or poorly treated wastewater, and loss of urban biodiversity	Urban Wetlands, Habitat Restoration, and Reconnecting Fragmented Riparian Ecosystems	Supports healthy watershed function, which regulates water flow and directly restores habitats for aquatic and terrestrial species

(continued)

Table 1 (continued)

Demo site (Pop.)	Key problems	NbS interventions	Co-benefits & rationale
Medjerda, Tunisia (2.2)	Flood risk in urban areas, sedimentation in urban waterways, and aging water infrastructure	Floodplain restoration, Riparian Buffer Zones/ Strips along watercourses	Slows surface runoff, trapping suspended sediment, and provides water storage
Konya, Turkey (3.2)	Groundwater overexploitation for agricultural and domestic supply; land subsidence risks; desertification; urban heat stress	Water Harvesting (e.g., rainwater harvesting, small-scale for irrigation) and Storage, Lake renaturalization around surface water bodies, Forest restoration and conservation activities	Maximizes the capture of rainfall to reduce reliance on groundwater or surface water during dry periods

and environment departments; and the need for co-stewardship models involving communities and private actors.

Theme 3: Uncovering Hidden Social Landscapes

Local knowledge revealed non-physical factors influencing the suitability of NbS in all the demo sites, such as historical social tensions in certain neighborhoods that affect site selection, cultural preferences for specific types of green spaces, and equity considerations in the distribution of NbS benefits.

5.2 *Quantitative Changes in NbS*

Overall, one-third of the preliminary NbS required modification after validation. Adjustments included:

- Spatial relocation of interventions (22% of cases);
- Changes in the NbS type or scale (45% of cases);
- Complete removal or replacement of proposals (33% of cases).

5.3 *Stakeholder Perspectives on Benefits and Trade-Offs*

Participants identified multiple co-benefits of NbS, including:

- **Ecological**: Enhanced biodiversity, improved water quality, and microclimate regulation;
- **Social**: Increased recreational access, improved mental health, and social cohesion;
- **Economic**: Reduced infrastructure costs, increased property values, and job creation.

However, trade-offs also arose, particularly between maximizing hydrological performance and addressing social equity. In several cases, such as the Jucar, Spain demo site, experts preferred the wetlands option over the floodplain reconnection due to their prior experience with these solutions in the basin. Although these solutions were slightly less technically optimal, they were preferred to ensure broader community acceptance.

6 Discussion: Barriers, Enablers, and Transferability of the Framework

Our findings underscore the value of participatory validation and highlight the systemic factors that influence NbS implementation. Beyond confirming the methodological robustness of the participatory approach, the analysis revealed a set of barriers that constrain the operationalization of nature-based solutions, including institutional fragmentation, limited financial resources, and uneven technical capacities across demonstration sites. At the same time, several enabling conditions emerged, such as strong local leadership, cross-sectoral collaboration, and the presence of supportive policy frameworks, which collectively enhance the feasibility of NbS adoption. Importantly, the study also sheds light on the transferability of the framework: while certain design principles and validation protocols appear broadly applicable across contexts, others require careful adaptation to local socio-ecological realities. This interplay between barriers, enablers, and context-specific adjustments underscores the need for flexible implementation strategies that can balance standardization and sensitivity to place-based dynamics.

6.1 Key Barriers to NbS Implementation

Discussions among the workshop participants revealed several systemic challenges that constrain the effective planning, implementation, and scaling of NbS at different levels. These challenges span institutional, financial, technical, and regulatory dimensions, and collectively limit the ability of local governments to fully realize the long-term benefits of NbS. Key issues identified include:

1. **Institutional Fragmentation**: Siloed departments hinder integrated planning.

2. **Short-Term Funding Cycles**: Misaligned with the long-term nature of NbS benefits.
3. **Technical Capacity Gaps**: Municipal staff often lack training in NbS design and maintenance.
4. **Regulatory Inconsistencies**: Planning codes may not accommodate innovative NbS.

6.2 Critical Enablers for Success

Workshop discussions also highlighted a set of enabling factors that can significantly improve the uptake, effectiveness, and long-term sustainability of NbS. These enablers operate across governance, institutional, and social dimensions and help overcome many of the structural barriers identified above. The most frequently cited enablers include:

1. **Presence of "Sampling" Actors**: Dedicated individuals within municipalities or communities drive progress.
2. **Pilot Projects and Demonstrations**: Small-scale success builds trust and political support.
3. **Hybrid Governance Models**: Partnerships between the public, private, and community sectors enhance viability.
4. **Participatory Co-Design Processes**: Early engagement fosters ownership and reduces conflict.

6.3 Transferability of the Framework

The proposed three-stage framework is suitable for tailoring to diverse contexts. Key transferability principles include:

- **Modularity**: Each stage can be adjusted based on data availability and local priorities.
- **Scalability**: The approach can be applied from the neighborhood to the city scale.
- **Inclusivity**: The workshop protocol can be adapted to engage multiple stakeholder groups, including marginalized communities.

The large fraction of NbS needing modification is a powerful indicator of the risks associated with top-down planning. This strongly argues for the institutionalization of participatory validation in municipal climate adaptation protocols.

7 Conclusion and Future Research Directions

This chapter demonstrates that participatory validation is not merely a supplementary step but rather a core component of effective NbS planning. Our Mediterranean case study reveals that technical proposals often require significant revision to align with local realities, as evidenced by the one-third modification rate. This underscores a critical lesson: urban resilience cannot be engineered from above; it must be co-created with those who live in and manage urban landscapes. The proposed framework is fully replicable, as it relies exclusively on freely and easily accessible datasets; consequently, the approach can be readily reproduced and applied to other case studies and comparable contexts.

The framework presented here entails a transferable model for integrating technical analysis with expert validation. This study advances urban resilience theory by operationalizing the concepts of co-production, adaptive governance, and emergent resilience. For practitioners, it provides a structured yet flexible toolkit for developing NbS that are both scientifically sound and socially legitimate.

7.1 Future Research Avenues

1. **Longitudinal Comparative Studies**: Monitor the performance, maintenance, and community perceptions of validated vs. non-validated NbS over time.
2. **Scalability of Participation**: Investigate how digital platforms can deepen and broaden engagement while maintaining deliberation quality.
3. **Economic Valuation of Co-Design**: Perform full-cost accounting to compare the lifecycle costs and benefits of participatory vs. conventional approaches.
4. **Justice and Equity in NbS**: Explore how participatory processes can be designed to amplify marginalized voices and ensure the equitable distribution of benefits.

In an era of climate uncertainty and rapid urbanization, the pursuit of resilient cities requires more than technical innovation; it must also address social innovation. Participatory validation of NbS represents a critical pathway toward urban futures that are sustainable, resilient, inclusive, and just.

Acknowledgements This work was supported by the OurMED PRIMA Program project funded by the European Union's Horizon 2020 Research and Innovation program under grant agreement No. 2222 and the PRIN Project CHANCES An integrated modeling approach for mitigating climate CHANge effects through enhanCEd weathering in Southern Italy (CUP: E53D23021850001).

References

Amico, A., & Currà, E. (2014). The role of urban built heritage in qualify and quantify resilience: Specific issues in Mediterranean city. *Procedia Economics and Finance, 18*, 181–190.

Ali, E., Cramer, W., Carnicer, J., Georgopoulou, E., Hilmi, N. J. M., Le Cozannet, G., & Lionello, P. (2022). *Cross-chapter paper 4: Mediterranean region.* In H.-O. Pörtner, D. C. Roberts, M. Tignor, E. S. Poloczanska, K. Mintenbeck, A. Alegría, M. Craig, S. Langsdorf, S. Löschke, V. Möller, A. Okem, & B. Rama (Eds.), *Climate change 2022: Impacts, adaptation and vulnerability. Contribution of Working Group II to the Sixth Assessment Report of the Intergovernmental Panel on Climate Change* (pp. 2233–2272). Cambridge University Press.

Bai, X., McPhearson, T., Cleugh, H. A., Nagendra, H., Tong, X., Zhu, T., & Zhu, Y. (2017). Linking urbanization and the environment: Conceptual and empirical advances. *Annual Review of Environment and Resources, 42*(1), 215–240.

Bush, J., & Doyon, A. (2019). Building urban resilience with nature-based solutions: How can urban planning contribute? *Cities, 95*, 102483.

European Commission (EC). (2015). *Towards an EU research and innovation policy agenda for nature-based solutions & re-naturing cities.* Publications Office of the European Union.

Gobster, P. H. (2010). Urban ecological restoration. *Nature and Culture, 5*(3), 227–230.

International Union for Conservation of Nature (IUCN). (2020). *Global standard for nature-based solutions: A user-friendly framework for the verification, design and scaling up of NbS.* IUCN.

Jetol, J. B., Ariffin, A., Ghani, A., Rafiuddin, & Dayana. (2024). *Sustainable urban development in Madrid, Hammarby Sjöstad, Melaka, Vancouver and Paris.*

Kindon, S., Pain, R., & Kesby, M. (2007). *Participatory action research approaches and methods: Connecting people.* Routledge.

Lafortezza, R., & Sanesi, G. (2018). Nature-based solutions: Settling the issue of sustainable urbanization. *Environmental Research, 172*, 394–398.

Lehmann, S. (2021). Nature in the urban context: Renaturalisation as an important dimension of urban resilience and planning. *Módulo Arquitectura CUC, 26*(1), 161–190.

Meerow, S., Newell, J. P., & Stults, M. (2016). Defining urban resilience: A review. *Landscape and Urban Planning, 147*, 38–49.

Nesshöver, C., et al. (2017). The science, policy and practice of nature-based solutions: An interdisciplinary perspective. *Science of the Total Environment, 579*, 1215–1227.

Norström, A. V., et al. (2020). Principles for knowledge co-production in sustainability research. *Nature Sustainability, 3*(3), 182–190.

Sarwar, A. N., Caramiello, C., Pugliese, F., Jomaa, S., Guelmami, A., Ronse, M., ... Manfreda, S. (2025). A framework for selecting nature-based solutions: Applications and challenges at the catchment scale. *Journal of Environmental Management, 394*, 127220.

UN DESA. (2018). *World Urbanization Prospects: The 2018 Revision.* United Nations.

Varshney, K., MacKinnon, M., Zari, M. P., Shanahan, D. F., Woolley, C., Freeman, C., & van Heezik, Y. (2024). Biodiverse residential development: A review of New Zealand policies and strategies for urban biodiversity. *Urban Forestry & Urban Greening, 94*, 128276.

Wamsler, C., et al. (2020). Environmental and climate policy integration: Targeted strategies for overcoming barriers to nature-based solutions and climate change adaptation. *Journal of Cleaner Production, 247*, 119154.

Yiu, C. Y. (2025). Urban political ecology in action: Community-based planning for sustainability and heritage in a high-density urban landscape. *Sustainability, 17*(8), 3726.

Cultivating Belonging: Everyday Aesthetics and Care Practices in Nature-Based Solutions

Elisabetta Di Stefano

Abstract In response to the ecological, social, and emotional challenges faced by contemporary cities, Nature-Based Solutions (NBS) are increasingly promoted as instruments for urban resilience. However, many implementations remain anchored to technocratic or cosmetic approaches that prioritize visual impact and environmental performance over lived experience and long-term care. This chapter proposes a relational and aesthetic re-framing of NBS grounded in everyday aesthetics, the aesthetics of care, and affective ecology. Moving beyond functionalist paradigms, it argues that urban greenery becomes meaningful and durable when it is embedded in shared, slow, and sensory-based practices that foster situated forms of belonging. Adopting a theoretical and interpretive methodology rooted in environmental humanities and aesthetic theory, the chapter examines co-creation not merely as a participatory design strategy, but as an aesthetic practice unfolding over time between activation and maintenance. Through the discussion of participatory artistic interventions and ordinary urban gardening practices, the study shows how repeated gestures of care contribute to the formation of shared atmospheres, stabilize relationships with place, and support affective engagement at the local scale. The chapter concludes by suggesting that the effectiveness of Nature-Based Solutions cannot be assessed solely in technical or visual terms, but depends on their capacity to cultivate shared practices of care and situated beauty, understood as emergent qualities of collectively inhabited and continuously maintained urban environments.

Keywords Everyday aesthetics · aesthetics of care · affective ecology · co-creation · Nature-Based solutions

E. Di Stefano (✉)
Department of Humanities, University of Palermo, Palermo, Italy
e-mail: elisabetta.distefano@unipa.it

L. Diana et al. (eds.), *Greening and Resilience*, Urban Sustainability,
https://doi.org/10.1007/978-981-92-0419-9_4

1 Introduction

Nature-Based Solutions (NBS) have become a central framework in European and international environmental policy, promoted as integrative responses to pressing societal challenges such as climate change adaptation, biodiversity loss, urban heat stress, disaster risk reduction, and declining human well-being. The European Commission defines NBS as living solutions inspired by, supported by, or copied from nature, designed to address societal challenges in resource-efficient and adaptable ways while simultaneously providing environmental, social, and economic benefits (European Commission, 2015). A complementary definition proposed by IUCN frames NBS as actions to protect, sustainably manage, and restore natural or modified ecosystems in ways that address societal challenges effectively and adaptively, while simultaneously providing human well-being and biodiversity benefits (Cohen-Schacham et al., 2016). Within Horizon 2020 and subsequent research agendas, NBS are positioned as instruments capable of reconciling ecological sustainability with economic competitiveness and job creation.

From a policy perspective, NBS are often framed as pathways for systemic transformation toward a more sustainable economic model. Maes and Jacobs (2017) define them as a transition toward the use of ecosystem services with decreased input of non-renewable natural capital and increased investment in renewable natural processes. This systemic interpretation situates NBS within a broader reconfiguration of the "energy mix" underpinning socio-economic systems, aiming to reduce trade-offs between ecosystem service production and long-term ecological resilience. Similarly, Eggermont et al. (2015) emphasize that NBS integrate science, policy, and practice in order to address societal challenges while enhancing biodiversity and sustaining social–ecological systems.

However, despite their rapid institutional uptake, the conceptual boundaries of NBS remain fluid. As Nesshöver et al. (2017) observe, the term overlaps with a range of existing frameworks, including ecosystem services, ecological engineering, ecosystem-based adaptation, and green infrastructure. While this flexibility encourages interdisciplinary dialogue, it also risks ambiguity, conceptual dilution, and the instrumentalization of nature as a technical fix or symbolic marker of sustainability. Eggermont et al. (2015) warn that, without coherent governance and ecological integration, NBS may devolve into fragmented or cosmetic interventions, disconnected from long-term socio-ecological transformation.

In urban contexts, these tensions become particularly visible. Many NBS initiatives focus primarily on measurable environmental performance (e.g., carbon sequestration, water regulation, temperature mitigation) or on visible green infrastructures such as green roofs, vertical greening systems, and urban forests. While such interventions may deliver important ecosystem services, their evaluation often privileges quantitative indicators and short-term outputs over long-term relational and experiential dimensions. Even when co-creation and participation are incorporated into NBS frameworks, they are frequently framed as governance mechanisms aimed at improving efficiency, social acceptance, or cost-effectiveness, rather than as transformative processes capable of reshaping collective habits and modes of inhabitation.

Within the expanding NBS literature, the aesthetic and affective dimensions of urban green interventions remain marginal within prevailing evaluative frameworks. Although cultural ecosystem services are acknowledged, the lived, sensory, and relational experience of urban nature is rarely examined as a constitutive factor in the durability and effectiveness of NBS. Concepts such as belonging, care, and co-creation are frequently invoked descriptively rather than analytically articulated.

To address this gap, the present chapter draws on theoretical resources from everyday aesthetics and the aesthetics of care (Di Stefano, 2023, 2024; Saito, 2007, 2022), the biophilia hypothesis (Kellert & Wilson, 1995), and affective ecology (Barbiero, 2017). These perspectives foreground embodied perception, affective attachment, and the formation of habits as central dimensions of human–environment relations. Everyday aesthetics shifts attention from exceptional or monumental design to ordinary, repetitive, and situated practices through which environments are perceived and sustained. The aesthetics of care emphasizes maintenance, attention, and responsibility as aesthetic–ethical attitudes embedded in daily life. The biophilia hypothesis highlights an intrinsic human tendency to affiliate with natural processes, while affective ecology stresses the psychophysical benefits and emotional stabilization generated by meaningful engagement with living systems. Together, these approaches provide an interpretive lens capable of complementing performance-based NBS frameworks by illuminating how sensory experience, temporal rhythms, and shared practices contribute to socio-ecological resilience.

In this perspective, NBS are understood not only as ecological infrastructures or governance instruments but as processes embedded in socio-ecological systems, where embodied practices and shared atmospheres play a constitutive role. Co-creation, in particular, is reinterpreted not merely as a participatory planning technique but as a temporal aesthetic process unfolding between activation and maintenance, between event and habit. This shift enables a move from evaluating NBS solely through environmental performance metrics toward considering their relational durability: the extent to which they cultivate stable forms of belonging, responsibility, and identification with place.

The contribution of this chapter is threefold. First, it systematizes the conceptual discussion of NBS by situating the proposed aesthetic–relational perspective within existing policy and research frameworks, clarifying both convergences and divergences. Second, it advances an operational interpretation of belonging, care, and co-creation by proposing qualitative analytical criteria applicable to urban NBS contexts. Third, through the analysis of two exemplary urban practices, it demonstrates how participatory artistic activation and everyday gardening can function as climate-resilient strategies that strengthen socio-ecological cohesion beyond measurable environmental outputs.

By integrating insights from environmental humanities into the contemporary debate on Nature-Based Solutions, this chapter argues that the effectiveness of urban greening initiatives ultimately depends not only on ecological functionality or economic viability, but on their capacity to generate lived, shared, and sustained forms of urban habitability.

2 Theoretical Framework: Aesthetic–Relational Foundations

Nature-Based Solutions have predominantly been framed within ecological economics, environmental engineering, and urban governance. Even when social dimensions are acknowledged, they are frequently articulated through concepts such as ecosystem services, stakeholder engagement, and participatory planning. While these perspectives have expanded the understanding of urban sustainability, they often remain anchored to performance-based paradigms and instrumental logics.

The aesthetic–relational framework proposed here does not reject these approaches, but seeks to complement them by addressing dimensions that remain comparatively under-theorized: embodied perception, temporal continuity, maintenance, and shared world-making. Instead of treating urban greenery as a technological or managerial device, this perspective interprets it as a lived process shaped by habits, affects, and relational practices. Three interrelated concepts structure this framework: belonging, care, and co-creation.

2.1 Belonging as Situated and Temporal Attachment

In urban resilience discourse, belonging is often equated with social inclusion or community participation. However, from an aesthetic–relational standpoint, it is a situated form of attachment emerging through repeated interaction and cyclical continuity.

Drawing on Dewey's conception of habit (Dewey, 1922), belonging can be understood as the stabilization of the relationship between organism and environment. Habits are not mechanical repetitions but modes of attunement that shape perception and action. Through cyclical engagement—planting, observing seasonal change, maintaining shared spaces—urban environments gradually become meaningful.

Everyday aesthetics (Saito, 2007) clarifies that aesthetic experience is not restricted to extraordinary events or designed spectacles. It is embedded in routine gestures and ordinary interactions. Belonging thus emerges not from the visual presence of greenery alone, but from embodied familiarity and shared practice.

From the perspective of affective ecology (Barbiero, 2017), such attachment also has stabilizing functions. Emotional bonds with living systems can mitigate alienation and ecological anxiety by fostering continuity and orientation. However, these bonds require temporal depth; they cannot be produced through isolated interventions.

In this framework, belonging becomes a qualitative dimension of evaluation. It can be explored through indicators such as continuity of use, repetition of shared practices, intergenerational participation, and the emergence of local narratives around a place. These criteria complement, rather than replace, ecological metrics.

2.2 *Care as Aesthetic Practice of Maintenance*

If belonging stabilizes attachment, care sustains it. In environmental discourse, care is often framed in ethical terms as stewardship or responsibility toward future generations. Within the aesthetics of care (Di Stefano, 2023; Saito, 2007, 2022), however, it is also understood as an embodied and perceptual practice.

Care manifests through maintenance: watering, pruning, composting, repairing. These repetitive gestures are frequently invisible in architectural representation, yet they are constitutive of inhabited environments. Maintenance introduces temporal depth into urban space by shifting attention from inauguration to continuity. What sustains a green intervention is not only its initial design but its ongoing responsiveness to seasonal change and ecological fragility.

From this perspective, maintenance is not secondary to design but integral to relational durability. Without sustained practices of care, even ecologically effective interventions may remain externally managed rather than collectively inhabited.

Operationally, care in NBS contexts can be explored through the presence of shared maintenance structures, visible embodied engagement in cultivation practices, adaptation to seasonal cycles, and long-term commitment beyond initial activation phases. Care thus functions both as an aesthetic attitude and as an evaluative dimension: it signals whether urban nature is merely installed or genuinely sustained through collective practice.

2.3 *Co-Creation: From Governance Tool to Aesthetic Creation*

In European policy discourse, co-creation is commonly defined as collaborative planning involving stakeholders and communities (European Commission, 2015, 2021). In this procedural sense, co-creation enhances legitimacy and management efficiency.

However, the aesthetic genealogy of creation suggests a broader meaning. As Tatarkiewicz (1980) shows in his history of aesthetic concepts, the idea of creation evolved from a strictly individual or divine act to a relational and processual phenomenon in modern artistic practice. Creation unfolds in time and involves interaction between author, work, and audience.

This relational dimension becomes explicit in Bourriaud's theory of relational aesthetics (Bourriaud, 2002), where the artwork produces situations of encounter instead of autonomous objects. Artistic meaning emerges through participation and interaction.

Seen from this perspective, co-creation in participatory art does not dissolve authorship into procedural consensus. Rather, it reconfigures it. The artist establishes a framework within which participants become co-creative agents. Creativity remains intentional yet distributed.

This distinction is crucial for understanding co-creation in NBS contexts. When reduced to governance technique, co-creation risks becoming administrative inclusion. When interpreted as aesthetic practice, it becomes shared world-making—an embodied and temporal process through which belonging and care are formed.

Co-creation can therefore be explored through qualitative dimensions such as distributed but structured agency, transformation of functional acts into shared experiential processes, temporal continuity beyond isolated events, and the emergence of collective meaning.

Synthesis

Together, belonging, care, and co-creation articulate an aesthetic–relational understanding of Nature-Based Solutions. They shift attention from infrastructural performance to durable relational embedding, from measurable outputs to sustained practices that shape how spaces are inhabited.

This framework operates alongside ecological evaluation models by introducing qualitative criteria concerned with experiential integration of NBS within urban life.

3 Methodology

3.1 Epistemological and Disciplinary Positioning

This study adopts a qualitative and interpretive approach situated within environmental humanities and philosophical aesthetics. Rather than measuring ecological performance or socio-economic outputs, the research investigates the aesthetic–relational dimensions through which Nature-Based Solutions are experienced, inhabited, and sustained over time.

Environmental humanities provide the broader framework by emphasizing the entanglement of ecological processes and cultural practices. Within this field, everyday aesthetics and the aesthetics of care offer conceptual tools to analyze habit formation, embodied participation, and maintenance as constitutive elements of urban environments. Affect theory further informs the approach by foregrounding pre-reflective and relational dimensions of experience.

This epistemological positioning implies that the study does not seek generalizable statistical findings. Instead, it develops an interpretive framework by illuminating underexplored experiential dimensions of NBS.

3.2 Research Design and Case Selection

The research employs a qualitative case-based methodology. Two urban practices are analyzed: the participatory artistic practices of Egle Oddo and the Orto Capovolto urban gardening project.

The cases are selected for their explicit engagement with ecological processes, participatory involvement of local communities, temporal continuity beyond one-off events, integration of cultivation and maintenance practices, and documented interaction between human and non-human agents. The aim is not to evaluate ecological efficiency but to examine how belonging, care, and co-creation manifest in concrete socio-ecological settings.

3.3 Analytical Framework

The analysis applies the conceptual categories developed in Sect. 2 as interpretive lenses. To translate abstract concepts into evaluative dimensions, the case studies are analyzed according to the following qualitative criteria: temporality (whether the intervention is episodic or continuous and whether it incorporates seasonal cycles), embodied engagement (the degree of physical involvement in cultivation and maintenance), relational density (the extent to which sustained interaction among human and non-human agents occurs), maintenance structure (how responsibility for long-term care is distributed), and atmospheric formation, understood in Böhme's (2017) sense as the production of affectively charged spatial presence. These dimensions allow the assessment of what may be termed *relational durability*: the extent to which an NBS intervention becomes integrated into everyday life through sustained practices and affective attachment.

3.4 Scope and Limits

This study does not provide quantitative data on environmental performance, carbon reduction, or economic impact. Nor does it claim universal applicability. Its contribution is exploratory and conceptual: to expand the evaluative vocabulary of NBS by introducing aesthetic–relational criteria that may complement existing ecological assessment tools.

4 Case Analysis

The following case studies illustrate two complementary modalities of aesthetic–relational Nature-Based Solutions. While Egle Oddo's practice foregrounds artistic activation and performative co-creation, Orto Capovolto exemplifies stabilization through everyday continuity. Together, they demonstrate how belonging, care, and co-creation operate in different yet interconnected ways.

4.1 Artistic Activation: Egle Oddo

Egle Oddo's participatory practices exemplify a form of co-creation that is explicitly structured within an artistic framework. Instead of presenting vegetation as a finished design intervention, her projects unfold as temporally oriented processes centered on cultivation, shared labor, and embodied interaction. Her research on "performative habitats" explicitly explores the intersection between artistic activation and vegetal processes (Oddo, 2021; Oddo & Adragna, 2022).

Temporality and Process

Oddo's interventions are not conceived as static installations. Even when initiated as events, they are oriented toward growth, seasonality, and ecological change. Germination, waiting, tending, and harvesting structure the experience. Temporality is not a background condition but an integral dimension of the work.

Embodied and Performative Engagement

Participation is grounded in tactile engagement with soil and seeds. As illustrated in Fig. 1, the cultivated plot becomes activated through collective gestures of planting and handling vegetal matter. The spatial setting is minimal, yet it operates as a relational device.

A significant element in the scene is the artist's distinctive attire. Oddo does not wear neutral working clothes but a patterned garment that visually differentiates the situation from ordinary agricultural labor. This detail signals that the activity is not merely functional gardening but an artistic action. The extraordinary is inserted into the ordinary. Cultivation becomes framed as aesthetic practice without losing its practical dimension.

This performative framing is crucial: co-creation here does not eliminate authorship. The artist establishes a conceptual and situational framework within which participants act. Creativity is distributed but not dissolved. The cultivated space functions as a temporary micro-urban laboratory where artistic intentionality and collective participation intersect. In this sense, the situation resonates with Fischer-Lichte's understanding of performance as an event that transforms both participants and the perceptual configuration of space through embodied co-presence (Fischer-Lichte, 2008).

Fig. 1 Egle Oddo, participatory cultivation practice, bratislava (Photo: Adam Šakový). The image illustrates co-creation as a performative interaction in which artistic authorship and collective participation coexist

Relational Density and Maintenance

Interaction unfolds not only among participants but between human and non-human agents. Soil conditions, plant growth, and seasonal variation shape decisions and rhythms. Maintenance practices—watering, monitoring, adapting—extend the artistic gesture beyond the initial activation. Care thus becomes visible as shared responsibility, reinforcing sustained attachment.

Analytical Synthesis

Through the lens developed in Sections 2 and 3, Oddo's practice demonstrates how artistic activation can operate as an alternative model of Nature-Based Solution. It generates temporal continuity through ecological process, embodied engagement through tactile participation, distributed yet structured agency, shared maintenance practices, and an aesthetic reframing of cultivation as collective world-making. In this configuration, co-creation is not a managerial instrument but an aesthetic practice in which artistic authorship and collective participation coexist. In this sense, artistic co-creation functions less as a fully stabilized everyday practice than as an activating threshold capable of opening processes that may later sediment into habitual forms of engagement.

4.2 Everyday Stabilization: Orto Capovolto

If artistic activation operates as a catalytic threshold, Orto Capovolto illustrates how such impulses can sediment into repetition and habitual continuity. Rather than emerging from an explicitly artistic framework, the project is rooted in collective urban gardening practices that gradually transform a shared space into a lived socio-ecological environment.

Temporality and Continuity

Unlike interventions structured around singular events, Orto Capovolto unfolds through sustained engagement. The garden operates according to seasonal rhythms, cultivation cycles, and recurring maintenance routines. Its temporal structure is not episodic but cyclical. Participants return regularly to plant, water, harvest, and adapt to changing environmental conditions. This repetition fosters familiarity and long-term attachment.

Temporality here is not staged as aesthetic performance but embedded in practice. The gradual transformation of the space—visible growth, soil preparation, composting—creates a shared temporal horizon in which ecological processes become perceptible and meaningful.

Embodied Engagement and Habit Formation

Participation in Orto Capovolto is grounded in routine bodily practices. Cultivation requires physical presence, manual work, and attention to vegetal needs. Over time, these gestures become habits in Dewey's sense: stabilized patterns of interaction that structure perception and action.

Through repeated engagement, participants develop a form of situated competence. The space becomes legible not only visually but through embodied familiarity—knowing when to water, how to prune, how to respond to seasonal variation. This embodied knowledge contributes to the formation of belonging as described in Sect. 2.

Relational Density and Shared Responsibility

The garden functions as a micro-community. Interaction occurs not only between individuals but also between human and non-human agents. Plants, soil conditions, weather variations, and growth cycles shape collective decisions and responsibilities.

Maintenance structures are distributed rather than centralized. Care is not delegated to external authorities but shared among participants. This shared responsibility reinforces durable relational embedding: the space persists because it is collectively sustained.

Atmospheric Formation and Identification

Beyond its functional role, Orto Capovolto generates a recognizable atmosphere. The garden becomes a site of encounter, conversation, and intergenerational exchange. The sensory qualities of the space—smell of soil, textures of leaves, visual transformation across seasons—contribute to affective attachment.

In this configuration, belonging is not produced through symbolic representation but through lived continuity. The garden gradually becomes part of participants' everyday routines and personal narratives.

Analytical Synthesis

Interpreted through the aesthetic–relational framework, Orto Capovolto demonstrates how everyday gardening can operate as a climate-resilient strategy grounded in relational durability. The project generates:

- temporal continuity through cyclical cultivation
- embodied engagement through routine practices
- distributed maintenance structures
- affective stabilization through shared atmosphere, and
- collective identification with place

Unlike artistic activation, which introduces a performative reframing of cultivation, Orto Capovolto exemplifies how ordinary practice itself can acquire aesthetic significance through sustained inhabitation. Together, the two cases illustrate complementary modalities of aesthetic–relational NBS: one initiated through artistic orchestration, the other stabilized through habitual continuity.

5 Discussion and Implications

5.1 Positioning within the NBS Debate

The two case studies analyzed above allow for a reframing of Nature-Based Solutions beyond their dominant performance-based interpretation. Current NBS evaluation models tend to prioritize quantifiable ecosystem services, cost-efficiency, and governance indicators. Even when participation is included, it is frequently assessed in terms of procedural inclusion or stakeholder engagement.

The aesthetic–relational framework proposed here does not contest ecological metrics but highlights a dimension that remains structurally underrepresented in prevailing models: relational durability. While ecosystem services measure functional outputs, relational durability concerns the extent to which an intervention becomes embedded in everyday practices and integrated into local narratives through sustained maintenance.

In this sense, the contribution of the present study lies in shifting the evaluative focus from infrastructural performance to experiential integration. Belonging, care, and aesthetic co-creation are not supplementary cultural benefits but conditions that influence the long-term resilience of urban green interventions.

5.2 Complementarity with Existing Evaluation Tools

Recent assessment tools—such as performance-based environmental indicators or biophilic design matrices (Wijesooriya et al., 2025)—offer valuable criteria for evaluating ecological functionality and spatial quality. However, they often privilege design attributes over lived processes.

The qualitative indicators proposed in this chapter—temporality, embodied engagement, distributed maintenance, relational density, and atmospheric formation—do not aim to replace existing frameworks. Instead, they can operate as complementary dimensions within interdisciplinary evaluation models.

For instance, temporal continuity may indicate whether an intervention extends beyond inaugural activation; shared maintenance structures may signal community integration; and embodied engagement may reveal whether users are passive beneficiaries or active co-producers of space. Such criteria allow planners and designers to assess not only whether an intervention works ecologically, but whether it becomes inhabitable and socially sustained.

5.3 Reframing co-Creation

A central theoretical contribution of this study concerns the reinterpretation of co-creation. Within policy discourse, co-creation often denotes participatory governance. By contrast, the aesthetic genealogy reconstructed here—drawing on Tatarkiewicz and Bourriaud—reclaims co-creation as a category of artistic and relational creativity.

The cases analyzed demonstrate that co-creation can function as a form of shared world-making rather than merely procedural inclusion. In Oddo's practice, artistic authorship structures participation without dissolving it into managerial consensus. In Orto Capovolto, habitual continuity transforms collective gardening into an ongoing aesthetic process.

This distinction clarifies that co-creation in NBS contexts should not be reduced to stakeholder consultation. When interpreted aesthetically, it becomes a temporal process through which belonging and care are cultivated. This process unfolds in phases: artistic activation may initiate shared perception and participation, but relational durability depends on the subsequent transformation of these moments into repeated practices embedded in everyday life. Artistic activation does not replace everyday practice; rather, it intensifies perception and symbolically reframes ordinary acts, enabling them to become consciously shared and eventually sedimented.

5.4 Limits and Future Directions

This study does not provide quantitative evidence of ecological or economic impact. Its scope is conceptual and exploratory. Future research could integrate the proposed qualitative criteria into interdisciplinary assessment models, combining relational indicators with environmental performance data. Moreover, comparative studies across diverse urban contexts would be necessary to evaluate the broader applicability of the aesthetic–relational framework.

6 Concluding Remarks

Nature-Based Solutions have emerged as a central paradigm in contemporary environmental policy, yet their evaluation remains predominantly anchored to ecological performance and governance efficiency. This chapter has argued that such frameworks, while indispensable, are insufficient to account for the experiential and relational dimensions that shape the long-term resilience of urban green interventions.

By introducing an aesthetic–relational perspective grounded in everyday aesthetics, the aesthetics of care, and the aesthetic genealogy of co-creation, the study has proposed a complementary evaluative vocabulary centered on belonging, maintenance, and shared creativity. These concepts redirect attention from activation to continuity, and from participation as procedure to co-creation as shared world-making.

Through the analysis of artistic activation in Egle Oddo's practice and the everyday stabilization exemplified by Orto Capovolto, the chapter has demonstrated that urban greening initiatives acquire durability not solely through ecological functionality, but through sustained practices that generate attachment, responsibility, and identification with place.

The effectiveness of Nature-Based Solutions, therefore, depends not only on what they produce environmentally, but on how they are lived, maintained, and collectively shaped. Integrating aesthetic–relational criteria into NBS evaluation frameworks may contribute to a more comprehensive understanding of urban resilience—one that recognizes habit, care, and shared creativity as structural components of sustainable transformation.

References

Barbiero, G. (2017). Ecologia affettiva. Come trarre benessere psicofisico dalla natura.
Böhme, G. (2017). The aesthetics of atmospheres. Routledge.
Bourriaud, N. (2002). Relational aesthetics. Les Presses du réel.

Cohen-Schacham, E., Walters, G., Janzen, C., Maginnis, S. (2016). Nature-based solutions to address global societal challenges. IUCN.
Dewey, J. (1922). Human nature and conduct: An introduction to social psychology. Holt & Co.
Di Stefano, E. (2023). Estetica urbana. Atmosfere e artificazione degli spazi della città.
Di Stefano, E. (2024). Caring for the landscape: From participatory art to everyday aesthetics. *Studi di Estetica, 52*(2), 22–34.
Eggermont, H., Balian, E., Azevedo, J. M. N., Beumer, V., Brodin, T., Claudet, J., Fady, B., Grube, M., Keune, H., Lamarque, P., Reuter, K., Smith, M., van Ham, C., Weisser, W. W., & Le Roux, X. (2015). Nature-based solutions: New influence for environmental management and research in Europe. GAIA–ecological perspectives for. *Science and Society, 24*(4), 243–248.
European Commission (2015). Towards an EU research and innovation policy agenda for nature-based solutions and re-naturing cities. Publications Office of the European Union.
European Commission (2021). Evaluating the impact of nature-based solutions: A handbook for practitioners. Publications Office of the European Union.
Fischer-Lichte E (2008). The transformative power of performance: A new aesthetics. Routledge.
Kellert, S.R, Wilson, E.O. (1995). The biophilia hypothesis. Island Press.
Maes, J., & Jacobs, S. (2017). Nature-based solutions for Europe's sustainable development. *Conservation Letters, 10*(1), 121–124.
Nesshöver, C., Assmuth, T., Irvine, K. N., Rusch, G. M., Waylen, K., Delbaere, B., Haase, D., Jones-Walters, L., Keune, H., Kovacs, E., Ledoux, L., Raymond, C. M., van Dijk, J., Vijver, M., & Wilkinson, M. (2017). The science, policy and practice of nature-based solutions: An interdisciplinary perspective. *Science of the Total Environment, 579*, 1215–1227.
Oddo, E. (2021). Evolutionary gardens and performative habitats. Ruukku–Studies in Artistic Research 16.
Oddo, E., Adragna, L. (eds) (2022). Performative habitats between art, philosophy, and science. Postmedia Books.
Saito, Y. (2007). Everyday aesthetics. Oxford University Press.
Saito, Y. (2022). Aesthetics of care: Practice in everyday life.
Tatarkiewicz, W. (1980). A history of six ideas: An essay in aesthetics. Martinus Nijhoff.
Wijesooriya, N., Markauskaite, L., & Brambilla, A. (2025). Biophilic quality matrix: A tool to evaluate the biophilic quality of a building during early design stage. *Cleaner Production Letters, 8*, 1–14.

Nature-Based Solutions for the Regeneration and Resilience of Smart Villages: Design, Strategies and Case Studies in Mediterranean Inner Areas

Luisa Lombardo

Abstract This research investigates the strategic role of Nature-Based Solutions (NbS) in the regeneration, resilience, and adaptive transformation of rural settlements into Smart Villages, with a particular focus on Mediterranean inner areas. While NbS have been extensively applied in dense urban contexts, their deployment in heritage-rich rural territories remains limited, despite these areas' potential as experimental laboratories for ecological and socio-cultural innovation. The study adopts a qualitative, comparative, and place-based approach, examining how NbS can enhance environmental performance, reinforce local identity, and support sustainable development in small settlements facing demographic decline, infrastructural fragility, and climate-related pressures. The Madonie region in Sicily serves as the primary case study, representing a context where traditional architecture, ecological sensitivity, and socio-economic challenges intersect. Through a systematic review of scientific literature and a critical assessment of recent public interventions, the research identifies both effective design strategies and recurring shortcomings, including material incoherence, insufficient microclimatic mitigation, and limited integration of vegetation within the built environment. Findings indicate that even small-scale greening interventions—such as pocket gardens, vegetated pedestrian routes, and context-sensitive green roofs—can substantially improve thermal comfort, biodiversity, and the usability of public spaces when aligned with local morphology and cultural heritage. The study further analyses six European case studies selected for their replicability, documented outcomes, and relevance to heritage-rich rural contexts, providing comparative insights for the application of NbS in the Madonie region. To support evidence-based planning and replicability, a Smart Village Performance Index is introduced, integrating quantitative indicators across four dimensions: community engagement, governance effectiveness, heritage rehabilitation, and digital innovation. Overall, the research demonstrates that resilient and sustainable Smart Villages emerge from approaches that harmonize ecological functionality,

L. Lombardo (✉)
Department of Architecture, University of Palermo, Palermo, Italy
e-mail: luisa.lombardo01@unipa.it

L. Diana et al. (eds.), *Greening and Resilience*, Urban Sustainability,
https://doi.org/10.1007/978-981-92-0419-9_5

cultural coherence, and community-driven innovation, positioning NbS as essential drivers for the regeneration of Mediterranean inner areas.

Keywords Smart Villages · Nature-Based Solutions · Rural Resilience · Green Infrastructure · Climate-resilient planning · Rural Mediterranean landscapes

1 Introduction

Mediterranean inner rural regions are increasingly facing a convergence of environmental, social, and economic pressures. Climate change, demographic decline, and territorial abandonment threaten both natural landscapes and cultural heritage, challenging the long-term sustainability of these fragile territories. Yet, these areas possess unique ecological assets, historic settlements, and resilient local identities, offering opportunities for innovative strategies that reconcile preservation, resilience, and socio-environmental innovation. In this context, Nature-Based Solutions (NbS) (Hongpeng, 2023) have emerged as strategic instruments capable of transforming small rural settlements into adaptive and sustainable Smart Villages (Alhari & Fajrillah, 2022). NbS leverage natural processes and ecosystem services to enhance environmental performance, regulate hydrological cycles, and support biodiversity (Solheim & Capobianco, 2021). Beyond ecological benefits, they provide psychological and social advantages by reconnecting communities with nature and fostering human well-being. Conceptually, NbS are co-created interventions integrating natural features into landscapes and settlements, addressing complex societal challenges such as flooding, pollution, biodiversity loss, and reduced quality of life. By embedding ecological functionality into human habitats, NbS contribute to more resilient, sustainable, and adaptive socio-ecological systems.

Although widely studied in urban contexts, NbS remain underexplored in rural and heritage-rich territories, where proximity to natural environments allows interventions that can preserve ecological integrity (Pultrone, 2022). Small towns and rural settlements, while vulnerable due to demographic decline, infrastructural gaps, and climate-related risks, offer ideal conditions for low-impact, context-sensitive interventions. Their coherent spatial structures, strong local identities, and integrated landscape-architecture systems enable NbS to be implemented in ways that are both ecologically effective and culturally meaningful. The Smart Village paradigm provides a compelling framework for operationalizing NbS in rural contexts (Frantzeskaki et al., 2019). By integrating digital innovation, participatory governance, and environmental stewardship, Smart Villages foster sustainable local development while safeguarding ecological and cultural landscapes. This paradigm represents a holistic approach in which technological, environmental, and social dimensions converge to enhance resilience, quality of life, and territorial identity. This study investigates the integration of NbS within the Smart Village framework, focusing on Mediterranean inner areas where landscape conservation, architectural heritage, and environmental fragility intersect (Rahoveanu & Serban, 2022). The

Madonie region in Sicily serves as the primary case study, illustrating how targeted NbS interventions can reinforce local resilience, strengthen community cohesion, and enhance environmental and cultural performance.

The primary aim of this research is to develop a methodological framework for integrating NbS into the adaptive regeneration of small rural settlements, particularly within the Smart Village approach (Wolff & Rauf, 2023). Specifically, it examines how NbS can improve environmental comfort, support ecological resilience, and elevate quality of life while remaining sensitive to landscape identity, heritage significance, and spatial specificity. Despite growing literature on NbS and Smart Cities, operational frameworks tailored to small rural settlements with high heritage value remain scarce. This chapter addresses this gap by critically analyzing existing rehabilitation practices and planning strategies, identifying principles capable of guiding sustainable, context-aware interventions. Focusing on the Madonie region enables a rigorous evaluation of NbS viability in territories where climate change, socio-economic fragility, and territorial abandonment converge, demonstrating that such areas can become fertile grounds for innovation that is both ecologically sound and culturally respectful (Borham & Croxford, 2024). Building on doctoral research, including an atlas of innovative technological solutions for rural heritage, the study introduces practical tools—such as performance indices—to support decision-making, guide conservation authorities, and expand the repertoire of context-appropriate NbS interventions in well-preserved rural settlements. The Smart Village Performance Index helps municipalities, villages, and communities identify applicable NbS, draw comparisons across settlements, and align interventions with recognized design principles.

This chapter contributes three original insights: (i) a critical reinterpretation of Nature-Based Solutions in heritage-rich rural contexts; (ii) a comparative assessment of European NbS practices through the Smart Village lens; and (iii) the development of a Smart Village Performance Index tailored to Mediterranean inner areas. By systematically assessing the integration of NbS in heritage-rich rural settlements through a combined qualitative and quantitative approach, this study produces a replicable framework applicable across Mediterranean inner regions.

2 Background Knowledge & Methodology

This study adopts a qualitative, comparative, and literature-based methodology, drawing on scientific publications, operational guidelines, and critical analyses of existing projects. No primary data were collected; instead, insights were derived from the comparative analysis of selected cases, allowing the identification of transferable Nature-Based Solutions (NbS) strategies suitable for heritage-rich rural contexts. Rather than engaging directly with communities or deploying monitoring technologies, the research examined interventions already implemented in diverse territorial settings, generating lessons by analogy or contrast. This approach facilitated the identification of strategies aligned with the spatial, environmental, and cultural characteristics of small settlements, providing guidance for future interventions.

Six European case studies were selected using a structured scoring system based on multiple criteria: (i) relevance to rural or peri-urban contexts; (ii) documented environmental, social, and technological outcomes; (iii) replicability potential; and (iv) compatibility with heritage-sensitive settlements. While these cases offer representative insights, their selection is constrained by data availability, and findings should be interpreted as indicative rather than exhaustive. Particular attention was given to European contexts where water scarcity, heatwaves, and soil erosion intersect with complex settlement patterns and historical legacies. Although primary field measurements were not conducted in the Madonie region, the strategies identified are transferable by analogy, though their on-site effectiveness requires validation. Differences in reporting and data availability across case studies may also influence comparative assessments.

The six selected projects illustrate how Smart Villages can leverage NbS to enhance environmental sustainability, social engagement, and technological innovation. They include: Eco Street, Ober-Grafendorf, Austria (2015); Johann-Nepomuk-Vogl Square, Vienna, Austria (2019–2020); Rainwater Management, Pula, Croatia (2015); Grün statt Grau, Osnabrück, Germany (2019); Leipziger BlauGrün 416, Leipzig, Germany (2019–2024); and The Little Forest, Zichow, Germany (2020). These cases demonstrate that NbS can act as effective drivers of sustainable and resilient settlements while respecting cultural and environmental constraints (Cohen-Shacham et al., 2019).

2.1 Eco Street, Ober-Grafendorf, Austria (2015)

Eco Street (Fig. 1) represents a best-practice NbS for sustainable urban water management and local climate adaptation. Implemented on Buchengasse (100 m × 7 m), the project reduces impervious surfaces, filters stormwater, and channels it into green areas designed for infiltration and groundwater recharge. The innovative Draingarden® system, developed by Zenebio GmbH, combines engineered substrates and selected vegetation to ensure natural filtration, water retention for plants, and reduce drainage load. Funded by the EU C3-Alps project and implemented collaboratively with universities, local authorities, and private companies, the project requires minimal maintenance while providing continuous ecological benefits. Outcomes include microclimate improvement, increased biodiversity, reduced irrigation and purification costs, enhanced landscape quality, and improved resident comfort. Eco Street has demonstrated high replicability, having been extended to multiple streets across Austria.

Fig. 1 Ober-Grafendorf Ecostreet, Austria

Fig. 2 Johann-Nepomuk-Platz, Vienna, Austria

2.2 *Johann-Nepomuk-Vogl Square, Vienna, Austria (2019–2020)*

This Project Implements the "Sponge City" Concept in a 2700 m^2 Dense Urban Area (Fig. 2), Integrating Park Spaces, Market Areas, and Children's Recreational Zones. Participatory Design Involved Residents, Shopkeepers, and Local Authorities, Enabling Stormwater Infiltration from Paved Surfaces and Roofs, while Using Water from Splash Pads for Tree Irrigation. The Project Combines Climate Adaptation, Sustainable Water Management, and Social Enhancement, Improving Urban Microclimate, Quality of Life, and Social Inclusivity. Effective Replication Requires Multidisciplinary Planning and Participatory Engagement

2.3 *Rainwater Management, Pula, Croatia (2015)*

Targeting the Šijana watershed and Re Tomislav Square (Fig. 3), this intervention combines retention ponds, raingardens, infiltration trenches, and artificial wetlands to reduce runoff peaks and enhance water quality. Public spaces were redesigned to foster pedestrian access and community interaction. Benefits include urban heat

Fig. 3 The Šijana watershed and King Tomislav Square, Pula, Austria

Fig. 4 Greenroofs solutions in Osnabrück, Germany

mitigation, enhanced water quality, climate resilience, and social-environmental gains. Replicability in other Croatian cities depends on stakeholder involvement and ongoing monitoring.

2.4 Grün Statt Grau, Osnabrück, Germany (2019)

The "Grün statt Grau" project (Fig. 4) mitigates summer heat and flood risks through extensive green roofs, complemented by an interactive green roof registry, financial incentives, and citizen guidelines. Outcomes include improved microclimate, increased biodiversity, flood risk reduction, thermal insulation, and aesthetic enhancement. The project has been successfully replicated in Münster, demonstrating the importance of digital and financial tools in scaling NbS.

2.5 Leipziger BlauGrün 416, Leipzig, Germany (2019–2024)

This project integrates blue-green infrastructure (BGI) in Leipzig's new 416 district (Fig. 5), including green roofs, tree trenches, retention areas, and multifunctional green spaces. Community engagement through forums and workshops is central,

Fig. 5 Leipzig BlauGrün 416, Eutritzsch

while project funding (€2.8 million) was provided by the Federal Ministry of Education and Research. Objectives encompass climate resilience, urban microclimate improvement, water management, and enhanced quality of life. The approach is replicable in other urban contexts with similar environmental and social conditions.

2.6 *The Little Forest, Zichow, Germany (2020)*

The Little Forest (Fig. 6) creates micro-forests (200–2000 m^2) with dense native plantings, enhancing biodiversity and climate adaptation. The "Forest of Diversity" includes 33 species and 2500 plants, achieving 99% survival in its first summer. Benefits include climate change mitigation, air quality improvement, CO_2 sequestration, temperature and noise regulation, and recreational opportunities. Community participation fosters environmental education and active citizenship. Initially crowdfunded, the project follows the Miyawaki method and is highly replicable internationally.

These case studies highlight key criteria for assessing the relevance and adaptability of NbS in fragile territorial systems, including integration with local ecological and social conditions, multifunctionality, and community involvement. Table 1 provides a comparative overview of the six selected European Nature-Based Solutions (NbS) case studies analyzed in this research. It summarizes key features, objectives, implemented strategies, environmental and social outcomes, and replicability potential. This comparative framework allows for the identification of transferable

Fig. 6 The Little Forest, Uckermark

lessons and best practices that can inform the adaptation of NbS interventions in heritage-rich rural contexts such as the Madonie region, highlighting similarities, differences, and context-specific considerations across diverse settlements.

Table 2 provides a comparative overview of the selected European case studies, highlighting differences in NbS typologies, scales of implementation, community engagement, and measurable outcomes. This comparison informs the adaptation of strategies to the Madonie context. Cases were selected using a structured scoring system based on relevance to rural contexts, evidence of social and environmental outcomes, replicability potential, and compatibility with heritage-sensitive settlements. A comparative table (Table 2) summarizes key features, highlighting lessons transferable to the Madonie region.

The analysis then focused on the Madonie region in northern Sicily, a preserved rural environment facing depopulation, infrastructural decay, and heritage preservation challenges. Encompassing 26 small municipalities within a UNESCO Global Geopark, the area combines traditional architecture, environmental sensitivity, and limited technological infrastructure. Spatial, morphological, and climatic data were assessed alongside recent public interventions (Junaidi & Othman, 2025). Within the NbS framework, small-scale interventions can enhance heritage adaptability and environmental performance without compromising natural landscapes (Fig. 7). Observed challenges include inadequate materials, poor coherence between new and historical designs, and limited usability of public spaces. To ensure methodological robustness, the study introduces quantitative indicators for long-term monitoring across four dimensions: (i) Community engagement—active associations, workshop participation, volunteer continuity; (ii) Governance effectiveness—project implementation rates, decision-making speed, open data availability; (iii) Heritage rehabilitation—rehabilitated built heritage area, proportion of reused traditional materials, number of cultural events tied to local identity; (iv) Digital and technological innovation—broadband coverage, digital public services, ICT uptake. Indicators include baseline and target values for longitudinal monitoring. Aggregated results form a Smart Village Performance Index (SVPI) (Garndhi et al., 2020), providing a comprehensive tool to guide adaptive strategies and assess improvements. By integrating quantitative metrics with qualitative insights, this framework offers a replicable model for research and practical application in European inner areas (Anastasiou & Manika, 2021).

3 Climate-Resilient Strategies and Implications at the City Scale, a Performance Index

The Smart Village Performance Index (SVPI) proposed in this study provides a structured framework for evaluating and implementing Nature-Based Solutions (NbS) at both city and settlement scales. Its primary aim is to guide municipalities and planners

Table 1 Comparative overview of European Nature-Based solutions in rural and Heritage-Rich settlements

Case study	NbS type	Objectives	Scale	Community engagement	Outcomes	Replicability
Eco street, Austria	Draingarden®	Water management, microclimate	Street	Medium	High	High
Johann-Nepomuk, Austria	Sponge City	Water, social	Square	High	High	Medium-high
Pula, Croatia	Rainwater management	Flood, social	District	Medium	Medium	Medium
Grün statt Grau, Germany	Green roofs	Heat, flood	Neighborhood	High	High	High
Leipzig BlauGrün 416	Blue-green	Climate, water, social	District	High	High	Medium-high
Little Forest, Germany	Micro-forest	Biodiversity, climate	Site	High	High	High

Table 2 Smart village performance index for NbS, evaluative hypothesis of NbS solutions

NbS category	Objective	Performance indicators	Environmental impact	Social impact	Economic impact	Technical feasibility	Maintenance requirement
Urban green (parks, gardens, street trees)	Microclimate mitigation, CO_2 sequestration	Summer temperature reduction, CO_2 capture	High	High	Medium	High	Medium
Green roofs	Thermal insulation, energy reduction, stormwater management	Energy consumption reduction, water infiltration	Medium-high	Medium	Medium	Medium	High
Green walls	Acoustic and thermal insulation, vertical biodiversity	Noise reduction, increase in plant species	Medium	Medium	Medium	Medium	High
Permeable pavements	Stormwater management, flood risk reduction	Water infiltration volume, flood reduction	High	Medium	Medium	High	Medium
Reforestation and urban/rural forests	Biodiversity, soil stabilization	Number of species, vegetation cover	High	Medium	Low-medium	Medium	Medium
Green infrastructure (bioretention, swales, rain gardens)	Stormwater management, flood prevention	Water infiltration volume, flood reduction	High	Medium	Medium	Medium	Medium
Urban/ peri-urban agriculture	Food security, local economy	Food production, job creation	Medium	High	High	Medium	Medium

(continued)

Table 2 (continued)

NbS category	Objective	Performance indicators	Environmental impact	Social impact	Economic impact	Technical feasibility	Maintenance requirement
Wetlands and marshes	Water filtration, habitat creation	Water quality, biodiversity	High	Medium	Low	Medium-low	High
Ecological Corridors & Habitat Connectivity	Biodiversity conservation	Ecological connectivity, number of species	High	Medium	Low	Medium	Medium

Fig. 7 Villages in the Madonie UNESCO Geopark, Palermo

in systematically identifying existing NbS, defining objectives, and assigning quantitative performance indicators to each intervention. By simultaneously assessing environmental, social, and economic impacts, the index ensures alignment with the three pillars of sustainability, while also considering technical feasibility and maintenance requirements (Campisi & Lombardo, 2025). The SVPI enables planners to prioritize NbS interventions according to local heritage constraints, microclimatic conditions, and community needs. By aligning indicators with site-specific characteristics, it supports decisions that maximize ecological and social benefits while respecting cultural landscapes. This integrated approach facilitates informed choices for interventions in urban fabrics, heritage-rich rural settlements, or mixed contexts. SVPI complements existing planning instruments, such as the R.I.E. (Reduction of Building Impact) parameter, allowing administrators to assess current sustainability conditions, simulate alternative NbS scenarios, and anticipate long-term effects on resilience, heritage conservation, and public usability.

Each NbS category is linked to specific objectives, performance indicators, and expected outcomes. For instance, urban green spaces, street trees, and pocket parks primarily address microclimate mitigation and CO_2 sequestration, while green roofs and walls improve thermal insulation, reduce energy demand, and enhance vertical biodiversity. Permeable pavements, rain gardens, and reforestation enhance stormwater management and flood resilience, whereas wetlands and ecological corridors improve water filtration, biodiversity, and habitat connectivity.

3.1 Rationale for Indicator Weighting

The selection and weighting of SVPI indicators are based on both the objectives of NbS and the characteristics of heritage-rich rural settlements, such as those in the Madonie region. Each indicator captures a key sustainability dimension: environmental performance, social engagement, economic impact, technical feasibility, and maintenance requirements. Indicators were defined through literature review, analysis of European best practices, and contextual adaptation. A three-level scale—High (3), Medium (2), Low (1)—reflects relative effectiveness based on ecological, social, and economic outcomes, as well as technical feasibility. Thresholds were determined from documented evidence and observed performance in selected European cases. Interventions showing clear, measurable improvements in multiple dimensions received "High," moderate or context-dependent results "Medium," and limited or uncertain impacts "Low." Environmental indicators, such as temperature reduction, stormwater management, and biodiversity enhancement, are weighted according to their ecological impact. Social indicators—including community participation, volunteer engagement, and public usability—emphasize local involvement. Economic indicators consider solutions that generate income or reduce operational costs, while technical feasibility and maintenance reflect the municipality's capacity to implement and sustain interventions. For example, pocket gardens and vegetated pedestrian routes are rated high for environmental and social impact due to their effectiveness in microclimate regulation and community engagement; medium for economic impact, given modest but tangible benefits; and medium to high for technical feasibility and maintenance, considering the use of local materials and volunteer involvement. Aggregating weighted scores produces a comprehensive, replicable metric, enabling evidence-based prioritization of NbS interventions. This ensures that the interventions contributing most to ecological, social, and cultural resilience are clearly identified while highlighting technical or economic constraints.

The SVPI assigns relative scores for environmental, social, and economic benefits, as well as technical feasibility and maintenance. High-performing interventions maximize ecological and social benefits while remaining technically feasible and manageable over time. The framework also highlights opportunities to replace less sustainable or outdated interventions with nature-based alternatives.

3.2 Application of SVPI to Madonie Park

Table 3 presents a contextualized application of the SVPI to the Madonie Park, integrating quantitative assessment with qualitative insights on local conditions, heritage conservation, and landscape coherence. This enables evidence-based planning that is context-sensitive, socially acceptable, and technically feasible.

The SVPI analysis for Madonie Park highlights which interventions deliver the highest combined benefits. Urban green spaces score high in environmental and social

Table 3 Application of SVPI to NbS in the madonie park: contextualized evaluation and qualitative insights

NbS category	Objective	Performance indicators	Environmental impact	Social impact	Economic impact	Technical feasibility	Maintenance requirement	Notes on madonie context
Urban green (parks, gardens, street trees)	Microclimate mitigation, CO_2 sequestration	Summer temperature reduction, CO_2 capture	High	High	Medium	High	Medium	Local parks reduce heat, involve volunteers; medium economic impact due to limited tourism revenue
Green roofs	Thermal insulation, energy reduction	Energy consumption reduction, water infiltration	Medium	Medium	Medium	Medium	High	Limited applicability on historic buildings; high maintenance due to traditional materials
Green walls	Acoustic and thermal insulation, vertical biodiversity	Noise reduction, increase in plant species	Medium	Medium	Medium	Medium	High	Applicable on selected walls; supports biodiversity but limited social engagement
Permeable pavements	Stormwater management, flood risk reduction	Water infiltration volume, flood reduction	High	Medium	Medium	High	Medium	Effective in small squares; installation requires municipal support
Reforestation / urban-rural forests	Biodiversity, soil stabilization	Number of species, vegetation cover	High	Medium	Low-medium	Medium	Medium	High ecological benefit potential; limited economic return; terrain can limit technical feasibility

(continued)

Table 3 (continued)

NbS category	Objective	Performance indicators	Environmental impact	Social impact	Economic impact	Technical feasibility	Maintenance requirement	Notes on madonie context
Green infrastructure (bioretention, swales, rain gardens)	Stormwater management, flood prevention	Water infiltration volume, flood reduction	High	Medium	Medium	Medium	Medium	Suitable for public squares and school yards; requires moderate municipal and volunteer engagement
Urban/ peri-urban agriculture	Food security, local economy	Food production, job creation	Medium	High	High	Medium	Medium	Can engage communities and schools; small-scale local economic benefits; feasible on peri-urban plots
Wetlands and marshes	Water filtration, habitat creation	Water quality, biodiversity	High	Medium	Low	Medium-low	High	High ecological value; limited accessibility; long-term maintenance may require regional support
Ecological Corridors & Habitat Connectivity	Biodiversity conservation	Ecological connectivity, number of species	High	Medium	Low	Medium	Medium	Enhances regional biodiversity; limited direct social or economic benefit; requires coordinated management

dimensions due to their role in microclimate regulation and community engagement. Green roofs and walls are constrained by heritage building limitations. Green Infrastructure and Urban/Peri-Urban Agriculture provide moderate benefits with strong social engagement potential, while Wetlands and Ecological Corridors offer substantial ecological gains but lower economic or social impact. By integrating quantitative and qualitative data on environmental, cultural, and social conditions, the SVPI supports climate-resilient regeneration and the prioritization of interventions that maximize ecological, social, and cultural resilience in heritage-rich rural settlements.

4 Discussion and Implications

The analysis of the Madonie region underscores both the potential and the limitations of implementing Nature-Based Solutions (NbS) in small-scale rural contexts. Greening interventions, when carefully aligned with local morphology, material culture, and spatial traditions, can enhance thermal comfort, regulate surface runoff, and improve the experiential quality of public spaces. These benefits are particularly relevant in heritage-rich settlements experiencing demographic decline, infrastructural fragility, and climate-related pressures (Despotović & Joksimović, 2020). However, the study also revealed a series of challenges in recent implementations. Several projects failed to fully exploit the ecological and social potential of NbS due to cost-driven choices, reliance on standardized planning models, or insufficient attention to environmental and cultural coherence. Recurring issues included the use of non-traditional materials, lack of shading and ventilation in public spaces, and weak integration between vegetation and the built environment. Such design shortcomings have contributed to deteriorating microclimatic conditions—particularly during extreme heat events—and have reduced the liveability of certain areas, especially for vulnerable populations.

The findings highlight that NbS can be effective even at small scales if they are conceived within a broader, context-sensitive strategy. Interventions such as green roofs and walls applied to public buildings using appropriate local materials can simultaneously reduce overheating and reinforce architectural integration with the surrounding landscape. Likewise, the transformation of abandoned lots into pocket gardens or the creation of vegetated pedestrian corridors can significantly improve both social interaction and environmental performance in villages facing depopulation and infrastructural decay. The application of the Smart Village Performance Index (SVPI) to the Madonie context provided quantitative support for these observations, indicating that urban green spaces and pocket gardens score highest across environmental, social, and economic dimensions, thus offering strong contributions to both regeneration and resilience objectives. For instance, urban green spaces scored high for summer temperature reduction and CO_2 sequestration (environmental impact), while also maximizing community engagement (social impact) and demonstrating manageable maintenance requirements, highlighting their replicable benefits in heritage-rich rural settlements.

Lessons learned from these European projects, combined with SVPI assessments, inform the adaptation of NbS strategies in the Madonie context, emphasizing small-scale, low-impact interventions compatible with heritage conservation and local ecological conditions. Crucially, the success of NbS is not determined solely by technical or ecological performance; their effectiveness is deeply contingent on resonance with local identity, traditions, and community aspirations. Compatibility between ecological functions and architectural language emerges as a decisive factor for social acceptance, long-term sustainability, and functional resilience in heritage-rich rural settings. Rather than prescribing prescriptive interventions, this study proposes a conceptual framework, operationalized through SVPI, to guide planners, designers, and decision-makers in evaluating when and how NbS can be successfully deployed in similar contexts (Scott et al., 2016). The evidence suggests that rural planning should move away from standardized, one-size-fits-all approaches and instead embrace strategies informed by local materials, microclimatic dynamics, spatial traditions, and socio-cultural values. By doing so, NbS can become a catalyst for the adaptive regeneration of small settlements, harmonizing environmental performance, cultural coherence, and community-driven innovation. These findings demonstrate that NbS interventions contribute to the regeneration of heritage-rich villages by restoring public spaces, enhancing local identity, and reinforcing cultural continuity, while simultaneously improving resilience by mitigating climate risks, enhancing adaptive capacity, and sustaining essential socio-ecological functions.

5 Conclusion

This research contributes to the evolving discourse on Nature-Based Solutions (NbS) by shifting attention from their conventional large-scale urban applications to the specific needs, opportunities, and constraints of rural and inner areas. By examining the Madonie region through the lens of the Smart Village paradigm (Lombardo, 2026), the study demonstrates that small-scale, context-sensitive interventions—ranging from green roofs and walls to pocket gardens and vegetated pathways—can play a pivotal role in enhancing environmental resilience, spatial quality, and social well-being, even in settlements with limited financial and technological resources (Sarabi et al., 2019). The findings highlight that the effective implementation of NbS in rural areas requires a deep understanding of local identity, historical material cultures, and spatial morphology. Coherent material choices, sensitive design strategies, and a forward-looking vision that treats natural and cultural resources as mutually reinforcing assets are key to achieving both ecological functionality and aesthetic integration. In this context, Smart Villages are not passive recipients of innovation but active agents capable of shaping place-based strategies that address climate adaptation, ecological regeneration, and the strengthening of community cohesion (Edgars & Una, 2020). Moreover, this research emphasizes the importance of multi-dimensional evaluation frameworks that integrate environmental, social, and economic sustainability alongside technical feasibility and maintenance requirements. The Smart Village Performance Index (SVPI) operationalizes this framework, providing a replicable, quantitative tool that allows planners and decision-makers to prioritize NbS interventions, evaluate their contribution to both regeneration and resilience, and align them with local heritage constraints, socio-cultural values, and ecological contexts. By embedding these considerations into planning processes, interventions can be effective, socially accepted, and culturally resonant, avoiding the pitfalls of standardized, cost-driven solutions that fail to respect local landscapes or community expectations. The Madonie case demonstrates that even small-scale NbS interventions can generate cumulative positive effects, including microclimate improvement, biodiversity enhancement, and increased usability and attractiveness of public spaces—particularly in depopulated or infrastructure-degraded contexts. When embedded in a broader strategy of ecological and social compatibility, these interventions restore environmental functions while fostering local ownership, pride, and active participation, reinforcing the potential of rural territories to innovate without compromising their identity. Importantly, this research provides three original contributions. First, it offers a critical reinterpretation of NbS specifically within heritage-rich rural contexts, highlighting the interplay between ecological functionality and cultural identity. Second, it presents a comparative assessment of European NbS practices through the Smart Village lens, identifying transferable strategies and lessons for small rural settlements. Third, it develops the Smart Village Performance Index (SVPI) tailored to Mediterranean inner areas, providing a replicable and quantitative tool to guide planning, evaluate interventions, and support evidence-based decision-making. Collectively, these contributions advance operational knowledge for the sustainable transformation of rural territories through NbS, bridging theory and practice in heritage-sensitive environments. By providing a structured, evidence-based evaluation through the SVPI, this framework not only prioritizes effective NbS interventions but also directly informs strategies for the regeneration and resilience of heritage-rich rural settlements, ensuring that ecological, social, and cultural benefits are maximized in a context-sensitive and replicable manner. Ultimately, the sustainable future of rural regions depends on their capacity to reconcile tradition and innovation, integrating heritage conservation, ecological restoration, and adaptive planning into a cohesive territorial strategy (Lafortezza et al., 2018). Grounded in this perspective, NbS become powerful instruments of transformation (Gerli & Marco, 2022), capable of unlocking resilience, connectivity, and vitality in small communities, while demonstrating that environmental sustainability, cultural integrity, and social well-being are not competing goals, but mutually reinforcing pillars of a climate-resilient, regeneration-oriented, and inclusive rural future.

Acknowledgments This research was founded by the European Union EU Next Generation Italian PRIN/PNRR 2022 Project titled "Greenwork: an interdisciplinary framework for urban health and urban resilience enhancement based on greening strategies on buildings and open spaces" - Project Code MUR P2022MNKEC 003—Sector: ERC SH7—CUP: B53D23029580001 - PRJ-1358.

References

Alhari, M. I., & Fajrillah, A. A. N. (2022). Enterprise architecture: A strategy to achieve e-government dimension of smart village using TOGAF ADM 9.2. *International Journal on Informatics Visualization, 6*(2), 540–545. https://doi.org/10.30630/joiv.6.2-2.1147

Anastasiou, E., & Manika, S. (2021). Territorial and human geography challenges: How can smart villages support rural development and population inclusion? *Social Sciences*. https://doi.org/10.3390/socsci10060193

Borham, O., & Croxford, B. (2024). Biomimetic strategies for sustainable resilient cities: Review across scales and City systems. *Biomimetics*. https://doi.org/10.3390/biomimetics9090514

Campisi, T., & Lombardo, L. (2025). DESIGNING WITH WATER strategies, protocols, and projects in architecture and for cities. *AGATHÓN, 18*, 196–211.

Cohen-Shacham, E., Andrade, A., Dalton, J., et al. (2019). Core principles for successfully implementing and upscaling nature-based solutions. *Environmental Science & Policy, 98*, 20–29. https://doi.org/10.1016/j.envsci.2019.04.014

Despotović, A.; Joksimović, M. (2020). Demographic revitalization of Montenegrin rural areas through the smart village concept, Agriculture and Forestry, https://doi.org/10.17707/AgricultForest.66.4.10

Edgars, P.; Una, K. (2020). *Smart and sustainable local communities in global covid-19 pandemic conditions*, Landscape Architecture and Art, https://doi.org/10.22616/j.landarchart.2020.17.09

Frantzeskaki, N., McPhearson, T., Collier, M. J., et al. (2019). Nature-based solutions for urban climate change adaptation: Linking science, policy, and practice communities for evidence-based decision-making. *Bio Science, 69*(6), 455–466. https://doi.org/10.1093/biosci/biz042

Garndhi, S. & Wibowo, S. & Ebardo, R.; (2020). A New Performance Index for Evaluating Smart City Projects.

Gerli, P.; Marco, J. (2022). What makes a smart village smart?: People, Process and Policy, doi:https://doi.org/10.1108/TG-07-2021-0126.

Hongpeng, F. (2023). A comprehensive review of nature-based solutions: Current status and future research. *AIMS Environmental Science., 10*, 677–690. https://doi.org/10.3934/environsci.2023037

Junaidi, A., & Othman, M. S. B. (2025). Smart villages: A systematic review of trends, models, and metrics. *Cogent Social Sciences*. https://doi.org/10.1080/23311886.2025.2492833

Lafortezza, R., Chen, J., van den Bosch, C. K., et al. (2018). Nature-based solutions for resilient landscapes and cities. *Environmental Research, 165*, 431–441. https://doi.org/10.1016/j.envres.2017.11.038

Lombardo, L. (2026). Smart villages: Tools and strategies for sustainable development in inner rural areas. Springer. (SpringerBriefs in Applied Sciences and Technology). ISBN 978–3–032-16031-7; https://link.springer.com/book/9783032160300

Pultrone, G. (2022). *Combining the Ecological and Digital Transitions: Smart Villages for New Scenarios in the EU Rural Areas*, Springer, doi:https://doi.org/10.1007/978-3-031-06825-6_260.

Rahoveanu, M. M. T.; Serban, V. (2022). *Perspectives on Smart Villages from a Bibliometric Approach*, Sustainability, doi:https://doi.org/10.3390/su141710723.

Raymond, C. M., Frantzeskaki, N., Kabisch, N., et al. (2017). A framework for assessing and implementing the co-benefits of nature-based solutions in urban areas. *Environmental Science & Policy, 77*, 15–24. https://doi.org/10.1016/j.envsci.2017.07.008

Sarabi, S. E., Han, Q., Romme, A. G. L., et al. (2019). Key enablers of and barriers to the uptake and implementation of nature-based solutions in urban settings: A review. Resources, 8(3), 121. doi:https://doi.org/10.3390/resources8030121.

Scott, M., Lennon, M., Haase, D., et al. (2016). Nature-based solutions for the contemporary city: Insights from practice in Fingal. *Ireland. Planning Theory & Practice, 17*(2), 267–300. https://doi.org/10.1080/14649357.2016.1158907

Solheim, A., & Capobianco, V. (2021). Implementing nature-based solutions in rural landscapes: Barriers experienced in the PHUSICOS project. *Sustainability*. https://doi.org/10.3390/su13031461

Wolff, E., & Rauf, H. A. (2023). Nature-based solutions in informal settlements: A systematic review of projects in southeast Asian and Pacific countries. *Environmental Science & Policy*. https://doi.org/10.1016/j.envsci.2023.04.014

Urban Green Spaces for the Eco-Energy Balance: A Spatial Framework to Integrate Carbon Sequestration and Renewable Energy Systems in Local Planning

Stella Pennino, Valerio Martinelli, and Romano Fistola

Abstract Urban areas are pivotal to climate change mitigation and adaptation, acting as both primary greenhouse gas emitters and key arenas for transformative socio-ecological solutions. This study operationalizes the concept of *Urban Eco-Energy Balance* by proposing a replicable geospatial framework to jointly assess and coordinate the development of Urban Green Spaces (UGSs) and local renewable energy systems. The primary objective is to provide a Decision-support Tool for Public Administrations to evaluate existing *Eco-Energy Deltas*—the differential between local CO_2 emissions and UGSs' carbon sequestration—and to identify optimal spatial configurations for enhanced greenery and rooftop photovoltaic (PV) deployment. The methodology relies on open-access data and is structured in two phases: the spatial assessment of current *Eco-Energy Deltas* at the Census Section's scale, and the identification of priority zones for intervention through a Transformation Map and a Priority Matrix. The framework is applied to the city of Naples, Italy, employing species-specific allometric equations to estimate current carbon sequestration, and rooftop PV suitability according to national regulatory criteria. Results reveal strong spatial disparities in eco-energy performance across the city. North-western districts show high existing sequestration capacity, while eastern industrial peripheries display significant, currently untapped potential for PV-based Power Hubs and Renewable Energy Communities. The Priority Matrix classifies urban areas into Green Priority, Energy Priority, Conservation and High-Efficiency zone, translating complex information into actionable planning categories. By integrating UGSs development and distributed renewable energy systems, the framework helps bridge the gap between high-level climate strategies and local planning practice, supporting the transition toward climate-neutral and resilient urban metabolisms.

S. Pennino (✉) · V. Martinelli · R. Fistola
Department of Civil, Building and Environmental Engineering, University of Naples Federico II, Naples, Italy
e-mail: stella.pennino@unina.it

L. Diana et al. (eds.), *Greening and Resilience*, Urban Sustainability,
https://doi.org/10.1007/978-981-92-0419-9_6

Keywords Urban green spaces · Carbon sequestration · Renewable energy systems · Urban eco-energy balance · Climate change mitigation · Spatial planning · Geographic information systems

1 Introduction: Cities, Green Spaces and the *Urban Eco-Energy Balance*

Cities account for a great proportion of global energy consumption and greenhouse gas (GHG) emissions (IEA, 2024; IPCC, 2023). GHG stem primarily from urban energy systems sustaining essential infrastructure, such as transport, buildings and industry (Hillman & Ramaswami, 2010). Moreover, urban areas concentrate main drivers of emissions—population, technology, lifestyles and consumption patterns (Chavez & Sperling, 2017). Therefore, cities have a unique potential to drive the ecological and energetic transitions through spatial reconfiguration and technological innovation (Frantzeskaki et al., 2017). As key contributors to global warming and leaders in its mitigation, cities are priority arenas to experiment with transformative solutions.

Urban green spaces (UGSs) play a central role in climate change mitigation, due to the innate capacity to absorb atmospheric CO_2 and multiple co-benefits. UGSs act as carbon sinks, capturing and storing carbon within above- and below-ground biomass (Shadman et al., 2022). Beyond direct sequestration potential, UGSs indirectly reduce GHG by providing local shading and cooling, thereby mitigating the urban heat island effect and lowering cooling energy demand (Chen & You, 2020). They also deliver a range of ecosystem services—air pollutant removal, stormwater management, noise abatement—while providing accessible recreational spaces for urban residents (Chen et al., 2020; Fistola, 2023). Nevertheless, UGSs cannot entirely offset GHG: net-zero balance requires systemic transformations across all provisioning systems (Seto et al., 2021).

Consequently, various studies have proposed to integrate nature-based solutions and local renewable energy systems to foster climate-neutral urban environments (Chen et al., 2025; Liu et al., 2023). The increasing accessibility of rooftop photovoltaics (PV) supports this transition by utilizing underused urban surfaces to generate renewable energy (Wiginton et al., 2010) and offset residual emissions by locally implementing *Power Hubs* (Jing et al., 2020). Both strategies—UGSs and rooftop PV systems—are intrinsically linked to the spatial configuration of the city, which has been shown to be a critical determinant of GHG emission patterns (Zhang et al., 2021).

Building on existing literature, this study develops a geospatial framework to evaluate the contribution of UGSs and PV systems to the *Urban Eco-Energy Balance*. The framework introduces an operational methodology that enables Public Administrations to evaluate existing *Eco-Energy Deltas* and explore the optimal spatial combinations of enhanced UGSs and PV rooftops. The methodology is tested through an application to the city of Naples, southern Italy, within a GIS environment, providing insights into its feasibility and replicability.

2 Background Knowledge: UGSs and Renewable Energy Systems in Urban Planning

2.1 Urban Green Spaces in the Urban System: Current Definitions and Applications

Urban management of natural resources and green spaces planning have long been objects of research and policy debate (Trupiano & Fistola, 1993). In recent years, renewed attention to these themes has emerged due to the urgency to mitigate global climate change (De la Sota et al., 2019), and the growing need to adapt to its escalating impacts (Gill et al., 2007). UGSs provide ecological and social benefits through the delivery of Ecosystem Services (ES) and sustainability functions, acting as integrated solutions to multiple urban challenges (Farkas et al., 2023; Ugochukwu Kanayo Ashinze et al., 2024). Consequently, growing research has focused on developing innovative approaches for integrated planning and management of UGSs (Freundt et al., 2020), to preserve and enhance the ES they supply (Montis et al., 2025). In this study we refer to a conceptual framework that explicitly relate ES to cities' sub-systems, acknowledging the city as a complex socio-ecological system (Fistola, 2023). Despite broad recognition of these benefits, a comparative policy analysis by Baycan-Levent and Nijkamp (2009) reveals that UGSs often remain undervalued in both planning and management practices.

One of the challenges and emerging frontiers in UGSs research is the quantification of carbon sequestration and storage performed (Dong et al., 2024). This process relies on the use of tree biomass equations to estimate their carbon storage (McPherson et al., 2016). Accurate biomass estimation requires including both Exceeding Ground Biomass (EGB), comprising trunk, branches and foliage, and Beneath Ground Biomass (BGB), which encompasses the root system (Kumar et al., 2023). These parameters can be assessed through a range of methods, including LiDAR-derived metrics (Gülçin & Bosch, 2021), remote sensing techniques (Huang et al., 2024), allometric models and species-specific tree growth equations (McPherson et al., 2016). This study builds upon existing literature by applying and adapting allometric parameters and models to evaluate both the actual carbon storage and the annual carbon sequestration potential of UGSs.

2.2 *Renewable Energy Systems: Urban Planning Challenges to Their Deployment*

2015's Paris Agreement marked an increase in practice-oriented research to maintain global warming within the 1.5 °C threshold (Hwang et al., 2025) and provided impetus for the worldwide development of renewable energy (Rosokhata et al., 2021). However, energy policies lacked cross-sectoral planning and failed to mainstream principles of energy justice (Akrofi et al., 2022). In urban contexts, this gap is particularly evident as the diffusion of renewable energy systems is both a technological and a spatial governance issue, shaped by land-use regulations, landscape protection, environmental constraints and distributive equity concerns (Aguilar et al., 2025).

The relative ease of PV systems installation, combined with technological advances and decreasing production costs, generated an economy of scale and led to the global diffusion of solar power (Tyagi et al., 2024). Compared to other renewable sources that require substantial infrastructural investment, PV systems offer a flexible and cost-efficient option for small- and medium-scale energy generation in cities (Hachem-Vermette, 2022; Pan et al., 2022). Their lightweight design enables the productive reuse of rooftops, underutilized plots and vacant land (Jiang et al., 2020). Nevertheless, the suitability of PV installations within urban fabric strongly depends on morphological and ownership patterns (e.g. public VS private rooftops) and the alignment between energy planning and urban regulatory framework (Costa & Campanini, 2024).

Many European countries have translated EU directives into national and local frameworks to facilitate the energy transition. This includes simplifications in authorization procedures and the identification of *Renewable Acceleration Areas* where renewable projects are prioritized while environmental safeguards are retained (EU, 2024). In Italy, however, the deployment of urban renewable energy systems remains constrained by regulatory and urban planning barriers, particularly concerning heritage, environmental and landscape protection, as well as restrictions related to seismic and hydrogeological risks (e.g. protected areas, Natura 2000 sites, zones of agricultural significance) (Martinelli, 2025b). In response, Aruta et al. (2024) have proposed an innovative GIS-based methodology for designing sustainable energy districts, conceptualizing the city as a distributed *Power Hub*. This study connects to this line of work by introducing a method that integrates urban planning criteria into the urban design of renewable energy systems.

2.3 *The* **Urban Eco-Energy Balance***: Pilot Cases of Net-Zero Energy Districts*

Cities are complex systems where UGSs and renewable energy systems are interdependent components of a broader urban metabolism. Achieving the net-zero carbon outcome at the city scale demands systemic transformation (Seto et al., 2021) across multiple sub-systems—buildings, spatial planning, policy, finance and energy infrastructures (Duan & Kim, 2023).

Growing interests in urban-scale renewable energy systems have led to novel planning paradigms, and several European cities have already pledged to achieve net-zero emissions by 2030 (NetZeroCities, n.d.). Numerous pilot projects of Net-Zero Energy Districts (NZEDs) are leading the large-scale transition to sustainable energy systems (Martinelli, 2025a). A NZED is commonly defined as an urban area in which the total annual energy consumption of buildings is balanced by on-site renewable energy generation (Sameti & Haghighat, 2018). Such equilibrium is achieved through the optimization of district energy systems that efficiently provide power, heating and cooling, reducing primary energy demand (Yang et al., 2015), and transmission losses (Sameti & Haghighat, 2017). NZEDs generally combine renewable energy generation, nature-based solutions (NBSs) and smart systems for real-time energy management (Komninos, 2022). Positive Energy Districts (PEDs) are urban areas capable of producing more renewable energy than they consume annually (Derkenbaeva et al., 2022). To do so, PEDs adopt a multi-level systemic approach that extends beyond building-level efficiency, integrating production, storage, management, exchange and users' engagement through advanced spatial planning and innovative governance models (Tuerk et al., 2021). The application of spatial multi-criteria approaches, and GIS-based workflows, enable to identify suitable zones for energy-positive interventions, highlighting trade-offs between energy performance, urban morphology and socio-spatial equity (Alpagut et al., 2021). Moreover, Renewable Energy Communities (RECs) further shifted the focus from where to install energy systems to how to equitably distribute energy benefits within communities (Bosone et al., 2023; EU, 2024).

This study builds upon these advances, extending the principles of NZEDs from the district to the city scale through the concept of the *Urban Eco-Energy Balance*. It provides a systematic assessment framework, supporting decision-makers in the strategic coordination of renewable energy systems and UGSs development toward a balanced, net-zero urban metabolism.

3 Methodology: Urban Green Spaces and Power Hubs for the *Eco-Energy Balance*

This study proposes a geospatial framework that operationalizes the concept of *Urban Eco-Energy Balance*. It is structured in two parts: (A) assessment of *Urban Eco-Energy Deltas*, by comparing the current CO_2 emissions and UGSs' carbon sequestration; and (B) a Decision-making Support System to spatially combine strategic UGSs enhancement and rooftop PV deployment.

3.1 *Part A: Spatial Assessment of* Urban Eco-Energy Deltas

1. *Estimation of CO_2emissions at the Minimum Territorial Unit's scale*

The first step is the spatial estimation of CO_2 emissions. Raw data on CO_2 emissions at the city scale constitute the input data. Emissions are then downscaled from municipal data to Minimum Territorial Units (MTUs) by subdividing them over all the MTUs based on their land-use type. MTUs' area, perimeter or population are timely employed as proxy weights in the application of a proportionality formula. The emissions per MTU E_i were estimated as:

$$E_i = E_{TOT} \times \frac{W_i}{\sum W_i}$$

E_{TOT} is the total emission of the sector and W_i is the weighting parameter for the MTU i. We obtain a spatial dataset describing total and area-normalized CO_2 emissions (tCO_2/ha) for each MTU, setting the baseline for the *Eco-Energy Balance* analysis. The GIS environment enables the application of a combined formula in Python programming language which includes all weighting factors coupled to the land-use type. The result can then be visualized through a categorized map.

2. *Assessment of the Carbon Sequestration potential of existing UGSs*

In the second step we map the UGSs and we clip them across the MTUs. We then employ two parallel procedures to estimate the UGSs' Carbon Sequestration (CS) based on vegetation type. For tree-covered areas we evaluate the total biomass with species-specific allometric equations:

$$Total\ Biomass = Exceeding\ Ground\ Biomass(EGB) + Beneath\ Ground\ Biomass(BGB$$
$$\text{with } EGB = D_b \times H \times \rho \text{ and } BGB = 0.26 \times EGB$$

where D_b is the diameter at breast height, H the tree height and ρ the wood density.

Each UGS' surface is then divided by the crown diameter to derive the number of present trees. Therefore, we can estimate the carbon storage of the entire area.

$$n_T = \frac{Green\ Area}{Tree\ Crown\ Area}$$
$$Carbon\ Storage_{TOT}[tC] = (0.50 \times Total\ Biomass) \times n_T$$

For treeless or grass-covered areas, carbon storage values were extracted from literature benchmarks and adjusted for local growth rates (Tabacchi et al., 2011). To assess their total carbon storage, these benchmark density values were multiplied by each specific green space's area. Finally, CS capacity is expressed as annual CO_2 absorption using the formula below, where 3,67 is the chemical conversion factor from C to CO_2:

$$Carbon\ Sequestration\left(\frac{tCO_2}{year}\right) = (Carbon\ Storage_{TOT} \times Growth\ Rate) \times 3,67$$

3. *Assessment of current CO_2 avoided via PV rooftop systems*

In the third step we map existing PV rooftop systems and spatialize the avoided CO_2 emissions across the corresponding MTUs. The avoided CO_2 is estimated by applying a constant coefficient to the PV surfaces identified. This proxy value is derived from the product of the average annual energy yield for the case study area and the emission factor of the national electricity mix (Caputo, 2019). The calculation follows the standard energy output formula:

$$E = H \times \eta \times PR$$

$$CO_2\ Avoided\left(\frac{tCO_2}{year}\right) = Suitable\ Area\left(m^2\right) \times E\left(\frac{kWh}{m^2 year}\right) \times E_f\left(\frac{tCO_2}{kWh}\right)$$

where E is the annual energy potential, H is the annual solar irradiance (kWh/m^{2}year), η is the module Efficiency (%), PR is the performance ratio and E_f is the national emission factor.

4. *Evaluation of Current Eco-Energy Deltas for MTU*

To evaluate the current *Eco-Energy Deltas,* we subtract from each MTU's annual CO_2 emissions the sum of its UGSs' annual CS and CO_2 avoided via PV rooftop systems. The resulting map identifies net-source zones—where emissions substantially exceed the combined sequestration and avoided emissions—and balanced or net-sink zones—where vegetation and PV systems already bring the unit to (or below) net equilibrium.

3.2 *Part B: A Spatial Decision-Making Support System for the* Eco-Energy Balance

5. *Assessment of potential UGSs' Carbon Sequestration enhancement*

The map of current CS capacity of existing UGSs (Part A, step 2) highlights areas with limited or negligible CS. Policymakers can identify priority intervention spaces to deploy greening strategies:

- Increasing tree density and canopy coverage in sparsely vegetated spaces;
- Selecting species with higher sequestration rates and adaptive resilience;
- Implementing green corridors or linear parks connecting fragmented areas;
- Introducing vertical greening elements to supplement limited horizontal space.

Within the GIS environment, policymakers can simulate the deployment of selected strategies and assess changes in the *Eco-Energy Deltas* employing different targeted greening alternatives.

Table. 1 Priority Matrix for strategic urban reconfiguration

Category	Geospatial condition	Recommended strategy (Policy response)
Green Priority Zones	High net CO_2 emissions + Low PV suitability	Urban afforestation, linear green parks, tree-lined streets, vertical greening, and biodiversity restoration
Energy Priority Zones	High net CO_2 emissions + High PV suitability	Establishment of Power Hubs: large-scale PV installations, development of RECs
Conservation Zones	Low net CO_2 emissions + High existing greenery	Implementation of protective regulations, building restrictions, and biodiversity enhancement measures
High-Efficiency Zones	Net Eco-Energy equilibrium from existing or potential PV generation	Integration into Smart Grid networks for energy export to deficit areas

Source Elaborated by the authors

6. *Assessment of potential PV rooftop systems deployment*

In this step we map the areas identified as *suitable* or *non-suitable* for renewable energy production, based on the criteria established under the Renewable Energy Directive (EU, 2024) and corresponding national decrees. These include land-use type, landscape constraints, protected areas, and technical suitability. We assess CO_2 emissions reduction considering PV rooftop deployment over the identified areas. The output, spatialized in a GIS environment, provides municipal authorities with a tool to identify where to prioritize incentives and expedite permitting procedures, thereby aligning renewable-energy projects with energy-justice policies.

7. *A Decision-making Support System: Transformation Map and Priority Matrix*

In conclusion, a Transformation map synthesizes outputs from previous phases. The comprehensive visualization supports the classification of MTUs into actionable policy categories according to the Priority Matrix (Table 1).

4 Application: The City of Naples in Italy

To validate the framework, we applied the methodology to the city of Naples in Italy. Naples is an ideal testbed due to the availability of detailed environmental and socio-economic datasets and the heterogeneous urban fabric. Moreover, the analysis builds upon the findings of previous studies on the same city (Aruta et al., 2024; Antuono et al., 2025). To provide meaningful results at the municipal scale, the MTU adopted is the Census Section, the smallest territorial unit defined by the Italian Institute of Statistics (ISTAT) for the decennial census.

4.1 Part A: Spatial Assessment of Urban Eco-Energy Deltas

1. *Estimation of CO_2emissions at the Census Section's scale*

Raw data at the city scale were obtained from the City of Naples' Emissions Monitoring Inventory (reference year 2019), adopted in 2022 within the city's commitments under the Covenant of Mayors (Comune di Napoli, 2022). It categorizes emissions into seven sectors: (i) municipal buildings, (ii) tertiary buildings, (iii) residential buildings, (iv) public lighting, (v) industry, (vi) public transport and (vii) private and commercial transport. Land use and population data were derived from the 2021 ISTAT territorial datasets (ISTAT, n.d.). The analysis highlights a high-emission area at the core of the city, where the most densely populated areas lie and UGSs are lacking (Fig. 1). Only few zero or very low emissions areas exist at the outskirts of the city, mostly corresponding to UGSs, brownfields or abandoned areas.

2. *Assessment of Carbon Sequestration potential of existing UGSs*

UGSs were obtained integrating two open datasets: (i) ISPRA's *Carta della Natura* (Bagnaia et al., 2017), which provides land cover type and habitat detail and (ii) *OpenStreetMap* Land Use data, which increases UGSs granularity at the intra-urban scale (Fig. 2). Allometric parameters were derived from the Italian National Forest Inventory (Gasparini et al., 2022).

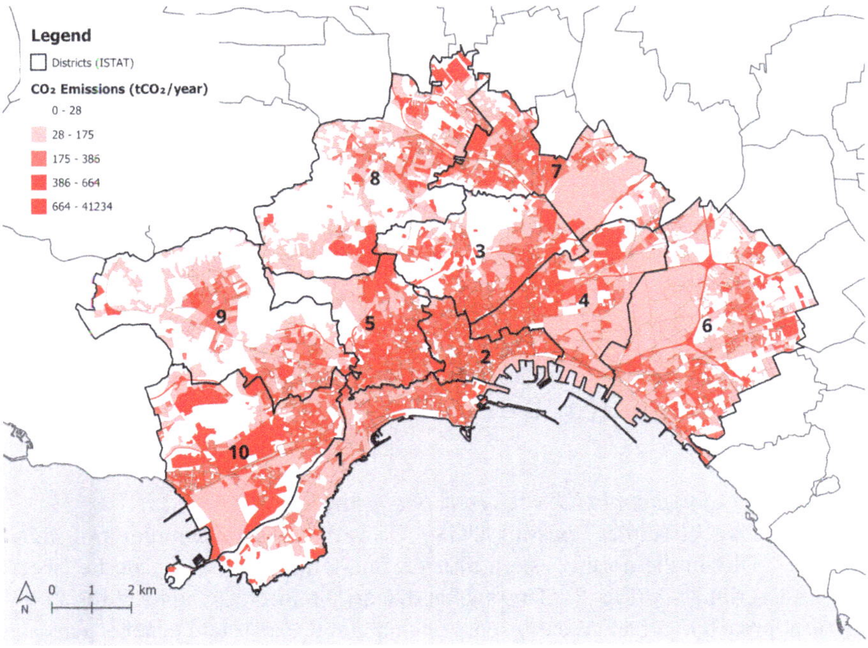

Fig. 1 Naples' CO_2 emissions at the census section's scale. *Source* Elaborated by the authors

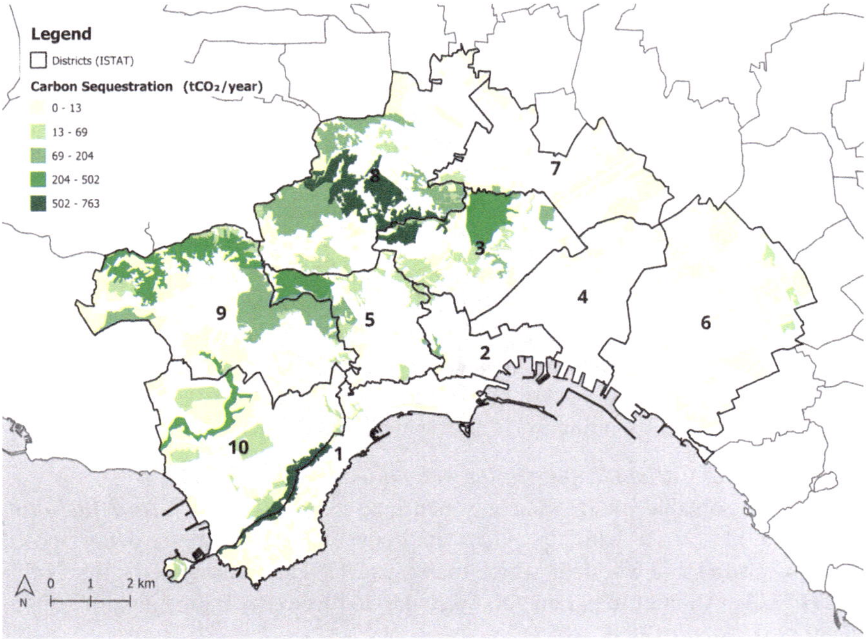

Fig. 2 Naples' Carbon Sequestration potential of existing UGSs. *Source* Elaborated by the authors

3. *Assessment of current CO_2avoided* via*PV rooftop systems*

   At present, no significantly extended PV rooftop system was identified in the city.

4. *Evaluation of current Eco-Energy Deltas for Census Section*

   The analysis highlights a concentration of unbalanced sections at the core of the city, and a general unbalance across the entire territory. Only few net-zero or very low emission areas exist, mainly in the North-Western part, corresponding to UGSs and hilly areas (Fig. 3).

4.2 *Part B: A Spatial Decision-Making Support System for the* Eco-Energy Balance

5. *Assessment of potential UGSs' Carbon Sequestration enhancement*

   The analysis of Naples' existing UGSs' CS revealed a concentration of highly effective UGSs in the north-western districts and a marked scarcity in the eastern industrial peripheries (Fig. 2). This spatial disparity reinforces the need for a polycentric approach to urban greening, ensuring equitable ecosystem benefits across all districts (Kabisch et al., 2016).

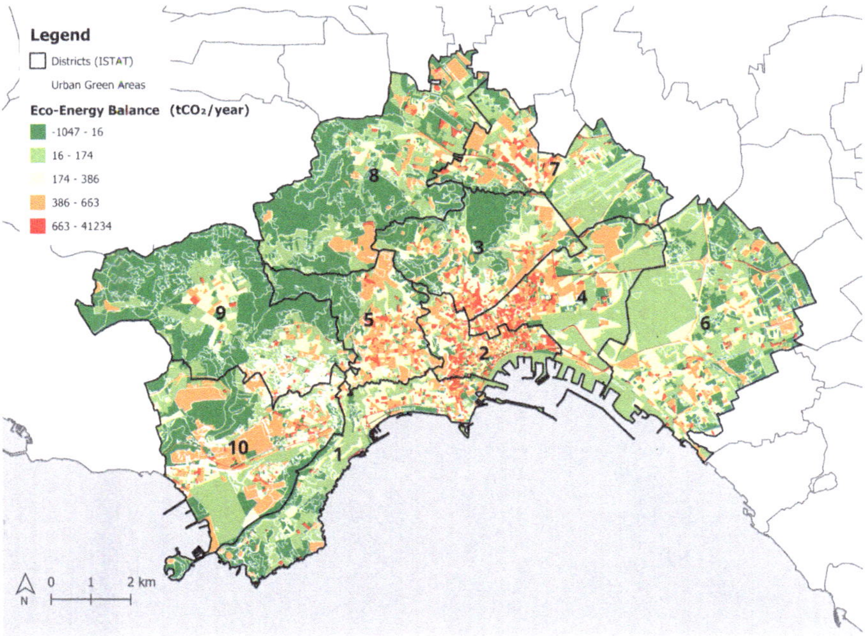

Fig. 3 Naples' current *Eco-Energy Deltas* by census section. *Source* Elaborated by the authors

6. *Assessment of potential PV rooftop systems deployment*

The *suitable areas* mapping reveals notable untapped potential in the eastern sector, dominated by disused or low-intensity industrial sites (Fig. 4). These areas offer great opportunities for PV rooftop systems deployment, particularly if considered through the lens of energy justice (Cumo et al., 2025), as they are characterized by socio-economic vulnerability and energy poverty.

7. *A Decision-making Support System: Transformation Map and Priority Matrix*

Naples' Transformation Map highlights the presence of three potentially high-efficiency *Power Hubs*: Naples Capodichino Airport, the Central Station and its surrounding area and the former ILVA industrial site in Bagnoli (Fig. 5). These areas are characterized by extensive disused or underutilized land and negligible green coverage. The map indicates that the deployment of large-scale PV rooftop systems in these areas—*suitable* according to current legislation—would represent, at present, the most effective strategy for reducing CO_2 emissions. Regarding existing UGSs, the analysis shows a positive contribution in terms of CS, although with significant variation depending on vegetation type. The map further identifies priority areas within the existing UGS network where targeted greening interventions could maximize overall CS performance.

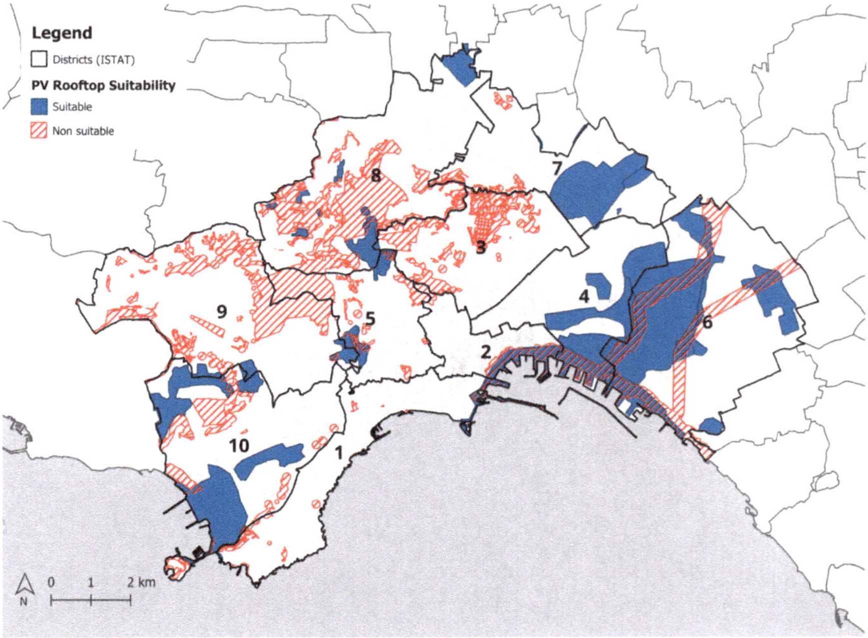

Fig. 4 *Suitable areas* for PV rooftop systems deployment. *Source* Elaborated by the authors

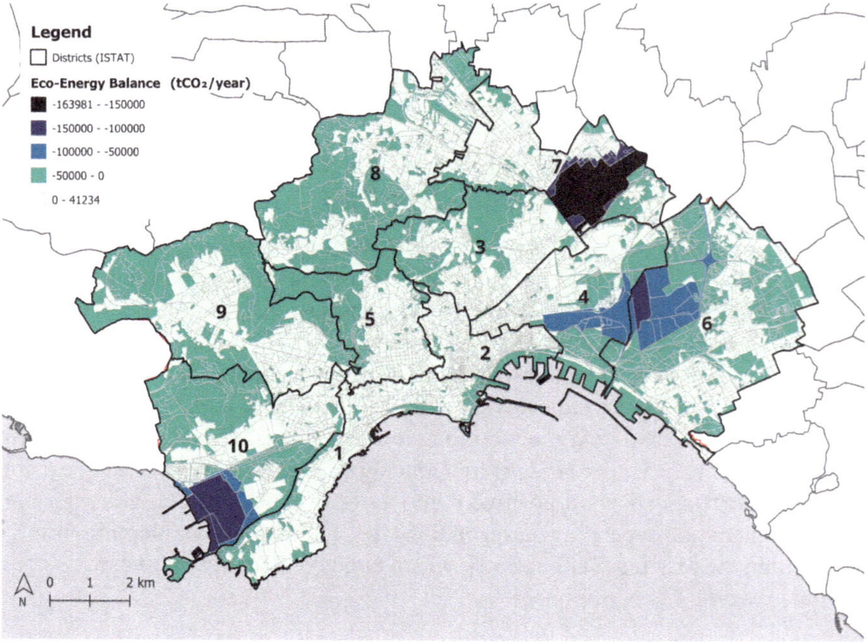

Fig. 5 Naples' Transformation Map for the *Eco-Energy Balance*. *Source* Elaborated by the authors

5 Discussion and Implications

The application of the framework to Naples confirms the results of previous studies on UGSs' CS (Antuono et al., 2025) and rooftop PV potential analysis (Aruta et al., 2024) in the city, adding an integrated perspective. The methodology extends the NZED concept from the district to the municipal scale through the *Urban Eco-Energy Balance*. The GIS environment is crucial as it allows the integration of heterogeneous open datasets, the use of Python scripts for advanced spatial analysis, and the visualization of outputs through maps. The Transformation Map and the Priority Matrix translate this information into defined categories, offering a practical Decision-making Support System for local climate governance and investment programming. The framework provides a replicable and operational approach to assess and improve the *Urban Eco-Energy Balance*. It supports evidence-based spatial planning and helps planners identify priority zones for greening, PV deployment and conservation. However, some limitations must be acknowledged. First, the analysis relies on open-access datasets, whose availability and quality may differ across municipalities. Second, CS estimates depend on general allometric coefficients and average growth rates, which could be refined using local field measurements and remote sensing. Third, PV potential is assessed without accounting for detailed architectural constraints—roof orientation, shading conditions, structural properties—resulting in conservative estimates. Future developments should address these limitations by refining input data. Moreover, integrating dynamic energy modelling and socio-economic indicators would better capture trade-offs between ecological goals, mitigation outcomes and energy justice over time.

6 Concluding Remarks

This study presented a geospatial methodology that enables municipalities to assess their *Urban Eco-Energy Balance* and to plan spatially targeted interventions to achieve climate neutrality. Tested in Naples, the framework combines the carbon sequestration of UGSs with renewable energy generation from rooftop PV systems, offering a comprehensive and replicable approach to spatial decision-making for climate mitigation. The resulting Transformation Map and Priority Matrix equip local administrations with an operational tool to define priority zones—whether for afforestation, renewable energy deployment or conservation—and to balance eco-energy goals through evidence-based spatial planning. Beyond its operational relevance, the methodology bridges the gap between high-level climate strategies and local planning tools, fostering the co-evolution of UGSs and RECs. The framework provides a blueprint for cities seeking to translate sustainability commitments into measurable, place-based actions toward neutrality and energy justice.

References

Aguilar, A. J., De La Hoz-Torres, M. L., Aguilar-Camacho, J., & Guerrero-Rivera, M. F. (2025). Assessment of energy efficiency and energy poverty of the residential building stock of the city of Seville using GIS. *Applied Sciences, 15*(12), 6438. https://doi.org/10.3390/app15126438

Akrofi, M. M., Okitasari, M., & Kandpal, R. (2022). Recent trends on the linkages between energy, SDGs and the Paris agreement: A review of policy-based studies. *Discover Sustainability, 3*(1), 32. https://doi.org/10.1007/s43621-022-00100-y

Alpagut, B., Romo, A. L., Hernández, P., Tabanoğlu, O., & Martinez, N. H. (2021). A GIS-Based multicriteria assessment for identification of positive energy districts boundary in cities. *Energies, 14*(22). https://doi.org/10.3390/en14227517

Antuono, G., Fistola, R., Cernicchiara, G., & D'Agostino, P. (2025). Urban forestry and environmental simulation for public spaces: An AR-supported workflow to a participatory experience. In O. Zerlenga, V. Marchetelli, & A. Gesse (Eds.), *Forest therapy: The challenge of the future for the design of living environments* (pp. 35–71). La Scuola di Pitagora.

Aruta, G., Ascione, F., Fistola, R., Iovane, T., Aruta, G., Ascione, F., Fistola, R., & Iovane, T. (2024). The city as a power hub for boosting renewable energy communities: A case study in Naples. *Sustainability, 16*(18). https://doi.org/10.3390/su16187988

Ashinze, U. K., Edeigba, B. A., Umoh, A. A., Biu, P. W., & Daraojimba, A. I. (2024). Urban green infrastructure and its role in sustainable cities: A comprehensive review. *World Journal of Advanced Research and Reviews, 21*(2), 928–936. https://doi.org/10.30574/wjarr.2024.21.2.0519

Bagnaia, R., Viglietti, S., Laureti, L., Giacanelli, V., Ceralli, D., Bianco, P. M., Loreto, A., Luce, E., & Fusco, L. (2017). Carta della Natura della Regione Campania: Carta degli habitat alla scala 1:25.000. ISPRA.

Baycan-Levent, T., & Nijkamp, P. (2009). Planning and management of urban green spaces in Europe: Comparative analysis. *Journal of Urban Planning and Development, 135*(1), 1–12. https://doi.org/10.1061/(ASCE)0733-9488(2009)135:1(1)

Bosone, M., Pirelli, B., & Vito, D. (2023). Evaluating energy communities: A new social and economic model for implementing the ecological transition. In O. Gervasi, O., et al. (Eds.), In *Computational Science and Its Applications—ICCSA 2023 Workshops. ICCSA 2023. Lecture Notes in Computer Science* (Vol. 14108). Springer. https://doi.org/10.1007/978-3-031-37117-2_19

Caputo, A. (2019). Fattori di emissione atmosferica di gas a effetto serra nel settore elettrico nazionale e nei principali Paesi Europei. ISPRA. Retrieved from: https://www.isprambiente.gov.it/files2019/pubblicazioni/rapporti/R_303_19_gas_serra_settore_elettrico.pdf

Chavez, A., & Sperling, J. (2017). Key drivers and trends of urban greenhouse gas emissions. In S. Dhakal & M. Ruth (Eds), *Creating low carbon cities* (pp. 157–168). Springer International Publishing. https://doi.org/10.1007/978-3-319-49730-3_14

Chen, R., & You, X. (2020). Reduction of urban heat island and associated greenhouse gas emissions. *Mitigation and Adaptation Strategies for Global Change, 25*(4), 689–711. https://doi.org/10.1007/s11027-019-09886-1

Chen, S., Wang, Y., Ni, Z., Zhang, X., & Xia, B. (2020). Benefits of the ecosystem services provided by urban green infrastructures: Differences between perception and measurements. *Urban Forestry & Urban Greening, 54*, Article 126774. https://doi.org/10.1016/j.ufug.2020.126774

Chen, T., Li, Y., Zhang, Y., Ji, H., Wang, X., He, J., Liu, Q., & Zhang, H. (2025). Coupling renewable energy with urban greening: Quantifying the sustainable development potential of photovoltaic-green roofs. *Sustainable Cities and Society, 133*, Article 106866. https://doi.org/10.1016/j.scs.2025.106866

Comune di Napoli. (2022). Deliberazione di Consiglio Comunale n. 34 del 5 agosto 2022 [City Council Resolution No. 34 of 5 August 2022]. https://trasparenza.comune.napoli.it/flex/files/1/9/9/D.9c1f1f4c913018ae55ec/Delibera_di_C.C._n._34_del_5_agosto_2022.pdf

Costa, V., & Campanini, F. (2024). Community-Centred energy planning: Within and beyond administrative borders. *Sustainability, 16*(5), 2049. https://doi.org/10.3390/su16052049

Cumo, F., Pennacchia, E., Maurelli, P., Rosa, F., & Zylka, C. (2025). Spaces, energy and shared resources: New technologies for promoting more inclusive and sustainable urban communities. *Energies, 18*(16). https://doi.org/10.3390/en18164410

De la Sota, C., Ruffato-Ferreira, V. J., Ruiz-García, L., & Alvarez, S. (2019). Urban green infrastructure as a strategy of climate change mitigation. A case study in northern Spain. *Urban Forestry & Urban Greening, Urban Green Infrastructure—Connecting People and Nature for Sustainable Cities, 40*, 145–151. https://doi.org/10.1016/j.ufug.2018.09.004

Derkenbaeva, E., Halleck Vega, S., Hofstede, G. J., & van Leeuwen, E. (2022). Positive energy districts: Mainstreaming energy transition in urban areas. *Renewable and Sustainable Energy Reviews, 153*, Article 111782. https://doi.org/10.1016/j.rser.2021.111782

Dong, L., Wang, Y., Ai, L., Cheng, X., & Luo, Y. (2024). A review of research methods for accounting urban green space carbon sinks and exploration of new approaches. *Frontiers in Environmental Science, 12*. https://doi.org/10.3389/fenvs.2024.1350185

Duan, Z., & Kim, S. (2023). Progress in research on net-zero-carbon cities: A literature review and knowledge framework. *Energies, 16*(17). https://doi.org/10.3390/en16176279

EU. (2024). DIRECTIVE (EU) 2018/2001 of the European parliament and of the council on the promotion of the use of energy from renewable sources. EU. Retrieved from https://eur-lex.europa.eu/legal-content/EN/TXT/PDF/?uri=CELEX:02018L2001-20240716

Farkas, J. Z., Hoyk, E., Morais, M. B. de, & Csomós, G. (2023). A systematic review of urban green space research over the last 30 years: A bibliometric analysis. *Heliyon, 9*(2). https://doi.org/10.1016/j.heliyon.2023.e13406

Fistola, R. (2023). Ecosystem services for the city as a complex system: A methodological proposal. *Sustainability, 15*(12). https://doi.org/10.3390/su15129318

Frantzeskaki, N., Castán Broto, V., Coenen, L., & Loorbach, D. (Eds.). (2017). *Urban sustainability transitions*. Routledge.

Freundt, M., Iwaszuk, E., Naumann, S., Davis, M., Gosk, I., Bergier, T., Jeleński, T., & Mederake, L. (Eds.). (2020). *Błękitno-zielona infrastruktura dla łagodzenia zmian klimatu w miastach: Narzędzia strategiczne*. Fundacja Sendzimira ; Ecologic Institute.

Gasparini, P., Di Cosmo, L., Floris, A., & De Laurentis, D. (Eds.). (2022). *Italian national forest inventory—methods and results of the third survey*. Springer. Retrieved from: https://www.inventarioforestale.org/it/pubblicato-il-volume-inventario-nazionale-delle-foreste-e-dei-serbatoi-forestali-di-carbonio-metodi-e-risultati-della-terza-indagine/

Gill, S. E., Handley, J. F., Ennos, A. R., & Pauleit, S. (2007). Adapting cities for climate change: The role of the green infrastructure. *Built Environment, 33*(1), 115–133. https://doi.org/10.2148/benv.33.1.115

Gülçin, D., & Bosch, C. C. K. van den. (2021). Assessment of above-ground carbon storage by urban trees using LiDAR data: The case of a university campus. *Forests, 12*(1). https://doi.org/10.3390/f12010062

Hachem-Vermette, C. (2022). Role of solar energy in achieving net zero energy neighborhoods. *Solar Energy Advances, 2*, Article 100018. https://doi.org/10.1016/j.seja.2022.100018

Hillman, T., & Ramaswami, A. (2010). Greenhouse gas emission footprints and energy use benchmarks for eight U.S. Cities. *Environmental Science & Technology, 44*(6), 1902–1910. https://doi.org/10.1021/es9024194

Huang, J., Song, P., Liu, X., Li, A., Wang, X., Liu, B., & Feng, Y. (2024). Carbon sequestration and landscape influences in urban greenspace coverage variability: A high-resolution remote sensing study in Luohe, China. *Forests, 15*(11). https://doi.org/10.3390/f15111849

Hwang, I., Jung, W., & Cho, K. (2025). Changes in research trends and topics on carbon neutrality before and after the 2016 Paris agreement. *IEEE Access, 13*, 133525–133545. https://doi.org/10.1109/ACCESS.2025.3583438

Intergovernmental Panel On Climate Change (Ipcc). (Ed.). (2023). Urban systems and other settlements. In *Climate change 2022—mitigation of climate change* (1st ed., pp. 861–952). Cambridge University Press. https://doi.org/10.1017/9781009157926.010

International Energy Agency (IEA). (2024). Empowering urban energy transitions: Smart cities and smart grids. Retrieved from https://iea.blob.core.windows.net/assets/00f7d520-d517-473d-b357-5adb43c4a57e/EmpoweringUrbanEnergyTransitions.pdf

ISTAT. (n.d.). Basi territoriali e variabili censuarie. Retrieved January 13, 2026, from https://www.istat.it/notizia/basi-territoriali-e-variabili-censuarie/

Jiang, M., Li, J., Wei, W., Miao, J., Zhang, P., Qian, H., Liu, J., & Yan, J. (2020). Using existing infrastructure to realize low-cost and flexible photovoltaic power generation in areas with high-power demand in China. *Iscience, 23*(12), 101867. https://doi.org/10.1016/j.isci.2020.101867

Jing, R., Hastings, A., & Guo, M. (2020). Sustainable design of urban rooftop food-energy-land Nexus. *IScience, 23*(11), 101743. https://doi.org/10.1016/j.isci.2020.101743

Kabisch, N., Frantzeskaki, N., Pauleit, S., Naumann, S., Davis, M., Artmann, M., Haase, D., Knapp, S., Korn, H., Stadler, J., Zaunberger, K., & Bonn, A. (2016). Ecology and society: Nature-based solutions to climate change mitigation and adaptation in urban areas: Perspectives on indicators, knowledge gaps, barriers, and opportunities for action. *Ecology and Society, 21*(2). https://doi.org/10.5751/ES-08373-210239

Komninos, N. (2022). Net zero energy districts: Connected intelligence for carbon-neutral cities. *Land, 11*(2). https://doi.org/10.3390/land11020210

Kumar, N., Nishtha, & Kumar, J. (2023). Estimation of carbon sequestration by some selected dominant tree species of Chandigarh University, Gharuan, Mohali. *Materials Today: Proceedings*. https://doi.org/10.1016/j.matpr.2023.03.210

Liu, H.-Y., Skandalos, N., Braslina, L., Kapsalis, V., Karamanis, D., Liu, H.-Y., Skandalos, N., Braslina, L., Kapsalis, V., & Karamanis, D. (2023). Integrating solar energy and nature-based solutions for climate-neutral urban environments. *Solar, 3*(3), 382–415. https://doi.org/10.3390/solar3030022

Martinelli, V. (2025a). Positive energy districts for urban energy transition: Regulatory challenges and implementation strategies. *TeMA—Journal of Land Use, Mobility and Environment, 18*(2), 293–297. https://doi.org/10.6093/1970-9870/12297

Martinelli, V. (2025b). Urban energy transition between regulatory evolution and scientific production: A bibliometric analysis. *TeMA—Journal of Land Use, Mobility and Environment, 18*(1), 131–141. https://doi.org/10.6093/1970-9870/11625

McPherson, E. G., Doorn, N. S. van, & Peper, P. J. (2016). Urban tree database and allometric equations. *Gen. Tech. Rep. PSW-GTR-253. Albany, CA: U.S. Department of Agriculture, Forest Service, Pacific Southwest Research Station, 86*, 253. https://doi.org/10.2737/PSW-GTR-253

Montis, A. D., Ledda, A., Serra, V., Manunta, A., Calia, G., Montis, A. D., Ledda, A., Serra, V., Manunta, A., & Calia, G. (2025). Urban green spaces and climate changes: Assessing ecosystem services for the municipality of Sassari (Italy). *Land, 14*(6). https://doi.org/10.3390/land14061308

NetZeroCities—European Climate, Infrastructure and Environment Executive Agency. (n.d.). Retrieved 13 January 2026, from https://cinea.ec.europa.eu/featured-projects/netzerocities_en

Pan, D., Bai, Y., Chang, M., Wang, X., & Wang, W. (2022). The technical and economic potential of urban rooftop photovoltaic systems for power generation in Guangzhou. *China. Energy and Buildings, 277*, Article 112591. https://doi.org/10.1016/j.enbuild.2022.112591

Rosokhata, A., Minchenko, M., Khomenko, L., & Chygryn, O. (2021). Renewable energy: A bibliometric analysis. *E3S Web of Conferences, 250*, 03002. https://doi.org/10.1051/e3sconf/202125003002

Sameti, M., & Haghighat, F. (2017). Optimization approaches in district heating and cooling thermal network. *Energy and Buildings, 140*, 121–130. https://doi.org/10.1016/j.enbuild.2017.01.062

Sameti, M., & Haghighat, F. (2018). Integration of distributed energy storage into net-zero energy district systems: Optimum design and operation. *Energy, 153*, 575–591. https://doi.org/10.1016/j.energy.2018.04.064

Seto, K. C., Churkina, G., Hsu, A., Keller, M., Newman, P. W. G., Qin, B., & Ramaswami, A. (2021). From Low- to net-zero carbon cities: The next global agenda. *Annual Review of Environment and Resources, 46*, 377–415. https://doi.org/10.1146/annurev-environ-050120-113117

Shadman, S., Ahanaf Khalid, P., Hanafiah, M. M., Koyande, A. K., Islam, Md. A., Bhuiyan, S. A., Sin Woon, K., & Show, P.-L. (2022). The carbon sequestration potential of urban public parks of densely populated cities to improve environmental sustainability. *Sustainable Energy Technologies and Assessments, 52*, Article 102064. https://doi.org/10.1016/j.seta.2022.102064

Tabacchi, G., Di Cosmo, L., Gasparini, P., & Morelli, S. (2011). Stima del volume e della fitomassa delle principali specie forestali italiane. Equazioni di previsione, tavole del volume e tavole della fitomassa arborea epigea. Consiglio per la Ricerca e la sperimentazione in Agricoltura, Unità di Ricerca per il monitoraggio e la pianificazione Forestale.

Trifiletti, E. G., & Iaccarino, M. (2020). Piano d'Azione per l'Energia Sostenibile (PAES): Relazione di monitoraggio al 15 maggio 2020. Servizio Controlli ambientali e attuazione PAES, Comune di Napoli. Retrieved from: https://static-www.comune.napoli.it/wp-content/uploads/2025/05/Relazione_monitoraggio_completa.pdf?mode=download

Trupiano, G., & Fistola, R. (1993). *Gestione delle risorse naturali e pianificazione degli spazi verdi*. Atena

Tuerk, A., Frieden, D., Neumann, C., Latanis, K., Tsitsanis, A., Kousouris, S., Llorente, J., Heimonen, I., Reda, F., Ala-Juusela, M., Allaerts, K., Caerts, C., Schwarzl, T., Ulbrich, M., Stosch, A., & Ramschak, T. (2021). Integrating plus energy buildings and districts with the EU energy community framework: Regulatory opportunities, barriers and technological solutions. *Buildings, 11*(10). https://doi.org/10.3390/buildings11100468

Tyagi, B., Bagui, D., Patil, A., Ramesh, K., Jena, A., & Chaudhary, M. (2024). Global perspectives on rooftop solar energy: A deep dive on how leading economies advance rooftop solar energy adoption. Council of Energy, Environment and Water (CEEW). https://www.ceew.in/sites/default/files/global-rts-issue-brief-clean-copy-final.pdf

Wiginton, L. K., Nguyen, H. T., & Pearce, J. M. (2010). Quantifying rooftop solar photovoltaic potential for regional renewable energy policy. *Computers, Environment and Urban Systems, 34*(4), 345–357. https://doi.org/10.1016/j.compenvurbsys.2010.01.001

Yang, Y., Zhang, S., & Xiao, Y. (2015). An MILP (mixed integer linear programming) model for optimal design of district-scale distributed energy resource systems. *Energy, 90*, 1901–1915. https://doi.org/10.1016/j.energy.2015.07.013

Zhang, H., Peng, J., Wang, R., Zhang, J., & Yu, D. (2021). Spatial planning factors that influence CO_2 emissions: A systematic literature review. *Urban Climate, 36*, Article 100809. https://doi.org/10.1016/j.uclim.2021.100809

Urban Green Infrastructure Performances Under Threat

Aude Zingraff-Hamed, Ayoub Hammadi, Hugo Le Pennec, and Mehdi Maanan

Abstract As global mean temperatures reach a + 1.4°C anomaly, urban environments are emerging as sentinels of climate change, experiencing localized thermal extremes that largely exceed global averages. This study evaluates the performance of Urban Green Infrastructure (UGI) under heat stress in Strasbourg, France, across 43 green spaces, and using NDVI and NDWI spectral indices as proxies for tree vitality, namely photosynthetic activity and hydric resilience, respectively. The results reveal a critical "performance gap" to functional collapse of the UGI. Results showed that forested areas maintain relatively high vitality, but UGI undergoes a functional collapse during sustained + 3°C local temperature anomalies and persistent system failure for 15 to 45 consecutive days under heat stress. At these tipping points, temperate species initiate defensive stomatal closure shifting the canopy's role from an active cooling system to a passive radiative barrier, start carbon starvation process endangering long-term tree health. To address the projected + 2.2°C to + 2.7°C scenarios by 2100, UGI needs an ecophysiological paradigm shift prioritizing large tree coverage, structural heterogeneity to protect biodiversity, and integrating Sponge City principles to safeguard city viability for human and nonhuman.

Keywords Urban greening · Nature-based solutions · One Health · Climate change adaptation · Ecosystem services · Nonhuman inclusion · Environmental justice

A. Zingraff-Hamed (✉) · A. Hammadi
Research Group Landscape Planning, Department of Spatial Planning, TU Dortmund University, Dortmund, Germany
e-mail: aude.zingraff-hamed@tu-dortmund.de

A. Zingraff-Hamed
Laboratoire Image Ville Environnement, Université de Strasbourg, Strasbourg, France

A. Zingraff-Hamed · H. Le Pennec
ENGEES National Institute for Water and Environmental Engineering, Strasbourg, France

A. Hammadi · M. Maanan
University Hassan II of Casablanca, Casablanca, Morocco

L. Diana et al. (eds.), *Greening and Resilience*, Urban Sustainability,
https://doi.org/10.1007/978-981-92-0419-9_7

Anthropogenic emissions link to temperature rises worldwide but especially in European cities. A wide range of anthropogenic activities emit greenhouse gases, e.g., energy production contributes up to 25–30% of CO_2 emissions yearly (IPCC, 2023), followed by other sectors such as agriculture and forestry (Tubiello et al., 2015). The anthropogenic increase of greenhouse gas concentrations is the radiative forcing of disrupted climate patterns and thermal escalation of global mean temperatures. In recent years, temperatures have consistently neared or exceeded + 1.5°C above pre-industrial time (CCCS, 2026), in 2025 with an increase of the land temperature average of 2.03 ± 0.17 °C (Berkeley Earth, 2026). The "warming gap" places Europe as the most warming continent, with a 2024 five-year average temperature approximately 2.3°C above the pre-industrial baseline, placing European ecosystems and human health under the highest pressure (WMO & Copernicus, 2024).

Global temperature increase endangers human wellbeing, health and potentially survival. Humans are a temperature-sensitive species and thermal increase causes physiological stress, first impacting the most vulnerable population, then impacting whole community health and productivity and finally potentially leading to increased morbidity and mortality. Europe counted in 2022, 2023 and 2024, respectively, 61,672, 47,490 and 62,775 heat-related deaths (Ballester et al., 2025) with Italy, Spain, France and Germany being the most affected. The 2003 heat wave in Europe was a turning point. It was the first heatwave with massive casualties, namely with over 70 000 heat-related deaths. It shattered the perception of Europe as a "safe" temperate zone. Furthermore, it questioned government preparedness for an extreme weather event. A year later, a landmark study published in *Nature* concluded that human-induced climate change will at least double the risk of 2003 heatwave calling for the development and deployment of adaptation plans (Scott et al., 2004). Studies have shown that heatwaves under emission scenarios will cause over 2 million deaths per year by 2100 (Masselot et al., 2025). Critical studies underscored the physiological limits of human survival in a warming world (e.g., Haines et al., 2006). In this context, climate protection and climate change adaptation became one of the top priorities of European cities.

While reducing emission remains a challenging target even more than 20 after this political turn, human strategy mostly relies on nonhuman performances in regulating climate processes, namely photosynthetic carbon sequestration and evaporative cooling. Indeed, based on the ecosystem services theory, vegetation provides essential ecosystem functions, such as microclimate regulation and atmospheric carbon sinks (Nowak & Crane, 2002), thereby maintaining Earth habitability. For example, Gill et al. (2007) showed that increasing tree cover by 10% in high-density urban areas can maintain surface temperatures at 1961–1990 levels during heat. In the wake of heatwave event, as in 2003, cities have transitioned toward strategies as nature-based solutions and urban green infrastructure (UGI) viewing nature no longer as a "decorative" luxury but as a "hybrid engineering" to be integrated into the urban fabric to perform at a larger scale than is technologically feasible, e.g., as with air conditioners and carbon scrubbers (Pauleit et al., 2017a).

UGI is a "strategically planned network of natural and semi-natural areas [...] designed to deliver essential ecosystem services" (EC, 2024). Unlike isolated parks,

UGI functions as a "living network" targeting biodiversity conservation, climate change adaptation, social cohesion and green economy. UGI are inherently multifunctional and deliver multiple ecosystem services (Pauleit et al., 2017a), namely benefits produced by nonhuman for human wellbeing and socio-cultural development, and are crucial life support area for complex ecosystem and ecosystem processes (Zölch et al., 2016). Finally, UGI acts as a critical interface for the regulation of basic biogeochemical cycles such as nitrogen and carbon cycles, and ecological integrity, e.g., habitat continuity and genetic flow. Consequently, the strategic deployment of UGI targets the maximization of vegetation performance to address global and systemic risks.

However, it is worthy to note that botanical systems are subject to biological constraints, including environmental stressors. UGI are composed of living organisms with specific "comfort zones." Their ability to regulate microclimate and store CO_2 depend on hydro-meteorological parameters (Pataki et al., 2011) as well as specific performance and overall health. In other words, their "service delivery" for climate protection and regulation is tied to fragile biological processes depending on the climate itself. Consequently, UGI have limit to tipping points that could trigger a collapse in the delivery of essential climate adaptation functions. Indeed, numerous scholars already formulate concerns about "climate-related 'collapse' of ecosystems" including the loss of critical ecosystem functions (e.g., Canadell & Jackson, 2021). This corresponds to an abrupt response of an ecosystem due to climate extreme disturbances. We adopt the "ecosystem collapse" definition build on previous scholars, mostly Keith et al., 2013, Lindenmayer et al., 2016, Cumming & Peterson, 2017, Bland et al., 2018 and Canadell & Jackson, 2021, and triggered it to the ecosystem functions. Accordingly, ecosystem functional collapse is defined as an abrupt and undesirable change in ecosystem performances, e.g., chlorophyll activity decrease.

Craig Allen and colleagues' seminal paper (Allen et al., 2010) rose first concerns about climate-related tree collapse, namely global heat-wave tree mortality in almost all forested ecosystems. This abrupt collapse of the forested ecosystem was the result of cumulative effect of drought and heat that weaken the species and led to intricate disturbance from diverse agents as pathogens. Within the ecosystem collapse definition, multiple and interacting drivers of collapse cause irreversible changes challenging resilience capacity of the ecosystems. In the case of chlorophyllin collapse, known drivers of this physiological collapse are threefold. First, under heat and drought, the increased atmospheric vapor pressure deficit forces plants to close their stomata to prevent dehydration. Without gas exchange capacity, there is no evapotranspiration capacity. This strategic temporary shutdown is a survival strategy but drastically reduces CO_2 fixation and halts evaporative cooling precisely when it is needed the most, namely during the peak of heat events (Yuan et al., 2019). Second, heat alters the photochemical process in the cells and damages the photosynthetic apparatus. This is compounded by the inactivation of heat-sensitive proteins (Wahid et al., 2007). This means that instead of fixing $CO_2$and cooling the air, the plant enters a survival mode. In other words, tree is starving and stopping service delivery. Finally, extreme heat can even lead to long-term vitality loss through mechanisms such as pollen sterility (Sehgal et al., 2023), undermining the long-term survival of vegetal communities.

Remote sensing has been broadly used for extensive and long-term monitoring of key ecosystem properties and ecosystem services, such as carbon stocks, and to document ecosystem collapse (Lloret & Batllori, 2021). Satellite-derived procedures successfully described processes at specific sites (Clark et al., 2004; McDowell et al., 2014) and across different climate regions, e.g., tropical forests (e.g., Verbesselt et al., 2016), temperate forests (e.g., Sommerfeld et al., 2018) and Mediterranean to semi-arid forests (e.g., Trichon et al., 2018). Remote sensing enables large-scale monitoring and the inclusion of indicators of vegetation vitality, e.g., the Normalized Difference Vegetation Index (NDVI). NDVI is the most widely used remote sensing indicator for assessing the quantity and health of vegetation. Variations of NDVI value within the vegetative development period (March to October) inform about photosystem disturbance. An abrupt decrease mirrors a photosystem II collapse. Consequently, NDVI variation can inform about tipping point for the development of early warning systems of vegetation functionality collapse (Liu et al., 2019; Rogers et al., 2018). The critical temperature of photosystem II disruption (T_{crit}) is one of the "tipping point" studied in the field of vegetation collapse. It is a "point of no return" at which leaves start to become irreversibly damaged and photosynthetic activity ceases (Hüve et al., 2011).

Ecosystem services derived from UGI constitute the backbones of city climate change mitigation and adaptation strategies. However, as biological entities, UGI have intrinsic physiological constraints and their performances result from physiological processes that are influenced by environmental stressors. While human bioclimatic heat stress has historically driven UGI policies, the biophysical limitations of the vegetation itself during heat remain under-considered in the context of urban design (Ahongshangbam et al., 2023). Despite the widespread promotion of UGI as a panacea for urban heat, current planning strategies often overlook these pitfalls. There is a significant knowledge gap regarding UGI resilience to tipping points, turning UGI from active climate regulators to passive or inactive components of the urban fabric. In this context, this study investigates the effect of heat on UGI service delivery capacity. Specifically, it aims to identify which UGI are performing the best when they are the most needed, namely during heat waves. This research seeks to provide better knowledge of these processes and a more realistic, physiology-based framework for preserving green infrastructure under future climate scenarios.

1 Material and Methods

The study investigates the impact of thermal heat on UGI vitality using spatiotemporal data extracted from sentinel (remote sensing) and meteorological data (Fig. 1).

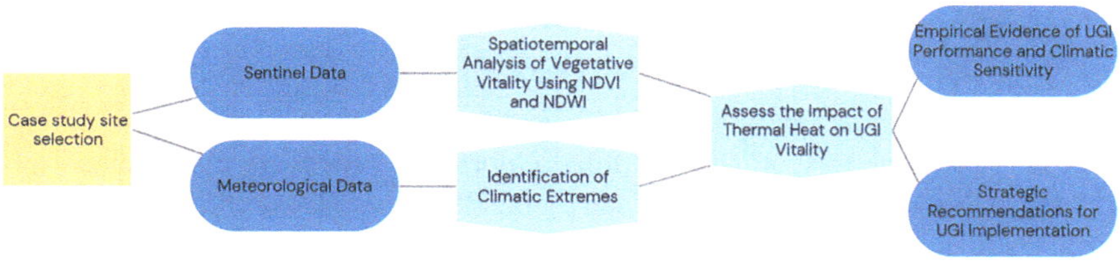

Fig. 1 Methodological approach

1.1 Case Study Site

Strasbourg (48°35′00″N,07°44′45″E) is the capital of the Grand Est region in north-eastern France, situated directly on the border with Germany (Fig. 2). It lies on the French side of the Rhine River, namely on the Upper Rhine Aquifer. It is the largest continuous groundwater reservoir in Europe holding an estimated 45 to 80 billion cubic meters of freshwater. Located between the Vosges Mountains and the Black Forest Mountains Strasbourg area was previously a wetland (Carbiener et al., 2021) and despite of massive drainage the groundwater remains reachable by linear vegetation, especially by trees (Loumagne et al, 1993), reducing during heat drought impact and enabling to investigate heat effect on UGI. Furthermore, Strasbourg is ranked the third greenest city in France, boasting 1,050 hectares of green space according to the French Green Cities Observatory (Observatoire des villes vertes de France, 2023). Furthermore, its climate is prone to extreme weather events, making it an ideal location for observing how green spaces react to heat stress.

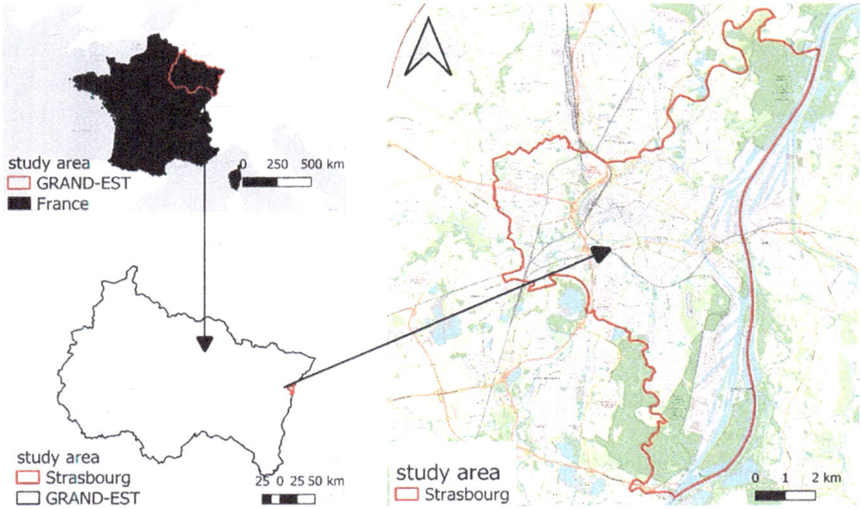

Fig. 2 Localization of the study area

1.2 Metrics

Plant vitality is reflected by the photosynthetic activity of trees. This biological process is facilitated by organelles called chloroplasts found in plant cells. These organelles contain chlorophyll, the green pigment that gives leaves their color. Consequently, the chlorophyll content provides information on the vegetation's capacity to perform photosynthesis—and thus, its direct vitality. There are several ways to measure the photosynthetic activity of vegetation: gas exchange measurements, fluorescence and spectral reflectance. The reflectance measurement is performed by remote sensing and allows large spatial and temporal data. This method is made possible by multispectral satellites, which continuously measure the Earth's spectral reflectance at very high resolutions.

The Normalized Difference Vegetation Index (NDVI) is the most widely used remote sensing indicator for assessing the quantity and health of vegetation. It acts as a "greenness" scale based on how plants interact with specific wavelengths of light (Rouse et al., 1974). While powerful, NDVI has few limitations: (a) in very dense forests (like the Rhine's riparian forests), the NDVI can "saturate," meaning it stays at 0.9 even if the biomass continues to increase, (b) in areas with sparse grass, the color of the soil can skew the results and (c) pixel resolution is 10m, meaning that isolated small trees are barely identified (Xu et al., 2022).

The Normalized Difference Water Index (NDWI) is a primary remote sensing indicator used to monitor the liquid water content in vegetation canopies and the presence of open water bodies. Developed to complement vegetation indices, it assesses the moisture levels and water stress of plants based on their interaction with short-wave infrared light (Gao, 1996). However, the NDWI also faces specific limitations: (a) in urban environments, certain man-made surfaces like asphalt or dark rooftops can produce signals that mimic water, leading to potential misclassification; (b) the index is highly sensitive to atmospheric conditions and high humidity can lead to wrong values; this occurs mostly in tropical areas; and (c) Short-Wave Infrared bands of Sentinel-2 often have a 20m resolution leading to a significant lack of spatial data coverage, e.g., to identify small ponds (Latuamury et al., 2025).

1.3 Vitality Data

Data were sourced from two ESA Sentinel-2 satellites. The first was launched on June 27, 2015, and the second on March 28, 2017. They continuously measure 13 spectral bands, ranging from the visible spectrum (400 nm–800 nm) and near-infrared (800 nm–1300 nm) to short-wave infrared (1300 nm–2500 nm). They provide a maximum spatial resolution of 10 m and a temporal resolution of 5 days. The satellites generate two measurement levels: L1C (Top-of-the-Atmosphere reflectance) and L2A (Surface Reflectance). While L2A data have only been systematically available since the launch of the second satellite in 2017, it is possible to process the L1C data available since 2015 converting them into L2A products.

1.4 Vitality Data Proceeding

COPERNICUS data collection was performed using Google Earth Engine (GEE) and Google Colab. To ensure a robust and consistent dataset for the longitudinal analysis of UGI in Strasbourg, specific filtering parameters were applied within the GEE environment. The selection aimed to balance maximum temporal coverage with high data quality. The following criteria were utilized: L1C (top of the atmosphere) image, cloud coverage lower than 50% and harmonized image. L1C-level data were processed on the 10 years temporal range of the study: 27/06/2015- 20/02/2024. The Python script was used for image filtering and extraction. 245 images were extracted and 108 were used for quality check. Python script was used for NDVI calculation on GEE.

1.5 Hydro-Meteorological Data Collection and Proceeding

Data were collected only for the vegetation growing season: April to October. Daily precipitation and temperature data were sourced from national weather agency (Météo France). The data used were sourced from the "Strasbourg Botanique" station which is the most central station in Strasbourg and provides the most accurate representation of the meteorological conditions affecting the UGI in Strasbourg City. The following daily parameters were collected: Precipitation (mm), minimum temperature (°C), maximum temperature (°C) and mean temperature (°C). Meteorological data were analyzed to identify yearly trends, detect anomalies and search for correlation with NDVI values. While most of the studies focus human response to heat waves and use night temperature to identify anomalies, this study is specific to thermal stress for trees and uses it as an indicator of plant physiology anomalies during 2-day long + 3°C over monthly average (10-year average).

Groundwater table level data were sourced from City Agency for environmental protection. Daily measurements were analyzed and showed no significant variation between 2014 and 2025.

1.6 UGI Map

UGI were extracted from the city map of the UGI. In total, 43 parks, 3 urban regional protected parks, 4900 community garden lots, 7 public spaces, 14 squares, 9 cemeteries, and all road greening (road trees), were studied. UGI were characterized according to their land cover properties, namely the percent of cover of 5 types of land cover: water, building, grass-land vegetation, shrubs and bushes, and forested (according to Darsow, 2023). Automatic mapping was performed on 2016, 2017, 2019, 2020 and 2022 to secure accurate mapping over time.

1.7 Data Analysis

The raster index images downloaded from GEE were subsequently analyzed using QGIS. Mean index value for each image and each specific green space was calculated to analyze chronological evolution of NDVI in link with hydro-meteorological conditions. To calculate the average pixel values within the boundaries of the shapefile (UGIs), the Zonal Statistics tool and the integrated QGIS Python console were utilized. Descriptive statistical data were performed to assess mean and atypical vegetation response to heat. Clustering analysis (Hierarchical Clustering on Principal Component) was performed to establish a typology of the best responding green infrastructure.

2 Climate-Related Stress and Implications at the City Scale

2.1 Overall Vitality of the UGI in Strasbourg

In 2020 the City of Strasbourg started a Canopy Plan to achieve 30% tree cover, satellite image analysis (2015–2025) enabled it to identify that the canopy index over the 10 years of the study was constantly around 26%.

The NDVI analysis on QGIS enables the display of the NDVI index for the city of Strasbourg (e.g., August 23, 2016). The values range from –0.414 (water) to 0.867 (vegetation). The lowest value corresponds to water bodies, followed by the built area. The light green to deep blue zones showed vegetated areas with varying levels of photosynthetic activity. The deep blue areas correspond to large forested areas (e.g., the three regional parks) as well as watered crops and gardens (Fig. 3).

NDVI mapping (Fig. 3) shows that the NDVI values are high throughout the regional parks, mirroring robust photosynthetic activity. However, the NDVI is not saturated as it does not reach 0,9. In comparison, city parks closer to the city center show heterogeneous NDVI values. Moreover, the totality of the urban green infrastructure recorded exhibit photosynthetic activity measurable by remote sensing. NDWI analysis shows different water stress among the various green infrastructures (Fig. 4). NDWI and NDVI respond similarly (Table 1).

Nature reserves, namely the regional parks, lead the chart, with NDVI values 20% higher than those of parks. Parks represent the second type of green infrastructure with the highest NDVI (avg: 0.505, max: 0.650) and NDWI (avg: 0.215, max: 0.392)

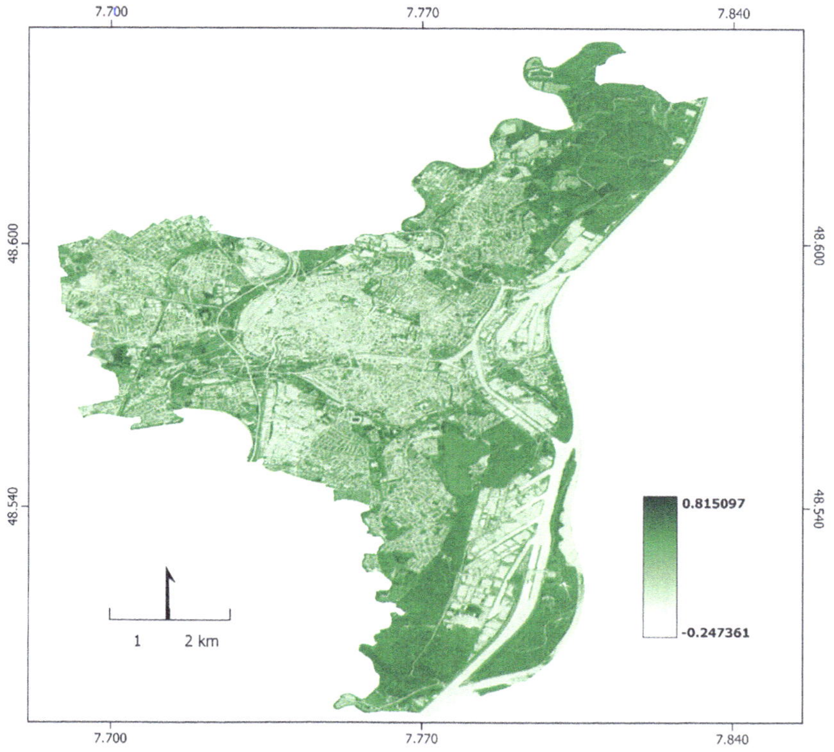

Fig. 3 NDVI map

results. However, among the 43 parks studied, significant disparities are observed in the results, which influence their photosynthetic capacity.

2.2 *Climate-Driven Anomalies*

Statistical analysis of the meteorological data showed that 2017 is the single year that can be considered as "heatwave free." 2021 started very fresh and wet but August hosted a strong heatwave. The plot of the mean NDVI per UGI type for each available image shows vegetation response to the heat (Fig. 5). Interestingly, the graphic shows abrupt decrease of photosynthetic activities followed by a partial recovery, meaning, followed by an increase of the NDVI values but staying below the former trend. Comparison with the climatic data show that these abrupt decreases of NDVI happened after two consecutive days of intense heat (+3°C). 2018 was particularly critic with a heat wave lasting weeks and reached between July 24 and August 8 the mark of 37°C. This graphic illustrates the intensity and frequency of the

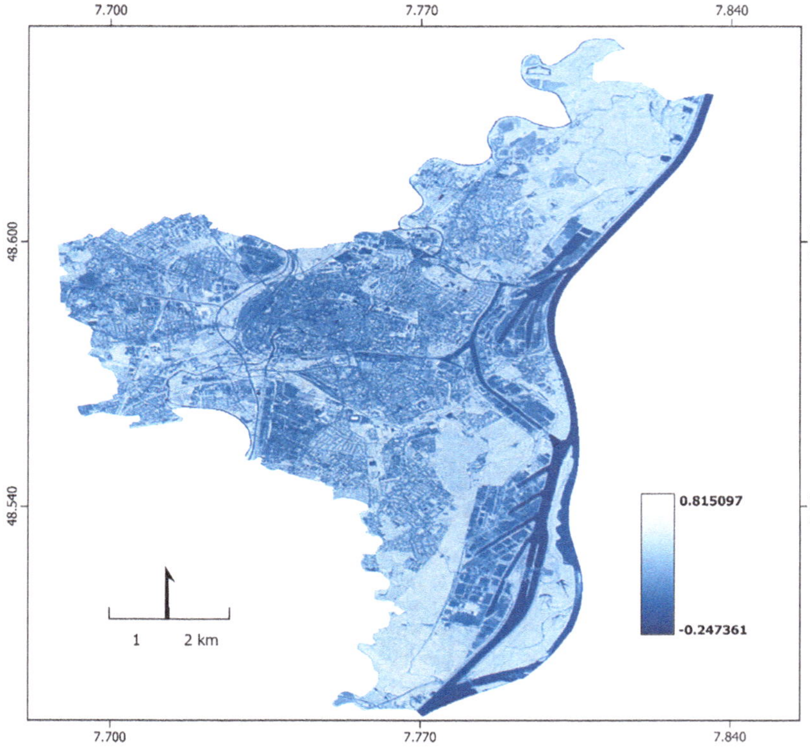

Fig. 4 NDWI map

Table 1 Mean NDVI and NDWI value per UGI type over the 10 years

	Nature reserves	Park	Garden	Place	Square	Cemetery
NDVI						
Medium	0.596	0.505	0.449	0.441	0.445	0.384
Max	0.791	0.647	0.603	0.584	0.589	0.506
NDWI						
Medium	0.215	0.162	0.139	0.146	0.125	0.075
Max	0.392	0.277	0.258	0.304	0.229	0.173

heat-related photosynthetic collapse of the vegetation in Strasbourg between 2015 and 2023.

The 19 anomalies do not impact green infrastructures in the same way. Due to their high baseline vitality, regional parks tend to be more resilient showing less sensitivity than other UGI (Fig. 3). NDWI analysis shows similar results than for NDVI analysis. However, the anomalies are more numerous. NDWI detects heatwave stress before NDVI as humid content decreases before the photosynthetic activity decreases. 2022

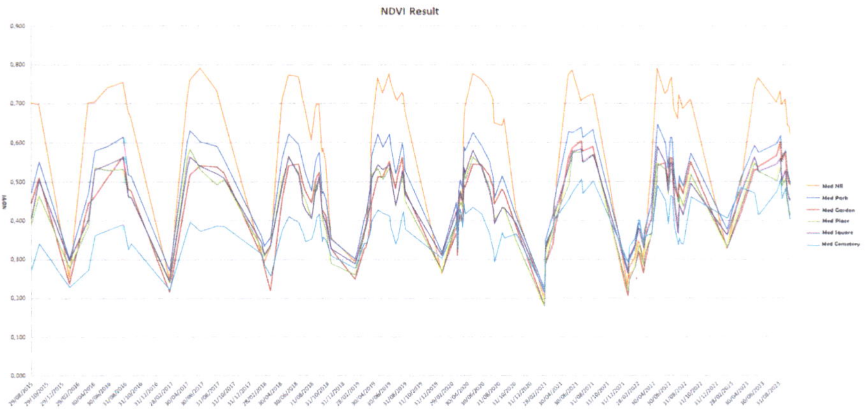

Fig. 5 Graphical result of the average NDVI index by green infrastructure type, with RN being the regional parks

is particularly interesting. In 2022, Strasbourg experienced an extremely hot and dry period, making it one of the most intense years for urban vegetation monitoring. 2022 was the warmest year ever recorded in France (at the time). Rainfall was down by 25% to 40% across the region. The forested areas overall better resist than other UGI types to multiple heatwaves. July was particularly severe and identified as "peak crisis" with a massive collapse of the vegetation chlorophyll activity. Here again the forested area better resisted this stress but recovered less than other UGI types. This reveals that forested areas better resist to peak temperature—2022 heat reached 38.9°C—but that the 45 consecutive days of heat irreversibly impacted the forested area vitality.

2.3 *The Higher the Tree Cover, the More Resilient Are Parks to Heat Stress*

While the regional reserves exhibit significantly higher vitality, the urban parks respond heterogeneously. This raises questions regarding the specific characteristics that enable higher or lower resilience within these spaces (Fig. 6).

Marguerite de Navarre Park, Sinbad Park and Chemin Cabichon, Brothers Grimm Park and Peter Pan Path and Pré Saint-Gall Park are the parks with the lowest NDVI and also the higher sensibility to heat. They exhibit degraded photosynthetic activity and a reduced capacity to cool the ambient air during heatwaves. The parks Kurgarten Park, Ciriers Park and the Muhlbach Urban Nature Park show results close to those of regional parks. These parks have the highest NDVI results, indicating the best photosynthetic capacities during heatwaves and maintaining high cooling potential.

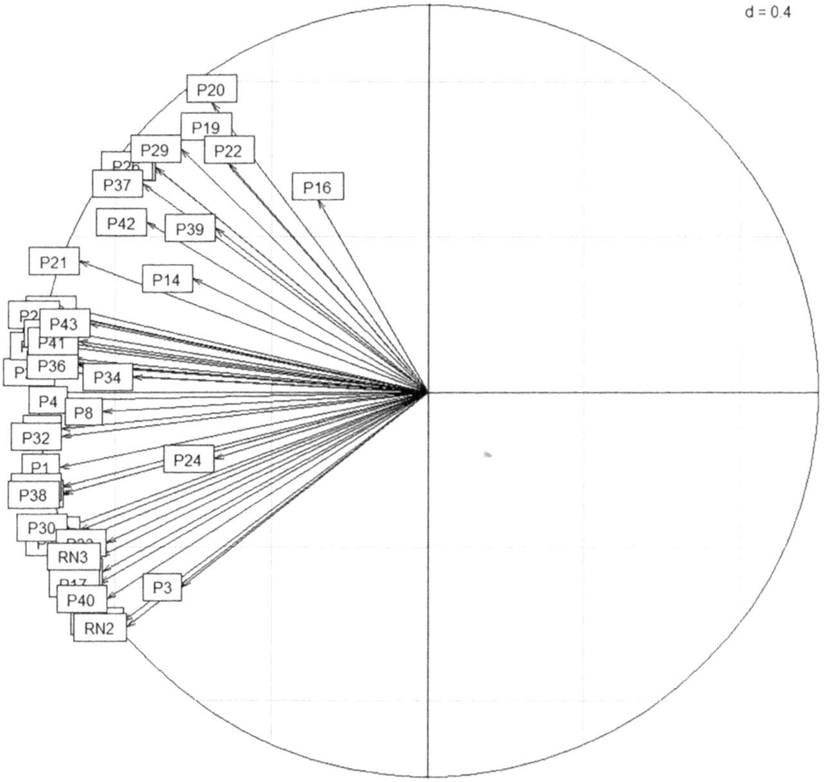

Fig. 6 Principal component analysis of the NDVI value of the different parks (including the regional parks—NR)

According to their vitality during heatwaves 4 groups of parks with similar responses can be identified.

Looking at the mean vegetation cover of the four park types shows that a minimum of 50% tree cover is necessary to achieve maximal resilience during heat. Less than 15% of tree cover led to major decrease of the chlorophyll activities and related cooling activities for themselves, surrounding vegetation, animals and humans. Less than 10% tree cover and isolated tree on grass muster led to a massive collapse of tree vitality during heatwaves.

3 Discussion and Implications

3.1 Consideration of Nonhuman

Our results indicated that the vegetation vitality and related cooling performance of Strasbourg's UGI is significantly compromised during sustained extreme heat events, particularly when temperatures exceed local norms by +3°C for more than two consecutive days. The study showed that while 15 consecutive days of heat cause abrupt decrease of photosynthetic activity, namely a vegetation functionality collapse, it allowed for recovery. It also showed that 45 consecutive days of heat hindered forested vegetation recovery. Photosynthetic damages are irreversible at least for the year. Long-term study should investigate the long-term impact of such seasonal collapse. The tipping point is between 15 and 45 consecutive days under extreme heat, but the study did not allow for more precise definition of the critical value.

These results are critical, because it showed that during a heatwave—when urban cooling demand is maximal for human wellbeing, productivity, health and survival—vegetative cooling capacity can and may undergo a functional collapse (Martínez-Villa et al., 2024).

Trees initiate a defensive stomatal closure to effectively stop evapotranspiration process and retain the minimal critical humidity in their most important organs. During these events, the UGI is not "just less effective"—it ceases to function as a cooling system and remains barely a shading structure. This represents a system failure at the most critical moment, namely when it is needed most. While urban vegetation usually cools down from its surrounding by 2°C to 4°C and surface such as asphalts by up to 20°C (Zölch et al., 2016), system failure of the UGI during heat can lead to catastrophic consequences for human health.

According to the literature, the abrupt cessations of photosynthetic activities, as showed in this study, trigger a secondary physiological crisis: starvation. Cessation of gas exchange terminates photosynthetic activity and the related CO_2 assimilation and glucose transformation (Ahongshangbam et al., 2023). Trees stop biomass accumulation and show a progressive decline of tree vitality and long-term health (Allen et al., 2010). Consequently, the long-term impact of such seasonal vegetation collapse endangers the UGI strategy as climate change adaptation and mitigation measure itself.

Of significant concern is the observation that despite a global mean temperature anomaly of approximately +1.4°C, none of the ten years within the study period were exempt from extreme thermal events. Every year recorded at least one +3°C local anomaly sustained for a minimum of two consecutive days. This disparity demonstrates that Strasbourg's urban microclimate is already undergoing localized temperature escalations that significantly exceed global averages. Such findings are critical for urban mitigation and adaptation strategies. While current global policies trajectory toward a +2.2°C to +2.7°C increase by 2100, the disproportionate frequency of localized extremes suggests that urban environments will reach physiological and infrastructural tipping points decades sooner than coarse-grained global models predict.

According to the literature, large and dense forested areas create their own internal "cool islands." Forested areas are 2°C to 10°C cooler than grass-land-dominated UGI (Ziter et al., 2019), keeping the trees themselves and surrounding vegetation below their own T_{crit} (Posch et al., 2022, Wang et al., 2015) and allowing them to stay longer as energy-free cooling machines for human wellbeing, health and survival (Drake et al., 2018). Furthermore, forested soils are organic matter rich and retain moisture better than isolated, compacted street-tree pits, providing water for transpiration during the first 48 hours of a heatwave. The study showed that for the temperature recorded between 2015 and 2023 in Strasbourg the heat waves were so intense that while the effect "cool island" exists, it is not sufficient to preserve tree functionality during 15 consecutive days of intense heat.

Beyond demonstrating that urban microclimates effectively "fast-forward" climate scenarios, this study reveals a profound vulnerability in UGI strategy. Within the last ten years and under the current +1.4°C global warming baseline, numerous +3°C events have already breached the T_{crit} of Strasbourg UGI. Extrapolating these findings to a +2.2°C–+2.7°C future scenario suggests a potential systemic and persistent collapse of UGI services in European cities, raising significant doubts regarding the long-term efficacy of current climate-mitigation strategies if they do not account for tree physiological limits.

3.2 Linkage of the UGI Objectives and Principles

It is essential to frame UGI not as a singular climate "fix" but as a multifunctional system. While the thermal regulation provided by the canopy is the focus of this study, UGI effectiveness is usually assessed based on the synergy between its four core pillars: climate adaptation and mitigation, biodiversity conservation, social cohesion and the green economy (Pauleit et al., 2017b).

Climate adaptation cannot be pursued in biological isolation. A forest canopy designed solely for rapid carbon fixation or shading and lacking biodiversity targets are highly vulnerable to systemic collapse via pests or specialized heat stress (Hansen & Pauleit, 2014). As observed in the + 3°C anomalies, stomatal closure and starvation are mitigated when the surrounding soil and ecosystem are healthy, e.g., for the regional parks. Biodiversity provides the functional redundancy necessary for the UGI to remain operational.

Furthermore, while common standards suggest 30% tree cover as a target, e.g., 3/30/300 rules, our study suggests that the 30% is the absolute minimal requirement in the 2100 scenario. Results suggest 60% of tree cover for the milestone UGI. However, a misconception would be an aggressive canopy plans that maximize tree cover to its maximal. Converting all available urban interstices into dense forested areas would paradoxically undermine local biodiversity. The ecological strength of the urban environment lies in its habitat mosaic. Indeed, many "weak" UGI for climate adaptation provide essential niches for heliophilic species and pollinators that cannot survive in closed-canopy forests. Furthermore, replacing open space with uniform forest cover can disrupt the migration of species that require heterogeneous landscapes, ultimately weakening the biological resilience that supports the trees themselves.

Moreover, social cohesion and related stewardship is a prerequisite for UGI longevity. Urban "foresting" that ignores the social utility of open space and community gardens indirectly the risk of UGI depletion. Furthermore, a resilient UGI should support a circular green economy to legitimate actions.

Future design must prioritize functionality under multiple scenarios instead of planting a monoculture of heat-tolerant species. Mix of species with different hydraulic strategies increase ecological value of the UGI and ensure performance capacity under heat. For example, if one biodiversity rich species reaches its T_{crit} and shuts down, others with higher tolerance will continue to provide cooling for the heat-sensitive and biodiversity critical tree.

3.3 *From the Planning to Actions*

Based on the empirical findings of this study, several strategic recommendations for UGI practice can be formulated. First, while contemporary benchmarks—most notably the 3/30/300 rule—advocate for a 30% municipal canopy cover (Konijnendijk, 2023), this research demonstrates that a 60% canopy threshold is the absolute minimum requirement for urban parks to maintain a functional "cool island" effect during extreme thermal events. This increased density is critical to delaying the onset of Tcrit, thereby extending the operational window of UGI before physiological collapse occurs. To ensure UGI remains a robust climate-mitigation instrument through 2100 with + 2.2°C to + 2.7°C target, a 60% canopy milestone should be adopted as a design standard. However, future implementation must transcend a purely arithmetic maximization of vegetative cover and also include biodiversity requirements. An indiscriminate, homogenized expansion that saturates all urban interstices risks the creation of ecological monoliths and will paradoxically undermine local biodiversity (Pauleit et al., 2017a).

Second, as climate projections converge on a 2.2°C to 2.7°C warming trajectory, UGI strategies must pivot toward an ecophysiological resilience framework. Adaptation is no longer just about surviving a drought; it is also about surviving together making the best use of diverse capacities. Current paradigms often prioritize "isohydric" (water-saving) species due to their high survival rates during acute moisture deficits. However, a resilient UGI must move beyond the "isohydric" bias, implementing a strategic taxonomic and functional mix that includes "anisohydric" (water-spending) and thermophilic species (high-performance heat sinker). The future strategy lies in assisted migration and combination introducing heat-adapted species from southern latitudes that possess higher T_{crit} values, ensuring that native species remain in a suitable habitat to remain functional. To support this functional diversity, urban planning must favor of structurally complex urban micro-forests (Pretzsch et al., 2015). Furthermore, UGI must be coupled with sponge city elements to ensure that when the + 3°C event hits, the trees have enough water "capital" in the soil to delay starvation process. Finally, to safeguard UGI functions, "strategic watering" should be considered as a critical maintenance protocol. Rather than a luxury, UGI needs to be considered as a critical infrastructure for which targeted water subsidies during peak thermal stress act as a "system saving" ensuring that the UGI continues to perform at scales that are technologically unachievable.

Finally, strategic UGI planning should be anchored into an ambitious and long-term city strategy. In this study, while municipal strategies report high planting volumes, the photosynthetic capacity and related services as cooling and CO_2 fixation reflected by indices as NDVI remains stagnant or even dips in the first decade of implementation. This phenomenon can be attributed to three critical factors. First, the administrative and implementation gap is a major limitation in temporal studies. Once a canopy plan is adopted many preliminary tasks as land acquisition and site preparation may slow down the implementation start (Dorst et al., 2019). Furthermore, seasonal planting windows and personal resources limit the implementation intensity. In this context, a 5-year study period is often too short to capture the positive trend of a new UGI policy. Second, in historic European cities like Strasbourg, the canopy plan needs to be highly ambitious to overcome the "zero-sum game" in the short term. Aging urban forests reach senescence simultaneously, requiring large-scale removals for public safety generating legacy gaps. Namely, when a mature tree with a 20m canopy diameter is replaced by three saplings, the immediate net loss in leaf area is over 90% (Nowak et al., 2013). Finally, in the context of remote sensing, even if the number of trees increases, the total pixel-based biomass on which NDVI is calculated decreases until the saplings' growth and crowns mature. While fast growing trees can take 10–15 years to reach a 10m bride crown, slow growing trees may need over 40 years to reach this size (Pretzsch et al., 2015). These findings underscore temporal delay to achieve coverage and thermal targets. In this regard, afforestation efforts in cities necessitate ambitious expansion of the canopy plans, namely hyper-proportional planting targets to counteract the inevitable tree deficit during canopy turnover.

3.4 Limitations

Despite the robust spectral evidence provided by the longitudinal analysis of Strasbourg's UGI, several limitations must be acknowledged. First, the spatial resolution of Sentinel-2 (10m) introduces a "mixed pixel" bias; in dense urban fabrics, a single pixel often aggregates the reflectance of a sapling crown, its surrounding soil and adjacent impervious surfaces, which can lead to an underestimation of early-stage canopy development and "diluted" NDVI values, e.g., for road greening. Second, while NDVI and NDWI are excellent proxies for photosynthetic activity and leaf water content, they do not directly inform about tree health. Third, in dense urban context, building shadowing and urban canyon wind patterns also positively influence tree vitality during heat waves. Fourth, the 10-year study window, while significant for policy review, remains a "snapshot" in the context of biological time. The full success of current planting and UGI strategies may not be detectable via remote sensing for another two decades. Finally, this study did not account for sub-pixel irrigation practices or specific soil volume variations across the 43 parks, both of which are critical "invisible" variables that determine whether a tree reaches its T_{crit} or maintains its cooling function during + 3°C anomalies. Finally, the study result and robustness are limited by the limited number of parks included in the study.

4 Conclusion

This study demonstrated that the future of urban climate change adaptation and mitigation strategies hinges on a fundamental transition from quantitative canopy targets to a qualitative ecophysiological strategy. The study showed that urban microclimates are frequently breaching the critical thermal limits of temperate species and revealed a looming "service collapse" that current mitigation plans may not fully anticipate. Cities rely on UGI performance to remain habitable. To avoid UGI functional collapse, UGI planning must shift toward structural complexity and density and UGI implementation should be more ambitious. By moving away from isolated, vulnerable tree alignments and embracing strong dense urban forests and multifunctional open spaces, cities can secure UGI performances under weather extreme thermal anomalies. The minimal canopy cover milestone of 60% is not merely a density target, but also a biodiversity target ensuring that the city remains a livable sanctuary for human and nonhuman.

Acknowledgements The authors wish to express their sincere gratitude to the Federation of Research in Environment and Sustainability (FERED) of the University of Strasbourg for the generous funding of the "Breathe Blue" project, which made this research possible. We are also deeply indebted to the Eurométropole of Strasbourg (EMS) for providing the critical spatial datasets and technical guidance throughout the study. Their collaboration and expertise in urban forestry management were essential in aligning this scientific inquiry with the practical challenges of municipal greening strategies. We also thank the Climate Agency of Eurométropole of Strasbourg for their insights into the local canopy plan and their commitment to evidence-based climate adaptation. Finally, we thank the city's technical services and regional park manager for their support.

References

Ahongshangbam, J., Kulmala, L., Soininen, J., Frühauf, Y., Karvinen, E., Salmon, Y., Lintunen, A., Karvonen, A., & Järvi, L. (2023). Sap flow and leaf gas exchange response to a drought and heatwave in urban green spaces in a Nordic city. *Biogeosciences, 20*, 4455–4475. https://doi.org/10.5194/bg-20-4455-2023

Allen, C. D., Macalady, A. K., Chenchouni, H., Bachelet, D., McDowell, N., Vennetier, M., Kitzberger, T., Rigling, A., Breshears, D. D., Hogg, E. H., Gonzalez, P., Fensham, R., Zhang, Z., Castro, J., Demidova, N., Lim, J.-H., Allard, G., Running, S. W., Semerci, A., & Cobb, N. (2010). A global overview of drought and heat-induced tree mortality reveals emerging climate change risks for forests. *Forest Ecology and Management, 259*(4), 660–684. https://doi.org/10.1016/j.foreco.2009.09.001

Ballester, J., Quijal-Zamorano, M., Méndez Turrubiates, R. B., Janoš, T., Shartova, N., Gallo, E., Beltrán Barrón, N. D., Peyrusse, F., & Smith, P. (2025). Heat-related mortality in Europe during 2024 and health emergency forecasting to reduce preventable deaths. *Nature Medicine, 31*(1), 4065–4074. https://doi.org/10.1038/s41591-025-03954-7

Berkeley Earth. (2026). Global temperature report for 2025. https://berkeleyearth.org/global-temperature-report-for-2025/

Bland, L. M., et al. (2018). Developing a standardized definition of ecosystem collapse for risk assessment. *Frontier Ecology and the Environment, 16*(1), 29–36. https://doi.org/10.1002/fee.1747

Canadell, J. G., & Jackson, R. B. (2021). Ecosystem collapse and climate change: An introduction. In J. G. Canadell & R. B.Jackson (Eds.), *Ecosystem collapse and climate change. Ecological studies* (Vol 241). Springer. https://doi.org/10.1007/978-3-030-71330-0_1

Carbiener, R., Schmitt, L., & Schnitzler, A. (2021). *Rhin vivant: Histoire du fleuve, des poissons et des hommes* (p. 228). La Nuée Bleue.

Clark, D. B., Soto Castro, C., Alfaro Alvarado, L. D., & Read, J. M. (2004). Quantifying mortality of tropical rain forest trees using high-spatial-resolution satellite data. *Ecological Letter, 7*, 52–59.

Copernicus Climate Change Service. (2026). Global climate highlights 2025. European Centre for medium-range weather forecasts. https://climate.copernicus.eu/global-climate-highlights-2025

Cumming, G. S., & Peterson, G. D. (2017). Unifying research on social–ecological resilience and collapse. *Trends in Ecology & Evolution, 32*(9), 695–713. https://doi.org/10.1016/j.tree.2017.06.014.ElsevierLtd

Darsow, A. (2023). *Effect of river morphological changes on riparian vegetation quality* (Master's thesis, Technical University of Munich). Mediatum.

Drake, J. E., Tjoelker, M. G., Vårhammar, A., Medlyn, B. E., Reich, P. B., Pfautsch, S., Pollard, W. H., & Barton, C. V. M. (2018). Trees tolerate an extreme heatwave via sustained transpirational cooling and thermal tolerance. *Global Change Biology, 24*(6), 2390–2402. https://doi.org/10.1111/gcb.14037

Dorst, H., van der Jagt, A., Raven, R., & Runhaar, H. (2019). Structural conditions for the implementation of nature-based solutions: A status quo analysis. *City and Environment Interactions, 1*, Article 100002. https://doi.org/10.1016/j.cities.2021.103283

European Commission. (2024). Green infrastructure. Retrieved May 22, 2024, from https://environment.ec.europa.eu/topics/nature-and-biodiversity/green-infrastructure_en

Gao, B. C. (1996). NDWI—A normalized difference water index for remote sensing of vegetation liquid water from space. *Remote Sensing of Environment, 58*(3), 257–266. https://doi.org/10.1016/S0034-4257(96)00067-3

Gill, S. E., Handley, J. F., Ennos, A. R., & Pauleit, S. (2007). Adapting cities for climate change: The role of the green infrastructure. *Built Environment, 33*(1), 115–133. https://doi.org/10.2148/benv.33.1.115

Haines, A., Kovats, R. S., Campbell-Lendrum, D., & Corvalán, C. (2006). Climate change and human health: Impacts, vulnerability and public health. *Public Health, 120*(7), 585–596. https://doi.org/10.1016/j.puhe.2006.01.002

Hansen, R., & Pauleit, S. (2014). From multifunctionality to multiple ecosystem services? A conceptual framework for multi-functional landscape development. *Landscape Online, 37*, 1–33. https://doi.org/10.3097/LO.201437

Hüve, K., Bichele, I., Rasulov, B., & Niinemets, U. (2011). When it is too hot for photosynthesis: Heat-induced instability of photosynthesis in relation to respiratory burst, cell permeability changes and H_2O_2 formation. *Plant, cell & environment, 34*(1), 113–126. https://doi.org/10.1111/j.1365-3040.2010.02229.x

Intergovernmental Panel on Climate Change (IPCC). (2023). Climate change 2023: Synthesis report. In H. Lee & J. Romero (Eds.), *Contribution of Working Groups I, II and III to the Sixth Assessment Report of the Intergovernmental Panel on Climate Change*. https://doi.org/10.59327/IPCC/AR6-9789291691647

Keith, D. A., Rodríguez, J. P., Rodríguez-Clark, K. M., Nicholson, E., Aapala, K., Alonso, A., Asmussen, M., Bachman, S., Basset, A., Barrow, E. G., & Benson, J. S. (2013). Scientific foundations for an IUCN red list of ecosystems. *PLoS One, 8*(5). https://doi.org/10.1371/journal.pone.0062111

Konijnendijk, C. C. (2023). Evidence-based guidelines for greener, healthier, more resilient neighbourhoods: Introducing the 3–30–300 rule. *Journal of Forestry Research, 34*(3), 821–830.

Latuamury, B., Imlabla, W. N., Sahusilawane, F., & Marasabessy, H. (2025). Spatial patterns of NDVI vegetation greenness index and NDWI water content index using sentinel 2A imagery. *IOP Conference Series: Earth and Environmental Science, 1527*(1), Article 012002. https://doi.org/10.1088/1755-1315/1527/1/012002

Lindenmayer, D., Messier, C., & Sato, C. (2016). Avoiding ecosystem collapse in managed forest ecosystems. *Frontiers in Ecology and the Environment, 14*(10), 561–568. https://doi.org/10.1002/fee.1434

Liu, Y., Kumar, M., Katul, G. G., & Porporat, A. (2019). Reduced resilience as an early warning signal of forest mortality. *Nature Climate Change, 9*, 880–885.

Lloret, F., & Batllori, E. (2021). Climate-Induced global forest shifts due to heatwave-drought. In J. G. Canadell & R. B. Jackson (Eds.), *Ecosystem collapse and climate change. Ecological studies* (vol. 241). Springer. https://doi.org/10.1007/978-3-030-71330-0_7

Loumagne, C., Normand, M., Vidal-Madjar, D., Ottle, C., & Louahala, S. (1993, May 17–19). Etat hydrique du sol: Mesures ponctuelles et mesures satellitaires [Paper presentation]. In *Journées du GSTS changements d'échelle dans les modèles de l'environnement et de la télédétection, Strasbourg, France* (pp. 65–71).

Martínez-Villa, J. A., Paquette, A., Feeley, K. J., Morales-Morales, P. A., Messier, C., & Durán, S. M. (2024). Changes in morphological and physiological traits of urban trees in response to elevated temperatures within an Urban Heat Island. *Tree physiology, 44*(12), tpae145. https://doi.org/10.1093/treephys/tpae145

Masselot, P., Ballester, J., Quijal-Zamorano, M., Méndez Turrubiates, R. B., Janoš, T., & Shartova, N. (2025). Heat-related mortality in Europe during 2024 and health emergency forecasting to reduce preventable deaths. *Nature Medicine, 31*(1), 4065–4074. https://doi.org/10.1038/s41591-025-03954-7

McDowell, N. G., Coops, N. C., Beck, P. S., Chambers, J. Q., Gangodagamage, C., Hicke, J. A., Huang, C. Y., Kennedy, R., Krofcheck, D. J., Litvak, M., & Meddens, A. J. (2014) Global satellite monitoring of climate-induced vegetation disturbances. *Trends in Plant Sciences, 20*(2), 114-123. https://doi.org/10.1016/j.tplants.2014.10.008

Nowak, D. J., & Crane, D. E. (2002). Carbon storage and sequestration by urban trees in the USA. *Environmental Pollution, 116*(3), 381–389. https://doi.org/10.1016/S0269-7491(01)00214-7

Nowak, D. J., Greenfield, E. J., Hoehn, R. E., & Lapoint, E. (2013). Carbon storage and sequestration by trees in urban and community areas of the United States. *Environmental Pollution, 178*, 229–236. https://doi.org/10.1016/j.envpol.2013.03.019

Observatoire des villes vertes. (2023). Palmarès 2023 des villes les plus vertes de France. Unep - Les Entreprises du Paysage & Hortis. https://www.observatoirevillesvertes.fr/

Pataki, D. E., Carreiro, M. M., Cherrier, J., Grulke, N. E., Jennings, V., Pincetl, S., Pouyat, R. V., Whitlow, T. H., & Zipperer, W. C. (2011). Coupling biogeochemical cycles in urban environments: Ecosystem services, green solutions, and misconceptions. *Frontiers in Ecology and the Environment, 9*(1), 27–36. https://doi.org/10.1890/090220

Pauleit, S., Zölch, T., Hansen, R., Randrup, T. B., & van den Bosch, C. K. (2017a). Nature-based solutions and climate change—Four shades of green. In N. Kabisch, H. Korn, J. Stadler & A. Bonn (Eds.), *Nature-based solutions to climate change adaptation in urban areas: Linkages between science, policy and practice* (pp. 29–49). Springer. https://doi.org/10.1007/978-3-319-56091-5_3

Pauleit, S., Zölch, T., Hansen, R., Rall, E. L., & Werner, R. (2017b). *Urban green infrastructure: Concepts and planning*. Oxford University Press. https://doi.org/10.1093/acrefore/9780199389414.013.93

Posch, B. C., Hammer, J., Atkin, O. K., Bramley, H., Ruan, Y. L., Trethowan, R., & Coast, O. (2022). Wheat photosystem II heat tolerance responds dynamically to short-and long-term warming. *Journal of Experimental Botany, 73*(10), 3268–3282. https://doi.org/10.1093/jxb/erac039

Pretzsch, H., Biber, P., Uhl, E., Dahlhausen, J., Rötzer, T., Caldentey, J., Koike, T., van Con, T., Chavana-Bryant, C., Konijnendijk, C. C., Roloff, A., Schuster, C., & Rahman, M. A. (2015). Crown size and growing space requirement of common tree species in urban centres, parks and forests. *Urban Forestry & Urban Greening, 14*(3), 466–479. https://doi.org/10.1016/j.ufug.2015.04.006

Rogers, B. M., Solvik, K., Hogg, E. H., Ju, J., Masek, J. G., Michaelian, M., Berner, L. T., & Goetz, S. J. (2018). Detecting early warning signals of tree mortality in boreal North America using multiscale satellite data. *Global Change Biology, 24*, 2284–2304.

Rouse, J. W., Haas, R. H., Schell, J. A., & Deering, D. W. (1974). Monitoring vegetation systems in the great plains with ERTS. In *Third Earth Resources Technology Satellite-1 Symposium* (Vol. 1, pp. 309–317). NASA SP-351.

Sehgal, H., Sharma, R. A., & Joshi, M. (2023). Transcriptomic profiling of desert tree Prosopis cineraria under heat stress reveals potential role of multiple gene families in its high thermotolerance. *Frontiers in Plant Science, 14*, Article 1271644. https://doi.org/10.3389/fpls.2023.1271644

Stott, P., Stone, D., & Allen, M. (2004). Human contribution to the European heatwave of 2003. *Nature, 432*, 610–614. https://doi.org/10.1038/nature03089

Sommerfeld, A., et al. (2018). Patterns and drivers of recent disturbances across the temperate forest biome. *Nature Communication, 9*, 4355.

Trichon, V., Hiernaux, P., Walcker, R., & Mougin, E. (2018). The persistent decline of patterned woody vegetation: The tiger bush in the context of the regional Sahel greening trend. *Global Change Biology, 24*, 2633–2648.

Tubiello, F. N., Salvatore, M., Ferrara, A. F., House, J., Federici, S., Rossi, S., Biancalani, R., Condor Golec, R. D., Jacobs, H., Flammini, A., Prosperi, P., Cardenas-Galindo, P., Schmidhuber, J., Sanz Sanchez, M. J., Srivastava, N., & Smith, P. (2015). The contribution of agriculture, forestry and other land use activities to global warming, 1990–2012. *Global Change Biology, 21*(7), 2655–2660. https://doi.org/10.1111/gcb.12865

Verbesselt, J., Umlauf, N., Hirota, M., Holmgren, M., Van Nes, E. H., Herold, M., Zeileis, A., & Schffer, M. (2016). Remotely sensed resilience of tropical forests. *Nature Climate Change, 6*, 1028–1032.

World Meteorological Organization & Copernicus Climate Change Service. (2024). State of the climate in Europe 2023. WMO-No. 1350. https://library.wmo.int/idurl/4/68835

Wahid, A., Gelani, S., Ashraf, M., & Foolad, M. R. (2007). Heat tolerance in plants: An overview. *Environmental and Experimental Botany, 61*(3), 199–223. https://doi.org/10.1016/j.envexpbot.2007.02.001

Wang, Y., Bakker, F., de Groot, R., Wörtche, H., & Leemans, R. (2015). Effects of Urban Green Infrastructure (UGI) on local outdoor microclimate during the growing season. *Environmental Monitoring and Assessment, 187*(12), Article 732. https://doi.org/10.1007/s10661-015-4943-2

Xu, Y., Yang, Y., Chen, X., & Liu, Y. (2022). Bibliometric analysis of global NDVI research trends from 1985 to 2021. *Remote Sens., 14*, 3967. https://doi.org/10.3390/rs14163967

Yuan, W., Zheng, Y., Piao, S., Ciais, P., Lombardozzi, D., Wang, Y., Ryu, Y., Chen, G., Dong, W., Hu, Z., Jain, A. K., Jiang, C., Zhu, Q., Holmgren, M., Fu, Y., & Wen, J. (2019). Increased atmospheric vapor pressure deficit reduces global vegetation growth. *Science Advances, 5*(8), eaax1396. https://doi.org/10.1126/sciadv.aax1396

Ziter, C. D., Pedersen, E. J., Kucharik, C. J., & Turner, M. G. (2019). Scale-dependent interactions between tree canopy cover and impervious surfaces reduce daytime urban heat during summer. *Proceedings of the National Academy of Sciences, 116*(15), 7575–7580. https://doi.org/10.1073/pnas.1817561116

Zölch, T., Maderspacher, J., Wamsler C., & Pauleit, S. (2016) Using green infrastructure for urban climate-proofing: An evaluation of heat mitigation measures at the micro-scale. *Urban Forestry & Urban Greening, 20*, 305-316. https://doi.org/10.1016/j.ufug.2016.09.011

Developing Composite Indicators to Evaluate the Feasibility of Green Walls in Urban Environments

Mariantonietta Ruggieri, Gianna Agrò, and Ornella Giambalvo

Abstract Urban areas across the world are increasingly confronted with complex environmental challenges, including climate change, air pollution, reduced biodiversity, and the intensification of urban heat island effects. Within the broader framework of nature-based solutions, green walls represent a promising strategy to enhance urban resilience, improve microclimatic conditions, promote ecological sustainability, and contribute to healthier and more liveable cities. However, planning and implementing green wall systems requires robust, multidimensional assessment tools capable of integrating heterogeneous information—technical, ecological, economic, social, and regulatory. This study proposes an integrated methodological framework for evaluating the feasibility of green walls through the construction of a Composite Indicator (CI). Building on the statistical theory of CI development, our work deals with a structured approach to indicator selection, data transformation, weighting, and aggregation. The methodology is applied to a case study involving buildings on the University of Palermo campus, with technical feasibility indicators representing structural, environmental, regulatory, and spatial characteristics. Different transformation methods (ranking, min–max normalisation, z-score standardisation) and weighting schemes (equal weights, expert-based weights, and aggregation rules) are compared to assess their influence on the resulting CI. The method used follows the Analytical Hierarchy Process (AHP) that is a decision-making framework that simplifies complex choices by organising them into a hierarchy of goals, criteria, and alternatives. A sensitivity analysis is conducted using Kendall's tau, Spearman's rho, and Pearson's correlation coefficients to evaluate the robustness of building rankings under different methodological scenarios. This paper offers both methodological insights and practical implications for urban sustainability planning. The results show that while rankings are generally highly correlated, the CI is sensitive to methodological choices, emphasising the need for transparency, justification of assumptions, and robustness checks. The proposed framework is adaptable to different contexts and provides urban planners and policymakers with a scientifically informed tool for decision-making regarding vertical greening interventions.

M. Ruggieri (✉) · G. Agrò · O. Giambalvo
Department of Economics, Business and Statistics, University of Palermo, Palermo, Italy
e-mail: mariantonietta.ruggieri@unipa.it

L. Diana et al. (eds.), *Greening and Resilience*, Urban Sustainability,
https://doi.org/10.1007/978-981-92-0419-9_8

Keywords Green walls · Feasibility assessment · Composite indicators · Analytic hierarchy process · Nature-based solutions · Sustainable cities

1 Introduction

In recent decades, cities have faced the challenge of reconciling intensive urban development with environmental sustainability and social well-being. Rising temperatures, increasing frequency of heatwaves, deteriorating air quality, and loss of biodiversity highlight the urgency of rethinking how buildings and open spaces are designed, retrofitted, and managed (Abid et al., 2025; Sanju, 2025).

Among several nature-based solutions, green walls (also referred to as living walls or vertical gardens) have attracted growing attention due to their potential to mitigate the urban heat island effect, enhance thermal comfort, improve air quality, support urban biodiversity, and contribute to more livable and aesthetically pleasing environments (Perini & Rosasco, 2013; Shi et al., 2025). Nevertheless, installing and maintaining green walls within complex urban systems is not a trivial task. Technical constraints, ecological conditions, economic costs, social acceptance, and regulatory requirements interact in ways that are often difficult to assess using simple, single-dimensional indicators.

This complexity motivates the use of CIs, which aggregate multidimensional information into a single metric to support decision-making. Properly constructed CIs can help identify the most suitable buildings or neighbourhoods for green wall interventions, compare alternative design scenarios, and guide the allocation of limited financial resources.

The aim of this work is twofold. First, it introduces a comprehensive evaluation framework for green wall feasibility that integrates technical, ecological, economic, social, and regulatory dimensions. Second, it discusses the methodological challenges associated with CI construction using the Analytic Hierarchy approach, including transformation of heterogeneous indicators, weighting schemes, and aggregation rules, and illustrates how these choices may affect the results.

The proposed framework is applied to a case study on the University of Palermo campus, where all buildings are assessed according to several technical feasibility indicators. In this paper, we also introduce the concept of CI, as a Transformability Index designed to capture the adaptive capacity of building façades and surrounding spaces to host and sustain green walls over time.

2 Background Knowledge

2.1 Green Walls and Urban Resilience

Green walls are a specific type of vertical greenery system that can be installed on existing façades or integrated into new buildings. They contribute to several ecosystem services: shading and insulation, microclimate regulation, air filtration, noise attenuation, habitat provision, and aesthetic enhancement (Perini & Rosasco, 2013; Shi et al., 2025). From an urban resilience perspective, they can reduce vulnerability to heatwaves and improve the livability of dense districts.

Successful examples from cities such as Milan (e.g. Bosco verticale), Paris, Singapore, Hong Kong, and Los Angeles show that, when well designed and maintained, green walls can become symbolic projects that combine environmental benefits with social and economic value (Shi et al., 2025). However, these flagship projects often rely on specific technical and financial conditions that may not be easily replicable.

2.2 The Need for Multidimensional Assessment

A comprehensive literature review and consultations with experts from architecture, botany, psychology, and sustainability studies were conducted to define the relevant dimensions for evaluating green wall feasibility.

The decision to install a green wall on a specific building involves multiple questions:

- Technical: Is the structure able to support additional loads? Is the façade geometry suitable? Are irrigation, drainage, and access for maintenance feasible?
- Ecological: Will the selected plant species thrive under local climatic conditions and exposure? Can the intervention meaningfully contribute to biodiversity and microclimate regulation?
- Economic: Are installation and maintenance costs sustainable? Are there potential energy savings or property value gains that can offset these costs? (Perini & Rosasco, 2013)
- Social: How will users and residents perceive the intervention? Does it offer educational or health-related benefits? (Kosorić, 2024)
- Regulatory: Are there planning constraints, heritage protections, or other regulatory issues that might restrict or delay implementation?

When these dimensions are considered in isolation, decision-making risks being fragmented. CIs provide a measure to synthesise heterogeneous information into a single, interpretable metric, while still allowing the underlying structure to be examined.

The case study for Palermo focuses on some technical feasibility indicators, which constitute a robust basis for initial assessment.

2.3 Composite Indicators: Principles and Challenges

Composite indicators are widely used in socio-economic and environmental assessment (e.g. human development, sustainability indices). Their construction typically involves the following stages (Aiello & Attanasio, 2004):

1. Conceptualisation: defining the phenomenon to be measured and its main dimensions.
2. Indicator selection: choosing simple indicators that represent each dimension.
3. Transformation: rendering indicators comparable through ranking, normalisation, or standardisation.
4. Weighting: assigning relative importance to each indicator or dimension.
5. Aggregation: combining transformed and weighted indicators into a single CI.

Each stage involves methodological choices and assumptions, often characterised by some degree of subjectivity. If not made explicit, these choices can undermine the transparency and credibility of the resulting indicators (OECD & JRC, 2008). A key message of this work is that sensitivity analysis is essential to examine how robust the CI is to alternative, yet plausible, methodological decisions. In this paper, a multicriteria approach is used for building indicators because it allows a more realistic representation of the complexity of the phenomena to be measured, by integrating different dimensions (economic, social, environmental, and institutional) into a single coherent framework. Instead of reducing everything to a single variable, it makes it possible to make the criteria explicit, collect expert judgements, and assign weights in a transparent way, making it clearer why an indicator takes a given value and facilitating discussion and scientific and policy legitimacy.

3 Methods

3.1 Overall Formulation of the Composite Indicator

The backbone of our methodology is a CI expressed as follows:

$$CI = f(T_1(X_1), T_2(X_2), \ldots, T_k(X_k)),$$

where:

- CI is the composite indicator,
- X_i are variables or elementary or simple indicators ($i = 1, \ldots, k$),
- T_i are transformation functions applied to the indicators in order to place them on a common comparable scale,
- $f(\cdot)$ is an aggregation function that combines the transformed indicators in order to obtain a single CI.

In the simplest and most common case, a weighted additive model is used:

$$CI = \sum_{i=1}^{k} w_i T(X_i),$$

where w_i are non-negative weights summing to one. This formulation is easily interpretable but requires that indicators be expressed on a common scale and that the compensability implied by addition is acceptable (i.e. a poor performance in one indicator can be offset by a good performance in another).

3.2 Data and Technical Feasibility Indicators

For our case study, the evaluation dimensions were carefully defined through a comprehensive review of the existing literature (Kosorić, 2024; Perini & Rosasco, 2013; Saberib, 2021; Shi et al., 2025), analysing established approaches and criteria relevant to assessing the feasibility of green wall installation. This process involved identifying key indicators that reflect structural, environmental, regulatory, and spatial suitability, ensuring a multidimensional assessment aligned with project objectives. The selection of these indicators was guided by their relevance and prevalence in academic and technical sources, such as recent studies and expert consensus (Lee et al., 2025; Li et al., 2023; Shi et al., 2025), to capture the complexity of feasibility considerations. Additionally, extensive consultations with the multidisciplinary research team of the project "Greenwork: an interdisciplinary framework for urban health and urban resilience enhancement based on greening strategies on buildings and open spaces", including experts in building design, landscape architecture, botany, biology, psychology, and statistics, were conducted to refine and validate the chosen criteria. These collaborative exchanges integrated academic insights with practical expertise and sector best practices, enhancing the robustness and applicability of the evaluation framework. The selected indicators are listed as follows:

- Year of construction.
- Façade construction technology (Load-bearing masonry = 1, Mixed tuff/ reinforced concrete masonry = 2, Reinforced concrete = 3, Pre-stressed concrete = 4, Steel structure = 5).
- Load-bearing capacity.
- Regulatory compliance and permits (Yes = 1, Not = 0).
- Condition of the façade (Very poor = 1, Poor = 2, Sufficient = 3, Good = 4, Excellent/Recently renovated = 5).
- Building height (m).
- Eight indicators related to the usable façade area for different orientations (N, S, W, E, NW, NE, SW, SE).

- Accessibility (%) for installation and maintenance.
- Ground-level-related areas (m^2) relevant for anchoring, logistics, and safety.
- Water supply (Yes = 1, Not = 0).

These indicators form a 20×17 data matrix, where rows correspond to buildings and columns to indicators. They are primarily technical, but many of them are also proxies for economic, ecological, and social aspects (e.g. façade condition affects both structural safety and aesthetics; accessibility affects maintenance costs and safety conditions).

3.3 Transformation Functions

Since the indicators are measured on heterogeneous scales (metric, ordinal, and binary), appropriate transformation functions are required to ensure comparability.

We consider the following approaches:

1. Rank transformation. Values of each indicator are replaced by their position in the ordered series. This is simple to implement and robust to outliers, but it only preserves ordinal information and does not reflect absolute distances between units.
2. Normalisation to [0, 1].

   Max normalisation:

$$T(X) = \frac{X}{\max(X)}.$$

   Min–max normalisation:

$$T(X) = \frac{X - \min(X)}{\max(X) - \min(X)}.$$

These transformations differ because the first takes into account only the maximum value of the distribution. So, the largest value becomes 1 (or—1 if negative), and other values are scaled proportionally relative to that maximum. In the second, otherwise, each value is rescaled using both the minimum and the maximum of the data. Then, the minimum becomes 0, the maximum becomes 1, and all other values lie strictly between 0 and 1. When all data are positive, the only difference between the two transformations is linked to the minimum value.

Normalisation and standardisation affect the distribution of transformed indicators, particularly when indicators exhibit skewness or outliers. Min–max normalisation tends to compress differences when there are extreme values, whereas z-score standardisation emphasises deviations from the mean.

3. Standardisation (z-scores):

$$T(X) = \frac{X - \mu}{\sigma},$$

where μ and σ are the sample mean and standard deviation. Standardisation centres indicators at zero and scales them by their variability.

All transformations eliminate measurement units, thus enhancing comparability across indicators. However, they also embed different assumptions about the relative influence of dispersion and extreme values, which need to be examined in the sensitivity analysis.

3.4 The Analytic Hierarchy Process for the Weighting Method

Weights w_i can be determined using expert judgement, statistical procedures, or a combination of both (Karagiannis & Karagiannis, 2025). In this study, we adopt the Analytic Hierarchy Process (AHP) proposed by Saaty (1990), which is widely used in multi-criteria decision-making (Belton & Stewart, 2002).

The Analytic Hierarchy Process (AHP) for indicator construction is useful because it clearly structures complex problems by breaking them down into hierarchies of objectives, dimensions, and sub-indicators, making the link between the final indicator and its components transparent.

AHP is based on pairwise comparisons of indicators. This approach allows for a systematic and consistent estimation of the weights of individual indicators (including qualitative ones), checks the consistency of expert judgements, and produces a weighting scheme that is easy to explain and defend in scientific and policy contexts (Zhao & He, 2022).

For each pair (i, j), experts express the relative importance of the indicator i over indicator j using a discrete scale, typically 1 to 9 (with reciprocals for less important indicators), as follows:

Scale value	Interpretation
1	Equal importance
3	Moderate importance of *i* over *j*
5	Strong importance
7	Very strong importance
9	Extreme importance
2, 4, 6, 8	Intermediate values

In our application, values are generated in the range of 1 to 5, ensuring reciprocal symmetry (i.e., if $a_{ij} = 3$, then $a_{ji} = 1/3$), while respecting the ordering suggested by experts.

All judgements are arranged in a reciprocal matrix A of order k, where k is the number of indicators:

$$A = \left[a_{ij}\right] = \begin{bmatrix} 1 & a_{12}a_{13} \ldots & a_{1k} \\ \vdots & \vdots\vdots & \vdots \\ \frac{1}{a_{1k}} & \frac{1}{a_{2k}}\frac{1}{a_{3k}} \ldots & 1 \end{bmatrix}$$

Weights can be obtained as follows:

- the geometric mean of each row of A matrix:

$w_i = \left(\prod_{j=1}^{k} a_{ij}\right)^{1/k}$;

- the principal eigenvector w associated with the largest eigenvalue $\lambda_{\max}$ of A, as originally proposed by Saaty, obtained by solving the following:

$Aw = \lambda_{\max}$ w.The geometric mean has the advantage of being non-compensatory at the level of pairwise comparisons, while the eigenvector method is more directly linked to the theoretical formulation of AHP. In both cases, a normalisation step ensures that $\sum_i w_i = 1$.

AHP also allows the evaluation of consistency and coherence in expert judgements. Consistency refers to the logical transitivity of preferences (if A > B and B > C, then A > C), while coherence relates to the overall reasonableness and non-contradictory nature of judgements (Lin et al., 2024).

3.5 Aggregation and Composite Indicator

Once transformed indicators and weights have been obtained, they are aggregated into the composite indicator:

$$CI = \sum_{i=1}^{k} w_i T(X_i).$$

This global CI aims to capture the adaptive capacity and long-term sustainability of green walls on a given building. Conceptually, the CI emphasises indicators linked to the following:

- structural and façade adaptability (possibility of reinforcement, modular systems),
- flexibility of irrigation and drainage systems,
- potential for integrating future technological or ecological upgrades,
- maintenance accessibility and safety over time.

The CI can be constructed as a separate CI focusing on a subset of indicators, or as a dimension-specific sub-index within the overall framework. In both cases, it provides planners with additional information about the futureproofing of the investment.

3.6 Sensitivity Analysis

To ensure the robustness of the method used for constructing the Compatibility Index (CI), we perform a structured sensitivity analysis (Blancas & Contreras, 2025). This analysis includes the following:

- Varying the transformation functions (ranks, normalisation, standardisation).
- Comparing different weighted schemes (equal weights vs. AHP-derived weights).
- Exploring different aggregation rules, where feasible (purely additive vs. less compensatory options).

For each scenario, we compute the CI for all buildings and derive the corresponding ranking. The stability of rankings is assessed using correlation measures between different scenarios: Kendall's tau, Spearman's rho, and the Pearson correlation coefficient.

High correlation values indicate that the CI is relatively insensitive to methodological choices, whereas lower values highlight dimensions or buildings that warrant closer scrutiny.

In the analysis, we apply multiple standardisation methods to examine their influence on the CI results. The choice of each method is typically guided by the nature of the indicators. These methodological alternatives are included primarily for sensitivity analysis purposes, to assess the robustness of the rankings under different data transformations, as well as different weighting schemes and aggregation rules.

Beyond the methodological considerations, the core contribution of our framework lies in its capacity to support practical design and planning decisions. The CI provides an integrated view that helps identify buildings with high potential for green wall implementation, informing strategic planning, resource allocation, and stakeholder communication.

4 Case Study: University of Palermo Campus

The proposed framework is applied to twenty buildings located on the University of Palermo campus. The campus offers a controlled yet realistic urban environment, characterised by buildings of varying ages, construction technologies, and functional uses.

The most recent five buildings were constructed in 2000 using reinforced concrete. The oldest four buildings date back to between 1948 and 1950. Additionally, one building, constructed in 1990, is made of steel. According to experts, twelve out of the twenty buildings are in excellent or good condition, while only three are in poor condition. The buildings are generally not very tall, averaging around 15 m in height; only two structures, the oldest ones, reach or exceed 30 m. The load-bearing capacity, measured in kNm (kilonewton meters), averaging above 0.72, indicates that the structures can withstand high loads without failure due to bending.

Specifically, Building 1 is a retirement residence that also houses the Regional Office for University Student Rights. Building 2 hosts the Orientation and Tutoring Centre, while Building 3 contains the Student Secretariat. Building 11 accommodates the University's General IT Services. Building 20 serves as an educational hub. The remaining buildings are dedicated to various departments, including Agriculture, Engineering, Architecture, Economics/Statistics/Business, Biology, Physics, Chemistry, and Humanities, each equipped with their respective degree programmes and classrooms.

The data matrix provides a comprehensive description of the selected buildings, highlighting key features such as age, construction technology, functional purpose, façade orientation, and existing structural conditions. These buildings differ in height, architectural style, and renovation status, enabling a representative assessment of feasibility across diverse typologies within the campus. The data matrix summarises these attributes, facilitating an understanding of the heterogeneity within the sample and its implications for the feasibility of green walls.

4.1 Data Collection

For each building, the 17 technical feasibility indicators listed in Sect. 3.2 were collected through the following:

- direct inspection of façades and surrounding open spaces,
- consultation of building documentation (structural drawings, refurbishment records),
- interaction with technical staff and members of the interdisciplinary project "Greenwork: an interdisciplinary framework for urban health and urban resilience enhancement based on greening strategies on buildings and open spaces".

The data were assembled into a 20×17 matrix, forming the empirical basis for the CI calculations.

4.2 *Application of the Methodology*

The methodological steps were implemented as follows:

1. Transformation: each indicator was transformed using alternative schemes (ranks, min–max normalisation, z-score standardisation).
2. Weighting: AHP was applied to derive weights reflecting expert opinions on the relative importance of indicators for green wall feasibility. Reciprocal matrices were generated on a 1–5 scale, ensuring compliance with expert rankings and reciprocal symmetry. Weights were calculated using both geometric mean and eigenvector methods, then normalised.
3. Aggregation: transformed indicators were combined into the CI using equal weights (unweighted scenarios) and AHP weights (weighted scenarios).
4. Ranking and sensitivity analysis: buildings were ranked according to the CI under each scenario. Rank correlations (Kendall, Spearman, Pearson) were used to assess stability and identify buildings whose position is highly sensitive to methodological choices.

5 Results and Discussion

5.1 *Results*

Using the transformation functions outlined in Sect. 3.3, we conducted a sensitivity analysis on rankings derived from both unweighted and weighted CIs. Tables 1 and 2 present the corresponding correlation matrices for the unweighted and weighted rankings, respectively.

The sensitivity analysis highlights several key findings:

1. Correlation between scenarios. Overall robustness of rankings.

   Kendall's tau, Spearman's rho, and Pearson's coefficient confirm that rankings are highly correlated across methodological choices, but not identical. This indicates that decisions based solely on a single CI and a single methodological option may be misleading, and that scenario-based interpretation is preferable. Correlations between rankings obtained under different transformation functions are generally high, indicating that the relative ordering of buildings is fairly stable. Nevertheless, some units change position, especially when moving from unweighted to weighted schemes, revealing borderline cases whose feasibility is more uncertain.

Table 1 Correlation matrices for rankings from unweighted CIs

	Rank	Max	Minmax	Scale
(A) Spearman/ Pearson correlation				
Rank	1.000	0.914	0.988	0.976
Max	0.914	1.000	0.923	0.904
Minmax	0.988	0.923	1.000	0.983
Scale	0.976	0.904	0.983	1.000
(B) Kendall correlation				
Rank	1.000	0.778	0.947	0.894
Max	0.778	1.000	0.810	0.799
Minmax	0.947	0.810	1.000	0.926
Scale	0.894	0.799	0.926	1.000

Table 2 Correlation matrices for rankings from weighted CIs

	Rank	Max	Minmax	Scale
(A) Spearman/ Pearson correlation				
Rank	1.000	0.883	0.947	0.974
Max	0.883	1.000	0.949	0.884
Minmax	0.947	0.949	1.000	0.965
Scale	0.974	0.884	0.965	1.000
(B) Kendall correlation				
Rank	1.000	0.746	0.820	0.905
Max	0.746	1.000	0.841	0.757
Minmax	0.820	0.841	1.000	0.873
Scale	0.905	0.757	0.873	1.000

2. Impact of transformation choices.

   The properties of the transformations are reflected in the rankings. When variables with different measurement scales are present, as in our case, the rank transformation was preferred (see Table 3), as its results are similar to those of other transformations (see Tables 1 and 2), while providing more robust and stable statistical analyses compared to using the original values.
3. Effect of weighting.

   Applying AHP weights leads to substantial variations in the CI for some buildings, especially those that perform unevenly across indicators. Buildings that score very well in highly weighted criteria (e.g. structural integrity, accessibility, water supply) may climb in the ranking, while others with more balanced but mediocre performance may drop. This underscores the importance of transparent and well-justified weighting schemes.

Table 3 Rankings derived from Unweighted and Weighted CIs for Rank transformation function

Unweighted CIs		Weighted CIs	
Buildings	Score	Buildings	Score
B20	20.0	B20	20.0
B4	19.0	B5	19.0
B14	18.0	B14	18.0
B16	17.0	B4	17.0
B6	16.0	B6	16.0
B5	15.0	B16	15.0
B17	13.5	B17	13.5
B19	13.5	B19	13.5
B18	12.0	B18	12.0
B7	11.0	B7	11.0
B8	10.0	B8	10.0
B9	9.0	B9	9.0
B1	8.0	B1	8.0
B12	7.0	B13	7.0
B13	6.0	B15	6.0
B3	5.0	B12	5.0
B15	4.0	B10	4.0
B10	3.0	B3	3.0
B11	2.0	B11	2.0
B2	1.0	B2	1.0

4. Aggregation and Composite indicator.

   The additive CI identifies a currently preferred building (referred to as Building 20 in this notation) as the most feasible candidate for a green wall installation. At the same time, the CI provides complementary information on which buildings are better suited for future adaptation and long-term maintenance, even if they are not top ranked in the current CI.

The results demonstrate that, despite some variability introduced by different methodological choices, the overall ranking of buildings remains relatively stable. This stability underscores the utility of the CI as a decision-support tool for urban planners and policymakers.

It is noteworthy that the buildings occupying the top positions in the ranking share certain common characteristics: they are of more recent construction, dating back to the 1990s or 2000s, whereas the others were built around the 1960s. Most of these structures are made of reinforced concrete, and their facades are in good condition.

Our analysis highlights that certain buildings consistently perform well across various scenarios, indicating their suitability for green wall interventions. Conversely, buildings with rankings highly sensitive to methodological choices should be examined further to understand the underlying factors influencing their feasibility.

5.2 Discussion

The results demonstrate that CIs, when carefully constructed and accompanied by sensitivity analysis, can be powerful tools for prioritising green wall interventions. The framework is as follows:

- Replicable: it can be applied to other campuses or districts by collecting a comparable set of indicators.
- Adaptable: new indicators and dimensions (e.g. detailed lifecycle costs, more refined ecological metrics) can be incorporated as data become available.
- Transparent: methodological choices are explicit, and their impact is quantified.

  From a broader urban resilience perspective, this work contributes to bridging the gap between quantitative indicator construction and applied sustainability planning. Many existing CIs in practice are developed without sufficient attention to their mathematical and statistical properties. Our approach emphasises that:
- The conceptual framework must guide indicator selection and transformation,
- weights should be grounded in both expert knowledge and, where appropriate, statistical evidence,
- sensitivity analysis is essential to avoid overconfidence in a single ranking.

  At the city scale, the proposed framework can support the development of strategic greening plans, helping local governments to:
- identify priority buildings and pathways for vertical greenery,
- align interventions with regulatory frameworks and funding schemes,
- communicate decisions to stakeholders using intuitive yet rigorous metrics.

6 Concluding Remarks

The CI framework facilitates transparent decision-making by clearly illustrating how different criteria and assumptions impact the prioritisation of sites. This approach enhances confidence in the planning process and supports the design of resilient, sustainable urban greening strategies. This study introduces a comprehensive, multi-dimensional framework for assessing the feasibility of green walls in urban environments using composite indicators (CIs). By integrating indicator transformation, AHP-based weighting, additive aggregation, and systematic sensitivity analysis, the approach offers a methodologically robust yet accessible tool for practitioners.

The case study conducted at the University of Palermo campus demonstrates that, although the rankings of buildings are generally highly correlated across different methodological choices, they are nonetheless influenced, above all, by the weighting process. This underscores the importance of viewing CIs as tools for structured comparison rather than definitive judgements.

The development of the CI emphasises the need to consider not only the current feasibility of green walls but also their adaptive capacity and long-term sustainability.

Ultimately, the proposed framework enhances the robustness and transparency of decision-making processes related to green infrastructure investments, thereby supporting the transition toward more resilient and sustainable urban environments.

In conclusion, our study presents a flexible and transparent approach for evaluating green wall feasibility through composite indicators. While methodological choices, such as data transformation and weighting, may influence specific rankings, overall consistency affirms the utility of the CI as a reliable decision support tool. Importantly, the framework highlights the significance of sensitivity analysis and stakeholder engagement to ensure outcomes are credible and contextually relevant.

Future research will aim to refine the indicators, expand the framework to incorporate additional dimensions (e.g. lifecycle costs and ecological metrics), and conduct simulation studies on AHP matrices to evaluate the coherence and consistency of expert judgements more thoroughly.

Acknowledgements The authors gratefully acknowledge the contributions of the interdisciplinary research team of the project "Greenwork: an interdisciplinary framework for urban health and urban resilience enhancement based on greening strategies on buildings and open spaces", as well as the technical staff of the University of Palermo for their support in data collection and on-site inspections.

References

Abid, M. A., Shahzad, F., Saeed, M., & Zain, M. (2025). Sustainability and green building practices in the UK: A mini review of economic, environmental, and socio-psychological impacts.

Aiello, F., & Attanasio, M. (2004). *How to transform a batch of simple indicators to make up a unique one? Atti del Convegno SIS, Bari* (pp. 327–338).

Belton, V., & Stewart, T. J. (2002). *Multiple criteria decision analysis: An integrated approach*. Springer.

Blancas, F. J., & Contreras, I. (2025). Global SDG composite indicator: A new methodological proposal that combines compensatory and non-compensatory aggregations. *Sustainable Development, 33*(1), 158–176.

Karagiannis, R., & Karagiannis, G. (2025). Distance-based weighting methods for composite indicators, with applications related to energy sustainability. *International Transactions in Operational Research, 32*(1), 478–501.

Kosorić, V. (2024). Social acceptance of urban farming in and on buildings. In *The vertical farm: Scientific advances and technological developments* (pp. 57–81). Taylor & Francis Group.

Lee, K., Lim, H., Hwang, J., & Lee, D. (2025). Development of building benchmarking index for improving gross-floor-area-based energy use intensity. *Energy and Buildings, 328*, Article 115103.

Li, H., Johra, H., de Andrade Pereira, F., Hong, T., Le Dréau, J., Maturo, A., Wei, M., Liu, Y., Saberi-Derakhtenjani, A., Nagy, Z., & Marszal-Pomianowska, A. (2023). Data-driven key performance indicators and datasets for building energy flexibility: A review and perspectives. *Applied Energy, 343*, Article 121217.

Lin, S-M., Hung, C. C. W., Chien, K-H, Hu, H-L., Shiue, F-J., Lee, H-Y. (2024). A hierarchical model of accelerating factors to promote urban renewal and reconstruction of unsafe and old buildings. *Heliyon, 10*, (e30857). https://doi.org/10.1016/j.heliyon.2024.e30857

OECD & JRC. (2008). *Handbook on constructing composite indicators: Methodology and user guide*. OECD Publishing.

Perini, K., & Rosasco, P. (2013). Cost–benefit analysis for green façades and living wall systems. *Building and Environment, 70*, 110–121.

Saaty, T. L. (1990). How to make a decision: The analytic hierarchy process. *European Journal of Operational Research, 48*(1), 9–26.

Saberib, E. T. M. H. (2021). Feasibility study of development of vertical green spaces (Green Roof) to achieve an ecological city; case study: District 5 of Isfahan.

Sanju, P. (2025). The role of urban green spaces in mitigating urban heat Island effect amidst climate change. *Research Journal of Chemistry and Environment, 29*(1), (1), 75–84.

Shi, X., Shi, C., Tablada, A., Guan, X., Cui, M., Rong, Y., Zhang, Q., & Xie, X. (2025). A review of research progress in vertical farming on façades: Design, technology, and benefits. *Sustainability, 17*(3), 921.

Zhao, P., & He, P. (2022). Construction and analysis of public management system indicators based on AHP (Analytic Hierarchy Process). *Journal of Sensors*, 5686855.

Urban Green Spaces Environmental Performance during Heatwave Events: Remote Sensing and GIS Mapping for Assessing the Temporal Response and Support NbS Climate-Resilient Design

Sabrina Puzone, Maria Fabrizia Clemente, and Enza Tersigni

Abstract Nature-Based Solutions (NbS) represent a strategic approach to addressing the impacts of climate change in urban settlements, particularly in relation to the intensification of heatwaves. As living systems, NbS do not perform statically, but are characterised by a dynamic behaviour that varies over time depending on climatic conditions, the physiological state of vegetation and the morphological characteristics of the urban context. This makes it necessary to move beyond specific assessment and design approaches, adopting methodologies that explicitly integrate the temporal dimension. This study proposes a conceptual and operative approach based on the integration of open-source satellite remote sensing data into GIS systems, to support the vulnerability analysis of urban green spaces, in terms of environmental performances, during heatwaves for decision makers' support in climate-resilient urban design. The conceptual framework is based on the four phases of Disaster Risk Management (preparedness, absorption, response and adaptation), applied to the multi-temporal analysis of environmental performance indicators of NbS. The test is conducted in the district of Soccavo (Naples), analysing the heatwave of July 2022 using the Land Surface Temperature (LST), Normalised Difference Vegetation Index (NDVI) and Normalised Difference Moisture Index (NDMI) indicators, derived from Google Earth Engine processing and integrated into a GIS environment. The results highlight different responses according to the different types of urban greenery over time, confirming the role of NbS as dynamic infrastructure.

Keywords Climate-resilient design · Urban green spaces · Heatwave · Environmental performances · Remote sensing

S. Puzone (✉) · M. F. Clemente · E. Tersigni
DiARC – Department of Architecture, University of Federico II, Naples, Italy
e-mail: sabrina.puzone@unina.it

L. Diana et al. (eds.), *Greening and Resilience*, Urban Sustainability,
https://doi.org/10.1007/978-981-92-0419-9_9

1 Thematic Framework

In the context of the constantly evolving relationship between society and the environment, nature-based solutions (NbS) play a strategic role in responding to the challenges of climate change mitigation and adaptation, biodiversity protection and ensuring human well-being (O'Hogain & McCarton, 2018; Buckley et al., 2024). These challenges are interdependent and require a systemic approach, considering that excessive urbanisation has led to deep changes in growth and development patterns, with an increase in impervious surfaces and a consequent decrease in green areas, combined with the growing frequency and intensity of extreme climate events (IPCC, 2023).

NbS have the potential to address both climate mitigation and adaptation challenges in a multi-hazard perspective and at relatively low cost, while offering multiple additional benefits for people and nature. Recent studies have shown that solutions such as increasing tree cover and designing green infrastructure and corridors contribute significantly to reduce urban surface temperatures and improve microclimatic comfort (Kabisch et al., 2017; Frantzeskaki, 2019), as well as reduce flood risk and absorb climate-changing gases caused by air pollution. This occurs through two key processes: evaporative cooling and shading (Shashua-Bar et al., 2011), which reduce the heat load caused by densely built-up areas, bringing additional benefits in terms of water resource management, air quality and improved resident well-being (Skarbit et al., 2024).

To enable a more informed choice and definition of nature-based strategies and solutions that maximise the benefits of risk adaptation and mitigation, the problem needs to be addressed through a multidisciplinary approach (Potschin et al., 2015). Over time, different perspectives on risk management and reduction have emerged, shifting the focus from approaches geared exclusively towards post-event management to an understanding of the need for holistic and systemic approaches. This has led to the gradual integration of Climate Change Adaptation (CCA) and Disaster Risk Reduction (DRR) approaches within the broader framework of Disaster Risk Management (DRM) (Mitchell & van Aalst, 2008). The DRM is a dynamic system encompassing the components of prevention (DRR), response, recovery and adaptation (CCA), in a continuous cycle oriented towards resilience, understood as the ability of systems to anticipate, absorb, adapt and transform in response to critical events (UNISDR, 2005; Barroca, 2018).

In line with this theoretical framework, understanding the vulnerabilities of urban systems requires a dynamic and temporal analysis linked to the evolution of hazardous phenomena. Urban green systems are not static components of the settlement; their microclimatic effectiveness and mitigation capacity vary over time in relation to the weather conditions, the physiological state of vegetation and the morphological characteristics of the context.

Time cannot be considered a simple contextual parameter, but rather takes on the role of an analytical variable allowing a better interpretation of the intensity, duration and persistence of climate impacts, as well as the capacity of urban systems to react, adapt and transform (Linkov & Trump, 2019; Wen et al., 2023; Ren et al., 2023).

1.1 Temporal Assessment of Nature-Based Solutions During Heatwaves

Among extreme weather events, heatwaves are a paradigmatic example, configured as dynamic events characterised by an initial phase, a period of maximum intensity and a subsequent attenuation phase (Lo et al., 2021). Each of these phases corresponds to different levels of thermal stress and variable responses from urban systems, both in technological-environmental and socio-economic terms.

Within climate resilience frameworks, the temporal dimension takes on a structuring role, as it allows the urban response to be read through the four phases of DRM (Fig. 1) of preparedness (T_0), absorption (T_1), response (T_2) and adaptation (T_3) (Clemente & Puzone, 2026).

Applying a temporal approach to the study of climate impacts on urban settlement systems, and to urban open spaces, allows to move beyond assessments based on single moments of observation, by introducing a perspective that includes all the different phases of the phenomena. In this conceptual framework, urban vegetation and open spaces take on a pivotal role, serving not only as structural components of the settlement system, but also as dynamic assets that enhance the city's capacity to respond to and withstand extreme events.

Urban greenery is characterised by highly variable behaviour over time, influenced by weather conditions, the physiological state of plant species, water availability and relationships with the surrounding built environment (Moncada-Morales et al., 2025). During a heatwave, these components do not respond uniformly but show different mitigation and adaptation capacities at different stages of the event, highlighting that their microclimatic effectiveness cannot be adequately evaluated through static observations.

For decades, research has focused on understanding the ecological mechanisms that govern the structure and functions of plant communities and their ability to

Fig. 1 Temporal phases of the extreme event (elaboration of the authors)

respond to stress conditions (Lortie et al., 2004). However, gaps remain in the operational translation of this knowledge into urban design and environmental monitoring processes (Bush & Doyon, 2019).

In this context, it is necessary to develop conceptual and operative approaches that can integrate the temporal dimension into the assessment of the environmental vulnerabilities of open spaces, through the use of indicators capable of describing their behaviour in the different phases of an extreme event to support decision makers. This research aligns with this perspective by proposing a GIS-based approach supported by open-source remote sensing data, which analyses the urban green spaces' environmental performance during heatwave events to support decision makers in climate-resilient design.

2 Remote Sensing for Environmental Performance Monitoring

Remote sensing is a science that allows qualitative and/or quantitative information to be obtained about an object, area or phenomenon using data acquired by an instrument that is not in direct physical contact with it. It exploits the possibility of detecting the absorption, emission, reflection and transmission of electromagnetic energy by a given object, allowing humans to extend their observation capabilities beyond their field of vision.

In the environmental field, the application of remote sensing technology in vegetation monitoring has been an area of considerable research in the past decades (Carlson et al., 2007). In particular, the collection of information through satellite remote sensing represents a valid methodology for multi-scale and multi-temporal monitoring, allowing the collection of information on the dynamics and behaviour of vegetation in complex urban contexts subject to climate change, overcoming the limitations of in-situ measurements. Furthermore, thanks to the growing availability of new technologies and innovative open-source tools, the scientific community has access to spatially continuous and frequent satellite images (de Araujo Barbosa et al., 2015).

During the Second World War, colour infrared film began to be used to detect the health of vegetation, based on reflectance, i.e. the amount of reflected sunlight that decreases in the near-infrared wavelength band and increases in the true red band (Ustin & Gamon, 2010). Compared to film, digital images extend the measurable spectral range, with digital sensors offering superior automation in data management, providing detailed information on changes in the shape, condition and density of vegetation (Estoque, 2020). These characteristics have facilitated the development of methods and models for monitoring environmental performance and measuring the effectiveness of nature-based solutions.

The era of global digital satellite imagery began in 1972 with the launch of the first Earth observation satellite programmes, including the Earth Resources Technology

Satellite (ERTS 1), later renamed Landsat (Triendl, 1979). From that moment on, remote sensing gradually took on a central role in mapping land cover and studying vegetation systems, thanks to the possibility of combining spatially contiguous and temporally repeatable observations (Wang et al., 2010; Parra, 2022). Over time, various types of satellite sources have become established, with highly variable characteristics according to differentiated spatial, temporal and spectral resolution.

Several studies have analysed satellite sources in different fields of application, stating that the most widely used are Landsat 8, followed by MODIS and Sentinel-2 (Reinermann et al., 2020; Mahrad et al., 2020; Li et al., 2020). Landsat 8 and Sentinel-2 represent a category of environmental monitoring satellites equipped with multispectral sensors capable of acquiring data with resolutions of tens of metres and a temporal frequency of several weeks. These sensors allow vegetation to be characterised by distinguishing between vegetated surfaces, bare soil and artificial surfaces, thanks to the acquisition of information in selected spectral bands of the visible (VIS), near-infrared (NIR) and short-wave infrared (SWIR) (Drusch et al., 2012). MODIS belongs to the category of global coverage meteorological and climate satellites, designed to acquire data with high daily temporal resolution, albeit at coarser spatial resolutions, on the order of kilometres (Zhang et al., 2015). These instruments have opened up new perspectives for the study of seasonal vegetation dynamics, enabling the analysis of plant system responses to climate variations and extreme events (Li et al., 2020).

2.1 Google Earth Engine and GIS for Environmental Performance Assessment

Within the evolutionary framework outlined above, satellite remote sensing is now not only a data acquisition technology, but also a structural component of knowledge systems for environmental and urban analysis aimed at supporting urban design. Indicators derived from satellite images no longer represent exclusively biophysical proxies but take on a key role in the assessment of environmental performance and in the functional interpretation of urban greenery in relation to extreme weather events. However, the exponential increase in the availability of satellite data highlights the need for tools capable of managing large datasets in a systematic, reproducible and scalable perspective.

In this context, cloud-based platforms for geospatial analysis have emerged, representing a natural evolution of remote sensing applications, enabling the transformation of extensive Earth Observation archives into operational knowledge systems. It is within this trajectory that Google Earth Engine (GEE) is configured as a technological infrastructure useful for monitoring environmental performance and vulnerability assessment of open spaces, thanks to the combination of open-source Earth Observation archives and cloud-based computing capabilities (Gorelick et al., 2017).

GEE provides direct, free access to historical satellite data series from various missions, enabling the processing of multispectral and thermal images over large spatial areas and different time horizons using Google's JavaScript and Python programming interfaces (Kwok, 2018; Velastegui-Montoya et al., 2023). These aspects are particularly relevant in studies on the impacts of phenomena such as urban heatwaves, where the analysis of thermal dynamics requires consistent temporal comparisons between the pre-, during and post-extreme event phases.

The integration between GEE and GIS strengthens the operational nature of methods for assessing the impacts of hazardous phenomena, configuring geospatial as decision-support systems for climate-resilient urban planning and design. In particular, the possibility of combining thermal and vegetation indicators in a multiscale and multi-temporal perspective allows urban heat to be interpreted as a systemic phenomenon influenced by urban morphology, the distribution of green spaces and the materials characteristics. This integrated approach supports the identification of priority areas for intervention and definition of nature-based adaptation and mitigation strategies, aligning with the objectives of sustainable urban design and climate resilience.

3 Case Study Application: The Soccavo District (Naples)

Based on the defined conceptual research framework, the study was applied to the urban district of Soccavo, located in the IX Municipality of the city of Naples, in southern Italy.

Soccavo has a consolidated urban context characterised by the presence of the Metropolitan Park of the hills of Naples and the Vomero hill, which define its geographical boundaries to the N-NE, respectively, by high building density and by the fragmented and discontinuous presence of urban green spaces (Fig. 2).

The district is significantly exposed to various environmental risks such as volcanic eruptions and bradyseism, as well as extreme climatic phenomena such as urban overheating and rain floods (Turchi et al., 2023). The vulnerability of

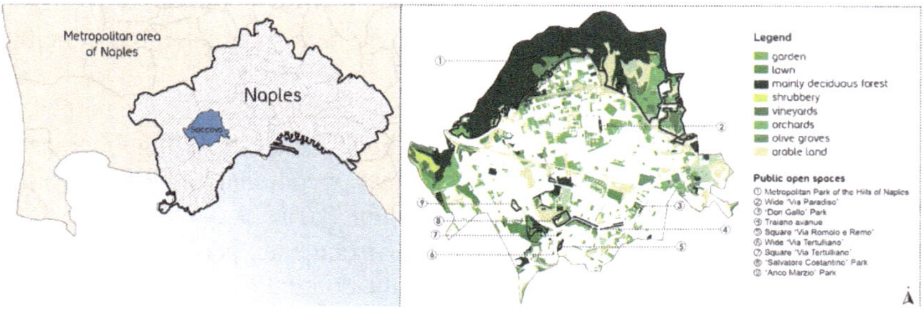

Fig. 2 Geographical location and vegetation system of the study area: the district of Soccavo (elaboration of the authors)

urban open spaces to heatwaves is therefore studied according to a multi-scale and multi-temporal approach.

The analysis was conducted considering the heatwave event that occurred from 21 to 31 July 2022. This reference scenario was identified through a combination of three factors (Apreda et al., 2019; D'Ambrosio et al., 2023): ΔT (day–night temperature difference) <10 °C, Tmax >30 °C and Tmin >20 °C;

The study introduces an in-depth analysis focused on open spaces and urban green systems. The vulnerability of open spaces is quantified through the use of environmental indices and indicators that are able to describe the complex conditions of the environment during the temporal phases of the extreme event.

The vulnerability analysis of open spaces in the Soccavo district to heatwaves is carried out through the GEE platform, that enabled the selection, pre-processing and export of satellite images in raster format for the study of Land Surface Temperature (LST) and the calculation of vegetation indices such as the Normalised Difference Vegetation Index (NDVI) and the Normalised Difference Moisture Index (NDMI). Data are acquired from Sentinel-2 for NDVI and NDMI, with a spatial resolution of up to 10 m, and from Landsat-8 for LST, with a spatial resolution of 30 m (Table 1).

Thermal data acquired allow for the analysis of the spatial distribution of surface temperatures and the identification of patterns of heat accumulation and dissipation in different urban fabrics. At the same time, multispectral data allow for the assessment of the vigour and response of vegetation cover under conditions of thermal stress, supporting the interpretation of the mitigating role of green spaces with respect to urban overheating phenomena (Cardone et al., 2024).

Each indicator has been linked to the four stages of risk management, each of which corresponds to a specific time frame of the heatwave event:

- T_0: 11–20 July 2022;
- T_1: 21–31 July 2022;
- T_2: 01–10 August 2022;
- T_3: 11–20 August 2022.

The resulting raster images, referring to the four time phases of the framework (T_0–T_3), were exported in georeferenced format and subsequently imported into a Geographic Information System (GIS) for detailed spatial processing with the aim of analysing changes in urban green space performance over time.

Table 1 Index and input data characteristics

Index	Description	Satellite	Spatial resolution
LST	Thermal index of the radiative temperature of the land	Landsat 8	30 m
NDVI	Spectral index of vegetation	Sentinel-2	10 m
NDMI	Spectral index of vegetation water content	Sentinel-2	10 m

4 Result and Discussion

The analysis of the selected indicator was structured through a multistep process aimed at translating satellite-derived environmental information into interpretable and operational knowledge to support climate-resilient urban design.

First, a zonal statistical process was carried out, associating homogeneous spatial reference units—namely the 2021 ISTAT census track—with the average value of the pixels covering each polygon. This step enabled the identification of spatial patterns and criticalities at the district scale. Then, the resulting values were subjected to a thematic normalisation process according to the Jenks Natural Breaks method, allowing continuous raster information to be synthesised into comparable classes. Finally, the average of the values falling within the Soccavo district administrative boundaries was calculated to synthesise the environmental information, while maintaining the link with the spatial dynamics observed at lower scales (Table 2).

The results highlight a differentiated temporal response of open spaces during the heatwave event. LST, for example, shows a significant increase during the absorption phase (T_1), in which a high average value (47.30 °C) is recorded, followed by a

Table 2 Indicator variations in the four phases of the heatwave event.

	T_0_PREPARE	T_1_ABSORB	T_2_RESPOND	T_3_ADAPT
LST	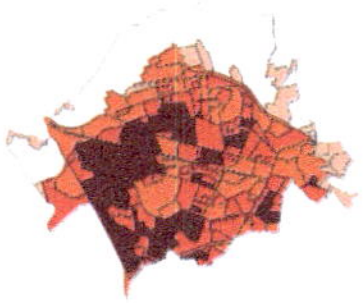	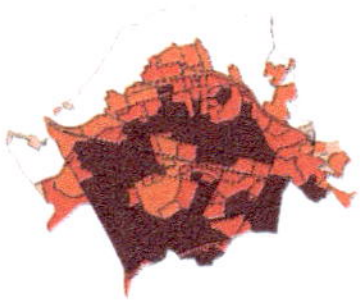	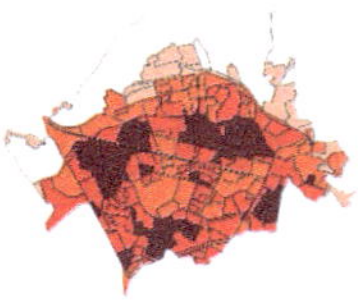	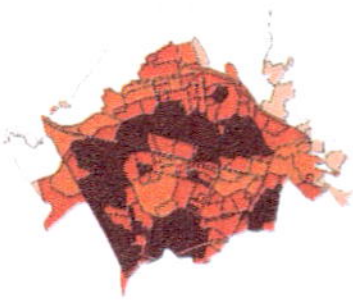
	43,44°C	47,30°C	42,08°C	37,94°C
NDVI				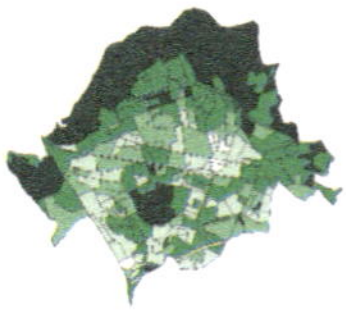
	-0.07	-0,10	0.08	0.03
NDMI				
	0.06	-0,09	-0.05	0.10

gradual reduction in the subsequent response (T_2) and adaptation (T_3) phases. Vegetation indicators show a complementary response. Both NDVI and NDMI decrease during the absorption phase, indicating a deterioration in the vigour of vegetation at maximum thermal stress, while a partial recovery of physiological conditions is observed in the subsequent phases. The results confirm that the environmental performances of urban greenery are not static, but evolve over time in relation to climatic conditions and thermal stress intensity.

To further explore these dynamics, a focused analysis was conducted on three representative types of urban green spaces within the study area, according to different morphological, dimensional and functional configurations (Fig. 3). The urban green areas are extracted from the topographic database provided by the Municipality of Naples.

The three focal points include: (1) a large peri-urban forest system (Metropolitan Park of the hills of Naples), (2) a public urban park (Salvatore Costantino" park) and (3) a series of smaller, fragmented green areas, such as private gardens.

The comparative analysis of the LST, NDVI and NDMI indicators in the four time phases (Table 3) highlights significantly different responses among the three systems, confirming that the response of urban greenery to heatwaves is strongly dependent on the type of open space and on its intrinsic capacity to absorb, respond and adapt to thermal stress.

The forest system (1) maintains the best overall performance throughout the entire event. Although surface temperatures increase during the absorption phase (T_1), LST values remain consistently lower than the other two systems and show a marked reduction in the adaptation phase (T_3). At the same time, the NDVI and NDMI

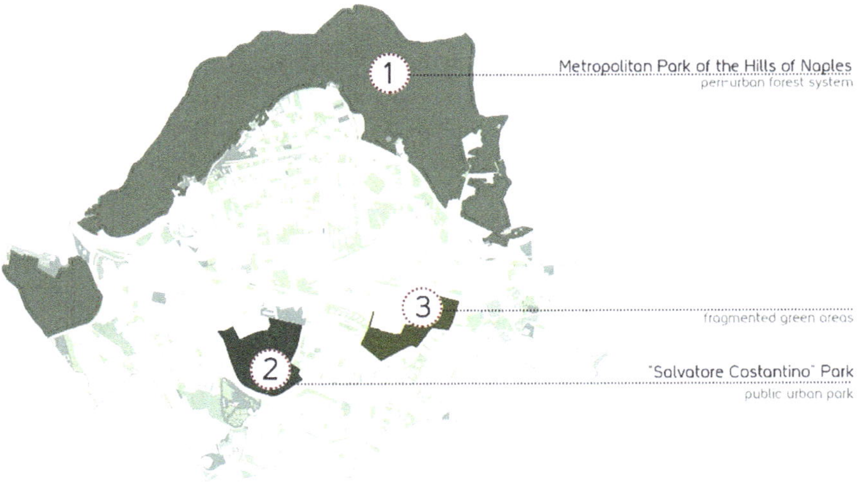

Fig. 3 Urban green spaces selected for the analysis (elaboration of the authors)

Table 3 Indicator variations in the four phases of DRM of green urban spaces selected

System	INDEX	T_0_PREPARE	T_1_ABSORB	T_2_RESPOND	T_3_ADAPT
1	LST	36,52 °C	41,74 °C	36,95 °C	33,31 °C
	NDVI	0.52	0.51	0.53	0.59
	NDMI	0.35	0.30	0.36	0.43
2	LST	42,22 °C	46,02 °C	40,81 °C	36,39 °C
	NDVI	0.34	0.34	0.37	0.38
	NDMI	0.09	0.06	0.07	0.17
3	LST	42,36 °C	46,45 °C	41,54 °C	36,71 °C
	NDVI	0.26	0.25	0.24	0.30
	NDMI	−0.02	−0.07	−0.03	0.04

show adequate stability during the event and a significant improvement after the event, indicating a high capacity for ecological resilience. This behaviour confirms the strategic role of large continuous green systems as infrastructures capable of ensuring persistent and long-term microclimatic benefits.

The public urban park (2), on the other hand, shows a significant increase in LST during the absorption phase (T_1), accompanied by a reduction in NDMI values, which indicates greater sensitivity to water stress. However, in the response phase (T_2) and especially in the adaptation phase (T_3), the system shows a reasonable recovery capacity, with decreasing surface temperatures and improving vegetation indices. This highlights the potential effectiveness of neighbourhood-scale green spaces in climate mitigation and adaptation, in addition to the recognised role in social terms, while also underlining their greater vulnerability compared to more extensive and ecologically continuous systems.

The third system, composed of small and fragmented green areas (3), highlights the most critical issues in terms of environmental performance. LST values persist across all phases, with a less marked reduction in the adaptation phase (T_3). At the same time, low NDVI and, above all, NDMI values indicate limited absorption and recovery capacity (T_2), suggesting structural vulnerability linked to fragmentation, reduced size and poor ecological continuity of these spaces. However, despite their inferior performance, the widespread distribution of these spaces still contributes to microclimatic mitigation at the local level, particularly if considered as part of a coordinated and integrated green network rather than as isolated elements.

The three cases analysed serve as examples, useful both for validating the methodological framework and for building a knowledge base to support the definition and selection of Nature-Based Solutions strategies, tailored to the characteristics of the urban context and climate resilience objectives. Beyond the empirical results, these findings call for a broader interpretative reflection on the role of methods and indicators in reading urban climate dynamics and urban green environmental performances.

The differentiated performance of urban green spaces typologies reflects not only biophysical or physiological factors, but also long-term processes of urbanisation, land fragmentation and governance of open spaces. Heatwaves act as systemic stressors that expose latent vulnerabilities embedded in contemporary urban forms. In particular, the limited adaptive capacity of small and fragmented green areas can be interpreted as the spatial outcome of incremental and non-systemic approaches to urban greening. These results align with theoretical frameworks that conceptualise climate change as a driver of urban transformation, highlighting the need to move beyond additive solutions towards integrated and infrastructural conceptions of urban greenery. Within this framework, urban green spaces emerge as dynamic climate infrastructures whose effectiveness depends on spatial continuity, ecological connectivity and integration within the built environment (Liu & Russo, 2021).

The temporal approach adopted in this study reinforces the idea that adaptation capacity cannot be adequately assessed through static indicators, but must be understood as a process unfolding across the different phases of extreme events. Consequently, the value of the proposed approach lies in bridging computational environmental assessment with urban climate adaptation theory, providing a basis for interpreting nature-based solutions not as isolated design interventions, but as structural components capable of supporting long-term resilience and transformation of urban settlements under climate change.

The proposed framework is grounded in open-source satellite-based indicators and cloud-based analytical platforms; its contribution does not lie in achieving high-resolution microclimatic accuracy. The use of medium-resolution satellite data inevitably introduces a degree of approximation, especially in heterogeneous urban contexts. However, this limitation should be understood as an intrinsic characteristic of approaches oriented towards district-scale strategic assessment and comparative and transferable evaluation. In this perspective, remote sensing-derived indicators function as relational and interpretative proxies, capable of revealing spatial patterns, temporal trajectories and differential responses of urban green systems, rather than as precise descriptors of local microclimatic conditions.

5 Concluding Remarks

The contribution proposed a conceptual and operative approach for analysing the vulnerability and resilience of urban open spaces to extreme heatwave events, based on the integration of satellite remote sensing, GIS and a temporal reading of the impacts of the phenomenon, in order to support decision makers in climate-resilient urban design. In line with the latest climate resilience and disaster risk management frameworks, the research highlighted how the introduction of the temporal dimension is a key element in overcoming static and point-in-time impact assessments, providing a more articulated understanding of urban dynamics during the different phases of the extreme event.

The multi-temporal analysis of the NDVI, NDMI and LST indicators made it possible to interpret the behaviour of urban greenery not only in terms of presence or quantity, but as an active system capable of responding in different ways to thermal stress conditions. The results of the test showed that different types of green spaces have heterogeneous environmental performance and mitigation capacities over time, highlighting the crucial role of vegetation structure, physiological state and morphological context in microclimatic regulation. In this sense, urban greenery emerges as a dynamic climate infrastructure, whose effectiveness varies between the different phases of preparation, absorption, response and adaptation to the extreme heat event.

From an operational point of view, the proposed framework represents a first step towards the development of a decision-support tool for climate-resilient urban design. The proposed approach allows complex environmental information to be translated into synthetic information useful for administrators, decision makers and designers, facilitating a spatial and temporal reading of urban vulnerabilities and guiding the selection of optimal nature-based adaptation and mitigation strategies.

Recognising the limitations associated with the spatial resolution of satellite data, future developments will integrate more detailed microclimatic and socio-demographic data within the GIS environment, enabling the spatial overlay of thermal stress patterns with population exposure, particularly for vulnerable groups (such as elderly people and children). The integration of exposure and social vulnerability factors will allow a shift from environmental vulnerability assessment towards an integrated evaluation of heatwave risk.

Moreover, future prospects will focus on extending the analytical assessment to the integration of ex-ante and ex-post design scenario simulations. In this sense, simulation models could be used to test how the Soccavo district would respond to different NbS, for example, considering an increase in the number of trees, with species in line with those currently effective in the Metropolitan Park of the Hills of Naples. The simulation would provide quantitative information on potential temperature reductions, supporting the comparison of alternative nature-based solution strategies and strengthening the role of the framework as a bridge between environmental analysis and climate-resilient design.

In conclusion, the study confirms that understanding the temporal behaviour of urban greenery and open spaces is a fundamental step towards building more resilient urban settlements. By bridging remote sensing-based environmental assessment, urban climate adaptation theory and project-oriented perspectives, the proposed framework contributes to rethinking nature-based solutions not as isolated technical measures, but as structural components of urban settlements.

Acknowledgements This study was carried out within the RETURN Extended Partnership and received funding from the European Union Next-GenerationEU (National Recovery and Resilience Plan—NRRP, Mission 4, Component 2, Investment 1.3—D.D. 1243 2/8/2022, PE0000005).

References

Apreda, C., D'Ambrosio, V., & Di Martino, F. (2019). A climate vulnerability and impact assessment model for complex urban systems. *Environmental Science & Policy, 93*, 11–26.

Barroca B (2018). Vulnerability, urban design and resilience. *Natural Hazards: Risk Assessment and Vulnerability Reduction 1.*

Buckley, Y. M., Austin, A., Bardgett, R., Catford, J. A., Hector, A., Iler, A., & Mariotte, P. (2024). The plant ecology of nature-based solutions for people, biodiversity and climate. *Journal of Ecology, 112*(11).

Bush, J., & Doyon, A. (2019). Building urban resilience with nature-based solutions: How can urban planning contribute? *Cities, 95.*

Cardone, B., Di Martino, F., Mauriello, C., & Miraglia, V. (2024). A GIS-based framework to analyze the behavior of urban greenery during heatwaves using satellite data. *ISPRS International Journal of, Geo-Information,13(11)*, 377.

Carlson, K. M., Asner, G. P., Hughes, R. F., Ostertag, R., & Martin, R. E. (2007). Hyperspectral remote sensing of canopy biodiversity in Hawaiian lowland rainforests. *Ecosystems, 10*(4), 536–549.

Clemente, M. F., & Puzone, S. (2026). The temporal dimension in climate adaptation and mitigation strategies and solutions aimed to increase urban integrated resilience. *Città, Territori, Progetti*, 214–229. https://doi.org/10.36253/contest-16544

D'Ambrosio, V., Di Martino, F., & Miraglia, V. (2023). A GIS-based framework to assess heatwave vulnerability and impact scenarios in urban systems. *Scientific Reports, 13*(1), 13073.

de Araujo Barbosa, C. C., Atkinson, P. M., & Dearing, J. A. (2015). Remote sensing of ecosystem services: A systematic review. *Ecological Indicators, 52*, 430–443.

Dorst, H., Van der Jagt, A., Raven, R., & Runhaar, H. (2019). Urban greening through nature-based solutions–key characteristics of an emerging concept. *Sustainable Cities and Society, 49.*

Drusch, M., Del Bello, U., Carlier, S., Colin, O., Fernandez, V., Gascon, F., et al. (2012). Sentinel-2: ESA'S optical high-resolution mission for GMES operational services. *Remote Sensing of Environment, 120*, 25–36.

Estoque, R. C. (2020). A review of the sustainability concept and the state of SDG monitoring using remote sensing. *Remote Sensing, 12*(11).

Frantzeskaki, N. (2019). Seven lessons for planning nature-based solutions in cities. *Environmental Science & Policy, 93*, 101–111.

Gorelick, N., Hancher, M., Dixon, M., Ilyushchenko, S., Thau, D., & Moore, R. (2017). Google earth engine: Planetary-scale geospatial analysis for everyone. *Remote Sensing of Environment, 202*, 18–27.

Grange, V. (2016). LA CANOPEE in Paris—An innovative architecture, a challenge for engineering. *Structural Engineering International, 26*(3), 235–241.

IPCC (2023): Summary for policymakers. In: Climate change 2023: Synthesis report. Contribution of working groups I, II and III to the sixth assessment report of the intergovernmental panel on climate change [Core writing team, H. Lee and J. Romero (eds.)]. IPCC, Geneva, Switzerland. pp. 1–34.

Li, J., Pei, Y., Zhao, S., Xiao, R., Sang, X., & Zhang, C. (2020). A review of remote sensing for environmental monitoring in China. *Remote Sensing, 12*(7), 1130.

Linkov, I., & Trump, B. D. (2019). Resilience and governance. In *The science and practice of resilience* (pp. 59–79). Springer International Publishing.

Liu, O. Y., & Russo, A. (2021). Assessing the contribution of urban green spaces in green infrastructure strategy planning for urban ecosystem conditions and services. *Sustainable Cities and Society, 68*, 102772.

Lo, S. H., Chen, C. T., Russo, S., Huang, W. R., & Shih, M. F. (2021). Tracking heatwave extremes from an event perspective. *Weather and Climate Extremes, 34, 100371.*

Lortie, C. J., Brooker, R. W., Choler, P., Kikvidze, Z., Michalet, R., Pugnaire, F. I., & Callaway, R. M. (2004). Rethinking plant community theory. *Oikos, 107*, 433–438.

Kabisch, N., Korn, H., Stadler, J., & Bonn, A. (2017). *Nature-based solutions to climate change adaptation in urban areas: Linkages between science, policy and practice*. Springer Nature.
Kwok, R. (2018). Ecology's remote-sensing revolution. *Nature, 556*(7699), 137–138.
Mahrad, B. E., Newton, A., Icely, J. D., Kacimi, I., Abalansa, S., & Snoussi, M. (2020). Contribution of remote sensing technologies to a holistic coastal and marine environmental management framework: A review. *Remote Sensing, 12*(14), 2313.
Mitchell, T., & van Aalst, M. (2008). Convergence of disaster risk reduction and climate change adaptation. *A review for DFID, 44*, 1–22.
Moncada-Morales, G. A., Verichev, K., López-Guerrero, R. E., & Carpio, M. (2025). A global review of vegetation's interaction effect on urban heat mitigation across different climates. *Urban Science, 9*(9), 361.
Nagendra, H. (2001). Using remote sensing to assess biodiversity. *International Journal of Remote Sensing, 22*(12), 2377–2400.
O'Hogain, S., & McCarton, L. (2018). Nature-based solutions. In *A technology portfolio of nature based solutions: Innovations in water management* (pp. 1–9). Springer International Publishing.
Parra, L. (2022). Remote sensing and GIS in environmental monitoring. *Applied Sciences, 12*(16), 8045.
Potschin, M., Kretsch, C., Haines-Young, R., Furman, E., Berry, P., & Baró, F. (2015). *Nature-based solutions*. OpenNESS Ecosystem Service Reference Book.
Reinermann, S., Asam, S., & Kuenzer, C. (2020). Remote sensing of grassland production and management—A review. *Remote Sensing, 12*(12), 1949.
Ren, Y., Liu, J., Zhang, T., Shalamzari, M. J., Arshad, A., Liu, T., et al. (2023). Identification and analysis of heatwave events considering temporal continuity and spatial dynamics. *Remote Sensing, 15*(5), 1369.
Shashua-Bar, L., Pearlmutter, D., & Erell, E. (2011). The influence of trees and grass on outdoor thermal comfort in a hot-arid environment. *International Journal of Climatology, 31*(10).
Skarbit, N., Unger, J., & Gál, T. (2024). Evaluating the impact of heat mitigation strategies using added urban green spaces during a heatwave in a medium-Sized City. *Sustainability, 16*(8).
Triendl EE (1979). Landsat image processing. Advances in digital image processing: Theory, application, implementation.: Springer US. pp. 165–175.
Turchi, A., Lumino, R., Gambardella, D., & Leone, M. F. (2023). Coping capacity, adaptive capacity, and transformative capacity preliminary characterization in a "multi-hazard" resilience perspective: The Soccavo district case study (City of Naples, Italy). *Sustainability, 15*(14), 10877.
UNISDR (2005). Hyogo framework for action 2005–2015: Building the resilience of nations and communities to disasters.
Ustin, S. L., & Gamon, J. A. (2010). Remote sensing of plant functional types. *New Phytologist, 186*(4), 795–816.
Velastegui-Montoya, A., Montalván-Burbano, N., Carrión-Mero, P., Rivera-Torres, H., Sadeck, L., & Adami, M. (2023). Google earth engine: A global analysis and future trends. *Remote Sensing, 15*(14), 3675.
Wang, K., Franklin, S. E., Guo, X., & Cattet, M. (2010). Remote sensing of ecology, biodiversity and conservation: A review from the perspective of remote sensing specialists. *Sensors, 10*(11), 9647–9667.
Wen, J., Wan, C., Ye, Q., Yan, J., & Li, W. (2023). Disaster risk reduction, climate change adaptation and their linkages with sustainable development over the past 30 years: A review. *International Journal of Disaster Risk Science, 14*(1), 1–13.
Zhang, B., Zhang, L., Xie, D., Yin, X., Liu, C., & Liu, G. (2015). Application of synthetic NDVI time series blended from Landsat and MODIS data for grassland biomass estimation. *Remote Sensing, 8*(1), 10.

Green Roofs: A Preliminary Computational Fluid Dynamics Approach

Felipe Rezende da Costa

Abstract Rapid urbanisation has reduced permeable surfaces in cities, leading to increased stormwater runoff volumes, higher flow velocities and greater discharges into urban drainage systems. These problems are further intensified by climate change, which has increased both the frequency and intensity of extreme rainfall events. In this context, green roofs have emerged as an effective solution, providing both rainwater infiltration and retention areas as well as social areas, offering environmental, economic and social benefits. Despite their great potential, accurately representing the hydrological behaviour of green roofs remains a challenge due to the complex interactions between precipitation and the structure of green roofs. This study presents a preliminary numerical investigation of the hydrodynamic response of a green roof system subjected to heavy rainfall based on a physical model. Rainfall was modelled using a volume fraction approach to represent water inflow into the domain, while the substrate layer was represented as a porous medium governed by a Darcy–Forchheimer formulation, allowing the simulation of infiltration and underground flow processes. The simulations were performed using the open-source computational fluid dynamics software OpenFOAM in a two-dimensional domain representing a longitudinal section of a green roof system. Rainfall with a duration of 30 minutes and an intensity of 60 mm/h was applied, corresponding to a high-intensity event. The results showed that although the model demonstrated stability and the ability to represent some aspects of the hydraulic response of green roofs, and thus showed promise, further refinement of the model is still necessary, such as the implementation of a soil water storage model.

Keywords Nature-based solutions · Stormwater management · Climate change adaptation · Urban ecology · Green infrastructure · CFD

F. R. da Costa (✉)
Department of Civil, Building and Environmental Engineering, University of Naples Federico II, Naples, Italy
e-mail: felipe.rezendedacosta@unina.it

L. Diana et al. (eds.), *Greening and Resilience*, Urban Sustainability,
https://doi.org/10.1007/978-981-92-0419-9_10

1 Introduction

Cities are expanding ever more rapidly, whether in terms of population, land area, resource consumption or pollutant emissions. This growth puts severe pressure on biodiversity, ecological systems and human well-being, with consequences both within cities and far beyond their boundaries (Mori et al., 2025; Qiao et al., 2025). An important change to lessen the impacts of this growth is to begin adapting cities to local ecosystems (Fig. 1), rather than simply changing the original environment to serve the process of metropolitan expansion (Shi & Moser, 2021).

The main difficulty for this implementation is the inefficiency of the systems applied, which end up ignoring cultural aspects and the value of nature and fail to consider the indirect impacts of cities (Kabisch et al., 2016; Castelo et al., 2023). Therefore, the future sustainability of cities depends on reintegrating nature into urban planning in order to recognise its multiple values (Mansur et al., 2022; Hölscher et al., 2023).

A sign of this failure is the fact that even with projects to increase green areas, the urbanisation process has led to a decrease in most cities worldwide (Radford & James, 2013). The loss of infiltration areas leads to an increase in the discharge, velocity and volume of water conveyed into urban sewer systems (Chen et al., 2017). If this were not enough to aggravate stormwater management problems, climate change has

Fig. 1 Integration of buildings and nature in the facilities of the Guadalajara Zoo, Mexico (*Source* Photo taken by the author)

Fig. 2 Green roof located in the facilities of the Guadalajara Zoo, Mexico (*Source* Photo taken by the author)

increased the frequency and intensity of extreme rainfall events (Myhre et al., 2019). In this context, the use of green solutions represents a reliable approach to couple engineering improvements with interesting environmental and socioeconomic co-benefits (Olabi et al., 2025). Green roofs (Fig. 2) take advantage of the retention and infiltration capacity of different layers of vegetation and soil, while providing economic, social and ecological benefits.

A green roof is composed of many layers with different functions that ensure the effectiveness of the system. The most visible layers are the surface layers, the vegetation and the growth layer, also known as substrate, which promotes plant growth, support, nutrients and water. On the lower side, a filter layer prevents the soil particles from entering and clogging the drainage system. Then, a drainage layer ensures adequate water flow, avoiding its accumulation. At a deeper level, there is a root barrier to protect the main structure. At last, before the roof itself, a waterproofing membrane protects the building ceiling from water infiltration (Mihalakakou et al., 2023; Stein et al., 2012).

The composition of the substrate is a crucial factor since soil texture influences root penetration, impacting plant stability and water infiltration (Harper, 2024). Nevertheless, understanding the hydrodynamics of these structures is very challenging, given the variable rainfall actions and the heterogeneous configurations of green roofs. In addition, testing is difficult for two reasons. The first is the cost of testing different soil compositions and constructing these structures, since green roofs tend to have a higher installation cost than conventional roofs and only become economically viable in the long term (Manso et al., 2021). The second is the time required, since after testing, the soil will no longer be in its initial condition and will no longer be able to generate the standard response in real cases.

To solve this problem, computer simulations prove to be a good solution. Computer simulations and modelling have long been powerful tools for reducing costs and, also combined with other methods, have shaped the industry for many years (Johnson et al. 2005). In addition to enhancing, optimising and shortening the product development cycle in various sectors (Gibson et al. 2004). By creating a model that can accurately represent the impacts of the substrate, the financial problem would be solved, as it would only be necessary to change its parameters. Meanwhile, time constraints would be resolved by performing more than one simulation at the same time and in succession.

This study aims to fill this gap. To do this, a preliminary two-dimensional model was assembled in OpenFOAM software. It consists of a domain that represents both the structure of a green roof and the atmosphere above it, so that, in addition to being able to compute the phenomena below the ground, it can also compute any possible surface runoff. This makes it possible to generate a performance database of different green roofs more quickly and cheaply than by constructing different structures.

2 Methodology

Accurate prediction of the hydraulic response of green roofs is essential for urban water management. Numerical simulations can be a powerful tool for this purpose, but their reliability depends on model validation against physical experiments. In this study, a full-scale green roof model was used to provide reference data for evaluating the simulation results.

2.1 Representation of Rain and Soil

To model the rainfall, one method is to set a fraction phase α of water to the total volume. It represents the amount of water within a computational cell, thus defining the volume convertible into rain flow. However, this would be for each water droplet diameter, so the rain flow rate needs to be multiplied by the fh factor, which is the probability density of droplets with a diameter less than or equal to d (Best, 1950), resulting in Eq. 1

$$\alpha = R^{*}fh/V \tag{1}$$

where R is the intensity of rain and V is the terminal velocity.

A rainfall intensity of 60 mm/h was set for simulations, corresponding to a violent rainfall (World Meteorological Organization [WMO], n.d.) for a duration of 30 min. A droplet diameter of 3 mm of a homogeneous distribution (fh = 1) was also selected, which results in a terminal velocity of approximately 8 m/s (Gunn & Kinzer, 1949).

However, at this point, a barrier is created for computational reasons, represented by the Courant number. The Courant number is a dimensionless value that measures how far a fluid travels across a computational grid cell, which is the division of the domain into parts so that the flow can be calculated. This is done by considering the distance travelled by the fluid particles in a single time step, linking flow velocity, time step size and mesh size. The time step is the small, discrete interval of time advanced in each calculation cycle.

Since, for simulation stability reasons, the Courant number must be less than or equal to 1, there are two options. The first would be to increase the size of the parts of this grip, which generates inaccuracy in the simulation, which would be unfeasible. The second would be to maintain short timesteps, which would increase the computational demand, making it impossible to run the simulations on ordinary computers.

Therefore, and since the objective of developing this model is to provide access to green roof simulations to a wide audience, another approach was adopted. Since speed would be the problem, the solution was to reduce the speed while maintaining the flow rate, thus increasing the amount of water. To maintain low speed even with the effect of gravity, a drag (Eq. 2) was implemented to counteract gravity.

$$D = \rho^{*} g / \mu^{*} U \tag{2}$$

where ρ is the density of a fluid, g is the gravity, μ is the viscosity of the fluid and U is the velocity, resulting in a drag of 5.87E+11.

For soil modelling, a porous medium was chosen, given its capability to effectively model the absorption of the flow energy. The advantage of this choice also relies on not needing to singularly model the grains, making the mesh generation easier with related shorter computational efforts. The momentum sink represented by Eq.3 is modelled.

$$Sm = d^{*} \mu^{*} U + 1/2^{*} f^{*} \rho^{*} U^{2} \tag{3}$$

in which d is the Darcy coefficient, μ is the viscosity of the fluid, U is the velocity, f is the Forchheimer coefficient and ρ is the density of the fluid. The Darcy coefficient represents the viscous resistance of the media, estimated by Eq. 4.

$$d = \left[150(1 - \varepsilon)^{2}\right] / \left(dp^{2*} \varepsilon^{3}\right) \tag{4}$$

where ε is the porosity (void fraction) and dp is particle diameter, whereas the Forchheimer coefficient represents the inertial resistance calculated by Eq. 5. However, it is worth noting that since the flow inside the soil is very slow due to the low porosity, this resistance cannot affect the result because the low velocity greatly reduces this

Table 1 Darcy–Forchheimer parameters

Soil	Porosity (ε)	Particle diameter (dp)	Solid Volume Fraction	Porous Coefficients	
				d	f
Sand	0,47	1,50E-04	0,53	1,80E+10	5,96E+04

absorption, as modelled by Eq. 2.

$$f = [7,5(1 - \varepsilon)]/\left(dp^{*}\varepsilon^{3}\right) \quad (5)$$

For the implementation of the porous medium, a sandy soil was chosen and the values for this type of soil are specified in Table 1.

2.2 *Simulation Setup*

The domain consists of a 2D representation of a longitudinal cut of the physical green roof with atmosphere above and a reservoir under it. This cut results in a bidimensional domain with a positive height of 0.1 m that consists of 10 cm representing the region of runoff; this way, the model will be able to compute any runoff that can occur. Besides, there is a negative depth to represent the substrate depth of 10 cm plus 10 cm to represent the reservoir under the structure to guarantee a smooth drainage. Thus, having all the results with negative depth being below ground level. Being 2.00 m long, the origin of the coordinate axis is located on top of the substrate and in the corner of the domain so that all coordinates in the XY plane are positive (Fig. 3). The red arrow represents the x-axis, the yellow (normal to the image) the y-axis and the green the z-axis.

Figure 3. shows the full computational mesh (a) used for the simulation, consisting of 13,000 hexahedral cells. This three-dimensional format was necessary because the software used (OpenFOAM) performs two-dimensional simulations through a three-dimensional domain of unit value on one axis. Local refinement is inside the soil region to accurately resolve hydraulic head gradients and porous resistance effects. Inflation layers are generated along the atmosphere and reservoir regions to guarantee smooth transition zones.

The boundary condition for all boundaries in the XZ planes was "empty", which applies a condition where the solver does not compute anything across these faces and represents a 2D cut. The YZ planes above the soil and the one with the reservoir at x = 0.0 m were configured as a zero-gradient region, to avoid return flows and work as outlets. The domain top was set as an inlet region with a partial inflow of water and air to simulate rainfall. The planes YZ at the soil region, the one with the reservoir at x = 2.0 m and the base of the domain were set with a no-slip condition. Finally, two porous media were used using parameters of a porous Darcy–Forchheimer medium.

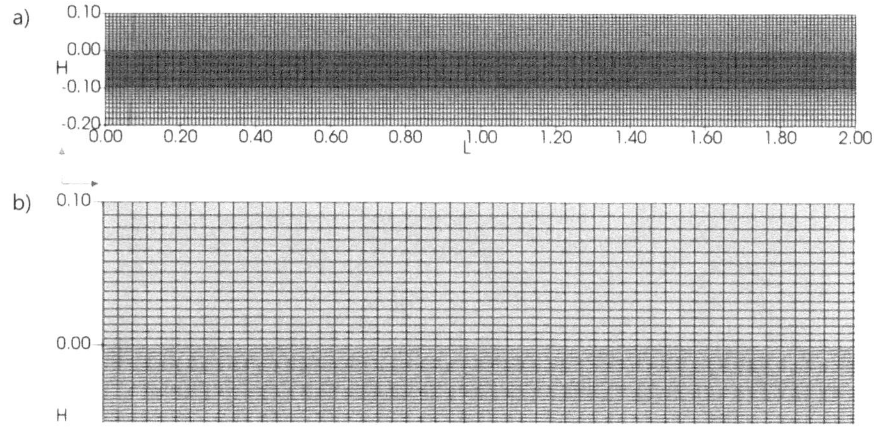

Fig. 3 Full domain mesh (**a**) and detailed view near the soil–atmosphere interface (**b**)

One to simulate the soil region and one with different values above it to represent the less computation demanding method.

To run the simulation, the open-source software OpenFOAM version 2412 (OpenCFD, 2025) was used. A non-turbulence model was chosen. The flow will be mostly two-dimensional and without recirculation zoning, which removes the need for more robust models. Thus, second-order numerical schemes were used for modelling. The interFoam module was chosen to generate the mesh and discretise the numerical schemes and governing equations, consisting of a solver for 2 incompressible, isothermal.

3 Results

Figure 4 shows the cumulative water depth over time obtained from numerical simulations (a) and the physical model experiment (b). The dashed black line represents the applied rainfall, while the solid lines correspond to the cumulative outflow from the green roof system, blue being from the simulation and orange with dots from the physical model.

Figure 4.b shows the effect that green roofs have on rainwater retention, where almost all the rainwater was still stored within the structure even after the end of the storm. This is the main factor for quantifying the effect of the structure on stormwater management because if the soil is capable of retaining it, all stored water can be discharged into the sewer system after the most critical moment, which is during the event. The foregoing can prevent the system from becoming overloaded and potential flooding when it is most strained. Meanwhile, Fig. 4.a shows that the model has not yet been able to represent this retention. However, it was able to represent infiltration since, as in the real case, there was no surface runoff. This factor is also important

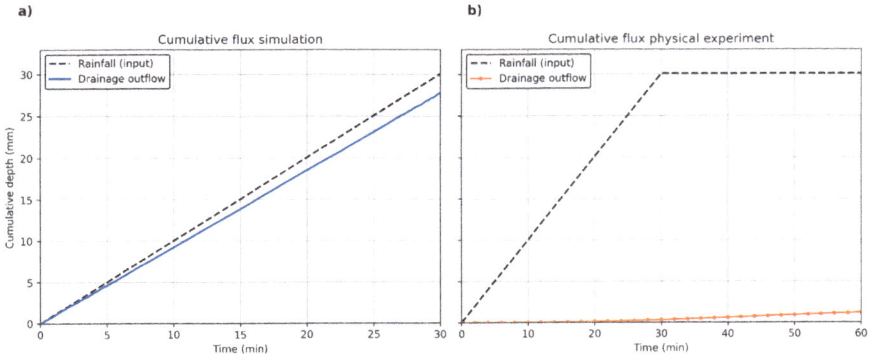

Fig. 4 Simulation (**a**) and physical experiment (**b**) cumulative flux comparison

because if water does not infiltrate the soil, it will be transported through surface runoff and flows to the sewerage. This leads to the same result as an impermeable zone.

To better understand what is missing and how to better represent the hydraulic response of the green roof, it is important to understand and analyse water storage, as this factor has yet to be accounted for. Figure 5 shows the amount of water retained in the soil during the simulation, showing an increase over the course of the simulation, but with some occasional drops and a maximum value of 0.52 mm.

4 Discussion and Implications

The slightly lower accumulated outflow in the simulation, whereas in the physical model experiment there was a large difference, may reflect simplifications in the hydraulic properties of the soil. This indicates that, although the numerical model captures the general dynamics of green roof runoff, water storage capacity is not yet being represented. Therefore, the model does not yet represent runoff in soil that has many particularities, something that the porous medium alone was unable to represent.

The retention effect cannot be simplified or ignored, since this deficit leads to an underrepresentation of the usefulness of green roofs as instruments of resistance to the effects of climate change. Making it necessary to implement a water storage model, like with the application of the Richards equation that represents the movement of water in unsaturated soils (Richards, 1931).

However, despite this problem, the implementation of the porous medium showed promising results as it still allowed water to drain. Since there was no surface runoff, the water was able to infiltrate the soil, which shows that even for such high resistance values, the high porosity did not generate a wall effect that could prevent water from entering the region.

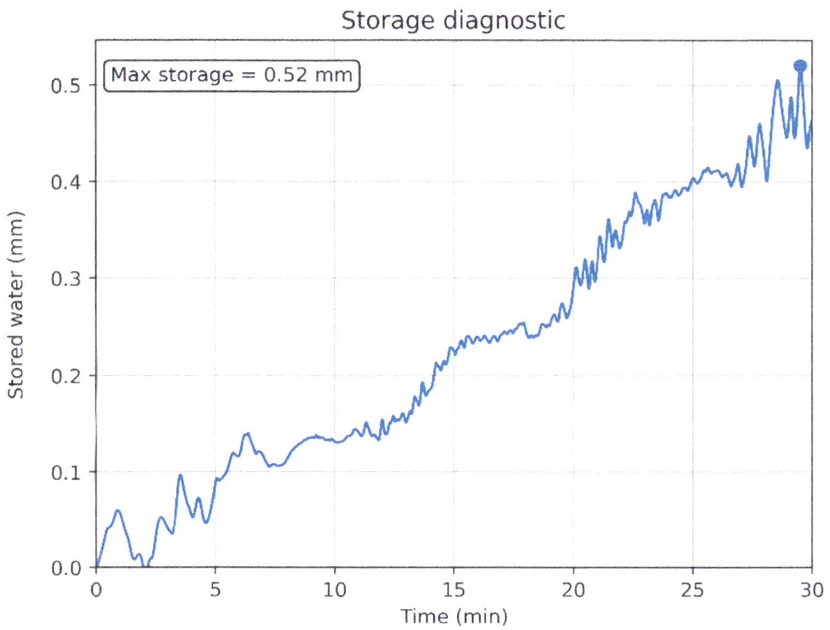

Fig. 5 Water stored inside the soil

Another important future step is the model's ability to simulate soil behaviour after the rain has stopped. This proves to be a challenge, as there is usually an abrupt halt in physical experiments due to the use of constant input. This sudden stop can cause instability in the model, which can generate errors such as water reflux from the outlets. One possible solution is to implement a gradual end to the event, more similar to real cases, which results in gentler pressure changes in the system.

It is important to understand the behaviour after the rain has stopped because water continues to drain through the substrate. This drainage usually decreases gradually, not instantly, and thus understanding the shape of this recession curve reflects internal storage processes. This behaviour is essential for understanding important parameters of green roof efficiency, such as peak runoff timing, drainage efficiency and the risks of prolonged soil saturation. Without post-rain observation, only the input is measured, not the system's response.

5 Concluding Remarks

Green roofs can increasingly be adopted as a sustainable solution for stormwater management in urban environments, storing and delaying the discharge of water into the urban drainage system. The issue is that the hydraulic response of green roofs is influenced by processes such as preferential flow paths, vegetation effects and spatial heterogeneity. These factors are complex and difficult to represent in numerical models. However, despite these limitations, representing groundwater storage through computational models remains essential. Therefore, even if the model does not fully reproduce all the hydraulic details of a green roof system, it still creates the possibility for advances in research, providing a basis that can be improved upon.

The model showed good initial results, as the porous medium can represent water infiltration into the soil and thus begin to fill the gap in hydraulic models of green roofs. In continuing the research, the next step is to implement additional parameters to represent the microscopic relationships between soil pores and water. Through this implementation, the model will be able to represent not only infiltration but also water retention, thus fully representing green roofs.

Acknowledgements The author acknowledges the financial support provided through a doctoral scholarship from the University of Naples Federico II. The construction of the physical green roof model was supported by funding from the European Union. The author also acknowledges the support of the Department of Civil, Building and Environmental Engineering of the University of Naples Federico II for providing laboratory facilities and technical assistance during the experimental work.

References

Best, A. C. (1950). The size distribution of raindrops. *Quarterly Journal of the Royal Meteorological Society, 76*(327), 16–36. https://doi.org/10.1002/qj.49707632704

Castelo, S., Amado, M., & Ferreira, F. (2023). Challenges and opportunities in the use of nature-based solutions for urban adaptation. *Sustainability, 15*(9), 7243. https://doi.org/10.3390/su15097243

Chen, J., Theller, L., Gitau, M. W., Engel, B. A., & Harbor, J. M. (2017). Urbanization impacts on surface runoff of the contiguous United States. *Journal of Environmental Management, 187*, 470–481. https://doi.org/10.1016/j.jenvman.2016.11.017

Gibson, I., Campbell, I., & Gao, Z. (2004). A comparative study of virtual prototyping and physical prototyping. *International Journal of Manufacturing Technology and Management, 6*(6), 503–522. https://doi.org/10.1504/IJMTM.2004.005931

Gunn, R., & Kinzer, G. D. (1949). The terminal velocity of fall for water droplets in stagnant air. *Journal of Atmospheric Sciences, 6*(4), 243–248.

Harper, O. (2024). Assessing the influence of soil composition on plant growth and development in USA. *American Journal of Physical Sciences, 2*(1), 61–72. https://doi.org/10.47604/ajps.2665

Hölscher, K., Frantzeskaki, N., Collier, M. J., Connop, S., Kooijman, E. D., Lodder, M., McQuaid, S., Vandergert, P., Xidous, D., Bešlagić, L., Dick, G., Dumitru, A., Dziubała, A., Fletcher, I., Garcia-Espina Adank, C., González Vázquez, M., Madajczyk, N., Malekkidou, E., Mavroudi, M., et al. (2023). Strategies for mainstreaming nature-based solutions in urban governance capacities in ten European cities. *npj Urban Sustainability, 3*(1), Article 54. https://doi.org/10.1038/s42949-023-00134-9

Johnson, F. T., Tinoco, E. N., & Yu, N. J. (2005). Thirty years of development and application of CFD at Boeing commercial airplanes. *Seattle. Computers & Fluids, 34*(10), 1115–1151. https://doi.org/10.1016/j.compfluid.2004.06.005

Kabisch, N., Frantzeskaki, N., Pauleit, S., Naumann, S., Davis, M., Artmann, M., Haase, D., Knapp, S., Korn, H., Stadler, J., Zaunberger, K., & Bonn, A. (2016). Nature-based solutions to climate change mitigation and adaptation in urban areas: Perspectives on indicators, knowledge gaps, barriers, and opportunities for action. *Ecology and Society, 21(2), Article 39.* https://doi.org/10.5751/ES-08373-210239

Manso, M., Teotónio, I., Matos Silva, C., & Oliveira Cruz, C. (2021). Green roof and green wall benefits and costs: A review of the quantitative evidence. *Renewable and Sustainable Energy Reviews, 135, Article 110111.* https://doi.org/10.1016/j.rser.2020.110111

Mansur, A. V., McDonald, R. I., Güneralp, B., Kim, H., Puppim de Oliveira, J. A., Callaghan, C. T., Hamel, P., Kuiper, J. J., Wolff, M., Liebelt, V., Martins, I. S., Elmqvist, T., & Pereira, H. M. (2022). Nature futures for the urban century: Integrating multiple values into urban management. *Environmental Science & Policy, 131*, 46–56. https://doi.org/10.1016/j.envsci.2022.01.013

Mihalakakou, G., Souliotis, M., Papadaki, M., Menounou, P., Dimopoulos, P., Kolokotsa, D., Paravantis, J., Tsangrassoulis, A., Panaras, G., Giannakopoulos, E., & Papaefthimiou, S. (2023). Green roofs as a nature-based solution for improving urban sustainability: Progress and perspectives. *Renewable and Sustainable Energy Reviews, 180*, 113306. https://doi.org/10.1016/j.rser.2023.113306

Mori, E., Di Lorenzo, T., Viviano, A., Jakovljević, T., Marra, E., Moura, B. B., Garosi, C., Manzini, J., Ancillotto, L., Hoshika, Y., & Paoletti, E. (2025). Under pressure: Environmental stressors in urban ecosystems and their ecological and social consequences on biodiversity and human Well-being. *Stress, 5*(4), 66. https://doi.org/10.3390/stresses5040066

Myhre, G., Alterskjær, K., Stjern, C. W., Hodnebrog, Ø., Marelle, L., Samset, B. H., Sillmann, J., Schaller, N., Fischer, E., Schulz, M., & Stohl, A. (2019). Frequency of extreme precipitation increases extensively with event rareness under global warming. *Scientific Reports, 9*(1), 16063. https://doi.org/10.1038/s41598-019-52277-4

Olabi, A. G., Shehata, N., Issa, U. H., Mohamed, O. A., Mahmoud, M., Abdelkareem, M. A., & Abdelzaher, M. A. (2025). The role of green buildings in achieving the sustainable development goals. International Journal of Thermofluids, *25*, Article 101002. doi:https://doi.org/10.1016/j.ijft.2024.101002.

OpenCFD (2025) OpenFOAM v2412 user guide. https://www.openfoam.com/documentation/user-guide

Qiao, W., Xie, Y., Liu, J., & Huang, X. (2025). The impacts of urbanization on carbon emission performance: New evidence from the Yangtze River Delta urban agglomeration. *China. Land, 14*(1), 12. https://doi.org/10.3390/land14010012

Radford, K. G., & James, P. (2013). Changes in the value of ecosystem services along a rural–urban gradient: A case study of greater Manchester, UK. *Landscape and Urban Planning, 109*(1), 117–127. https://doi.org/10.1016/j.landurbplan.2012.10.007

Richards, L. A. (1931). Capillary conduction of liquids through porous mediums. *Physics, 1*(5), 318–333. https://doi.org/10.1063/1.1745010

Shi, L., & Moser, S. (2021). Transformative climate adaptation in the United States: Trends and prospects. *Science, 372*(6549), eabc 8054. https://doi.org/10.1126/science.abc8054

Stein, J., McLaughlin, J., & Bendernagle, L. (2012). Rooftop control of stormwater: A sustainable source control strategy in NYC using blue roof and green roof applications. *Water Resources IMPACT, 14*(4), 10–12. http://www.jstor.org/stable/wateresoimpa.14.4.0010

World Meteorological Organization. (n.d.). *Aviation—hazards—precipitation.* Retrieved September 20, 2025, from https://community.wmo.int/en/activity-areas/aviation/hazards/precipitation

Multicriteria Evaluation of the Technical Capacity of the Green Walls of the Existing Building: Application to the Campus of the University of Palermo

Mariantonietta Ruggieri, Marco Bellomo, Luisa Lombardo, Tiziana Campisi, and Simona Colajanni

Abstract Urban areas are increasingly exposed to climate-related pressures, including rising temperatures, air pollution, and the intensification of urban heat island effects, which undermine environmental quality, public health, and resilience. Nature-Based Solutions (NBS) are recognised as effective strategies to address these challenges, but implementation on existing buildings is often limited by regulatory uncertainties and the lack of operational tools for early-stage decisions. Green walls offer significant potential in dense contexts where horizontal green space is scarce, yet their feasibility depends on structural, geometric, regulatory, and infrastructural constraints. This chapter introduces an indicator-based multicriteria framework for the preliminary assessment of green wall feasibility on existing buildings. Seventeen technical and contextual indicators—covering constructive and structural characteristics, regulatory conditions, geometric and operational aspects, and enabling infrastructures—are normalised and aggregated into a composite feasibility index, supporting transparent comparison between buildings and the identification of priority candidates for vertical greening. The methodology is applied to the University of Palermo campus, characterised by diverse building typologies, façade technologies, orientations, and accessibility conditions. The resulting feasibility profiles

---

M. Ruggieri
Department of Economics, Business and Statistics, University of Palermo, Palermo, Italy
e-mail: mariantonietta.ruggieri@unipa.it

M. Bellomo · L. Lombardo · T. Campisi · S. Colajanni (✉)
Department of Architecture, University of Palermo, Palermo, Italy
e-mail: simona.colajanni@unipa.it

M. Bellomo
e-mail: marco.bellomo01@unipa.it

L. Lombardo
e-mail: luisa.lombardo01@unipa.it

T. Campisi
e-mail: tiziana.campisi@unipa.it

L. Diana et al. (eds.), *Greening and Resilience*, Urban Sustainability,
https://doi.org/10.1007/978-981-92-0419-9_11

highlight that suitability for green walls emerges from the combined behaviour of multiple prerequisites rather than a single dominant factor. The framework does not prescribe specific technologies but provides a flexible technical layer that can be integrated into different decision-making schemes and adapted to other urban contexts, offering a scalable tool to support the integration of green walls as NBS within broader climate adaptation and sustainable campus development strategies.

Keywords Multicriterial framework · Green walls · Urban resilience and sustainable cities · Technical methodology feasibility · Transformability index · Greening application

1 Introduction

Urban environments are increasingly exposed to complex and interrelated challenges linked to climate change, environmental degradation, air pollution, and the intensification of the urban heat island effect. These phenomena directly affect not only environmental quality but also public health, social well-being, and the long-term resilience of cities. In response to these pressures, Nature-Based Solutions (NBS) have gained growing attention in both research and policy agendas as adaptive strategies capable of enhancing urban resilience while delivering multiple co-benefits across environmental, social, and economic dimensions (Alves et al., 2024; Bayissa et al., 2025). Among NBS, green walls—also referred to as vertical greening systems or vegetated façades—represent a particularly promising strategy due to their capacity to transform otherwise inert building envelopes into active environmental infrastructures (Fig. 1). By integrating vegetation into the vertical surfaces of buildings, green walls can contribute to microclimatic regulation, air quality improvement, biodiversity support, and enhanced thermal and visual comfort. These interlinked benefits and feedbacks are summarised in the conceptual framework shown in Fig. 1, which maps how Nature-Based Solutions can simultaneously address environmental risks and social vulnerabilities at different urban scales. Moreover, their application is especially relevant in dense urban contexts where horizontal green space is limited, making vertical surfaces a strategic resource for urban greening (Marsaglia, 2024; Xi et al., 2022).

Despite their recognised potential, the implementation of green walls remains uneven and often confined to isolated pilot projects. One of the main barriers to their widespread adoption lies in the absence of structured decision-support tools capable of guiding designers, planners, and public authorities in selecting suitable buildings and appropriate technological solutions. Existing studies frequently focus on individual performance aspects—such as energy savings, cooling effects, or aesthetic impact—without integrating technical feasibility, economic sustainability, ecological performance, and social acceptance within a unified evaluation framework. This fragmentation limits the scalability and replicability of green wall interventions, particularly in public or institutional contexts.

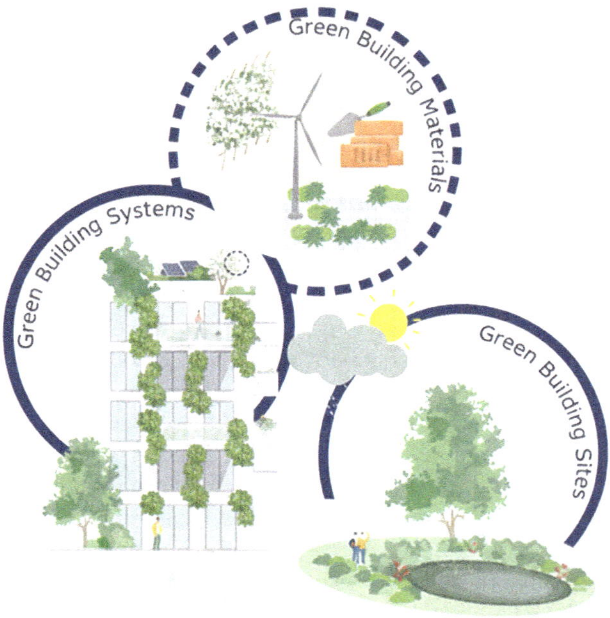

Fig. 1 Conceptual framework of the co-benefits generated by Nature-Based Solutions at building and district scale, highlighting how vertical greening strategies can enhance thermal comfort, biodiversity, urban liveability, and overall climate resilience

In this regard, multicriteria decision-making approaches offer a valuable methodological basis for addressing the inherent complexity of NBS evaluation. By allowing heterogeneous indicators to be structured, weighted, and aggregated, multicriteria frameworks enable transparent comparison between alternative solutions while explicitly accounting for stakeholder priorities. However, their application to green wall feasibility assessment at the urban or campus scale remains relatively underexplored, especially when social and ecological dimensions are considered alongside technical and economic factors (Barker et al., 2024; Kwak et al., 2025).

This chapter addresses this gap by proposing a holistic multicriterial framework for assessing the feasibility of green wall implementation in urban environments. The framework integrates technical, environmental, economic, and social indicators into a composite evaluation index designed to support evidence-based decision-making. The methodology is applied to the University of Palermo campus, a complex and heterogeneous urban system comprising multiple building typologies and a high density of users. Through this application, the chapter demonstrates how green walls can be strategically planned as scalable NBS, supporting urban resilience, climate adaptation, and sustainable campus development.

The contribution proposes a multicriteria framework oriented towards the preliminary screening of technical feasibility, aimed at supporting the selection and prioritisation of candidate buildings on the basis of structural prerequisites, regulatory constraints, and operational conditions (Ebrahimnejad et al., 2017). Nature-Based Solutions (NBS) applied to the existing building stock often require decisions to be made in the presence of multiple and potentially conflicting objectives (e.g. technical constraints, operational requirements, planning priorities, and regulatory limitations). Under such conditions, multicriteria decision-making methods (MCDM/MCDA) provide a robust methodological reference, as they allow heterogeneous information (quantitative and qualitative) to be integrated within a coherent and transparent evaluation framework. A multicriteria approach makes it possible to (i) structure the set of relevant factors into explicit indicators, (ii) make different alternatives comparable through normalisation procedures, and (iii) aggregate results in a traceable manner, making the effect of weighting choices explicit. This type of setting is particularly useful when the objective is not to estimate the final performance of a technology with high precision, but rather to support a preliminary and comparative selection of candidate buildings, reducing the arbitrariness of choices and making the process replicable.

In this contribution, technical feasibility is defined as the possibility, at an early stage, to consider a building suitable for green wall implementation on the basis of prerequisites and constraints that can be verified through survey data and available technical documentation (Geneletti, 2007). More specifically, technical feasibility is articulated into four main groups of aspects: (1) structural compatibility and constructive characteristics of the façade; (2) regulatory and permitting constraints that may prevent or condition the intervention; (3) geometric and operational conditions influencing installation and maintenance (e.g. height, accessibility, usable surface); and (4) availability of enabling facilities and infrastructures (e.g. water supply and service areas). Consistent with the objective of supporting planning and prioritisation, the proposed evaluation must be understood as an early-stage screening: it does not replace detailed design checks (structural, plant-related, or technological), but rather orients the selection of buildings on which to concentrate more in-depth analyses at subsequent stages.

2 Methodology

This study adopts a multicriterial evaluation framework to assess the feasibility of green wall implementation in urban environments, integrating technical, environmental, economic, and social dimensions within a unified decision-support structure. The methodological approach is designed to support early-stage planning and strategic decision-making, particularly in complex institutional contexts where multiple objectives and constraints coexist.

2.1 Case Study Application: University of Palermo Campus

The proposed indicator-based framework was applied to the University of Palermo (UNIPA) campus, selected as a representative urban environment characterised by a high degree of morphological, functional, and temporal heterogeneity (Fig. 2).

The campus hosts academic buildings, laboratories, administrative facilities, and collective spaces, distributed across a large area and characterised by varying construction periods, façade technologies, and exposure conditions. This diversity makes the campus particularly suitable for testing a feasibility-oriented approach to green wall implementation at a multi-building scale. The campus includes both historical and more recent buildings, offering a wide range of construction typologies and regulatory conditions. This variability allows the framework to capture different levels of structural compatibility, accessibility, and constraint, which are decisive factors in the preliminary selection of buildings suitable for vertical greening interventions. In addition, the high daily pedestrian flow and the presence of open spaces between buildings amplify the potential social and environmental impact of green wall applications.

Each building was treated as an independent evaluation unit. Data were collected through on-site surveys and analysis of available technical documentation, focusing

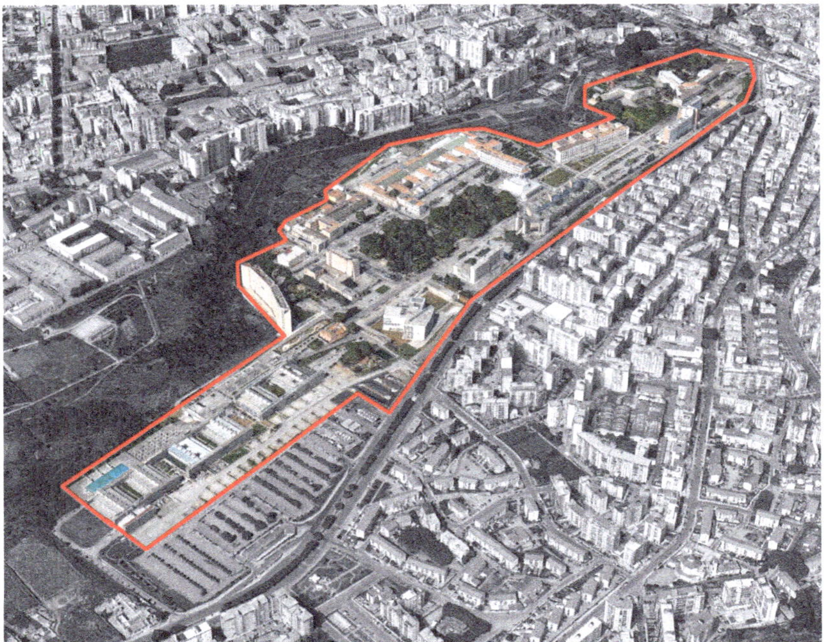

Fig. 2 Aerial view of the University of Palermo Campus, illustrating its integration within the surrounding urban fabric and the extent of the analysed area

on parameters directly related to the defined indicators. These included construction year, façade construction technology, estimated structural capacity, building height, state of conservation, and the percentage of usable façade surface for each orientation. Façade exposure was explicitly considered through separate indicators to reflect the differentiated microclimatic potential associated with solar radiation and prevailing wind conditions in the Mediterranean context (Bellomo et al., 2025). Accessibility was assessed by estimating the proportion of façade surface that could be safely reached for installation and maintenance operations, taking into account surrounding open spaces, existing access systems, and operational constraints. The availability of ground-level areas and water supply infrastructure was also verified, as these elements directly influence the feasibility of irrigation systems and long-term maintenance strategies. Figure 3 shows the plan of the University of Palermo campus with the buildings selected as priority candidates for different greening strategies (vertical green walls, entrance buffer areas, pergolas, terraces, and parking-related solutions). These buildings were subsequently documented through the standardised survey sheet described in Sect. 2.1, in order to derive the full set of technical feasibility indicators (I_1–I_{17}) used in the following analysis.

All collected data were organised into a structured dataset and processed using the normalisation and aggregation procedures described in the previous section. By applying a consistent indicator set and weighting scheme across the selected buildings, the framework enabled a comparative assessment of green wall feasibility at the campus scale. The resulting composite scores support the identification of priority buildings for pilot interventions and provide a strategic overview of the potential role of green walls within the broader sustainability and resilience agenda of the University of Palermo campus.

Fig. 3 Mapping of the buildings within the University of Palermo campus identified as priority targets for greening interventions, highlighting the different types of proposed measures (vertical greening systems, entrance buffer areas, pergolas, terraces, and parking-related solutions)

2.2 Multicriterial Framework Structure

The methodological approach adopted in this study is based on a set of technical and contextual indicators specifically designed to evaluate the feasibility of green wall implementation on existing buildings. Rather than structuring the evaluation through a hierarchical requirements tree, the framework directly operates at the indicator level, allowing a faster and more operational assessment suitable for early-stage screening and comparative analysis. The selected indicators focus primarily on constructive characteristics, regulatory feasibility, geometric constraints, exposure conditions, and infrastructural availability, which are considered decisive factors in determining whether a building is suitable for vertical greening interventions. This approach reflects the objective of identifying priority buildings for pilot applications rather than evaluating detailed performance at the system level (Kandakoglu et al., 2022; Rey, 2004).

The selection of indicators was carried out by combining evidence from the literature on recurrent requirements and constraints for the implementation of vertical greening systems on existing buildings, operational requirements relevant for preliminary screening (installation and maintenance), and the availability and verifiability of data through on-site surveys and technical documentation. The objective is not to estimate in detail the energy or environmental performance of the system, but rather to transparently identify prerequisites and enabling conditions that make a building a suitable and prioritizable candidate at an early stage. Seventeen indicators were defined, combining quantitative and qualitative/binary variables, in order to provide an operational description of the technical feasibility of implementing green walls on existing buildings (Table 1). Quantitative indicators were normalised to ensure comparability across different units of measurement, while qualitative indicators were converted into numerical scores using predefined discrete values (Abbas et al., 2023).

The data underlying these indicators were collected through a standardised survey sheet, compiled for each building/external area. The sheet includes geographical localisation (territorial context, coordinates, elevation, and climatic data), administrative localisation (building name, year of construction, ownership, and planning constraints), current land use, external accessibility and surrounding services, the position of the building within the urban fabric or the campus, typological and morphological data, dimensional data, functional and technological characteristics, state of conservation, the possible presence of existing greening strategies, and a section dedicated to qualitative evaluations and a transformability index. The survey sheet is organised into standardised sections specifically designed to support the compilation of the indicator set I_1–I_{17}:

- Geographical localisation: includes the territorial context of the building (province, municipality, district, address), geographical coordinates (latitude, longitude, elevation above sea level), and main climatic reference data.

Table 1 Indicator set adopted for the technical feasibility assessment of green wall implementation, including indicator type, unit or scale, and evaluation and normalisation method

Indicator ID	Indicator description	Indicator type	Unit/scale	Evaluation and normalisation method
I_1	Year of construction	Quantitative	Year	Min–max normalisation
I_2	Façade construction technology	Qualitative	Discrete classes	Ordinal scoring (e.g. load-bearing masonry = 0)
I_3	Structural capacity	Quantitative	kN·m (or equivalent)	Min–max normalisation
I_4	Regulatory and permitting feasibility	Binary	Yes/no	Yes = 1; no = 0
I_5	State of façade conservation	Qualitative	Ordinal scale	Ordinal scoring (very poor = 0)
I_6	Building height	Quantitative	m	Min–max normalisation
I_7	Usable façade surface—North orientation	Quantitative	%	Min–max normalisation
I_8	Usable façade surface—South orientation	Quantitative	%	Min–max normalisation
I_9	Usable façade surface—West orientation	Quantitative	%	Min–max normalisation
I_{10}	Usable façade surface—East orientation	Quantitative	%	Min–max normalisation
I_{11}	Usable façade surface—North-west orientation	Quantitative	%	Min–max normalisation
I_{12}	Usable façade surface—North-east orientation	Quantitative	%	Min–max normalisation
I_{13}	Usable façade surface—South-west orientation	Quantitative	%	Min–max normalisation
I_{14}	Usable façade surface—South-east orientation	Quantitative	%	Min–max normalisation
I_{15}	Accessibility for installation and maintenance	Quantitative	%	Min–max normalisation

(continued)

Table 1 (continued)

Indicator ID	Indicator description	Indicator type	Unit/scale	Evaluation and normalisation method
I_{16}	Ground-level service	Quantitative	m^2	Min–max normalisation
I_{17}	Water supply availability	Binary	Yes/no	Yes = 1; no = 0

- Administrative localisation: reports the denomination of the building, year of construction, ownership, and key information on planning constraints (volume, building density, possible landscape, hydrogeological and seismic constraints).
- Current land use: specifies the prevailing functions hosted in the building (classrooms, laboratories, offices, services, collective spaces, other functions), which help to characterise the use context and the intensity of occupancy.
- External accessibility and surrounding services: records the number and type of pedestrian and vehicular accesses, the width of the main road, the presence of parking areas and equipped green spaces, and the availability of nearby services and of public transport in the vicinity.

Based on the data collected through this survey sheet, the values of the 17 indicators I_1–I_{17} reported in Table 1 were systematically derived for each building, forming the baseline dataset for the subsequent normalisation steps and for the construction of summary result tables. Structural and technological information feed the indicators related to the constructive context (I_1–I_3, I_5–I_6); checks on planning and regulatory constraints feed the regulatory feasibility indicator (I_4); measurements of façade surfaces by orientation, obtained from dimensional data and elevation drawings, allow the indicators I_7–I_{14} to be calculated; information on accesses and external spaces is used to estimate operational accessibility (I_{15}); and finally, data on service areas and water infrastructures feed the indicators I_{16}–I_{17}.

The systematic compilation of the survey sheet for all the analysed buildings therefore produces, for each unit of analysis, a complete set of 17 raw values, organised into a dataset that underpins the subsequent normalisation phases and the construction of synthesis tables of the results:

- General data: building ID, prevailing function, year of construction (I_1), building height (I_6), and main typological information.
- Structure and façade: façade construction technology (I_2), estimated structural capacity (I_3), and state of conservation (I_5), derived from technical documentation and visual inspection.
- Regulatory framework: outcome of the regulatory/permitting feasibility check (I_4), obtained through the assessment of planning constraints and consultation with the competent heritage and planning authorities, coded as a binary variable (1 = feasible, 0 = not feasible).

- Façade surfaces: estimation, for each considered orientation (I_7–I_{14}), of the percentage of façade surface that is effectively usable for the installation of green walls, based on geometric survey and on the presence of openings, obstacles, and interferences with other elements.
- Operational accessibility: assessment of the percentage of façade surface that can be safely reached for installation and maintenance operations (I_{15}), compiled on the basis of on-site surveys and analysis of the surrounding open spaces.
- Ground-level facilities and infrastructures: measurement of available service areas at ground level (I_{16}) and verification of the presence of a usable water supply network (I_{17}), the latter coded as a binary indicator.

The systematic compilation of this survey sheet for all campus buildings makes it possible to derive, in a consistent and comparable way, the full set of indicators I_1–I_{17} for each unit of analysis, thereby creating the database on which the subsequent normalisation procedures and, where applicable, multicriteria aggregation are applied.

For the sake of replicability, each indicator was associated with: (i) a data source (on-site survey, technical documentation, direct observation), (ii) a calculation or coding rule (formula or discrete criterion), and (iii) an output expressed in units/scale consistent with Table 1. Quantitative indicators (e.g. year of construction, height, percentage of usable façade surface, accessibility, service areas) were derived from surveys and/or processing of geometric data and expressed in their respective units of measurement; they were then made comparable through min–max normalisation to the interval. Binary indicators (e.g. regulatory feasibility, water availability) were coded as 1 when the condition was satisfied and 0 otherwise. Qualitative/ordinal indicators (e.g. façade technology, state of conservation) were converted into discrete scores according to predefined criteria and, where necessary, rescaled to the same interval to ensure consistency within the multicriteria aggregation.

Façade orientation is treated through multiple indicators, each corresponding to a specific exposure (N, S, E, W, NE, NW, SE, SW). This allows the framework to capture the differentiated climatic and solar conditions associated with each orientation, which are particularly relevant in Mediterranean contexts (Perrucci et al., 2022). To limit possible distortions due to the presence of multiple orientation-related indicators (I_7–I_{14}), the weighting system was defined so that the overall weight assigned to the "façade/orientation" group remains comparable with that of the other families of indicators (structural, regulatory, operational, and infrastructural). In addition, a sensitivity check was carried out by comparing the ranking obtained with an alternative scenario in which the sub-orientations (NE, NW, SE, SW) receive lower weights than the main orientations. Compared to contributions mainly focused on the definition and construction of composite indices, this chapter places the emphasis on the operationalisation of the indicator set: survey procedures, coding rules, and calculation criteria required to make the application replicable across a portfolio of existing buildings in a screening phase.

On the basis of the data collected through the survey sheet, the values of the 17 indicators I_1–I_{17} reported in Table 1 were systematically derived for each building and organised into a single dataset, which constitutes the basis for subsequent normalisation and multicriteria analysis. Rather than providing a definitive multicriteria ranking, the resulting indicator dataset is conceived as a structured basis that can be used within different decision-support schemes (such as composite indices, scenario-based weighting, or clustering procedures), depending on the specific objectives and stakeholder preferences. Binary indicators were coded as 1 (condition satisfied) or 0 (condition not satisfied), while ordinal qualitative indicators were converted into numerical values based on predefined evaluation criteria. Once normalised, all indicators were aggregated using a weighted summation approach to obtain a composite feasibility score for each building. The weighting system reflects a predefined evaluation scenario aligned with the objectives of the study and the specific characteristics of the case study context. The aggregation allows buildings to be ranked according to their overall suitability for green wall implementation, supporting the identification of priority candidates for pilot interventions (Pons-Valladares et al., 2023; Xidonas et al., 2022).

This indicator-based methodology provides a flexible and replicable framework that can be easily adapted to other urban contexts by modifying indicator definitions, normalisation thresholds, or weighting schemes, without altering the overall evaluation logic.

3 Results

The indicator-based framework was applied to all the buildings of the University of Palermo campus, with the aim of describing and comparing their technical feasibility profile for potential green wall implementation. For each building, the raw values collected through the survey sheets were first organised into a coherent dataset covering the full set of indicators I_1–I_{17}. Quantitative indicators were normalised using a min–max transformation, expressed as, which scales values between 0 and 1 while preserving relative differences within the dataset. Qualitative indicators were translated into numerical values using predefined ordinal scores, while binary indicators—such as regulatory feasibility and water supply availability—were coded as 1 when the condition was satisfied and 0 otherwise. The resulting normalised matrix provides, for each building, a multi-indicator technical feasibility profile, combining structural and constructive parameters—such as year of construction (I_1), façade construction technology (I_2), structural capacity (I_3), regulatory feasibility (I_4), state of façade conservation (I_5), and building height (I_6)—with technological and architectural indicators related to usable façade surface by orientation (I_7–I_{14}), accessibility (I_{15}), availability of ground-level service areas (I_{16}), and water supply (I_{17}). This representation makes it possible to identify, at a glance, buildings that simultaneously satisfy key prerequisites (e.g. sufficient structural capacity and positive regulatory conditions) and exhibit favourable façade and infrastructural conditions for vertical greening. Instead of prescribing a single aggregated feasibility score or a definitive ranking, the normalised indicator set is intended to function as a flexible input layer for different multicriteria decision-making procedures. Depending on the chosen weighting scheme or decision rule, the same indicator matrix can support alternative evaluations—such as composite feasibility indices, scenario-based prioritisation, or clustering of buildings with similar feasibility patterns—while maintaining full transparency on the contribution of each indicator. From a purely descriptive point of view, the indicator profiles already highlight clear differences among campus buildings. Some buildings combine favourable regulatory conditions, adequate structural performance, full availability of water supply, and consistently high percentages of usable façade surfaces across several orientations, thus emerging as robust candidates for future pilot interventions. Others show more unbalanced profiles, where good performance in isolated indicators is offset by critical weaknesses in structural, regulatory, accessibility, or infrastructural dimensions. This variability confirms that technical feasibility is not driven by a single dominant parameter, but by the combined behaviour of multiple prerequisites and enabling conditions, and illustrates the added value of using a structured indicator set as a basis for subsequent multicriteria analysis rather than relying on ad-hoc or purely qualitative judgements.

4 Conclusion

This paper proposed an indicator-based framework for the preliminary characterisation of the technical feasibility of green wall implementation in complex urban environments. By focusing on a structured yet operational set of indicators, the methodology allows heterogeneous building characteristics—structural, geometric, regulatory, and infrastructural—to be collected, organised, and normalised within a unified and transparent data structure. The application to the University of Palermo campus demonstrates how the framework can be effectively used to build a consistent feasibility dataset at the campus scale, supporting early-stage planning and the identification of buildings that present favourable technical conditions for potential pilot interventions. The analysis of the indicator profiles confirms that technical feasibility is not determined by a single dominant factor, but by the combined behaviour of prerequisite conditions—such as structural capacity and regulatory feasibility—and enabling factors, including façade usability, accessibility, and the availability of water and service areas. Buildings that simultaneously satisfy these key prerequisites and exhibit favourable façade and infrastructural conditions emerge as promising candidates for further investigation, while cases characterised by critical weaknesses in one or more dimensions highlight where additional studies or alternative strategies may be required. Rather than prescribing specific technological solutions or a single aggregated feasibility index, the proposed framework provides a flexible technical layer that can be embedded within different multicriteria decision-making approaches. The normalised indicator set can be used to construct composite indices, explore alternative weighting scenarios, or cluster buildings according to their feasibility patterns, while preserving full transparency on the contribution of each indicator. In this sense, the framework is particularly suited to institutional and campus-scale contexts, where multiple buildings must be evaluated coherently under shared sustainability objectives and limited resources.

The methodology is inherently flexible and transferable. Indicator definitions, normalisation thresholds, and the choice of aggregation rules can be adapted to different climatic contexts, planning priorities, or policy goals without altering the underlying structure of the framework. This adaptability makes the approach applicable beyond the specific case study, offering a replicable technical basis for supporting the integration of green walls as Nature-Based Solutions within broader urban resilience and climate adaptation strategies. Future developments should focus on extending the indicator set to explicitly include environmental performance and social dimensions, and on coupling the technical feasibility dataset with fully integrated multicriteria models and stakeholder-driven weighting schemes. Combined with monitoring data and post-implementation evaluation, this would further strengthen the link between strategic planning and the operational implementation of vertical greening interventions in sustainable urban environments.

Acknowledgements This research was funded by the European Union EU Next Generation Italian PRIN/PNRR 2022 Project titled "Greenwork: An interdisciplinary framework for urban health and urban resilience enhancement based on greening strategies on buildings and open spaces"—Project Code MUR P2022MNKEC 003—Sector: ERC SH7—CUP:B53D23029580001—PRJ-1358.

References

Abbas, A. E., & Hupman, A. C. (2023). Scale dependence in weight and rate multicriteria decision methods. *European Journal of Operational Research*. https://doi.org/10.1016/j.ejor.2022.12.038

Alves, R. A., dos Santos, M. M., Rudke, A. P., Francisquetti Venturin, P. R., & Martins, J. A. (2024). Site selection for nature-based solutions for stormwater management in urban areas: An approach combining GIS and multi-criteria analysis. *Journal of Environmental Management, 359*. https://doi.org/10.1016/j.jenvman.2024.120999

Barker, A., Garcia-Blanco, G., Garcia, I., & Aguirre-Such, A. (2024). The role of strategic planning in nature- based solutions (NBS) transformation: An evaluation of the green cities framework in mainstreaming NBS in 6 European countries. *Nature-Based Solutions, 6*, 100157. https://doi.org/10.1016/j.nbsj.2024.100157

Bayissa, Y., Srinivasan, R., Hunink, J., Nyolei, D., Moges, S., de Andrade Costa, D., Tadesse, D., Melesse, A., & Tilahun, S. (2025). Evaluating the potential of nature-based solutions to mitigate land use and climate change impacts on the hydrology of the Gefersa and Legedadi watersheds in Ethiopia. *Journal of Hydrology: Regional Studies, 57*. https://doi.org/10.1016/j.ejrh.2024.102130

Bellomo, M., & Colajanni, S. (2025). Multi-criteria assessment of vegetated walls: A scientific approach to nature-based solutions. *Energy and Buildings, 344*, 115973. https://doi.org/10.1016/J.ENBUILD.2025.115973

Ebrahimnejad, R., Noori, O., & Deihimfard, R. (2017). Mitigation potential of green structures on local urban microclimate using ENVI-met model. *International Journal of Urban Sustainable Development, 9*(3), 274–285. https://doi.org/10.1080/19463138.2017.1370424

Geneletti, D. (2007). An approach based on spatial multicriteria analysis to map the nature conservation value of agricultural land. *Journal of Environmental Management, 83*(2), 228–235. https://doi.org/10.1016/j.jenvman.2006.03.002

Kandakoglu, M., Walther, G., & Amor, S. B. (2022). A robust multicriteria clustering methodology for portfolio decision analysis. *Computers and Industrial Engineering, 174*. https://doi.org/10.1016/j.cie.2022.108803

Kwak, Y., & Chen, S. (2025). Integrating seasonal climate variability and spatial accessibility in ecosystem service value assessment for optimized NbS allocation. *Urban Climate, 59*. https://doi.org/10.1016/j.uclim.2025.102314

Marsaglia, V. (2024). Technological greenery. Exploring cutting-edge solutions for performant greenery integration in building envelope design. *Energy and Buildings, 324*. https://doi.org/10.1016/j.enbuild.2024.114920

Perrucci, D. V., Hassan, M. M., & Baroud, H. (2022). Planning for temporary housing through multicriteria decision analysis. *Journal of Housing and the Built Environment*. https://doi.org/10.1007/s10901-022-09952-2

Pons-Valladares, O., Casanovas-Rubio, M. del M., Armengou, J., & de la Fuente, A. (2023). Approach for sustainability assessment for footbridge construction technologies: Application to the first world D-shape 3D-printed fiber-reinforced mortar footbridge in Madrid. Journal of Cleaner Production, 394. doi:https://doi.org/10.1016/j.jclepro.2023.136369.

Rey, E. (2004). Office building retrofitting strategies: Multicriteria approach of an architectural and technical issue. *Energy and Buildings, 36*(4), 367–372. https://doi.org/10.1016/j.enbuild.2004.01.015

Xi, C., Ding, J., Wang, J., Feng, Z., & Cao, S. J. (2022). Nature-based solution of greenery configuration design by comprehensive benefit evaluation of microclimate environment and carbon sequestration. *Energy and Buildings, 270*. https://doi.org/10.1016/j.enbuild.2022.112264

Xidonas, P., & Steuer, R. (2022). A multicriteria evaluation methodology for assessing the impact of COVID-19 in EU countries. *Decision Analytics Journal, 4*. https://doi.org/10.1016/j.dajour.2022.100123

Inclusive Paths for the Experimentation of Ecological Solutions. Application of Participatory Strategies at the UNIPA Campus

Simona Colajanni, Marco Bellomo, Luisa Lombardo, and Tiziana Campisi

Abstract Mediterranean university campuses represent critical urban microclimates increasingly vulnerable to extreme heatwaves and intense solar radiation, which compromise the thermal comfort and functionality of outdoor environments. This research evaluates the efficacy of Nature-Based Solutions (NBS) as catalysts for the environmental and social reactivation of neglected academic spaces, using the chapel courtyard at the University of Palermo (UNIPA) as a primary case study. Characterized by extreme microclimatic conditions—with summer surface temperatures exceeding 46 °C—the site serves as a testing ground for climate-responsive design. Adopting a Participatory Action Research (PAR) framework, the study analyzes the "GREENWORK//LAB" initiative, an interdisciplinary collaboration involving academic researchers, industrial stakeholders, and student associations. The methodology integrates quantitative environmental monitoring (in-situ thermal data collection) with qualitative socio-spatial surveys and user-perception assessments. The research led to the development of several modular, low-impact, and reversible design prototypes, including hexagonal shading systems with integrated brise-soleil, multifunctional units equipped with photovoltaic cells, and circular economy modules utilizing recycled materials. Empirical results demonstrate a significant temperature differential between shaded and exposed areas, validating the performance of vegetated shading and passive cooling in mitigating localized thermal stress. Beyond technical performance, the study assesses the impact of participatory processes on community stewardship and environmental awareness. The dissemination of findings at SHARPER 2025 (European Researchers' Night) further illustrates the potential of

S. Colajanni (✉) · M. Bellomo · L. Lombardo · T. Campisi
Department of Architecture, University of Palermo, Palermo, Italy
e-mail: simona.colajanni@unipa.it

M. Bellomo
e-mail: marco.bellomo01@unipa.it

L. Lombardo
e-mail: luisa.lombardo01@unipa.it

T. Campisi
e-mail: tiziana.campisi@unipa.it

L. Diana et al. (eds.), *Greening and Resilience*, Urban Sustainability,
https://doi.org/10.1007/978-981-92-0419-9_12

citizen science in urban greenery management. This research concludes that small-scale, user-centered interventions offer a scalable and holistic model for enhancing campus resilience. It advocates for an evolutionary governance model that integrates student-driven experimentation into permanent institutional spatial planning.

Keywords Greening solutions · Design for all · Greening applications · Participatory approaches · Architectural engineering workshop

1 Introduction

The implementation of greening strategies within university environments has gained significant traction in recent years, evolving into a core component of broader reflections on sustainable development, urban resilience, and environmental well-being (Alves et al., 2024; Bayissa et al., 2025). Within the consolidated framework of Nature-Based Solutions (NBS), academic campuses are increasingly recognized as "living laboratories"—strategic nodes where ecological interventions can simultaneously address climate adaptation, social inclusion, and educational innovation (Babí Almenar et al., 2021; Frantzeskaki, 2019). This paradigm shift views the university not merely as a site for knowledge dissemination, but as a dynamic testing ground for urban regeneration strategies that prioritize human comfort and ecological performance.

Mediterranean Universities, in particular, face pressing challenges due to the intensification of the Urban Heat Island (UHI) effect, characterized by recurrent heatwaves and high solar radiation. These factors severely compromise the thermal comfort and usability of outdoor spaces, which are essential for the social and academic life of the community. In such settings, greenery serves a dual purpose: it transcends aesthetic ornamentation to act as a climate-responsive infrastructure capable of mitigating localized heat stress through evapotranspiration and shading, while simultaneously stimulating sustainable behaviors and activating underutilized spatial assets (Frantzeskaki et al., 2019; Xi et al., 2022).

Despite this growing awareness, a gap persists between theoretical planning and practical application. Many campuses still present fragmented and inconsistent distributions of green infrastructure, often disconnected from spatial functionality and the actual needs of users (Bayissa et al., 2025). This condition is evident at the University of Palermo (UNIPA), where the deployment of green systems is uneven and frequently uncoordinated with the accessibility of open spaces. Several zones remain underused or completely abandoned, especially where vegetation is scarce or inappropriately integrated into the architectural and climatic context (Pueyo-Ros et al., 2025).

Among these neglected areas, the chapel courtyard emerges as a particularly emblematic case study. Despite its significant architectural and symbolic value, the site is defined by harsh microclimatic conditions, a lack of shading, and a total absence of urban furniture. These factors have led to its progressive abandonment, preventing

Fig. 1 The chapel courtyard at the UNIPA Campus in its current state. Representative view of the area showing high solar exposure; Mapping of critical thermal zones and experimentation of shading infrastructure.@Author2025

it from functioning as a meaningful public space within the campus fabric. Empirical preliminary surveys at this site recorded summer surface temperatures exceeding 46 °C, highlighting a critical thermal threshold that demands urgent intervention.

This research seeks to explore the potential of small-scale, user-centered greening strategies as a catalyst for reactivating such "orphan" university spaces. The project is rooted in the hypothesis that the integration of modular vegetated systems and environmentally responsive design elements can significantly improve the usability, comfort (Fig. 1), and social relevance of outdoor academic environments (Ascione et al., 2020; Marsaglia, 2024). To support a rigorous scientific investigation, the research adopts a participatory design methodology—integrated within the "GREENWORK//LAB" framework—to bridge the gap between analytical research and practical experimentation.

To address a robust framework for the study, the research pursues the following refined objectives:

- Microclimatic Diagnostic: To quantify and map the thermal stress of neglected campus nodes through empirical environmental monitoring. This phase involves the use of high-precision data loggers to establish a baseline of microclimatic criticalities, providing the evidence-based justification required for targeted Nature-Based Solutions.
- Performance Evaluation of NBS Prototypes: To test the technical efficacy and feasibility of low-impact, reversible greening structures. This objective focuses on measuring the temperature differentials generated by the prototypes to validate their role as effective passive cooling systems in Mediterranean climates.
- Participatory Governance and Social Inclusion: To evaluate a replicable co-design model involving students, researchers, and industrial partners like Green Habitat s.r.l. This objective aims to foster community ownership and environmental awareness, ensuring that the developed solutions are context-sensitive and inclusive of vulnerable groups (Chelli et al., 2025; Pinto et al., 2023).

By linking quantitative analysis, industrial collaboration, and public engagement, this research aims to offer a set of feasible and scalable interventions that may serve as a replicable reference for other Mediterranean academic institutions facing similar climatic and spatial degradation.

2 Methodology

The methodological framework adopted for this study follows a Participatory Action Research (PAR) approach, structured to integrate analytical investigation, industrial synergy, and experimental design. By involving students from architecture and design programs not merely as learners but as active researchers and co-designers, the methodology establishes a bridge between academic theory and the practical challenges of urban climate adaptation. The research was articulated in different distinct, yet interconnected, operational phases.

2.1 Site Selection and Environmental Diagnostics

The initial phase focused on identifying priority "orphan" spaces within the UNIPA Campus through a multi-criteria survey. The selection process was guided by three diagnostic parameters: (i) high microclimatic discomfort, (ii) evidence of spatial abandonment or underutilization, and (iii) absence of functional urban furniture.

To address a microclimatic diagnosis, a quantitative environmental monitoring campaign was conducted. The chapel courtyard was identified as the critical test site. Microclimatic data were collected using an HD32.3TC data logger, equipped with sensors for measuring air temperature, globe-thermometer temperature (for radiant heat analysis), and relative humidity. In-situ measurements were complemented by infrared thermography to map surface temperature variations across different materials (stone, asphalt, and shaded soil). These quantitative readings were integrated with qualitative tools, including user-perception surveys and spatial mapping of student flow, to provide a comprehensive baseline of the courtyard's environmental and social performance (Marsaglia, 2024).

2.2 Interdisciplinary co-Design and Industrial Collaboration

The experimental core of the research was the "GREENWORK//LAB" workshop, an interdisciplinary collaboration between the Department of Architecture, student associations (GIN), and the industrial partner Green Habitat s.r.l. This phase has been carefully structured to assess the efficacy of participatory methodologies. Specifically, it aims to test and evaluate how collaborative frameworks, by leveraging the localized knowledge of active participants, can drive the development of innovative,

Fig. 2 Participatory Action Research (PAR) approach: the workshop integrates analytical investigation, industrial synergy, and experimental design, engaging architecture and design students as active researchers and co-designers to bridge academic theory with practical urban climate adaptation challenges. @Author2025

context-sensitive design solutions that effectively respond to specific microclimatic and urban challenges (Fig. 2).

The workshop has been structured as an iterative cycle of analysis, design exploration, and critical review. The teams were organized into four working groups, each tasked with developing a specific design response to the environmental criticalities identified in Phase I. The presence of an industrial partner was crucial for ensuring the technical feasibility and material sustainability of the proposals. Discussions focused on the selection of "living materials" (native plant species) and eco-compatible construction systems (recycled wood, low-impact metals), aligning the design process with the principles of the circular economy.

2.3 *Prototyping and Technical Validation*

To address the performance evaluation aim, the students translated their conceptual ideas into technical prototypes. This phase utilized a "multi-scale" modelling approach:

- Digital Simulations: Three-dimensional modelling was used to simulate shading patterns and solar radiation impact during the summer solstice.
- Physical Mock-ups: 1:25 scale models were produced through 3D printing and analogue modelling to evaluate the modularity and reversibility of the structures (Fig. 3).
- Performance Benchmarking: Since full-scale long-term monitoring was not yet feasible for all modules, their expected cooling efficiency was inferred by cross-referencing the design geometries with performance data documented in recent

Fig. 3 Conceptual prototypes and technical details of the student-led NBS interventions. Clockwise from top left: Hexagonal modular seating (Group 1); Multifunctional thermal buffer (Group 2); Photovoltaic-integrated green trellis (Group 3); Circular economy vertical partition (Group 4). @Author2025

NBS literature and previous experiments conducted by the research team (Baez-Garcia et al., 2025; Tehrani et al., 2025).

2.4 Dissemination and Institutional Impact

The final phase aimed at scaling the results through public engagement. The presentation of prototypes at SHARPER 2025 (European Researchers' Night) served as a validation tool, allowing for real-time feedback from a broader audience of citizens and stakeholders (Fig. 3). This phase was fundamental for identifying the bureaucratic and administrative barriers that often prevent the transition from temporary student installations to permanent institutional spatial governance. By documenting these obstacles, the methodology provides a replicable roadmap for other Mediterranean institutions seeking to implement student-driven greening initiatives.

3 Results

The experimental phase of the research yielded two primary categories of results: empirical microclimatic data validating the need for intervention, and the development of four distinct Nature-Based Solution (NBS) prototypes designed to mitigate thermal stress and reactivate the social fabric of the UNIPA Campus.

3.1 *Empirical Microclimatic Analysis*

The environmental monitoring campaign carried out in the chapel courtyard provided quantitative evidence of the extreme thermal conditions characterizing the site. Using the HD32.3TC data logger during summer peak conditions, the research recorded surface temperatures exceeding 46.2 °C on unshaded pavement, a value that significantly exceeds the comfort threshold for outdoor permanence. In contrast, measurements taken in shaded areas or in proximity to existing vegetation revealed a significant temperature differential. Shaded zones maintained temperatures around 33.1 °C, demonstrating that passive cooling strategies could achieve a reduction of over 13 °C in surface heat. Air temperature measurements further confirmed this trend, showing localized cooling of 3–4 °C near vegetated points. Additionally, relative humidity levels increased by 5–7% in the immediate vicinity of the prototypes, contributing to a more favorable microclimate through evapotranspiration. These findings provided the empirical basis for the subsequent design phase, proving that vegetated shading is not merely an aesthetic choice but a thermal necessity for Mediterranean campuses (Fig. 4).

3.2 *Prototyping and Technical Solutions*

Following the diagnostic phase, the GREENWORK//LAB produced four modular design solutions (Table 1). Each proposal was evaluated based on its capacity for climate adaptability, spatial inclusivity, and material sustainability (Aduse-Poku et al., 2024).

Group 1 Hexagonal Adaptive System. Developed for a flexible utilization, this module utilizes a hexagonal geometry that allows for high spatial flexibility and aggregation (Fig. 3a). The structure, built from recycled pallets, incorporates a seating system integrated with flowerbeds. Its primary technical innovation is a modular brise-soleil made of wooden slats and interwoven ribbons, which creates a protected microclimate by filtering direct solar radiation (led by: Giovanni Scialabba, Alessandro La Rosa, Christian Barresi, and Sofia Scoma).

Group 2 Multifunctional Rectangular Buffer. This 200 x 100 x 202 cm module emphasizes spatial versatility. It features mobile components, including four stools

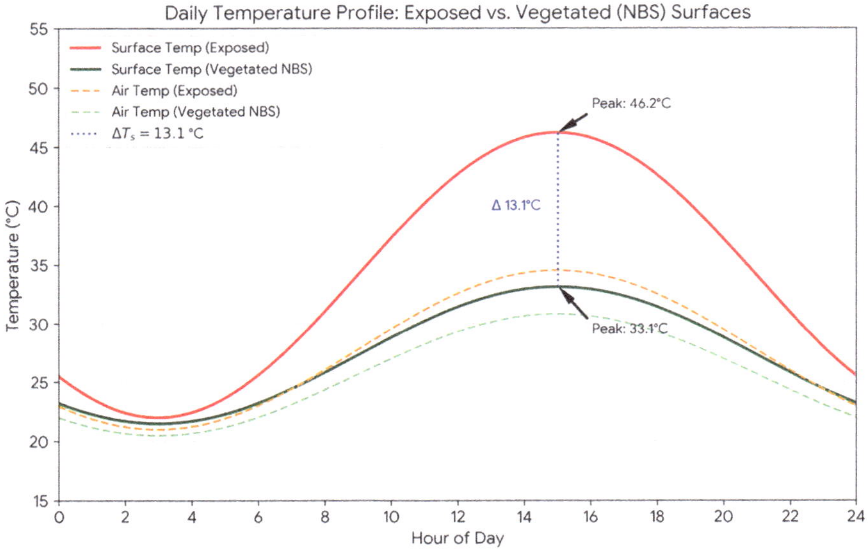

Measurement Parameter	Exposed (Baseline) Areas	Shaded/Vegetated (NBS) Zones	Recorded Differential (ΔT/ΔRH)
Surface Temperature (T_s)	46.2 °C	33.1 °C	-13.1 °C
Air Temperature (T_a)	34.5 °C	30.8 °C	-3.7 °C
Relative Humidity (RH)	32 %	38 %	+6 %
Radiant Heat (Globe Temp)	High Stress (40 °C)	Moderate/Low Stress	Significant attenuation

Fig. 4 Comparative analysis of surface temperatures during summer peak conditions. The graph demonstrates the 13 °C differential between fully exposed surfaces and the micro-environments created by the vegetated shading prototypes. @Author2025

Table 1 Comparative Analysis of the proposed Nature-Based Solutions (NBS) prototypes. The table synthesizes the technical specifications, design flexibility, and circular economy principles applied by each student group

Group	Design Concept	Key Technical Features	Sustainability & Circularity
Group 1	Hexagonal seating	Integrated brise-soleil (woven ribbons); modular aggregation.	Recycled pallets; integrated planters for native flora.
Group 2	Multifunctional unit	200 × 100 cm frame; adjustable boards (table/seat); climbing plant trellises.	Modular reversibility; high-durability timber.
Group 3	Tech-integrated seating	Photovoltaic cells for off-grid charging; vertical greenery on galvanized mesh.	Solar energy integration; low-impact galvanized steel & wood.
Group 4	Circular economy partition	220 cm vegetated screen; elevated base to reduce ground-heat conduction.	100% recycled pallet wood; focus on permeability and natural shading.

that can be stowed within the main frame and adjustable wooden boards that serve as either seats or work tables (Fig. 3b). The vertical trellises support climbing plants, which, combined with horizontal shading slats, function as a natural thermal buffer against intense solar gain (led by: Francesca Lombardo, Giuseppe Scionti, Chiara Cangemi, Annatea Scaffidi Muta, and Loide Giorgianni).

Group 3 Energy-Integrated Longitudinal Module. This proposal integrates NBS with renewable energy technology (Fig. 3c). The structure features a wooden trellis with galvanized mesh for vertical greenery, paired with integrated photovoltaic cells on the upper shading surface. This allows the furniture to function as an active off-grid charging station for students, linking environmental comfort with technological utility (led by: Giulio Accardo, Riccardo Drago, Patrizia Giacoma Ferreri, Giuliana Severina, and Ivan Sungkur).

Group 4 Circular Economy Vertical Screen. Centered on 100% recycled materials, this module (75 x 120 cm footprint) focuses on preventing ground-heat conduction through an elevated base (Fig. 3d). It features a 220 cm vertical partition with varying levels of permeability—partially solid and partially perforated—designed to balance the need for privacy and ventilation with high-performance shading (led by: Alice Spataro, Giorgia Siragusano, Lorenzo Giordano, and Daniele Di Fiore).

3.3 Educational and Social Outcomes

Beyond technical performance, the workshop demonstrated significant results in terms of community stewardship (Objective C). The participatory nature of the process enabled students to experience the full research-to-design pipeline. The solutions were not merely academic exercises but tangible contributions to the campus governance debate.

The presentation of these results at SHARPER 2025 served as a vital validation of the "Third Mission." By showcasing the physical prototypes and the related microclimatic data, the research team successfully engaged the public in a dialogue regarding urban greenery and citizen science (Fig. 5). This engagement highlighted that while technical efficiency is paramount, the social acceptance and "ownership" of green spaces by the university community are equally essential for the long-term success of urban regeneration projects.

Despite these achievements, the results also highlighted critical obstacles, such as institutional bureaucratic inertia and the lack of a permanent framework for integrating student-driven projects into official campus planning.

These findings suggest that the transition from experimental prototype to permanent fixture requires a formal evolution in spatial governance.

Fig. 5 Presentation of research results and prototypes during SHARPER 2025, fostering public engagement and institutional dialogue on urban resilience. @Author2025

4 Discussion and Conclusion

The findings of this research confirm that small-scale, user-centered Nature-Based Solutions (NBS) can serve as effective catalysts for the regeneration of neglected academic environments. In Mediterranean contexts, where thermal stress and intense solar exposure significantly reduce the usability of outdoor spaces, the transition from "residual" areas to "social hubs" requires a multi-scalar and interdisciplinary approach.

The chapel courtyard at UNIPA, previously perceived as an inhospitable and "orphan" space, was successfully reframed as a potential center for social interaction and environmental resilience. This transformation was achieved not through high-cost technological interventions, but through the sensitive application of climate-adaptation principles, material compatibility, and participatory design.

The empirical evidence—specifically the 13 °C temperature differential recorded during the study—validates the hypothesis that even minimal, low-impact interventions can profoundly alter the microclimatic performance and social perception of a site. Furthermore, the "GREENWORK//LAB" experience highlights the educational value of embedding real-world environmental challenges into the academic curriculum, fostering a new generation of designers equipped with a sense of responsibility toward spatial justice and climate resilience.

4.1 Study Limitations

Despite the positive outcomes, the researchers acknowledge several limitations that should be addressed in subsequent investigations:

- Temporal and Seasonal Constraints: The microclimatic data collection was primarily focused on summer peak conditions. A longitudinal study covering

all seasons would be necessary to assess the performance of the prototypes under diverse weather patterns, such as winter wind exposure or heavy rainfall.
- Durability and Maintenance: As the prototypes were constructed using low-tech and recycled materials (e.g., pallet wood), their long-term durability in an outdoor environment remains to be tested. The lack of a permanent maintenance protocol within the university's administrative structure poses a risk to the longevity of such interventions.
- Scale of Impact: While the prototypes effectively mitigated localized heat stress, their impact on the overall campus microclimate is limited by their small scale. Achieving a significant cooling effect at the urban level would require a systemic network of such modules.

4.2 *Future Outlook and Governance*

To overcome these limitations, future developments of this research will focus on the integration of IoT (Internet of Things) sensors within the modular units to provide real-time, open-access environmental data. This "Citizen Science" approach will allow for continuous monitoring and iterative refinement of the design solutions.

Furthermore, the study advocates for a shift in institutional spatial governance. The transition from experimental "living labs" to permanent campus fixtures requires overcoming the "bureaucratic inertia" identified during the workshop. This can be achieved by the following:

- Drafting a "Green Campus Protocol": A formal policy to streamline the implementation of student-led, low-impact interventions.
- Expanding Industrial Partnerships: Strengthening the synergy with partners like Green Habitat s.r.l. to scale the production of modular NBS.
- Institutionalizing Participation: Embedding participatory design processes into the permanent planning strategies of the University, ensuring that students and researchers are recognized as key stakeholders in the transformation of their own environment.

In conclusion, the UNIPA experience demonstrates that the regeneration of university spaces through greening strategies is both scientifically feasible and socially urgent. By linking analytical research, design experimentation, and community engagement, the initiative offers a holistic and replicable model for Mediterranean campuses, promoting a vision of the academic environment as a dynamic, resilient, and inclusive ecosystem.

5 Conclusion

This study demonstrates that the regeneration of Mediterranean academic spaces through participatory greening is both scientifically valid and socially urgent.

By bridging analytical research with industrial collaboration, the UNIPA campus can evolve into a resilient ecosystem.

The project emphasizes that, beyond technical performance, institutional commitment is essential to transform these experimental "living labs" into permanent pillars of university governance.

Acknowledgements This research was funded by the European Union EU Next Generation Italian PRIN/PNRR 2022 Project titled "Greenwork: An interdisciplinary framework for urban health and urban resilience enhancement based on greening strategies on buildings and open spaces"—Project Code MUR P2022MNKEC 003—Sector: ERC SH7—CUP:B53D23029580001—PRJ-1358.

References

Aduse-Poku, M., Rohrer, F., Winter, B., & Edelmann, H. G. (2024). Methodology for the quantification of the absorption potential of greenhouse - and pollutant gases by climbing plants used in façade greening; a case study on ivy (Hedera helix). *Environmental Advances*. https://doi.org/10.1016/j.envadv.2024.100568

Alves, R. A., Santos, M. M. dos, Rudke, A. P., Francisquetti Venturin, P. R., & Martins, J. A. (2024). Site selection for nature-based solutions for stormwater management in urban areas: An approach combining GIS and multi-criteria analysis. Journal of Environmental Management, 359. doi:https://doi.org/10.1016/j.jenvman.2024.120999.

Ascione, F., De Masi, R. F., Mastellone, M., Ruggiero, S., & Vanoli, G. P. (2020). Green walls, a critical review: Knowledge gaps, design parameters, thermal performances and multi-criteria design approaches. In Energies (Vol. 13, Issue 9). MDPI AG. doi:https://doi.org/10.3390/en13092296.

Babí Almenar, J., Elliot, T., Rugani, B., Philippe, B., Navarrete Gutierrez, T., Sonnemann, G., & Geneletti, D. (2021). Nexus between nature-based solutions, ecosystem services and urban challenges. *Land Use Policy, 100*. https://doi.org/10.1016/j.landusepol.2020.104898

Baez-Garcia, W. G., Simá, E., Chagolla-Aranda, M. A., Carreto-Hernandez, L. G., & Aguilar, J. O. (2025). Experimental evaluation of the thermal behavior of a green facade in the cold and warm seasons in a subtropical climate (Cwa) of México. *Journal of Building Engineering, 99*. https://doi.org/10.1016/j.jobe.2024.111627

Bakhshoodeh, R., Ocampo, C., & Oldham, C. (2023). Impact of ambient air temperature, orientation, and plant status on the thermal performance of green façades. *Energy and Buildings, 296*. https://doi.org/10.1016/j.enbuild.2023.113389

Bayissa, Y., Srinivasan, R., Hunink, J., Nyolei, D., Moges, S., de Andrade Costa, D., Tadesse, D., Melesse, A., & Tilahun, S. (2025). Evaluating the potential of nature-based solutions to mitigate land use and climate change impacts on the hydrology of the Gefersa and Legedadi watersheds in Ethiopia. *Journal of Hydrology: Regional Studies, 57*. https://doi.org/10.1016/j.ejrh.2024.102130

Chelli, A., Brander, L., & Geneletti, D. (2025). Cost-benefit analysis of urban nature-based solutions: A systematic review of approaches and scales with a focus on benefit valuation. In Ecosystem services (Vol. 71). Elsevier B.V. doi:https://doi.org/10.1016/j.ecoser.2024.101684.

Debele, S. E., Leo, L. S., Kumar, P., Sahani, J., Ommer, J., Bucchignani, E., Vranić, S., Kalas, M., Amirzada, Z., Pavlova, I., Shah, M. A. R., Gonzalez-Ollauri, A., & Di Sabatino, S. (2023). Nature-based solutions can help reduce the impact of natural hazards: A global analysis of NBS case studies. *Science of the Total Environment*. https://doi.org/10.1016/j.scitotenv.2023.165824

Frantzeskaki, N. (2019). Seven lessons for planning nature-based solutions in cities. *Environmental Science and Policy, 93*, 101–111. https://doi.org/10.1016/j.envsci.2018.12.033

Frantzeskaki, N., McPhearson, T., Collier, M. J., Kendal, D., Bulkeley, H., Dumitru, A., Walsh, C., Noble, K., Van Wyk, E., Ordóñez, C., Oke, C., & Pintér, L. (2019). Nature-based solutions for urban climate change adaptation: Linking science, policy, and practice communities for evidence-based decision-making. *Bio Science, 69*(6), 455–466. https://doi.org/10.1093/biosci/biz042

Kabisch, N., Korn, H., Stadler, J., & Bonn, A. (n.d.). *Theory and Practice of Urban Sustainability Transitions Natureebased Solutions to Climate Change Adaptation in Urban Areas*. Retrieved from http://www.springer.com/series/13408

Marsaglia, V. (2024). Technological greenery. Exploring cutting-edge solutions for performant greenery integration in building envelope design. *Energy and Buildings, 324*. https://doi.org/10.1016/j.enbuild.2024.114920

Morabito, M., Crisci, A., Gioli, B., Gualtieri, G., Toscano, P., Di Stefano, V., Orlandini, S., & Gensini, G. F. (2015). Urban-hazard risk analysis: Mapping of heat-related risks in the elderly in major Italian cities. *PLoS One, 10*(5). https://doi.org/10.1371/journal.pone.0127277

Pinto, L. V., Inácio, M., & Pereira, P. (2023). Green and blue infrastructure (GBI) and urban nature-based solutions (NbS) contribution to human and ecological Well-being and health. *Oxford Open Infrastructure and Health, 1*. https://doi.org/10.1093/ooih/ouad004

Pueyo-Ros, J., Riva, M., Gonzalvo-Henry, G., Guillaume-Ruty, S., & Comas, J. (2025). Knowledge base dataset on nature-based solutions for wastewater treatment and stormwater management. *Data in Brief, 60*. https://doi.org/10.1016/j.dib.2025.111469

Tehrani, A. A., Sobhaninia, S., Nikookar, N., Levinson, R., Sailor, D. J., & Amaripadath, D. (2025). Data-driven approach to estimate urban heat Island impacts on building energy consumption. *Energy, 316*. https://doi.org/10.1016/j.energy.2025.134508

Xi, C., Ding, J., Wang, J., Feng, Z., & Cao, S. J. (2022). Nature-based solution of greenery configuration design by comprehensive benefit evaluation of microclimate environment and carbon sequestration. *Energy and Buildings, 270*. https://doi.org/10.1016/j.enbuild.2022.112264

Multispecies co-Existence: Shaping an Urban Garden in Milan as a Ground for Human and More-than-Human Relationships

Eugenio Morello, Luca Lazzarini, Israa Mahmoud, and Asef Ayatollahi

Abstract Cities are increasingly confronted with the challenge of addressing biodiversity loss while undergoing profound processes of urban transformation. Despite the long-standing tradition of participatory planning and co-design in urban studies, nature is rarely included as an active stakeholder in deliberative processes, and tools to recognize the rights, agency, and needs of more-than-human species in the co-design and co-production of urban solutions remain limited. This gap is particularly evident in the design of urban green spaces, where biodiversity is often treated as an outcome rather than a participant in decision-making. This chapter aims to advance the understanding of multispecies co-existence as an approach to co-designing urban green spaces. It draws on a concrete case study: the co-design of an urban garden in Milan within a living-lab setting established in 2024 as part of the research activities of the National Biodiversity Future Center (NBFC). The approach experimented in the research has seen the combined use of co-design tools to support citizens and local stakeholders in recognizing the rights and stakes of more-than-human species, and of ecological and social monitoring tools, which enabled an assessment of the impacts of the co-design process on both the human and more-than-human species inhabiting the garden. The co-design process generated a learning environment grounded in the recognition of the needs of different species, which allowed the garden to transform into a multispecies arena in which biodiversity is interpreted as an active agent shaping the space. The findings reveal several challenges associated with co-designing for multispecies coexistence, including the crucial contribution of experts in guiding participants to recognize the needs of the species inhabiting the garden, or the presence of institutional representatives in the process which sometimes undermined the emergence of innovations. At the same time, the monitoring activities highlight the significant role of the co-designed garden in providing habitat

E. Morello · L. Lazzarini (✉) · I. Mahmoud · A. Ayatollahi
Laboratorio di Simulazione Urbana 'Fausto Curti', Department of Architecture and Urban Studies (DAStU), Politecnico di Milano, Milan, Italy
e-mail: luca.lazzarini@polimi.it

National Biodiversity Future Center (NBFC), Palermo, Italy

I. Mahmoud
e-mail: israa.mahmoud@polimi.it

L. Diana et al. (eds.), *Greening and Resilience*, Urban Sustainability,
https://doi.org/10.1007/978-981-92-0419-9_13

for specific species, while also revealing a high level of user recognition of the garden as a multispecies arena.

Keywords Multispecies · Urban biodiversity · Co-design · More-than-human · Public participation

1 Introduction

Despite their increasing prominence in public policy and urban planning discourses (Oke et al., 2021), nature and biodiversity continue to be framed predominantly through traditional and instrumental perspectives. As highlighted by Pascual et al. (2017), the integration of nature into policy frameworks often relies on a narrow, unidimensional lens rooted in a utilitarian understanding of human–nature relationships. This anthropocentric orientation conceptualizes nature primarily in terms of the benefits it provides to humans. As several authors have noted (Díaz et al., 2015; Lazzarini, 2024), such perspectives are closely linked to notions of well-being grounded in economic value, particularly those derived from the production or consumption of nature as a commodity. Within this framework, the contributions of nature to physical, mental, and social well-being are largely examined through health and social science approaches, reinforcing an instrumental view of nature in human life (Einzeberg, 2018).

Given this context, there is growing recognition in the debate that the succession of crises we have faced in recent years—referred to as a "poly-crisis" by Morin and Kern (1999)—demands an expanded understanding of how human and more-than-human worlds are intertwined (Whatmore, 2006), challenging the idea that humans are distinct from the complex biophysical systems we live within.

Looking at the urban planning field, Hernandez-Santin et al. (2023) proposed to consider biodiversity as a more-than-human citizen who inhabits a particular locality. In fact, most of the species living in our cities can be interpreted as *active agents*, capable of responding to and transforming their surroundings. A key implication of this perspective is that natural species may be considered potential stakeholders in policy-making processes; entities with whom designers, planners, and policy-makers can, in a sense, co-design. In the same vein, Houston et al. (2017: 194) have urged to develop a relational understanding of planning as «'*co-becoming*' in and for diverse multispecies communities, where humans, plants, soils, microbes, birds, fungi, insects, native and non-native animals shape urban landscapes and interactions and where we are more attentive to the storied contexts of multispecies places within cities and regions».

To embed nature as both a right holder and a stakeholder, the IUCN recognizes different approaches, from a source of inspiration to a formal governance role (van den Tillaar, 2025). While the nature-as-stakeholder perspective is increasingly being tested both in governance and civil-society initiatives and supported by toolkits designed for companies and non-governmental organizations seeking to engage with nature as a stakeholder, there have been far fewer attempts by public administrations to apply this concept within the policy domain. This gap limits the use of the concept as a principle for developing policies, plans, and projects that transform urban green spaces into multispecies arenas of coexistence between human and more-than-human species.

Despite this gap, a growing body of literature has examined the role of urban and community gardens as sites that enable multi-species coexistence. Urban gardens are frequently conceptualized as spaces of social and ecological production, particularly when issues of accessibility and spatial distribution are addressed through inclusive planning approaches (Huq & Deacon, 2025). In this context, community gardens have also been described as instruments of hybrid governance, operating at the intersection of top-down institutional frameworks and bottom-up civic initiatives, and relying on forms of public engagement oriented toward environmental sustainability. From an ecological perspective, urban gardens can generate measurable Ecosystem Services (ES)—such as pollination support, microclimate regulation, and stormwater retention—while also contributing to social-ecological memory that underpins community resilience and environmental stewardship (Barthel et al., 2010; Clarke et al., 2019). From a governance perspective, they function as arenas in which responsibilities for participation, environmental management, and care are negotiated between citizens and public authorities. These processes may simultaneously produce forms of empowerment and tension, which, as noted by Jacob and Rocha (2021), can play a critical role in shaping the longevity and stability of garden-based initiatives.

Within this broader frame, this chapter aims to deepen the understanding of multispecies coexistence as an approach for co-designing an urban green space by presenting and discussing a concrete case study, the co-design of an urban garden in Milan, Italy, through a living lab setting established in 2024 as part of the research activities of the National Biodiversity Future Center (NBFC). The main research questions underlying this contribution are the following:

- How can multispecies relationships be recognized and incorporated into urban planning, particularly in the design and management of urban green spaces?
- What tools, methods, and techniques can support the co-design of urban green spaces as multispecies arenas?
- What obstacles and challenges arise when scaling up multispecies co-design into existing governance and institutional frameworks?

In this study, we propose an approach that employs co-design methodologies to redesign an urban green space as a multispecies garden. The innovative contribution of our work mainly relates to the process dimension of the research, specifically in the combined use of co-design tools to support citizens and local stakeholders in recognizing the rights and stakes of more-than-human species, and of ecological and social monitoring tools, which enabled an assessment of the impacts of the co-design process on both human and more-than-human species inhabiting the garden.

This chapter is organized into four sections. The first develops the theoretical background on co-design for multispecies coexistence in urban contexts. The second introduces the case study and outlines the methodological approach adopted in the research. The third discusses the results by reflecting on three key lessons emerging from the co-design process. The chapter concludes with a fourth section, which highlights some implications related to planning for multispecies coexistence.

2 Theoretical Background: Co-design for Multispecies Entanglements

Co-design—often used interchangeably with co-creation and rooted in public participatory design and management—reframes end-users as creative partners across all stages of a design process, shifting roles and power relations from "user as subject" to "user as partner and part of the process". This turn is widely credited to Sanders and Stappers (2008), whose work documented the move from expert-led, user-centered approaches to collaborative, generative practices that aim to produce more democratic and sustainable outcomes. Participatory design's political heritage—emphasizing rights, power, and workplace democratization—remains salient as co-design expands to community development, policy, and urban planning domains. Recent syntheses explicitly connect co-design to sustainable development, arguing that when applied beyond products and services, it must confront inclusion, empowerment, and collaborative governance in urban planning (Mahmoud et al., 2025; O'Donnell et al., 2025).

Urban biodiversity research provides a nexus knowledge base for co-design and co-creation in cities. Seminal work by Nilon et al. (2017) established both the ecological distinctiveness and value of urban nature—from semi-natural remnants to heavily modified habitats—and called for integrating ecological knowledge into planning to maintain biodiversity amid urbanization. Subsequent urban ecology research consolidated patterns and processes across social-ecological systems and emphasized planning applications, ES, and human well-being in cities. Green infrastructure (GI) emerged as a planning concept that treats interconnected green–blue networks as critical urban systems for delivering multiple ES. Frameworks by Hansen and Pauleit (2014) operationalize GI's multifunctionality, highlighting how planners can assess network integrity, balance service supply and demand, and navigate trade-offs; later guides show how European cities mainstream GI through audits, indicators, and design for multifunctional parks. These developments dovetail with Nature-based Solutions (NbS), formally defined by IUCN (2020: 2) as "actions to protect, manage, and restore ecosystems that address societal challenges while providing biodiversity and human well-being benefits", a definition increasingly adopted for urban stormwater, heat, and health interventions.

A rapidly growing strand of literature reframes cities as ground for multispecies justice and more-than-human worlds (Raymond et al., 2025). In planning theory, Houston et al. (2017) argue that urban futures are co-produced by human and nonhuman agents and propose "multispecies entanglements" as a corrective to human exceptionalism, pushing planning to nurture conditions for multispecies flourishing and climate-adaptive justice. More-than-human geographies further develop relational ethics and materialist practices that attend to the "livingness" of urban worlds, challenging representational politics and inviting experimental engagements with urban wilds (Kirksey & Helmreich, 2010). Together, these literatures converge on a proposition: co-design for urban biodiversity and multispecies coexistence entails plural ways of knowing and valuing urban natures, operationalized through GI, NbS, and biodiversity-sensitive design, while embracing the realities of novel ecologies and the ethical project of more-than-human flourishing (Hinchliff et al., 2005).

3 Case-Study Presentation and Methodology

3.1 The Case-Study: From a Wasteland to a Biodiversity-Rich Garden

As part of the research activities of the Italian National Biodiversity Future Center (NBFC) [2022–2025] funded under the National Recovery and Resilience Plan, an urban living lab has been established in Milan to test co-design methods for a neighborhood park in the eastern periphery of the city, the Sorelle Mirabal Garden, with the goal of preserving and enhancing its biodiversity. The initiative was launched following a joint agreement between the Politecnico di Milano and Municipio 3 (local Council) of Milan in May 2023, centering on the "civic experimentation of urban biodiversity", as stated in the municipal resolution approved by the local authority (Municipio 3, Comune di Milano, 2023). The aim of the agreement is to collaboratively safeguard and enrich the existing park's biodiversity through participatory practices embedded in a living lab framework, fostering new knowledge and a renewed sensitivity toward nature conservation. The actions outlined in the municipal resolution include identifying a wild area ("area ad elevata naturalità"), separated from the rest of the garden but connected to the railway embankment to the west, and co-designing and installing urban furnishings and NbSs in the garden.

The Sorelle Mirabal Garden spans approximately two hectares and is located adjacent to a railway embankment, bordered by two elderly care facilities and surrounded by a mixed-use urban fabric where new large-scale residential developments are being built (Fig. 1). Once a wasteland and an illegal dumping site, the area was reclaimed in 2017 by a group of non-profit associations that secured the space through a public tender. The associations restored the existing and planted new vegetation and transformed the site into a shared community garden. Since then, the garden has hosted numerous projects, including the co-design of a therapeutic park for elderly people, aimed at evaluating the cognitive and social benefits of urban nature (Boffi et al., 2022).

In 2022, the loan-for-use agreement expired, resulting in the park's closure to the public. Over the subsequent years, the vegetation initially restored by the local

Fig. 1 Location of the Sorelle Mirabal Garden in Milan, highlighted by the red perimeter. *Source* Elaboration by the authors from Google Earth. Date: January 2026

associations continued to grow spontaneously, turning the space into a valuable site for observing biodiversity dynamics. The park currently hosts a variety of trees, including mulberries and cherry trees, as well as spontaneous species like *Acer negundo* and invasive ones such as *Ailanthus altissima* (Fumagalli et al., 2020). It is also habitat to a diverse range of vascular plants, shrubs, birds, insects, and small mammals, the latter of which use the nearby railway as a corridor to access the park.

3.2 The Methodology: Co-Design for Human and More-Than-Human Interaction

The co-design activities took place between May and September 2024. The process has involved a group of 20–25 participants, mainly citizens living in the neighborhoods (Lambrate and Ortica) where the park is located and was structured in four main workshops held in Milan [Fig. 2].

In the first workshop, following a brief presentation of the main research objectives, participants were guided to develop an initial definition of urban biodiversity, explore its benefits through the concept of ES (Ayatollahi & Morello, 2025), and identify which of these benefits could be enhanced within the Sorelle Mirabal Garden.

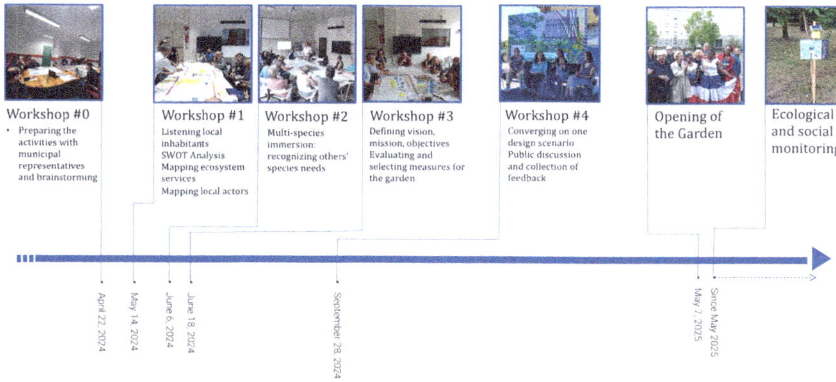

Fig. 2 Overview of the workshops and main activities carried out during the co-design process. Source: Elaboration by the authors. Date: February 4, 2026

Using a set of 28 cards divided into three categories, each representing a type of ecosystem service in urban areas, participants were invited to discuss the cards and select their top five ES with reference to the park. The most frequently selected services included heat reduction through increased green coverage, improved social interaction among residents, reduction of anxiety and depression, and control of disease and sanitary risks. These preferences helped to prioritize specific ES in the park's redesign. Subsequently, participants were called to share their preliminary knowledge and expectations about the garden, and this has served to establish its collaborative vision, mission, and objectives.

In the second workshop, a multispecies role-playing game was set up through the use of a set of cards representing a selection of animal and plant species which find their habitat in the park. Species were selected with the help of two local experts from the Natural History Museum and the Civic Aquarium of Milan, who joined the workshop, providing relevant insights on local biodiversity. The decision to include not only desirable species but also non-native and even invasive species was driven by the aim of guiding citizens' understanding of the complexity of biodiversity dynamics in urban environments, where harsh competition between species is often observed. In line with this approach, each participant working in pairs received two cards (one featuring an animal and the other a plant) and was asked to complete two boards. Each board contained blank fields to be filled in with information related to the species' needs (resources and habitats), their seasonal and daily trajectories, the potential interactions with other species, and the benefits and threats they pose to humans.

In the third workshop, the goal was to identify the design elements to be implemented in the garden. Once again, a set of cards was prepared, featuring a selection of NbS drawn from existing catalogues, along with various types of urban furniture essential for the public use of the garden. The cards were laid out on the table, and participants were invited to collaboratively discuss and decide which elements could be introduced into the park, ensuring alignment with the vision, mission, and objectives established during the first workshop.

In the fourth and final workshop, a public presentation and discussion of the activities conducted in the previous months was organized with the participation of institutional representatives, academics, and biodiversity experts during the Milano Green Week, a program of events about greening and sustainability promoted by the Municipality of Milan.

After a long preparation phase and some delays due to administrative issues, the garden was finally inaugurated on May 7, 2025. One of its most distinctive traits, resulting from the co-design process and related to the vision of the park as a «multispecies, multifunctional, collaborative and inclusive garden», is the creation of an area of high naturalness ("area ad elevata naturalità") located in the western section of the garden, near the railway. This area is enclosed by a light and porous metal fence, with limited human access, allowing nature to flourish spontaneously (Figs. 3, 4). Thanks to its wide mesh, the fence allows small mammals to pass through and allows park visitors to observe natural dynamics. Its educational role is further enhanced by a series of informational panels installed along its length, which explains the ecological value of the area. Experts from the Nature History Museum and the Civil Aquarium in Milan are monitoring several species in this space that were previously discussed during the workshop activities, such as the European green toad (*Bufotes viridis*), for which a dedicated pond was created and is now under continuous observation, and which became a sort of symbol of the park.

3.3 The Biodiversity and Social Monitoring

Biodiversity monitoring is supported by a range of sensors installed throughout the garden, especially in two portions, the area of high naturalness and another area located at the entrance of the garden. Several types of sensors measure soil volumetric water content, soil water potential, and carbon sequestration capacity. In addition, camera traps for wildlife video recording have been installed in various parts of the garden, particularly within the enclosed area. Furthermore, 11 *IoT Spectrum* sensors designed to measure the frequency and abundance of pollinators (*Diptera, Hymenoptera, and Coleoptera*) were installed thanks to an agreement with 3Bee, a company based in Milan specialized in environmental and biodiversity monitoring. Of these sensors, eight were placed within the garden—four of which in the area of high naturalness—while two additional sensors were placed outside the garden, in the surrounding neighborhood: one in the green courtyard of a nearby residential complex and another one in the garden of an elderly center, in order to highlight differences between the garden's performance and that of nearby urban areas. The *IoT Spectrum* sensors enable continuous, non-invasive detection and analysis of pollinator activity. The variables analyzed include pollinator abundance, clustering patterns, and average hourly frequency (buzzes/hour/year).

Fig. 3 Schematic plan of the garden, resulting from the co-design process. *Source* Elaboration by the authors. Date: April 30, 2025

Fig. 4 Views of the high biodiversity area in the Sorelle Mirabal Garden. *Source* Photos by the authors. Date: October 21, 2025

In parallel with biodiversity monitoring, a social monitoring activity was conducted through an online questionnaire administered to the visitors of the Garden. The survey was designed on the basis of a previous social monitoring activity conducted in 2019–2021 in Milan (Mahmoud et al., 2021). It began in September 2025 and, to date, 74 responses have been collected. The questionnaire aims to investigate visitors' preferences and attitudes toward nature and biodiversity and is organized into three main sections. The first section collects demographic data about the respondents, including age, gender, occupation, and place of residence. The second section addresses the respondents' relationship with the neighborhood, their understanding of and attitudes toward biodiversity. The third section focuses on the respondents' engagement with the Sorelle Mirabal Garden, including visitation frequency, preferred times of the day, activities carried out in the Garden, and perceptions of its biodiversity.

4 Challenges of co-Designing for Multispecies Coexistence

The co-design activities implemented in the Sorelle Mirabal Garden provide a few key insights regarding the acknowledgment of multispecies relationships in participatory processes, which can be summarized in three main aspects.

The first aspect concerns the ability of the co-design process to foster a learning environment grounded in reciprocal relationships between human and more-than-human species. This perspective moves beyond human ontological exceptionalism and challenges views of nature as separate from or subordinate to human society, instead emphasizing the multispecies entanglements, responsibilities, and practices of care that the garden can support. From this standpoint, both the vision of the garden as a "multispecies, multifunctional, collaborative, and inclusive garden" and its spatial and functional setting are understood as co-produced by human and more-than-human species. This relational framing provides the conceptual basis for local institutions to treat biodiversity not merely as a commodity providing benefits to humans but as an active agent shaping urban spaces and experimental trajectories, particularly in the context of climate change and ecological transition (Houston et al., 2017).

The second aspect highlights the crucial role of experts throughout the process. Across the three workshops, experts guided participants in reflecting on the needs, daily and seasonal trajectories, and micro-histories of the species inhabiting the garden. Prior to the co-design phase, the experts identified and mapped both native and invasive species present in the garden, recognizing the vulnerabilities and threats that characterize their habitats. During the workshops, they shared their knowledge with participants (including institutional representatives), encouraging them to empathize with the species and to appreciate their specific needs and trajectories, whose presence and agency were actively acknowledged within the participatory process (Hernandez-Santin et al., 2023). After the workshops, experts continued to play a critical role in mediating the dialogue between the citizens and the institutional representatives and in guiding local authority's intervention on restoring or improving the habitats of the garden—such as the pond for the European green toad, and designing and installing the facilities (e.g., the fence bordering the area of high naturalness). While institutional representatives and public officers from the local authority actively participated in the co-design process, their involvement helped align the local political agenda with the themes explored during co-design and engage them in the project's learning process. However, their participation also introduced specific challenges. On certain occasions, representatives drew upon existing power dynamics and political priorities to steer discussions in particular directions, which not only constrained the emergence of innovative ideas but also generated some tensions among workshop participants.

A final aspect concerns monitoring as a crucial component of co-design for multispecies co-existence. By combining ecological and biodiversity monitoring through sensors with social monitoring based on questionnaires, the project was able to build an integrated dataset and initiate an assessment of the impacts of the co-design process on both human garden users and more-than-human species. While the full results of the monitoring activity will require long-term data collection to generate robust scientific evidence, an initial key finding points to users' high level of recognition of the garden's vision as a multispecies arena, in which human and more-than-human species establish balanced patterns of coexistence. In this context, more than 70 respondents identified various elements of biodiversity and nature that positively affect their daily lives, including the provision of tranquil, calming spaces, increased awareness and knowledge of living beings, reduced noise, and improved air quality. Another equally important aspect is the broad social acceptance of the area of high naturalness where human access is restricted to protect species' habitats. In this case, in more than 60 responses, participants indicated strong or very strong positive feelings about the presence and value of an area of high naturalness in the garden. To confirm recognition of biodiversity and its dedicated space, 54 respondents indicated that, even after removal of the protective fence, they would respect the accessibility limitations to not harm the species living in that area. Furthermore, all respondents reported high vegetation diversity in the garden, with most floral diversity observed within the area of high naturalness. While 66 respondents indicated at least some benefit from the strong presence of biodiversity, 70 responses also addressed its negative aspects and, in broader terms, ecosystem disservices. These negative aspects primarily concern the presence of pests, allergens, and invasive species, as well as inadequate management (primarily trimming and care), especially of floral species. An overview of the findings indicates that biodiversity and green spaces, despite the presence of some disservices, are highly valued by the garden's residents and visitors because of the broad range of cultural and provisioning ES they provide.

The value of the area of high naturalness in terms of biodiversity is further supported by the ecological and biodiversity monitoring results, as the *Spectrum* sensors installed within the fenced area recorded higher levels of pollinator activity than those located both outside the area within the garden, and outside the garden in the neighborhood. Despite the measurements showing a low level of pollinator presence relative to similar-sized green spaces in urban areas, a comparison of pollinator presence between sensors installed inside and outside the area of high naturalness within the garden shows that the average pollinator presence is significantly higher inside the fenced area. For instance, from April to October 2025, a sensor located in the central part of the garden recorded 0.34 buzz/h/y, while a sensor installed on the northern side of the area of high naturalness recorded 0.80 buzz/h/y. While these results suggest a relative effectiveness of the area characterized by higher naturalness, the analysis of sensor data over longer and more comprehensive time periods will be crucial to consolidate robust trends and to develop a more complete understanding of pollinator activities. Based on the monitoring data currently available, aggregated sensor readings indicate the effectiveness of the high-biodiversity area remains limited, with only negligible differences in pollinator presence observed inside and outside the fenced area. This result might be primarily attributable to the small size of the garden and its relative isolation from other large green spaces in the surrounding neighborhood, as well as to microscale factors, such as the presence and distribution of specific plant species in the immediate vicinity of the sensors.

5 Conclusions

In the research presented in this chapter, we employed several instruments and tools to reflect upon and foster multispecies coexistence in the co-design of an urban green space. The research demonstrates the capacity of the co-design processes to foster a learning environment grounded in reciprocal relationships between human and more-than-human species. In addition, the integration of ecological and social monitoring tools enabled a systematic assessment of the impacts of the co-design process on both human and more-than-human communities inhabiting the garden.

In terms of ecological monitoring, while the findings indicate strong pollinator presence in the area of high naturalness, the relatively low overall performance of the garden suggests the need for further interventions oriented, for instance, to increase plant species diversity in the garden and strengthen its ecological connectivity with the nearby ecological corridors at both neighborhood and city scales. Looking at the social monitoring, the findings reveal that garden users exhibit a high level of awareness and recognition of the garden's vision as a multispecies arena, in which humans and more-than-human species engage in balanced patterns of coexistence. Moreover, users express strong respect for the non-human species inhabiting the garden, indicating the effectiveness of the co-design process in shaping shared values and practices of care.

In conclusion, we draw on the insights from the case study to highlight broader implications for planning toward multispecies coexistence.

First, there are epistemic and representational challenges: who gets to "speak for" non-humans, and how are their interests included in planning deliberations? More-than-human geographies show that extending rights-based frames to non-humans is insufficient without reworking relational ethics and embodied competences of care. Urban "cosmopolitics" experiments with practices that sense and negotiate the presence of urban wild geography yet also reveal how contested and unstable such politics remain. Planning scholarship therefore calls for new mediations (e.g., citizen science, ecological proxies, and design probes) that transduce multispecies needs into actionable spatial decisions (Whatmore, 2006). In the case of Sorelle Mirabal Garden, the experts from the Natural History Museum and Civic Aquarium played an advocacy role for the non-humans in the workshops.

Second, institutional and governance limitations still persist in the incorporation of multispecies needs and trajectories into policies and planning. While GI and NbS are now mainstream policy instruments, many city strategies lack mechanisms to assess human and more-than-human relationships with the multifunctionality of ES, monitor trade-offs, or coordinate actions across departments and scales. In the case study, the local authority of Milan (Municipio 3) relied more often on showcasing examples whereas this coordination has worked. Comparative studies in European cities found that strategic plans rarely detail how to enhance urban biodiversity or manage synergies and conflicts between the needs of residents and those of the more-than-human species. Moreover, although biodiversity-sensitive strategies are often promoted for climate and health, recent reviews indicate that explicit biodiversity goals remain secondary in many urban initiatives, raising risks of "green solutions" that underperform for non-human life (Tannous et al., 2024).

Third, environmental and multispecies justice issues complicate implementation. Urban greening can unintentionally trigger ecological gentrification and uneven distributions of benefits and burdens; practitioners frequently oscillate between frames of public health and biodiversity, with divergent implications for coexistence. Contributions from the literature on multispecies justice, such as Raymond et al. (2025), urge planners to redistribute attention and resources toward species typically marginalized by human-centric priorities and to design for safe passage, habitat heterogeneity, and temporal rhythms (e.g., nesting, migration) that exceed standard human use cycles (Hunold, 2020).

Fourth, conflict–coexistence dynamics require re-framing. Human–wildlife interactions in cities range from tolerance to conflict; turning conflict into coexistence demands attention to values, emotions, and social identity as much as to ecological mitigation. Multilevel approaches such as the co-design process of the garden—combining behavioral, expertise, and governance interventions—are recommended to move along the conflict-to-co-existence continuum. The more-than-human planning agenda recasts such work as cultivating shared urban worlds, not merely minimizing nuisance (Hobbs et al., 2006).

To conclude, we argue that the methods and tools tested in the case-study area of the Sorelle Mirabal Garden demonstrate a significant potential for replicability in supporting urban regeneration processes aimed at enhancing biodiversity, both in quantitative terms and in relation to social acceptance. Although the procedure developed within the community-level urban living lab is strongly place-based and site-specific, it can nevertheless be replicated in comparable urban contexts that share similar spatial, ecological, and social conditions. From a co-creation perspective, two phases are particularly relevant for replication. First, the systematic integration of social and ecological monitoring and co-monitoring activities throughout the process makes it possible to capture evolving relationships and interspecies dynamics over time. Second, a co-creation approach that explicitly attends to the representation of multispecies interests fosters awareness, sensitivity, and pro-biodiversity attitudes within local communities. When applied consistently across urban regeneration initiatives involving green areas, this approach can contribute to the accumulation of robust scientific data, while also informing the development of evidence-based policies that better integrate biodiversity considerations into urban transformation processes.

Acknowledgements This research has received funding from the Project "National Biodiversity Future Center—NBFC" funded under the National Recovery and Resilience Plan (NRRP), Mission 4 Component 2 Investment 1.4—Call for tender No. 3138 of 16 December 2021, rectified by Decree n.3175 of 18 December 2021 of Italian Ministry of University and Research funded by the European Union NextGenerationEU; Project code CN_00000033, Concession Decree No. 1034 of 17 June 2022 adopted by the Italian Ministry of University and Research, CUP, D43C22001250001.

Author Contributions The authors' contribution according to CRediT (Contributor Roles Taxonomy) is described as follows. Eugenio Morello (EM); Luca Lazzarini (LL); Israa Mahmoud (IM); Asef Ayatollahi (AA). Conceptualization: EM, LL, IM, AA; Methodology: EM, LL, IM, AA; investigation: EM, LL, IM, AA; Visualization: LL; writing—original draft: LL, IM; writing—review & editing: EM, LL, IM, AA; Supervision: EM.

References

Ayatollahi, A., & Morello, E. (2025). Scoping the emerging role of urban biodiversity in social impact assessment, a systematic review of regulating ecosystem services indicator types. *Frontiers in Sustainable Cities, 7*, 1623650. https://doi.org/10.3389/frsc.2025.1623650

Barthel, S., Folke, C., & Colding, J. (2010). Social–ecological memory in urban gardens—Retaining the capacity for management of ecosystem services. *Global Environmental Change, 20*(2), 255–265. https://doi.org/10.1016/j.gloenvcha.2010.01.001

Boffi, M., Pola, L. G., Fermani, E., Senes, G., Inghilleri, P., Piga, B., Stancato, G., & Fumagalli, N. (2022). Visual post-occupancy evaluation of a restorative garden using virtual reality photography: Restoration, emotions, and behavior in older and younger people. *Frontiers in Psychology, 13*, 927688. https://doi.org/10.3389/fpsyg.2022.927688

Clarke, M., Davidson, M., Egerer, M., Anderson, E., & Fouch, N. (2019). The underutilized role of community gardens in improving cities' adaptation to climate change: A review. *People, Place and Policy, 12*(3), 241–251. https://doi.org/10.3351/ppp.2019.3396732665

Díaz, S., Demissew, S., Carabias, J., et al. (2015). The IPBES conceptual framework—Connecting nature and people. *Current Opinion in Environmental Sustainability, 14*, 1–16. https://doi.org/10.1016/j.cosust.2014.11.002

Eizenberg, E. (2018). One landscape, multiple meanings: Revisiting contemporary discourses on Urban Community gardens. *Built Environment, 44*(3), 326–338. http://www.jstor.org/stable/45173839

Fumagalli, N., Fermani, E., Senes, G., Boffi, M., Pola, L., & Inghilleri, P. (2020). Sustainable co-design with older people: The case of a public restorative garden in Milan (Italy). *Sustainability, 12*(8), 3166. https://doi.org/10.3390/su12083166

Jacob, M., & Rocha, C. (2021). Models of governance in community gardening: Administrative support fosters project longevity. *Local Environment, 26*(5), 557–574. https://doi.org/10.1080/13549839.2021.1904855

Hansen, R., & Pauleit, S. (2014). From multifunctionality to multiple ecosystem services? A conceptual framework for multifunctionality in green infrastructure planning for urban areas. *Ambio, 43*(4), 516–529. https://doi.org/10.1007/s13280-014-0510-2

Hinchliffe, S., Kearnes, M. B., Degen, M., & Whatmore, S. (2005). Urban wild things: A Cosmopolitical experiment. Environment and Planning D: Society and Space, 23(5), 643–658. doi:https://doi.org/10.1068/d351t (Original work published 2005).

Hernandez-Santin, C., Amati, A., Bekessy, S., & Desha, C. (2023). Integrating biodiversity as a non-human stakeholder within urban development. *Landscape and Urban Planning, 232*(104678). https://doi.org/10.1016/j.landurbplan.2022.104678

Hunold, C. (2020). Urban Greening and Human–Wildlife Relations in Philadelphia: From Animal Control to Multispecies Coexistence? Environmental Values, 29(1), 67–88. https://www.jstor.org/stable/26870972.

Huq, F. F., & Deacon, L. (2025). A systematic review of community gardens and their role in urban food security and resilience. *Discover Sustainability, 6*(696). https://doi.org/10.1007/s43621-025-01628-5

Houston, D., Hillier, J., MacCallum, D., Steele, W., & Byrne, J. (2017). Make kin, not cities! Multispecies entanglements and 'becoming-world' in planning theory. *Planning Theory, 17*(2), 190–212. https://doi.org/10.1177/1473095216688042

Hobbs, R. J., Arico, S., Aronson, J., Baron, J. S., Bridgewater, P., Cramer, V. A., Epstein, P. R., Ewel, J. J., Klink, C. A., Lugo, A. E., Norton, D., Ojima, D., Richardson, D. M., Sanderson, E. W., Valladares, F., Vilà, M., Zamora, R., & Zobel, M. (2006). Novel ecosystems: Theoretical and management aspects of the new ecological world order. *Global Ecology and Biogeography, 15*(1), 1–7. https://doi.org/10.1111/j.1466-822X.2006.00212.x

IUCN. (2020). IUCN global standard for nature-based solutions: A user-friendly framework for the verification, design and scaling-up of NbS. *International Union for Conservation of Nature.* https://doi.org/10.2305/IUCN.CH.2020.08.en

Kirksey, S. E., & Helmreich, S. (2010). The emergence of multispecies ethnography. *Cultural Anthropology, 25*(4), 545–576. by the American Anthropological Association. As Published:. https://doi.org/10.1111/j.1548-1360.2010.01069.x

Lazzarini, L. (2024). Biodiversity as a key relational value for cities. In F. Calabrò, L. Madureira, F. C. Morabito, & M. J. Piñeira Mantiñán (Eds.), *Networks, Markets & People. NMP 2024* (*Lecture notes in networks and systems*) (Vol. 1187, pp. 62–71). Springer. https://doi.org/10.1007/978-3-031-74704-5_7

Mahmoud, I., Dubois, G., Liquete, C., & Robuchon, M. (2025). Understanding collaborative governance of biodiversity-inclusive urban planning: Methodological approach and benchmarking results for urban nature plans in 10 European cities. *Urban Ecosystems, 28*, 17. https://doi.org/10.1007/s11252-024-01656-5

Mahmoud, I. H., Morello, E., Vona, C., Benciolini, M., Sejdullahu, I., Trentin, M., & Pascual, K. H. (2021). Setting the social monitoring framework for nature-based solutions impact: Methodological approach and pre-greening measurements in the case study from CLEVER cities Milan. *Sustainability, 13*(17), 28. https://doi.org/10.3390/su13179672

Morin, E., & Kern, B. (1999). *Homeland earth: A manifesto for the new millennium.* Hampton Press.

Municipio 3, Comune di Milano (2023), *Linee guida per la manutenzione, la valorizzazione e l'apertura del Giardino comunale di via San Faustino*, Deliberazione di Giunta di Municipio 3 n. 67 del 24 maggio 2023.

Nilon, C. H., Aronson, M. F. J., Cilliers, S. S., Dobbs, C., Frazee, L. J., Goddard, M. A., O'Neill, K. M., Roberts, D., Stander, E. K., Werner, P., Winter, M., & Yocom, K. P. (2017). Planning for the future of urban biodiversity: A global review of City-scale initiatives. *Bioscience, 67*(4), 332–342. https://doi.org/10.1093/biosci/bix012

Oke, C., Bekessy, S. A., Frantzeskaki, N., Bush, J., Fitzsimons, J. A., Garrard, G. E., Grenfell, M., et al. (2021). Cities should respond to the biodiversity extinction crisis. *Npj Urban Sustainability, 1*(1), 1–4. https://doi.org/10.1038/s42949-020-00010-w

O'Donnell, M., Collier, M., Pineda-Pinto, M., Cooper, C., Nulty, F., & Castañeda, N. R. (2025). Redefining co-design for social-ecological research and practice: A systematic literature review. *Environmental Science & Policy, 164*, 103998. https://doi.org/10.1016/j.envsci.2025.103998

Pascual, U., Balvanera, P., Díaz, S., et al. (2017). Valuing nature's contributions to people: The IPBES approach. *Current Opinion in Environmental Sustainability, 26–27*, 7–16. https://doi.org/10.1016/j.cosust.2016.12.006

Raymond, C. M., Rautio, P., Fagerholm, N., Aaltonen, V. A., Andersson, E., Celermajer, D., Christie, M., Hällfors, M., Saari, M. H., Mishra, H. S., Lechner, A. M., Pineda-Pinto, M., & Schlosberg, D. (2025). Applying multispecies justice in nature-based solutions and urban sustainability planning: Tensions and prospects. *Npj Urban Sustainability, 5*(1), 2. https://doi.org/10.1038/s42949-025-00191-2

Sanders, E. B.-N., & Stappers, P. J. (2008). Co-creation and the new landscapes of design. *Journal of CoDesign, 4*(1), 5–18. https://doi.org/10.1080/15710880701875068

Tannous, J., Aderele, M. O., & Adetula, O. I. (2024). Assessing the contributions of nature-based solutions to urban biodiversity enrichment. In W. Leal Filho, G. J. Nagy, & D. Ayal (Eds.), *Handbook of nature-based solutions to mitigation and adaptation to climate change*. Springer. https://doi.org/10.1007/978-3-030-98067-2_42-1

Van den Tillaar, M. (2025), *Deep dive—How to engage nature as a stakeholder?* Community of Practice "Meaningful Stakeholder Dialogue", https://www.ser.nl/en/themes/irbc/meaningful-dialogue/community-of-practice (accessed on December 8th 2025).

Whatmore, S. (2006). Materialist returns: Practising cultural geography in and for a more-than-human world. *Cultural Geographies, 13*(4), 600–609. https://doi.org/10.1191/1474474006cgj377oa

Green Interventions at the Urban Scale: Approaches and Methods

The Phytopolis Strategy: Redesigning Barcelona, Copenhagen, and Paris Through Green Climate-Risk Infrastructure

Maicol Negrello

Abstract Urban climate risks increasingly manifest through compound heat–flood dynamics that exceed the coping range of conventional grey infrastructure and sectoral planning. Nature-based solutions (NBS) are therefore shifting from "add-on greening" towards urban-scale climate-risk infrastructure, where soil, vegetation, and blue–green networks are designed, financed, and governed as core public systems delivering adaptation, mitigation, and health co-benefits (European Commission, 2015; IPCC, 2022). This chapter develops a comparative analysis of three European frontrunner cities—Barcelona, Copenhagen, and Paris—focusing on (i) the unit of action used to scale interventions, (ii) the delivery and economic mechanisms enabling implementation, and (iii) the performance evidence and indicators used to justify, prioritise, and evaluate projects. Methods combine document analysis of municipal plans, regulatory instruments, and financing arrangements with a targeted review of peer-reviewed literature on urban health, climate adaptation, and NBS effectiveness. Barcelona's Eixample "green axes and squares" programme operationalises climate-risk infrastructure through a street-grid transformation (e.g., Consell de Cent and Girona) aligned with mobility and public-space redesign. Copenhagen's Cloudburst Management Plan structures investments by hydrological catchments and a co-financing model between municipality and utility (e.g., Enghaveparken; Sankt Kjelds Plads). Paris leverages regulatory reform (PLU bioclimatique), municipal greening programmes, and green bond finance (Paris Climate Bond), with emblematic cooling projects (e.g., Place de Catalogne; streets and school-adjacent greening). By synthesising scaling logics, governance and finance, and measurable outcomes, the chapter contributes an operational framework for NBS as climate-risk infrastructure, relevant to trends in urban policy, climate adaptation, and delivery models for large-scale implementation.

Keywords Nature-based solutions · Climate change adaptation · Urban resilience · Blue–green infrastructure · Delivery mechanisms · Urban health · Policy instruments

M. Negrello (✉)
Department of Architecture and Design (DAD), Politecnico di Torino, Torino, Italy
e-mail: maicol.negrello@polito.it

L. Diana et al. (eds.), *Greening and Resilience*, Urban Sustainability,
https://doi.org/10.1007/978-981-92-0419-9_14

1 Introduction

European cities are undergoing a rapid intensification of climate risks, with heatwaves and extreme rainfall among the most critical drivers affecting safety, public health, and the continuity of urban services. Recent European risk syntheses indicate an increase in intense precipitation across many areas, with a corresponding rise in urban flood risk, and further increases expected as global temperatures continue to rise (EEA, 2024). In parallel, thermal extremes are becoming more frequent and severe: heat stress is amplified by dense urban morphologies, high-thermal-mass materials, limited vegetation cover, and anthropogenic heat emissions, which together reinforce the urban heat island effect (IPCC, 2022; Santamouris, 2015). Crucially, these processes are increasingly taking the form of compound and cascading risks, in which heat, hydraulic overload, service disruptions, and health impacts interact through shared exposure pathways—impervious surfaces, ageing or undersized drainage networks, contested public space, and socio-spatial inequalities (EEA, 2025; IPCC, 2022). In many European contexts, the same physical drivers that increase thermal load—high levels of surface sealing, reduced evapotranspiration, deficits of living soil, and shading—also accelerate runoff generation during short, intense events, facilitating the transition from cloudbursts to pluvial flooding (EEA, 2024; Zölch et al., 2016).

Within this context, the design of urban space—and therefore urban design—is increasingly framed as a strategic lever that extends well beyond the production of singular, localised project outputs. From a systems perspective, the redesign of urban fragments can generate indirect and cumulative regenerative impacts across environmental (microclimate, hydrological functioning, ecological quality), social (health, comfort, perceived safety, equitable access to climate-protective public spaces), and economic dimensions (damage reduction, improved service reliability, and the creation of maintenance and management value chains) (European Commission, 2015; IPCC, 2022). In this perspective, Nature-based Solutions (NBS) should not be understood as a merely technical, "vegetal" add-on; rather, they become constitutive elements of urban design and of a renewed form of urbanity, in which nature, soil, and water actively shape the form, performance, and meaning of public space (Cohen-Shacham et al., 2016; European Commission, 2015).

This implies a shift in design paradigm: nature is treated as a true material of architectural and urban design, able to structure spatial composition while simultaneously reducing exposure to climate hazards. In other words, NBS become "infrastructure" not only because they provide measurable ecosystem services, but because they underpin replicable spatial typologies and devices—streets, squares, parks, and corridors—that enable the transition from isolated interventions to coherent, adaptive, and regenerative urban systems (European Commission, 2015). In line with a Nature-Based Regeneration perspective, nature is not an ornamental layer but a compositional and performance driver capable of reshaping the relationships between urban form, biodiversity, and users' well-being (Negrello, 2025).

A decisive issue nevertheless concerns scaling. Much of the evidence base for NBS derives from plot- or park-scale interventions, whereas urban climate risks propagate through interconnected networks and catchments. The operational question for planning and design is therefore how to define a unit of action that allows local interventions to be replicated and assembled into a coherent, city-wide risk-management system. Two additional challenges follow. First, delivery: large-scale transformation requires stable governance and financing arrangements capable of supporting implementation and long-term maintenance. Second, evidence: cities must justify priorities and investments through credible indicators and monitoring frameworks that demonstrate performance over time (European Commission, 2015; IPCC, 2022). Against this backdrop, this chapter adopts a comparative lens across three European urban contexts—Barcelona, Copenhagen, and Paris—to show how different climatic regimes and institutional-cultural settings can support different design approaches and policy devices for implementing NBS at the urban scale. The aim is not to rank "best practices", but to clarify: (i) which units of action enable scaling, (ii) which economic and institutional instruments make transformation feasible, and (iii) which indicators support accountability and learning, with direct implications for the quality of urban space and for the transition towards robust and regenerative climate-risk infrastructure.

2 Background Knowledge

Nature-based solutions (NBS) are commonly defined as actions inspired and supported by nature that address societal challenges while delivering benefits for biodiversity and human well-being (Cohen-Shacham et al., 2016; European Commission, 2015). In urban contexts, NBS intersect with blue–green infrastructure, sustainable drainage, and climate-adaptive public-space design; however, a key shift in contemporary European practice is their increasingly explicit positioning as climate-risk infrastructure—systems expected to achieve defined service levels (e.g., outdoor heat mitigation, runoff retention, peak-flow attenuation, flood routing, and recovery) and to remain reliable over decades, akin to conventional networks (IPCC, 2022). This infrastructural framing makes evidence central: rather than treating "more green" as a generic proxy for resilience, cities and researchers are progressively converging on indicator frameworks that link design decisions (soil volume, canopy structure, permeability, shading patterns, and connectivity) to measurable outcomes and co-benefits, while making trade-offs and distributional effects visible at neighbourhood scale (Raymond et al., 2017; Hölscher et al., 2023). The urban-health rationale has become particularly influential in strengthening the investment case for large-scale street and public-realm redesign: health impact assessment (HIA) and exposure modelling for Barcelona's Superblock approach show that reallocating street space away from motor traffic—combined with greening and the reduction of air pollution, noise, and heat exposure—can produce quantifiable health gains, including reductions in premature mortality, though results depend on assumptions, implementation quality, and the spatial distribution of benefits (Mueller et al., 2020; Nieuwenhuijsen et al., 2024). More recent assessments of Barcelona's "Eixos Verds" street-greening strategy further support this direction by modelling mental-health benefits associated with increased neighbourhood greenness and improved public-space conditions, strengthening the argument that NBS-enabled urban redesign can operate simultaneously as adaptation infrastructure and public-health intervention (Vidal Yañez et al., 2023). At the same time, the literature cautions that scaling requires more than technical performance: mainstream adoption depends on governance capacities that embed NBS into mandates, budgeting routines, procurement, and long-term stewardship, ensuring that maintenance and monitoring are treated as integral to infrastructure delivery rather than residual costs (Hölscher et al., 2023). Within this evolving evidence base, an emerging regenerative urban design perspective reframes NBS as a compositional and ecological agenda in which nature functions as a material of urbanism—structuring morphology, use and identity of streets, squares, and parks while supporting biodiversity and well-being beyond risk reduction alone (Negrello, 2025).

3 Climate-Resilient Strategies and Implications at the City Scale

This study adopts a comparative, multiple-case design aimed at explaining how urban-scale green interventions become deliverable climate-risk infrastructure through the alignment of spatial typologies, policy instruments, financing arrangements, and evaluation practices. A structured case-study protocol was used to ensure analytical comparability across Barcelona, Copenhagen, and Paris, focusing on implementation logic rather than descriptive narratives (Yin, 2018).

3.1 Case Selection and Evidence Base

The three cases were selected through purposive sampling as theoretically contrasting "programme architectures" for scaling nature-based solutions (NBS): street-network transformation (Barcelona), catchment-based cloudburst management (Copenhagen), and regulatory mainstreaming via bioclimatic planning (Paris). Selection was supported by the presence of (i) mature policy frameworks, (ii) identifiable flagship urban-design projects, and (iii) sufficient documentary and scientific material to reconstruct delivery mechanisms and performance evidence. The evidence base combines: plans and technical reports (City of Copenhagen, 2012; Ville de Paris, 2024a, 2024b), municipal programme documentation and delivery agency materials (Ajuntament de Barcelona, n.d.; BIMSA, n.d.; Ville de Paris, 2025), and peer-reviewed studies and synthesis reviews relevant to impact, governance, and evaluation (Nieuwenhuijsen, 2024; Salih & Báthoryné Nagy, 2024; Vidal Yañez et al., 2023; Whittaker & Jespersen, 2022).

3.2 *Analytical Coding: Instruments–Typologies–Evidence*

To operationalise the comparison, the study adopted a coding framework aligned with the cross-case matrix, structured around five categories: (1) the scaling unit ("unit of action"), defined as the smallest repeatable spatial and administrative unit enabling programme replication; (2) the key policy / plan frame (plans, binding rules and programme targets); (3) core delivery and economic mechanisms, including named governance arrangements, financing instruments, and maintenance responsibilities; (4) typological and flagship urban-design outputs, capturing both repeatable spatial forms and named projects; and (5) performance evidence / indicator maturity, describing the extent to which objectives are translated into auditable indicators through modelling, monitoring, reporting, and, where applicable, health impact assessment.

This structure enables the analysis to trace a pathway from institutional arrangements to repeatable design units and measurable outcomes. In Copenhagen, the scaling unit is organised around hydrological sub-catchments and flow routes, while the plan-level embedding of return-period standards supports risk-based prioritisation and portfolio sequencing (City of Copenhagen, 2012). Delivery is stabilised through utility–municipality co-delivery and co-financing enabled by national regulation (BEK nr 159, 2016), and its evidence profile prioritises service-level flood protection metrics and mapped flood-depth reduction (Whittaker & Jespersen, 2022). In Barcelona, the scaling unit is the corridor–node logic of Eixample green axes and intersection squares, which supports exposure-based evaluation; evidence maturity is strengthened by HIA approaches linking changes in greenness exposure to health outcomes (Vidal Yañez et al., 2023) alongside synthesis evidence on the Superblock model (Nieuwenhuijsen, 2024). In Paris, regulation-led delivery shifts the implementation locus towards plot-level compliance under the PLU bioclimatique, complemented by distributed public-space typologies (e.g., "rues végétalisées" and urban forests). This implies an evaluation approach centred on cumulative, citywide monitoring of canopy trajectories, permeability/open-ground proxies, and cool-space accessibility, supported by auditable programme reporting (Ville de Paris, 2024a, 2024b; Ville de Paris, 2025).

3.3 *Evidence Appraisal, Triangulation, and Limitation*

Evidence was triangulated across document types and scientific literature to reduce reliance on single-source claims. Peer-reviewed studies were prioritised when they explicitly reconstruct an evidence chain connecting intervention-induced exposure changes to outcomes. This includes the quantitative health impact assessment of Barcelona's Eixos Verds (Vidal Yañez et al., 2023) and a 2024 review synthesising the Superblock model's sustainability and health implications (Nieuwenhuijsen, 2024). For Copenhagen, the financing and governance of adaptation delivery were interpreted using peer-reviewed work on institutional and market conditions shaping long-horizon investment (Whittaker & Jespersen, 2022). Methodological claims about microclimate modelling uncertainty and sensitivity—relevant across all cases—were grounded in recent review evidence emphasising the importance of transparent assumptions and parameterisation (Salih & Báthoryné Nagy, 2024).

Two limitations are acknowledged. First, accounting conventions and reporting units differ across cities (e.g., municipal vs utility budgets; planned vs delivered outputs), constraining direct cost-effectiveness comparisons and motivating an emphasis on mechanisms and evaluation readiness. Second, causal attribution is inherently uneven: Barcelona's evidence base is relatively mature for health-related outcomes (Vidal Yañez et al., 2023), while Paris' regulation-led pathway produces more diffuse causality that demands longitudinal monitoring and compliance-oriented metrics (Ville de Paris, 2024a, 2024b). These limitations are treated not as weaknesses but as substantive findings regarding what different programme architectures make measurable—and when (Table 1).

3.4 Barcelona: Street-Grid Greening as Mobility–Health–Adaptation Infrastructure

Barcelona's recent urban strategy is notable for treating the street network itself as a primary surface for climate-risk infrastructure. The Superilla approach—operationalised in the Eixample through "green axes" (eixos verds) and redesigned intersections ("new squares")—uses the Cerdà grid as an advantage: repetition and standardisation allow a consistent design model to scale across multiple streets while maintaining local adaptability. In the official competition documentation, the model is articulated through a set of street "premises": pedestrian priority (limited destination traffic at low speed), continuous universal accessibility (single-level platform), asphalt removal, intensified greening at chamfered corners, rainwater management, and integration of energy features (Ajuntament de Barcelona, 2021).

Scaling Unit: The "Green Axis + Square" as Replicable Urban Design Module

Barcelona's scaling unit is neither the park nor the plot. It is a replicable street-grid module: a street segment re-urbanised as a "green axis", paired with nodal squares at key intersections. The 2021 programme identifies specific early transformations,

Table 1 Comparative matrix: scaling units, policies, delivery mechanisms, and evidence (Barcelona–Copenhagen–Paris)

City	Scaling unit ("unit of action")	Key policy / plan frame	Core delivery / economic mechanisms (named)	Flagship urban-design outputs (≥2)	Performance evidence / indicators (examples)
Barcelona	**Street-grid transformation unit** in the Eixample: *street segments as "green axes" + nodal intersections as "squares"* deployed through a network logic (replicable module)	**Superilla Barcelona** strategy and the **Eixample Green Axes & Squares** programme (implementation phase 2020–2023 for first set of axes/squares)	**EIB municipal framework financing**: **€95m (approved 2020)** to finance ~40 climate action/urban regeneration projects; **EIB loan €175m (signed 2025)** aligned with Barcelona's **Municipal Investment Plan (PIM) 2024–2027** supporting regeneration, green spaces, flood protection, and sustainable mobility	**Consell de Cent** "green axis"; **Girona** "green axis"; (additional first-phase axes: **Rocafort**, **Comte Borrell**) + **squares** at key intersections including **Consell de Cent × Rocafort**, × **Comte Borrell**, × **Enric Granados**, × **Girona** (Ajuntament de Barcelona, 2020–2023)	**Health/liveability rationale** used to justify public-space reallocation and greening: quantitative HIA evidence and superblock literature (Mueller et al., 2020; Nieuwenhuijsen, 2024); **mental-health HIA** for the "Eixos Verds" intervention (Vidal Yañez et al., 2023). Typical indicators: traffic reduction/ proximity exposure, greenness exposure, heat-risk reduction proxies, public-space reallocation metrics

(continued)

Table 1 (continued)

City	Scaling unit ("unit of action")	Key policy / plan frame	Core delivery / economic mechanisms (named)	Flagship urban-design outputs (≥2)	Performance evidence / indicators (examples)
Copenhagen	**Hydrological catchment + flow-path unit**: cloudburst corridors, retention nodes, streets-as-conveyance, and sewer interfaces organised by risk and conveyance; city subdivided into prioritised catchments; portfolio of ~300 projects over ~20 years	**Cloudburst Management Plan (2012)** (linked to the earlier **Climate Adaptation Plan**) adopting a service-level risk-management frame for extreme rainfall	**Utility–municipality co-delivery** (City of Copenhagen + **HOFOR**) funded via **water tariffs / drainage charges**; enabled by national **co-financing regulation** allowing utilities to co-finance surface projects: **BEK nr 159 (26/02/2016)** on wastewater utilities' co-financing of roof/ surface-water projects; long-term programme sequencing and annual prioritisation of project packages	**Enghaveparken Climate Park** (cloudburst storage, multifunctional park); **Sankt Kjelds Plads & Bryggervangen** (nature-based cloudburst corridor and street transformation); (early demonstrator often cited: **Tåsinge Plads** climate-adapted square)	**Service-level framing** (protection against extreme rainfall events) and project/programme performance metrics. Typical indicators: **retention capacity (m^3)** (e.g., park reservoirs), conveyance capacity, flood-depth/extent maps, avoided-damage logic, and cost-effectiveness (City of Copenhagen, 2012; C40, 2016). Programme-scale monitoring often reports the number/status of projects and delivered blue–green assets

(continued)

Table 1 (continued)

City	Scaling unit ("unit of action")	Key policy / plan frame	Core delivery / economic mechanisms (named)	Flagship urban-design outputs (≥2)	Performance evidence / indicators (examples)
Paris	**Regulatory + programme unit**: (i) parcel/plot rules under **PLU bioclimatique** (prescriptive standards for permeable/vegetated surfaces) + (ii) distributed public-space typologies (streets, school-adjacent streets, squares/urban forests) implemented via recurring municipal greening pipelines	**PLU bioclimatique** approved **20 Nov 2024** and entered into force thereafter; complementary city programmes for greening streets/squares (e.g., "rues végétalisées", "rues aux écoles")	**Paris Climate Bond (2015)**: **€300m** sustainability/green bond with an adaptation sub-portfolio (notably **20,000 trees** + **30 ha new parks**, deadline 2020); plus routine municipal capital programming for street greening and planting seasons (e.g., 2024–2025 cycle with large batches of streets/places)	**Place de Catalogne urban forest** (**470 trees**, > **4,000 m^2** forested area); **Rue Charles-Moureu** "rue-jardin" (target > **58%** vegetated surface); plus citywide **"rues aux écoles"** transformations as a scalable typology	**Accountability through targets and public reporting**: citywide baseline shares and trajectories—**31%** of Paris territory as nature/green surfaces, dropping to **21%** without Bois de Boulogne/Vincennes (Cour des comptes, 2024); policy targets such as **35% permeable vegetated surfaces** and "full soil" ambitions appear across planning documents and audits. Typical indicators: tree-planting progress (net/gross accounting), de-sealing area, permeable surface share, heat-risk proxies, and programme delivery pace; increasing demand for auditable KPIs and consistent definitions (Cour des comptes, 2024)

including the Consell de Cent corridor and additional axes such as Girona, Rocafort, and Comte Borrell, together with redesigned squares at intersections along Consell de Cent (e.g., Consell de Cent–Rocafort; –Borrell; –Enric Granados; –Girona) (Ajuntament de Barcelona, 2021). This matters because the unit is both morphological (street section + surface materials + planting logic) and governance-ready (a bounded scope for contracts, phasing, and local consultation).

Delivery and Economics: Municipal Investment + EIB Financing for Regeneration and Climate Action

Barcelona's ability to maintain continuity and scale relies on embedding green transformation within broader municipal investment and regeneration packages. The European Investment Bank approved a €95 million loan in 2020 to support around 40 projects, explicitly framing the investment around urban regeneration and climate-related outcomes (European Investment Bank, 2020). More recently, Barcelona reached an agreement on a €175 million EIB loan (announced March 2025) to fund urban regeneration projects linked to sustainable mobility, green spaces, and public infrastructures, aligned with the Municipal Investment Plan 2024–2027 and climate action priorities (Ajuntament de Barcelona, 2025). The key point is not only the volume of finance, but the portfolio logic: NBS and public-space redesign are funded as part of a multi-scheme investment narrative rather than as isolated greening projects.

Performance Evidence: From Liveability to Measurable Health Impacts

Barcelona is also one of the clearest European cases where health evidence is used to legitimise street transformation. Research on Superblocks has quantified potential reductions in premature mortality and produced multi-exposure models linking traffic, air pollution, noise, heat, and green space to health outcomes (Mueller et al., 2020). Recent synthesis further frames the Superblock approach as a replicable urban policy model with health and climate co-benefits, while also highlighting that results depend on implementation fidelity, exposure baselines, and behavioural assumptions (Nieuwenhuijsen, 2024). For urban-scale NBS delivery, this is crucial: street redesign becomes a public-health intervention, reinforcing the policy rationale for reallocating space from vehicles to canopy, permeable soils, and social use.

3.5 Copenhagen: Catchment-Based Cloudburst Urbanism and Utility Co-Financing

Copenhagen's strategy is shaped by a clear triggering event: the 2011 cloudburst and its high damage costs. The city responded by developing a long-term, city-wide programme: the Cloudburst Management Plan (CMP), adopted in 2012, designed to reduce flood impacts from extreme rainfall through a combined portfolio of surface blue–green solutions and underground conveyance (City of Copenhagen, 2012).

Scaling Unit: The Catchment, the Corridor, and the Retention Node

Where Barcelona's unit of action is morphological (street-grid modules), Copenhagen's is primarily hydrological. The CMP is structured around catchments, flow routes (cloudburst roads/boulevards), and retention spaces, enabling the city to prioritise projects where water can be safely stored, slowed, infiltrated, or conveyed to the harbour. This scaling logic makes climate-risk infrastructure legible as a network: streets can become controlled "rivers" during cloudbursts, while parks and squares act as temporary reservoirs. Official and semi-official documentation consistently describes a multi-decade programme with a large project pipeline (often reported as 300+ actions over ~20 years) (City of Copenhagen, 2012; Climate-ADAPT, 2013).

Delivery and Economics: The Co-financing Model (City + HOFOR) and Regulatory Enabling Copenhagen's Implementation Depends on the Institutional Architecture Linking Municipality and Utility

Delivery is widely documented as a shared investment between the City of Copenhagen and the Greater Copenhagen Utility (HOFOR), with private owners covering some property-level measures, while major conveyance and public-space components are financed through a combination of municipal resources and drainage charges / water fees (Climate-ADAPT, 2013). A pivotal enabling instrument is Denmark's utility co-financing regulation, which allows utilities to finance surface-level climate adaptation when it is cost-effective compared to purely sewer-based alternatives. The co-financing executive order is codified as BEK nr 159 of 26 February 2016 (Retsinformation, 2016). This mechanism matters because it unlocks financing for "hybrid" public-realm projects—parks, streets, and squares that deliver both flood protection and recreational value—while maintaining the economic discipline of water-sector regulation.

Flagship Projects as Performance Prototypes: Enghaveparken and Sankt Kjelds Plads

Two projects illustrate how the catchment-based scaling logic becomes designed form. Enghaveparken Climate Park retrofits a historic park into a flood reservoir with a documented cloudburst capacity of ~ 22,600 m^3, while preserving everyday recreational use (State of Green, 2021). Sankt Kjelds Plads and Bryggervangen is frequently presented as a landmark nature-based cloudburst project combining flood protection, biodiversity, and public space, delivered during 2015–2019 (SLA, 2019). Together, these prototypes demonstrate how Copenhagen operationalises "climate-risk infrastructure" through urban design typologies—retention parks, cloudburst streets, and distributed blue–green nodes.

3.6 Paris: Regulatory Scaling (PLU Bioclimatique), Programme Delivery, and Green-Bond Finance

Paris faces an acute constraint: a very dense and historically layered urban fabric with limited open ground. Regional audit work emphasises that Paris is highly exposed to heat due to low vegetation and dense built form, documenting that nature/green areas cover about 31% of the municipal territory when including the outer woods, and about 21% without them (Cour des comptes, 2024).

Scaling Unit: From "Project Sites" to a City-Wide Regulatory and Typological System

Paris's unit of action combines (i) regulatory scaling—rules and targets embedded in the PLU bioclimatique—and (ii) typological programmes applied across many sites (streets, school-adjacent streets, squares). The City states that the PLU bioclimatique (adopted 20 November 2024) commits to creating roughly 55 hectares of new green spaces, mainly through ~ 10 major parks, and a metropolitan green corridor/park structure of over 25 hectares in the north-east (Ville de Paris, 2024a). These commitments translate adaptation into land-use regulation, enabling greening not only through municipal projects but also via planning permissions and plot-based transformations. In parallel, Paris scales through programme typologies. The city's official communication for the 2024/2025 planting season states that 120 streets or squares would be vegetated, including streets transformed into child-friendly pedestrian priority routes ("rues aux enfants"), and highlights specific design targets such as Rue Charles-Moureu becoming a "rue-jardin" with > 58% of its surface vegetated (Ville de Paris, 2024b). This approach creates a replicable delivery pipeline: not one megaproject, but hundreds of smaller, typologically consistent transformations across neighbourhoods.

Delivery and Economics: Paris Climate Bond and Municipal Programmes

Paris is a leading European example of using green finance to support climate programmes. The Paris Climate Bond (2015) is widely documented as a €300 million green bond used to finance climate mitigation and adaptation initiatives (Climate Bonds Initiative & UNEP Inquiry, 2015; Climate-ADAPT, 2016). The adaptation component includes two explicit projects: planting 20,000 trees and creating 30 hectares of new parks, with implementation planned up to 2020 (Climate-ADAPT, 2016). Recent case-study work also notes continued bond activity and follow-on issuances, signalling that green bonds have become a recurrent financing channel within the city's climate investment strategy (ClimateFIT, 2024). At the same time, audit reporting underscores the importance—and difficulty—of maintaining transparent accounting, consistent indicators, and credible monitoring for large-scale greening targets, including progress towards 35% permeable vegetated surfaces and long-term tree-planting commitments (Cour des comptes, 2024).

Flagship Projects: Place de Catalogne and Street Greening as Heat-Risk Reduction

Paris's flagship "cooling" interventions illustrate how typological programmes crystallise into emblematic sites. The City reports that Place de Catalogne—previously highly mineral—has been transformed with 470 trees and more than 4,000 m^2 of urban forest (including a clearing), explicitly framed as an environmental asset and an urban cooling device (Ville de Paris, 2024c). This project is important analytically because it demonstrates the "square as cooling infrastructure" typology: a node that increases canopy, reduces radiant heat exposure, and expands permeable ground in a heat-vulnerable area.

On streets, Paris's planting-season programme highlights the conversion of ordinary corridors into "rue-jardin" and the expansion of school-adjacent pedestrian and greening programmes. By September 2025, the city reports 300 "rues aux écoles" (calmed streets near schools), with 100 already redesigned and vegetated (Ville de Paris, 2025). In risk-infrastructure terms, these interventions combine exposure reduction (traffic and heat) with increased permeability and shade in daily mobility environments.

4 Discussion and Implications

4.1 *What "Urban Scale" Means: Three Distinct Scaling Logics and Their Design Consequences*

The three cases demonstrate that "urban-scale greening" is not a single methodology but a family of scaling logics that operate through different *units of action* and therefore generate different design consequences, governance requirements, and evidence expectations. In other words, "urban scale" is not defined merely by the size of interventions, but by the way programmes create repeatable implementation units that can be aggregated into a coherent city-wide system.

In Barcelona, scaling is primarily *morphological*: the street section and intersection geometry become the main design platform through which NBS are delivered as climate-risk infrastructure. The Green Axes and Squares programme leverages the repetition of the Cerdà grid to standardise a street-grid module—green axis plus nodal square—capable of controlled variation while retaining a consistent infrastructural brief. This is particularly aligned with the practice of urban design because it translates NBS into compositional rules (platform continuity and universal accessibility, depaving, intensified planting at chamfered corners, and stormwater-sensitive surfaces), enabling a city-wide transformation to proceed through a series of bounded projects that remain legible to architects and urban designers (Ajuntament de Barcelona, 2021, 2022). The key implication is that morphological scaling creates a direct bridge between climate adaptation and the everyday public realm: streets become measurable exposure-management devices, not only mobility corridors.

In Copenhagen, scaling is primarily *hydrological*: the unit of action is not "a green street" in isolation, but a segment within a catchment-based network defined by flow paths, conveyance corridors, and retention nodes. The Cloudburst Management Plan makes climate-risk infrastructure legible as a system by linking each intervention to a functional role—safe conveyance, detention/retention, infiltration, or controlled discharge—benchmarked against explicit service levels for extreme rainfall events (City of Copenhagen, 2012). This logic is closer to infrastructure planning than to conventional streetscape improvement, yet it becomes urban design when the network is realised through multifunctional typologies: streets designed to behave as controlled "rivers" during cloudbursts and parks redesigned as reservoirs while remaining daily civic spaces (SLA, 2019; State of Green, 2021). The implication is that hydrological scaling increases the probability of risk reduction at the system level, but it also raises design demands: the public realm must be shaped for dual temporality—ordinary use and episodic extreme events—without compromising either. In Paris, scaling operates through a *regulatory–typological hybrid* suited to a highly constrained morphology. Rather than relying on a single dominant typology, Paris combines regulatory embedding (PLU bioclimatique commitments and rule-sets) with programme typologies delivered across many sites (rues végétalisées, school-adjacent calm streets, and flagship cooling nodes such as Place de Catalogne). This approach is pragmatic in a dense city with limited open ground:

regulation provides the long-term structural frame that extends greening beyond municipal projects, while repeated typologies deliver a near-term pipeline with measurable surface-change targets (Ville 2024b; de Paris, 2024a). The implication is that Paris' "urban scale" is achieved through cumulative, distributed implementation rather than through large contiguous green expansion, which in turn shifts the evaluation problem: performance must be captured through compliance, aggregation, and longitudinal monitoring rather than through single-site attribution.

4.2 Core Delivery and Economic Mechanisms: Why Implementation is not Reducible to "Good Design"

Across the cases, a central finding is that design ambition and technical plausibility are insufficient for scaling unless the programme architecture is stabilised through delivery and economic mechanisms. These mechanisms do not merely "fund" projects; they actively shape what can be designed, where it can be delivered, and whether it can be maintained. First, portfolio finance supports continuity by framing greening as essential public infrastructure within multi-year capital programmes rather than as a sequence of isolated projects. In Barcelona, European Investment Bank lending has been explicitly structured to support packages of urban regeneration and climate action, reinforcing the municipal ability to deliver street-based NBS as a coherent pipeline across budget cycles and political transitions (European Investment Bank,). The policy implication is that when NBS are financed as a portfolio, they can be delivered as a system—critical for network effects such as cooling corridors and city-wide exposure reduction. Second, utility co-financing can transform the public realm into stormwater infrastructure under regulated economic logic. Copenhagen's model shows how co-financing regulation allows utilities to finance surface-level adaptation measures when they are cost-effective compared to sewer-only alternatives, thereby unlocking investment in hybrid spaces that deliver both hydraulic protection and civic amenity (City of Copenhagen, 2014; Climate-ADAPT, 2013). The implication is not simply that "more money" becomes available; rather, the regulatory framework legitimises blue–green public-space typologies as part of the water-service system, overcoming the limitations of sectoral budgeting that often underfunds multifunctional design. Third, green bonds and auditability can increase both investment capacity and the pressure for measurable outcomes. Paris' climate bond is widely documented as a financing channel for climate-related programmes, including adaptation-relevant greening measures, but it also elevates the need for robust indicators, transparent accounting, and credible monitoring. Audit reporting on Paris' greening policy explicitly highlights the governance challenge of measurement consistency and maintenance capacity—precisely the issues that become unavoidable when greening is repositioned as climate-risk infrastructure rather than discretionary improvement (Climate-ADAPT, 2016; Cour des comptes, 2024). A practical implication is that finance instruments that demand reporting can indirectly drive methodological upgrades in monitoring and evaluation, provided the indicator system is well designed.

4.3 *Performance Evidence: Indicators, Credibility, and the "Regenerative" Dimension*

Evidence plays at least three roles across the cases: justification (why invest), prioritisation (where next), and accountability (did it work). Yet indicators are not neutral; they shape design decisions by determining what is counted, optimised, and politically defended. The three cases reveal distinct "indicator cultures" that correspond to their scaling logics.

Barcelona's evidence base leans towards health and liveability outcomes arising from traffic reduction and greening of daily environments. Multi-exposure modelling and health impact assessment have been used to estimate potential reductions in premature mortality and improvements in health determinants under large-scale Superblock implementation, linking air pollution, noise, heat, physical activity, and green-space exposure to outcomes (Mueller et al., 2020). Subsequent synthesis positions Superblocks as a replicable policy model but stresses that impact estimates depend on baseline exposures, behavioural assumptions, and implementation fidelity, implying that evaluation must be integrated into delivery rather than treated as post hoc validation (Nieuwenhuijsen et al., 2024). This matters for policy transfer: where the street network is the adaptation platform, exposure metrics and health-relevant indicators become central to credibility.

Copenhagen's evidence culture is more explicitly service-level and capacity oriented. The Cloudburst Management Plan frames performance in terms of system behaviour under extreme rainfall and the functional roles of corridors and retention nodes, supported by hydraulic mapping and prioritisation (City of Copenhagen, 2012). Prototype projects such as Enghaveparken further anchor evaluation in volumetric capacity (m^3) and controlled routing logic while making co-benefits legible through everyday park quality (State of Green, 2021). The implication is that when the unit of action is hydrological, the core accountability metrics necessarily include capacity, routing, and protection standards, yet this can be complemented by systematic measurement of co-benefits (thermal comfort, biodiversity, social use) to strengthen the multi-benefit claim.

Paris combines quantitative targets (hectares, streets vegetated, surface permeabilisation, tree planting) with explicit scrutiny over whether measurement, maintenance, and governance capacity can sustain delivery at scale (Cour des comptes, 2024). The PLU bioclimatique provides a regulatory mechanism to mainstream bioclimatic objectives, while programme typologies such as rues végétalisées and flagship cooling nodes like Place de Catalogne translate those objectives into visible interventions with measurable surface-change outputs (Ville de Paris, 2024a, 2024c). The implication is that regulation-led scaling shifts the evaluation challenge from "did this project work?" to "is the city's cumulative trajectory on track?", requiring longitudinal monitoring and compliance-oriented metrics.

From a regenerative urban design perspective, these indicator systems can be strengthened by integrating spatial–compositional and relational metrics alongside classic "counting" indicators. In practice, this means evaluating not only trees planted or hectares created, but the continuity of shade and cool routes, the accessibility of cooling infrastructure for vulnerable groups, and ecological connectivity across the network—because NBS performance depends on spatial configuration, user patterns, and maintenance trajectories as much as on nominal quantities (Ingaramo & Negrello, 2023; Negrello et al., 2023; Salih & Báthoryné Nagy, 2024). These additions are not decorative: they align indicator systems with the causal mechanisms through which NBS reduce exposure and produce co-benefits, thereby improving credibility and supporting better design decisions.

5 Concluding Remarks

In this chapter, urban-scale nature-based solutions (NBS) are treated as climate-risk infrastructure: interventions that should deliver defined service levels for reducing heat risk and managing pluvial flooding, while also generating co-benefits for health, liveability, and biodiversity. Looking across Barcelona, Copenhagen, and Paris, the comparison suggests that NBS scale well when three things come together: there is a clear unit of action that designers and administrations can repeat, a delivery and economic mechanism that makes implementation feasible over time, and an auditable indicator set that supports justification, prioritisation, and accountability (European Commission, 2015; IPCC, 2022; Salih & Báthoryné Nagy, 2024).

The cases do not point to a single template that can be copied everywhere. Instead, they show different—yet complementary—ways of organising delivery. Barcelona works through a morphological module (the green axis plus nodal square), turning adaptation aims into repeatable public-space typologies and drawing on health and exposure evidence to underpin investment choices and the reallocation of street space (Ajuntament de Barcelona, 2021, 2022; Mueller et al., 2020; Nieuwenhuijsen et al., 2024). Copenhagen operates through a catchment-network logic, where corridors, conveyance streets, and retention nodes form a system designed against explicit rainfall service levels; importantly, utility-enabled co-financing is a practical condition of scalability, not a secondary institutional detail (City of Copenhagen, 2012; Whittaker & Jespersen, 2022). Paris, by contrast, combines a regulatory–typological pipeline: adaptation requirements are embedded in the rule-set of urban production and reinforced through programme typologies deployed across many sites, which raises the stakes for monitoring, compliance, and credible reporting (Ville de Paris, 2024a, 2024b, 2024c; Cour des comptes, 2024; Climate-ADAPT, 2016).

For transferability, the chapter therefore argues for keeping the evaluation framework both rigorous and repeatable. Table 2 brings together a compact set of KPI domains—hydrological, thermal, health/exposure, and public-space use—supplemented by ecology and maintenance metrics that are necessary for long-term infrastructure reliability. The main point is straightforward: indicator systems should be built into programmes from the outset, so that modelling assumptions, definitions, responsibilities, and maintenance commitments remain transparent, comparable, and open to audit over time.

Table 2 Synthetic overview of performance indicators (KPI families) and their transferability

Indicator domain	Example KPIs (synthetic)	Typical method/ data	Transferability	Primary purpose
Hydrological (flood risk)	Retention/detention volume (m^3); peak-flow attenuation; flood depth/extent under design storms; % imperviousness reduced	Hydraulic/ hydrodynamic modelling; design-storm mapping; post-event flood mapping	High	Prioritisation + accountability
Thermal & climatic (heat risk)	Canopy cover (%); shade-hours / shade continuity; surface/air temperature reduction; cool-route accessibility (e.g., within 300 m)	Microclimate modelling + remote sensing; on-site sensors; GIS accessibility analysis	High	Justification + accountability
Health & exposure	Multi-exposure indicators (air/noise/ heat/green); HIA outputs (e.g., avoided premature mortality); well-being proxies	Health Impact Assessment; exposure modelling; epidemiological parameters	Medium–High	Justification (investment case)
Public-space use & liveability	Public space reallocation (m^2); pedestrian/cycle counts; dwell time; perceived safety/comfort	Before/after counts; observation; surveys; mobility data	High	Accountability + learning
Ecology & biodiversity	Habitat connectivity; biodiversity proxies; soil quality/volume; blue–green network continuity	Biodiversity monitoring; ecological network analysis; soil audits	Medium	Learning + co-benefit proof
Delivery & maintenance (reliability)	Plant survival & canopy growth; O&M budgets; compliance rates; asset-condition KPIs	Asset management logs; contract KPIs; compliance reporting	High	Accountability + durability

Key Results and Key Messages

- Urban-scale delivery relies on repeatable units of action: the *green axis + square* module (Barcelona), the *catchment/corridor + retention node* network (Copenhagen), and a *regulatory + programme typology* system (Paris).
- Delivery architectures shape what is realistically buildable: portfolio finance, utility co-financing, and green-bond/regulatory accountability influence siting, sequencing, and long-term upkeep.

- Evidence needs to be embedded in delivery: indicator systems should support investment justification, risk-based prioritisation, and long-term accountability, using consistent definitions and transparent assumptions.
- The most transferable KPI families are service-level hydrological metrics and broadly comparable thermal and public-space indicators (canopy development, shade continuity, permeability/de-sealing, access to cool spaces, and observed everyday use).
- Health-impact evidence strengthens the case for investment, but it travels poorly unless baseline exposures, behavioural assumptions, and equity/distributional effects are made explicit.
- A regenerative evaluation lens improves robustness by adding configuration-sensitive measures (connectivity; continuity of shade and cool routes) and reliability measures (maintenance performance; asset-condition KPIs) alongside simple output counts.

Taken together, these findings clarify when NBS move beyond symbolic greening and operate as measurable, transferable climate-risk infrastructure, capable of delivering durable risk reduction and associated co-benefits under accelerating climate pressures (European Commission, 2015; IPCC, 2022).

References

Ajuntament de Barcelona. (2021, March 3). *Guanyadors del concurs d'eixos verds i places (Superilla Barcelona)* [PDF]. Ajuntament de Barcelona. https://ajuntament.barcelona.cat/superilles/sites/default/files/Guanyadors_concurs_SuperillaBarcelona.pdf

Ajuntament de Barcelona. (2025, March 6). *Urban regeneration projects get 175 million euros from Europe* [News release]. Ajuntament de Barcelona. https://www.barcelona.cat/infobarcelona/en/projectes-de-regeneracio-urbana-reben-175-milions-deuros-deuropa_1490199.html

BIMSA (Barcelona d'Infraestructures Municipals, S.A.). (n.d.). *Creació de quatre nous eixos verds i quatre places a l'Eixample* [Project documentation]. https://www.bimsa.cat/actuacio/creacio-de-quatre-nous-eixos-verds-i-quatre-places-a-leixample/

Chambre régionale des comptes Île-de-France. (2024). *La politique de végétalisation de la Ville de Paris comme moyen d'adaptation au changement climatique: Rapport d'observations définitives* [PDF]. Cour des comptes. https://www.ccomptes.fr/sites/default/files/2024-05/IDR2024-22.pdf

City of Copenhagen. (2012). *Cloudburst management plan 2012* [PDF]. City of Copenhagen. https://climate-adapt.eea.europa.eu/en/metadata/case-studies/the-economics-of-managing-heavy-rains-and-stormwater-in-copenhagen-2013-the-cloudburst-management-plan/cloudburst_management_plan_2012.pdf

City of Copenhagen. (2014). Cloudburst Management Pays Off: Economics of Cloudburst and Stormwater Management in Copenhagen. Copenhagen: City of Copenhagen.

Climate-ADAPT (European Environment Agency). (2013). *The economics of managing heavy rains and stormwater in Copenhagen: The cloudburst management plan* (Case study). https://climate-adapt.eea.europa.eu/en/metadata/case-studies/the-economics-of-managing-heavy-rains-and-stormwater-in-copenhagen-2013-the-cloudburst-management-plan

Climate-ADAPT (European Environment Agency). (2016). *Climate bond financing urban adaptation actions to reduce heat stress in Paris, France* (Case study). https://climate-adapt.eea.europa.eu/en/metadata/case-studies/climate-bond-financing-adaptation-actions-in-paris

Climate Bonds Initiative & UNEP Inquiry. (2015). *Scaling green bond markets: Guide for the public sector*. Climate Bonds Initiative.

CLIMATEFIT. (2024). *Paris climate bond* [Case study report, PDF]. CLIMATEFIT Horizon Europe Project. https://climatefit-heu.eu/wp-content/uploads/2024/07/09_Paris-Climate-Bond.pdf

Cohen-Shacham, E., Walters, G., Janzen, C., & Maginnis, S. (Eds.). (2016). *Nature-based solutions to address global societal challenges*. IUCN. https://portals.iucn.org/library/sites/library/files/documents/2016-036.pdf

European Commission, Directorate-General for Research and Innovation. (2015). *Towards an EU research and innovation policy agenda for nature-based solutions & re-naturing cities* (Final report). Publications Office of the European Union. https://op.europa.eu/en/publication-detail/-/publication/fb117980-d5aa-46df-8edc-af367cddc202

European Environment Agency. (2024). *European climate risk assessment* (EEA Report 01/2024). European Environment Agency. https://www.eea.europa.eu/en/analysis/publications/european-climate-risk-assessment

European Environment Agency. (2025). *Europe's environment and climate: Knowledge for resilience, prosperity and sustainability* (EEA Report 11/2025). European Environment Agency. https://www.eea.europa.eu/en/europe-environment-2025/main-report

European Investment Bank. (2020, August 3). *EIB to provide €95 million to finance 40 climate action projects in Barcelona* [Press release]. https://www.eib.org/en/press/all/2020-215-eib-to-provide-eur95-million-to-finance-40-climate-action-projects-in-barcelona

European Investment Bank. (2025, March 6). *EIB and city of Barcelona sign €175 million loan for urban regeneration and climate strategies* [Press release]. https://www.eib.org/en/press/all/2025-125-el-bei-y-barcelona-firman-un-prestamo-de-175-millones-de-euros-para-financiar-proyectos-de-regeneracion-urbana

Hölscher, K., Frantzeskaki, N., Collier, M. J., Connop, S., Kooijman, E., Lodder, M., McQuaid, S., Vandergert, P., Xidous, D., Bešlagić, L., Dick, G., Dumitru, A., Dziubała, A., Fletcher, I., Garcia-Espina Adank, C., González Vázquez, M., Madajczyk, N., Malekkidou, E., Mavroudi, M., Loizou, E., et al. (2023). Strategies for mainstreaming nature-based solutions in urban governance capacities in ten European cities. *Abstract npj Urban Sustainability, 3*(1). https://doi.org/10.1038/s42949-023-00134-9

Ingaramo, R., & Negrello, M. (2023). "Surviving the city": Nature as an architecture design strategy for a more resilient urban ecosystem. In *Green infrastructure: Planning strategies and environmental design*. Springer.

IPCC. (2022). *Climate change 2022: Impacts, adaptation and vulnerability. Contribution of Working group II to the Sixth assessment report of the intergovernmental panel on climate change.* Cambridge University Press.

Mueller, N., Rojas-Rueda, D., Khreis, H., Cirach, M., Andrés, D., Ballester, J., Bartoll, X., Rueda, S., & Nieuwenhuijsen, M. J. (2020). Changing the urban design of cities for health: The superblock model. *Environment International, 134*, Article 105132. https://doi.org/10.1016/j.envint.2019.105132

Negrello, M. (2025). *Nature-based regeneration: Five stories of architecture that will change the world*. LetteraVentidue, Siracusa.

Negrello, M., Fonsdituri, C., Busca, F., & Ingaramo, R. (2023). Nature-based solutions for urban design: The case of Turin. *Officina, 43*, 30–39.

Nieuwenhuijsen, M. J., de Nazelle, A., Pradas, M. C., Daher, C., Dzhambov, A. M., Echave, C., et al. (2024). The superblock model: A review of an innovative urban model for sustainability, liveability, health and well-being. *Environmental Research, 251*, Article 118550. https://doi.org/10.1016/j.envres.2024.118550

Raymond, C. M., Frantzeskaki, N., Kabisch, N., Berry, P., Breil, M., Nita, M. R., Geneletti, D., & Calfapietra, C. (2017). A framework for assessing and implementing the co-benefits of nature-based solutions in urban areas. *Environmental Science & Policy*, 7715–24. https://doi.org/10.1016/j.envsci.2017.07.008

Retsinformation (Denmark). (2016). *Bekendtgørelse nr. 159 af 26/02/2016 om spildevandsforsyningsselskabers medfinansiering af kommunale og private projekter vedrørende tag- og overfladevand* [Executive Order]. https://www.retsinformation.dk/eli/lta/2016/159

Salih, K., & Báthoryné Nagy, I. R. (2024). Review evidence on urban green infrastructure and heat-mitigation modelling: Sensitivity, parameterisation and evaluation implications. *Urban Science, 8*(4), 220.

Santamouris, M. (2015). Analyzing the heat island magnitude and characteristics in one hundred Asian and Australian cities and regions. *Science of the Total Environment, 512–513*, 582–598. https://doi.org/10.1016/j.scitotenv.2015.01.060

SLA. (2019). *Sankt Kjeld's Square & Bryggervangen* [Project documentation]. https://www.sla.dk/cases/sankt-kjelds-square-and-bryggervangen/

State of Green. (2021). *Enghaveparken – Climate Park* [Project profile]. https://stateofgreen.com/en/solutions/enghaveparken-climate-park/

UNEP-CCC Knowledge Portal. (n.d.). *Climate bond financing adaptation actions in Paris*. https://c2e2.unepccc.org/kms_object/climate-bond-financing-adaptation-actions-in-paris/

Vidal Yañez, D., Pereira Barboza, E., Cirach, M., Daher, C., Nieuwenhuijsen, M. J., & Mueller, N. (2023). An urban green space intervention with benefits for mental health: A health impact assessment of the Barcelona "Eixos Verds" plan. *Environment International, 174*, Article 107880. https://doi.org/10.1016/j.envint.2023.107880

Ville de Paris. (2024a, November 22). *Les temps forts du Conseil de Paris de novembre 2024* (incl. PLU bioclimatique). https://www.paris.fr/pages/les-temps-forts-du-conseil-de-paris-de-novembre-2024-29311

Ville de Paris. (2024b, December 11). *La saison des plantations 2024/2025 est lancée: 120 nouvelles rues et places végétalisées !* https://www.paris.fr/pages/la-saison-des-plantations-2024-2025-est-lancee-120-nouvelles-rues-et-places-vegetalisees-29620

Ville de Paris. (2025, September 10). *Plus de 300 "rues aux écoles" dans Paris*. https://www.paris.fr/pages/57-nouvelles-rues-aux-ecoles-dans-paris-8197

Whittaker, S., & Jespersen, K. (2022). Stretching or conforming? Financing urban climate change adaptation in Copenhagen. *Buildings and Cities, 3*(1), 974–999. https://doi.org/10.5334/bc.238

Yin, R. K. (2018). *Case study research and applications: Design and methods* (6th ed.). Sage.

Zölch, T., Maderspacher, J., Wamsler, C., & Pauleit, S. (2016). Using green infrastructure for urban climate-proofing: An evaluation of heat mitigation measures at the micro-scale. *Urban Forestry & Urban Greening, 20*, 305–316. https://doi.org/10.1016/j.ufug.2016.09.011

Urban Greenery and Participation: Transforming Cities in Transition

Ferdinando Trapani

Abstract The case study examined in this chapter concerns the euPOLIS H2020 EU project (2025) in Palermo, located in a residual peripheral area that still preserves the remains of an important eighteenth-century garden, when the agricultural and monumental greenery of the Conca d'Oro was still present. The case highlights an experience of co-programming and co-design between the municipal administration and a network of environmental and cultural associations at local and regional levels, developed as part of academic research on eco-sustainable urban governance models in an international perspective. The chapter focuses on the participatory phase of the euPOLIS project in Palermo. The analysis examines the scientific approach adopted by the municipal administration to experiment with the direct involvement of citizens in the design choices for the new urban park. The outcomes of the euPOLIS project are analysed in policy terms, through comparison with other experiences of green space provision. The specificity of the case study lies in its location in an urban fringe area characterised by high-density housing, fragile social cohesion and the morphological and functional fragmentation of urban tissues. The conclusions highlight the strengths and weaknesses of participatory policies for the construction of urban green spaces. On the positive side, the case shows citizens' willingness and interest in obtaining green areas close to their places of residence. On the negative side, the delay between co-design hypotheses and implementation leads to a decline in attention and growing disaffection with participation among residents.

Keywords Urban planning · Green urbanism · Green policies · Participative planning process · Co-production · Co-design

F. Trapani (✉)
Department of Architecture, University of Palermo, Palermo, Italy
e-mail: ferdinando.trapani@unipa.it

L. Diana et al. (eds.), *Greening and Resilience*, Urban Sustainability,
https://doi.org/10.1007/978-981-92-0419-9_15

1 Introduction

Nature-based solutions (NBS) have emerged as a central concept in European urban policy, framing green and blue spaces as multi-functional responses to intertwined climatic, environmental and social challenges. Parks, greenways, wetlands and urban forests are increasingly mobilised not only to mitigate flood risk, urban heat islands and air pollution, but also to support social cohesion, public health and environmental justice.

Recent Horizon 2020 projects have played a prominent role in advancing this agenda. Initiatives such as proGIreg, URBiNAT, CLEVER Cities, JUSTNature and euPOLIS have experimented with co-created NBS in post-industrial districts, social housing neighbourhoods and fragmented peri-urban areas across Europe. They combine technical innovation with new forms of collaborative governance, often through living labs and other experimental platforms.

The euPOLIS project (Integrated NBS-based Urban Planning Methodology for Enhancing the Health and Well-being of Citizens) explicitly seeks to "return nature to cities" by integrating NBS and blue–green solutions (BGS) into planning processes with a strong focus on public health and well-being. The methodology is first tested in four front-runner cities (Belgrade, Łódź, Piraeus, Gladsaxe) and then transferred to four follower cities (Bogotá, Palermo, Limassol, Trebinje). Palermo, as a follower city, applies and adapts this approach to the residual areas of Villa Turrisi in the western part of the city.

Villa Turrisi forms part of the historic Conca d'Oro, a once-productive citrus landscape progressively fragmented by speculative urbanisation and infrastructure. The area is today characterised by high-density residential fabrics, major road axes, a tram line, the Passo di Rigano canal and the disused Palermo–Corleone railway, which is now being transformed into the Palermo–Monreale greenway. At the same time, it is the focus of long-standing civic mobilisation by the "Parco Villa Turrisi" association and other actors, advocating the creation of an urban park and denouncing illegal land-use changes.

The municipal master plan scheme (Comune di Palermo, 2023a) designates the broader Villa Turrisi area as an Urban Park, while the first operational step is the "Parco di Villa Turrisi – primo lotto", a public garden funded through the National Recovery and Resilience Plan (PNRR—Mission 5, urban regeneration). Within this context, the euPOLIS methodology is expected to (i) guide the design of NBS/BGS in and around the park, (ii) structure co-creation with citizens and stakeholders, and (iii) support the evaluation of impacts on health and well-being.

Within this broader experimental landscape, euPOLIS is one of several large-scale Horizon 2020 initiatives testing co-created NBS in European cities, alongside projects such as proGIreg, URBiNAT, CLEVER Cities and JUSTNature.

The chapter addresses three research questions:

- How is the euPOLIS NBS-based planning methodology translated into the specific socio-spatial and institutional context of Palermo as a follower city?

- In what ways does the Villa Turrisi project use NBS/BGS to address ecological fragmentation, climate–water risks and social vulnerabilities in a dense and functionally fragmented area?
- What forms of co-design and governance innovation emerge from the euPOLIS process, and how might they inform Italian debates on urban standards and green–blue infrastructure?

The analysis builds on international literature on NBS and collaborative governance and on a close reading of local documentation: euPOLIS project materials, municipal reports, the technical dossier of the Villa Turrisi Park project (Palermo, 2021) and the Reticula article by Liuzzo (2021), which interprets euPOLIS as a step towards the "bio-city". By treating euPOLIS as a European methodological framework and Villa Turrisi as its local demonstrator in Palermo, the chapter positions the case study at the intersection between experimental EU research and the ordinary tools of Italian urban planning. In line with the chapter template provided by the editors, the contribution is structured in six sections, from background to conclusions.

2 Background and Theoretical Framework

2.1 Nature-Based Solutions, Collaborative Governance and Care

Nature-based solutions (NBS) are commonly defined as actions that protect, restore or manage ecosystems in ways that address societal challenges while providing multiple co-benefits for biodiversity and human well-being. In cities, this includes interventions such as parks, greenways, wetlands, green roofs and urban forests, which can mitigate climate risks, enhance air quality and create spaces for recreation and social interaction. In the wider debate on climate adaptation, NBS are increasingly presented as multi-benefit measures that can counteract soil sealing, the loss of natural areas and urban densification, while delivering ecosystem services to urban residents (Frantzeskaki, 2019; Kabisch et al., 2017).

Implementing NBS in dense, fragmented urban contexts is not merely a technical task. It requires collaborative governance arrangements that bring together public institutions, knowledge organisations, civic initiatives and, in some cases, private actors. The collaborative governance literature emphasises joint decision-making forums where stakeholders share resources, negotiate goals and co-produce policy solutions (Ansell & Gash, 2008). In this perspective, NBS become both ecological and institutional innovations, because they demand new forms of co-ordination across sectors and scales.

In recent years, this perspective has been extended to the field of NBS and urban green infrastructure. Van der Jagt et al. (2019) propose a guiding framework for co-creating urban green infrastructure, highlighting the importance of living labs, multi-actor partnerships and iterative co-design processes. Hölscher et al. (2024)

discuss how co-production capacities can become embedded in urban governance structures, moving from one-off projects to more durable institutional arrangements. This literature resonates with an emerging discourse on care in urban planning: NBS are not only technologies for managing environmental risks, but also infrastructures for caring for both nature and people. They can support everyday practices of care (walking, playing, gardening, meeting) and contribute to repairing damaged socio-ecological relations, especially in marginalised neighbourhoods, while also raising questions of environmental justice and green gentrification (Anguelovski et al., 2022; Raymond et al., 2017).

2.2 International Reference Cases and Lessons for Policy Evaluation

Among the many European initiatives on NBS, four Horizon 2020 projects provide particularly relevant reference points for the Palermo case.

proGIreg (Productive Green Infrastructure for post-industrial urban regeneration) operates in post-industrial districts and brownfield sites (Dortmund, Turin, Zagreb, Ningbo), developing NBS that are "productive" in socio-economic terms (urban agriculture, green businesses, etc.) through local living labs.

URBiNAT focuses on "healthy corridors" in under-served social housing neighbourhoods (Porto, Nantes, Sofia), where co-created NBS connect fragmented open spaces and act as drivers for social cohesion and healthier lifestyles.

CLEVER Cities includes the Thamesmead area in London, a large social housing estate where NBS such as park improvements, rain gardens and street trees are used to address issues of health, safety and social isolation, through co-design, co-implementation and co-monitoring involving housing associations, NGOs, schools and "community researchers".

JUSTNature explicitly links NBS to environmental justice, focusing on the "right to ecological space" and testing City Practice Labs that address both climatic effectiveness and the fair distribution of ecological benefits in several European cities.

From these projects, three sets of lessons are particularly relevant for euPOLIS and for the Villa Turrisi case. First, NBS are most effective when they are conceived as corridors and networks that re-stitch fragmented urban fabrics, rather than as isolated green patches. Second, living labs and similar platforms are crucial for stabilising collaborative governance, going beyond episodic participation. Third, integrating health and justice objectives into Horizon 2020 NBS projects requires explicit attention to both distributive and procedural dimensions of environmental justice.

Table 1 provides a synthetic overview of these four Horizon 2020 projects, highlighting their urban contexts, governance set-ups and the specific lessons that will be used to interpret the euPOLIS case in Palermo along the five evaluation dimensions presented in the next section.

3 Methodology and Policy Evaluation Framework

The analysis adopts a policy analysis perspective, treating euPOLIS/Villa Turrisi as a policy device at the intersection of European research, national funding programmes and local planning. Empirically, it is based on:

- official euPOLIS documentation regarding the consortium's research work (euPOLIS, 2025);
- municipal materials and masterplans (Comune di Palermo, 2004, 2016, 2023a) including the euPOLIS project in Palermo (Comune di Palermo, 2023b);
- the technical dossier and planning report for the garden executive project (Palermo 2021);
- the detailed report and the programme of the co-creation workshop held at the "Cantieri Culturali della Zisa", public cultural district in Palermo, in September 2023 (Palermo, 2023b); and
- the Reticula article by Liuzzo (2021) on euPOLIS, NBS/BGS and the bio-city.

These sources are analysed through an evaluation framework derived from international literature and from euPOLIS conceptual documents. Following Liuzzo's interpretation of euPOLIS as a passage from the smart city to the bio-city, the framework is organised into five dimensions:

1. Ecological reconnection and defragmentation of natural capital: does the project contribute to reconnecting residual agricultural areas, the Passo di Rigano canal and the Palermo–Monreale greenway into a coherent green–blue corridor?
2. Blue–green infrastructure and climate–water resilience: to what extent are NBS/BGS used as alternatives or complements to grey infrastructure, in line with regional guidelines on hydraulic invariance and climate adaptation?
3. Public health and well-being outcomes: how are health and well-being objectives articulated and monitored, in connection with the wider euPOLIS evaluation protocol?
4. Co-design, inclusion and governance innovation: how is the co-creation process structured (online survey, workshop, GDPM), and which social groups and actors are involved?
5. Policy integration, learning and transferability: how does the Villa Turrisi experience feed into the re organisations of the master plan, the Palermo Green City Action Plan and the institutionalisation of new planning standards and indicators?

Each dimension is explored through document analysis and, where relevant, comparison with practices observed in the international projects mentioned above.

These H2020 projects constitute various attempts to address and solve the problems that the use of NBS had presented in various European cities already before the

Table 1 International H2020 NBS projects—Lessons for euPOLIS/Villa Turrisi

H2020 project	Urban context and NBS object	Governance/co-design set-up	Lessons learnt useful for the Palermo—Villa Turrisi case
proGIreg—Productive Green Infrastructure for post-industrial urban regeneration (progireg.eu)	Post-industrial neighbourhoods and brownfield sites in European and Chinese cities (Dortmund, Turin, Zagreb, Ningbo). Productive NBS (community gardens, greenhouses, riverbank renaturation, green corridors) to regenerate residual and degraded areas	Local Living Labs as stable platforms bringing together the municipality, universities, associations, businesses and citizen groups. NBS are co-developed and partly co-managed by civic and economic actors, with strong attention to socio-economic benefits and the productive use of green spaces	(1) Residual areas can be treated as "productive infrastructures" (not only ornamental parks), generating jobs, services and social inclusion (2) The Living Lab platform is crucial to stabilise co-design over time and avoid reducing it to isolated meetings. (3) Measuring social, economic and environmental benefits from the outset helps to politically legitimise NBS and to embed them into planning instruments

(continued)

Table 1 (continued)

H2020 project	Urban context and NBS object	Governance/co-design set-up	Lessons learnt useful for the Palermo—Villa Turrisi case
URBiNAT—Healthy corridors as drivers of social housing neighbourhoods' regeneration (CORDIS)	Deprived social housing neighbourhoods (Porto – Campanhã, Nantes Nord, Sofia—Nadezhda). Implementation of Healthy Corridors as collective spaces connecting fragmented parts of the city (parks, vacant lots, gardens, pathways)	Co-diagnosis and co-creation structured in phases (presentation, cultural mapping, participatory activities), with a Healthy Corridor Lab involving residents, associations, schools and social services. Strong attention to building new place identities and relations between neighbouring areas	(1) Green and blue corridors work as devices of physical and symbolic connection in fragmented contexts: not only "green spaces" but relational infrastructures (2) The co-diagnosis phase (understanding the neighbourhood together with residents before designing) is fundamental and distinct from simply involving people in a pre-defined project (3) The evaluation of impacts on social cohesion and neighbourhood perception must be integrated from the very beginning into policy design

(continued)

Table 1 (continued)

H2020 project	Urban context and NBS object	Governance/co-design set-up	Lessons learnt useful for the Palermo—Villa Turrisi case
CLEVER Cities—Thamesmead (London) (CLEVER Cities—Home)	Large social housing estate from the 1960s–1970s, dense and stigmatized, with wide but poorly qualified green areas. Distributed NBS: park upgrades, rain gardens, street trees, meeting and play spaces, "urban greening" interventions for climatic comfort and perceived safety	Multi-stage co-creation process (co-design, co-implementation, co-monitoring) involving the municipality, housing associations, NGOs (e.g., Groundwork), schools and residents. Use of community researchers and capacity-building pathways to turn residents into co-producers of knowledge and not only "consulted" stakeholders	(1) The quality of the process depends on the ability to move from co-design to co-implementation and co-monitoring, strengthening trust and sense of belonging (2) Instruments such as community researchers make it possible to reach groups that are usually excluded from participatory processes (3) When NBS are co-designed, they become levers to work on safety, health and neighbourhood perception, not only on environmental indicators
JUSTNature—Activation of NBS for a just low-carbon transition (JUSTNature)	Network of City Practice Labs (including Leuven, Munich, Merano/Bolzano, Chania, Gżira, Szombathely) that experiment with NBS for the low-carbon transition in urban neighbourhoods, with attention to air quality, thermal comfort and access to ecological space	The explicit objective is to guarantee a "just" transition, based on the principle of the right to ecological space (right to clean air, thermal comfort and access to green areas). NBS are co-designed with vulnerable stakeholders and integrated into city climate strategies	(1) NBS must be assessed in terms of climatic effectiveness and in terms of environmental justice (2) The concept of the right to ecological space provides a strong normative basis to green policies (3) Explicitly integrating NBS into climate–energy policies and urban plans reinforce the stability of decisions over time

COVID19 period (Almassy, 2018: 63–64), but many of these questions still appear today and are not completely resolved.

4 Case Study: euPOLIS and the Villa Turrisi Urban Park

4.1 Urban and Planning Context

The Villa Turrisi area is located between Viale Leonardo da Vinci and the neighbourhoods of Uditore and Passo di Rigano. Historically, it formed part of the Conca d'Oro agricultural landscape, whose progressive fragmentation has produced a mosaic of residual green spaces and built-up areas.

Figures 1 and 2 illustrate the area intended for the future urban park, showing the relationship between the residual green spaces, the dense built fabric and the main transport infrastructures.

In the municipal master plan scheme, Villa Turrisi is included within the system of urban parks and classified, for the first lot, as V3 "public green spaces". The area is also strategic for the emerging network of green–blue infrastructure, as it hosts a stretch of the Passo di Rigano canal and the former railway line to Corleone, now being reconverted into the Palermo–Monreale greenway.

The first step of the public garden executive project (Palermo, 2021) is financed through the PNRR as an urban regeneration intervention, with the declared goal of improving environmental quality, reconnecting ecological networks and providing accessible green space in a dense and socially vulnerable context. The project builds on years of civic mobilisation, which have helped to keep the area on the political and planning agenda.

In the most recent phase, the planning and implementation process has made an additional step forward. Municipal Council Resolution (no. 20 of 18 April 2024) gave effect to the variation request (approved by resolution on 10.3.2021), re-determining the land-use classification of an area previously designated as "public green spaces", "nursery school" and "compulsory schools" following the expiry of the expropriation constraint. Once the procedure for adopting the urban plan variation had been concluded, and in light of the critical issues affecting the implementation timetable of the PNRR-funded project, the municipal administration considered it necessary and urgent to approve the feasibility design for the project of new urban green park.

4.2 Palermo as euPOLIS Follower City

Within euPOLIS, Palermo is a follower city tasked with adapting and demonstrating the integrated NBS-based planning methodology developed in front-runner cities. Official documents present the methodology to "return nature to the city" and to

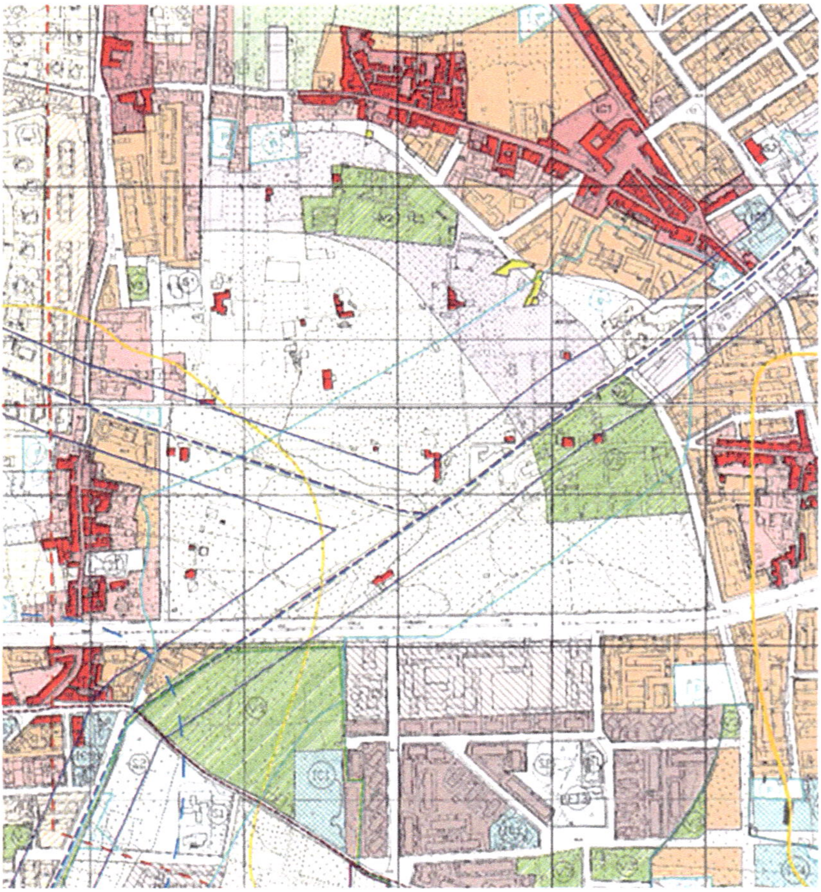

Fig. 1 Masterplan of the area intended for the future urban park of Villa Turrisi. Green: green areas; yellow: demolitions; blue: public services; red: historic urban fabric; orange and brown: residential areas; violet: productive areas. Blue lines indicate water channels. *Source* Comune di Palermo, 2004, tavv. 5007 and 5010 (map excerpt and author's elaboration)

address low environmental quality, fragmented public spaces and water stress through NBS/BGS that are co-created with citizens.

The Villa Turrisi Park is framed as a "giardino culturale" (cultural garden), where the relationship between humans and nature is narrated and performed through biodiversity observation, urban agriculture, sport and educational activities. At the same time, the project is explicitly linked to the broader Palermo Green City Action Plan and to the network of Natura 2000 sites, including Monte Pellegrino and the Valle dell'Oreto, which together define a metropolitan green–blue framework.

Fig. 2 Plan of the area intended for the future urban park of Villa Turrisi (Palermo, 2021)

4.3 *Co-Creation Process and GDPM*

The co-creation process combines online and in-presence activities. An initial online survey and information event introduce the project and collect general perceptions and expectations. On 21 September 2023, a co-creation workshop was held at the NOZ space (Cantieri Culturali alla Zisa), structured as a world café with thematic tables.

Participants include representatives of diverse groups: children, young people, working women and caregivers, elderly people, large low-income families, people with special needs, religious communities, neighbourhood associations and professionals. The thematic tables address:

- vegetation and NBS selection;
- comparison between two design alternatives ("Soluzione 1" and "Soluzione 2");
- desired activities and facilities;
- contextual analysis and perceived qualities; and
- environmental/urban, social/economic and health/well-being indicators.

A key methodological innovation is the introduction of the Goal Driven Planning Matrix (GDPM), explicitly socio-economic and gender-oriented, to structure the comparison of alternatives and link design options with multidimensional objectives. The GDPM is intended to translate participatory inputs into a transparent evaluation of trade-offs and synergies, guiding the selection and refinement of the preferred solution.

From a critical perspective, however, the co-creation process also reveals some limitations. Participation remains largely dependent on pre-existing active groups and networks, so that some categories, such as migrants or precarious workers, risk being under-represented. Time constraints linked to the PNRR funding calendar limit the possibility of iterative feedback rounds, and the technical complexity of the GDPM may reduce the transparency of decision-making for non-expert participants. These elements echo broader debates on the risks of merely procedural participation in NBS projects and confirm the need to stabilise more durable and inclusive governance arrangements.

4.4 Design Options and NBS/BGS

The technical dossier develops two main functional alternatives, both compliant with PNRR requirements and regional rules on hydraulic invariance. Shared elements include:

- a network of pedestrian and cycling paths connecting entrances, the greenway and internal spaces;
- areas for play, sport and community gardening, with attention to different age groups;
- biodiversity points and vegetated patches aimed at preserving remnants of the traditional orchard and enhancing habitat quality; and
- buffer vegetation along busy roads to mitigate noise and air pollution.

The alternatives differ mainly in the intensity and distribution of facilities and activities (e.g., outdoor markets, eco-business spaces and educational hubs), which are explicitly discussed with participants. Both incorporate NBS/BGS such as retention and infiltration areas for stormwater, permeable surfaces, vegetated swales and shaded areas designed to reduce heat stress. Preservation of an existing wooded strip near the main entrance is prioritised to maintain ecological continuity and microclimatic benefits.

Overall, the design reflects the euPOLIS ambition to substitute or hybridise grey infrastructure with nature-based systems, while responding to local demands for safety, accessibility and everyday use.

5 Discussion and Implications

5.1 Ecological Reconnection and Green–Blue Infrastructure

From an ecological perspective, the Villa Turrisi project represents a significant attempt to defragment natural capital in the western part of Palermo. By linking residual green spaces with the Passo di Rigano canal and the Palermo–Monreale greenway, the project contributes to a green–blue corridor that connects dense neighbourhoods with metropolitan ecological assets.

However, the current planning horizon—focused on the first PNRR-funded lot—only partially covers the potential corridor. The effectiveness of ecological reconnection will depend on future extensions, management of canal margins and integration with private green spaces. This partiality is typical of NBS projects embedded in fragmented funding cycles and underlines the importance of long-term planning beyond project boundaries.

5.2 Climate–Water Resilience and BGS

The project's strong alignment with BGS principles is a core strength. The explicit use of retention and infiltration systems, permeable surfaces and vegetated buffers respond to regional guidance on hydraulic invariance and to euPOLIS objectives on water stress and microclimate. Compared with front-runner cities, Palermo is at an earlier stage in terms of environmental sensor deployment and modelling; nevertheless, the design creates a physical basis for future monitoring and adaptive management.

5.3 Health and Well-Being: From Goals to Metrics

The Palermo experience clearly integrates health and well-being into the framing of the park. Desired functions and facilities emerging from the workshop—shaded seating, child-friendly spaces, areas for movement and quiet rest, community gardens—reflect a care-oriented approach to everyday life.

At the same time, health-related metrics and monitoring protocols are still under construction. Whereas some front-runner cities combine surveys, ethnographic observations and physiological measurements, Palermo currently relies more on ex ante indicator design than on systematic data collection. Strengthening this dimension could enhance the capacity to demonstrate benefits and support future policy decisions.

At this stage, quantitative evidence is limited to the number and typology of participants in the co-creation events and to the scoring of design alternatives within the GDPM and the indicator matrix, which remain largely internal to the municipal technical team. Making these semi-quantitative tools more explicit and publicly available would be an important step for future phases of the project, strengthening the empirical basis of the discussion and the robustness of conclusions.

5.4 Co-Design and Governance Innovation

In terms of co-design, the Palermo case is notable for the attention paid to inclusion and differentiated publics. The use of targeted invitations and multiple formats (online survey, workshop, thematic tables) aligns with best practice on NBS co-creation. The GDPM adds an original layer by linking participation with transparent multi-criteria evaluation.

Yet, on the governance side, the picture is more ambiguous. While the Environment Service leads the process and civic actors are actively involved, there is limited evidence of formalised co-governance arrangements (e.g., co-management agreements and permanent multi-actor platforms) akin to living labs in proGIreg or URBiNAT. Palermo appears to be in a transitional phase, where collaborative practices are emergent but not yet fully institutionalised.

5.5 Policy Integration and Learning

Finally, the integration of euPOLIS/Villa Turrisi into local planning and national policy frameworks is both promising and fragile. On the one hand, the project is firmly connected to the master plan revision, the Palermo Green City Action Plan and PNRR-funded interventions and has been recognised nationally as a "good territorial practice". On the other hand, the Italian planning system remains strongly anchored to quantitative standards (e.g., DM 1444/1968), and the translation of ecosystem-service and health-oriented indicators into binding norms is still incipient.

The Palermo case thus functions as a laboratory for hybridisation, where experimental metrics and collaborative practices interact with established regulatory frameworks. The extent to which these innovations will be mainstreamed depends on political continuity, administrative capacity and the ability to link euPOLIS outcomes to other policy arenas (housing, mobility and public health).

6 Concluding Remarks

For Italian urban planning, the case indicates that NBS can become a vehicle for re-thinking urban standards and policies around ecological connectivity, health and justice, particularly in dense and fragmented contexts. From a methodological viewpoint, several elements of the approach adopted in Palermo appear replicable in other cities: the combination of comparative learning from Horizon 2020 projects with a locally grounded policy analysis framework; the five-dimensional matrix that links ecological reconnection, blue–green infrastructure, health, governance and policy integration; and the use of the GDPM as a socio-economically and gender-sensitive tool for co-evaluating design alternatives. Transferring these elements to other contexts would require adaptation to local institutional settings and planning cultures, but the Palermo experience suggests that even follower cities with limited resources can use euPOLIS-like frameworks to negotiate more inclusive and health-oriented green infrastructure policies. Future research could deepen this analysis through comparative studies across follower cities, longitudinal monitoring of environmental and health outcomes, and closer investigation of how euPOLIS-inspired practices are embedded in municipal routines and national regulatory debates.

Acknowledgements This chapter was prepared as part of the project "ECO-SET, A Multidisciplinary Approach to Plan ECOsystem SErvices for Cities in Transition", funded by Ministry of Universities and Research, Projects of Relevant National Interest, 2022, code no. P202249BMJ. Institution: University of ROME "La Sapienza". Principal investigator: Carmelina Bevilacqua. Location of this contribution research: DSEAS, Department of Economic, Business and Statistical Sciences, University of Palermo.

References

Almenar, J. B., Rugani, B., Geneletti, D., Brewer, T., & Carnelli, F. (2021). Nexus between nature-based solutions, ecosystem services and urban challenges. *Ecosystem Services, 50*, Article 101341.

Anguelovski, I., Connolly, J. J., Cole, H., Garcia-Lamarca, M., Triguero-Mas, M., Baró, F., Martin, N., Conesa, D., Shokry, G., Pérez del Pulgar, C., Argüelles Ramos, L., Matheney, A., Gallez, E., Oscilowicz, E., López Máñez, J., Sarzo, B., Beltrán, M. A., & Martinez Minaya, J. (2022). Green gentrification in European and North American cities. *Nature Communications, 13*, 2313. https://doi.org/10.1038/s41467-022-31572-1

Ansell, C., & Gash, A. (2008). Collaborative governance in theory and practice. *Journal of Public Administration Research and Theory, 18*(4), 543–571. https://doi.org/10.1093/jopart/mum032

Frantzeskaki, N. (2019). Seven lessons for planning nature-based solutions in cities. *Environmental Science & Policy, 93*, 101–111. https://doi.org/10.1016/j.envsci.2018.12.033

Hölscher, K., Frantzeskaki, N., Kindlon, D., et al. (2024). Embedding co-production of nature-based solutions in urban governance: Emerging co-production capacities in three European cities. *Environmental Science & Policy, 152*, Article 103652. https://doi.org/10.1016/j.envsci.2023.103652

Kabisch, N., Bonn, A., Korn, H., & Stadler, J. (Eds.). (2017). *Nature-based solutions to climate change adaptation in urban areas: Linkages between science, policy and practice*. Springer. https://doi.org/10.1007/978-3-319-56091-5

Liuzzo, G. (2021). La questione urbana per euPOLIS. NBS e BGS: nuove regole e nuove pratiche urbane per la bio-city. *Reticula, 28*, 111–120. Istituto Superiore per la Protezione e la Ricerca Ambientale (ISPRA).

Raymond, C. M., Frantzeskaki, N., Kabisch, N., Berry, P., Breil, M., Nita, M. R., Geneletti, D., & Calfapietra, C. (2017). A framework for assessing and implementing the co-benefits of nature-based solutions in urban areas. *Environmental Science & Policy, 77*, 15–24. https://doi.org/10.1016/j.envsci.2017.07.008

van der Jagt, A. P. N., Smith, M., Ambrose-Oji, B., Konijnendijk, C. C., Giannico, V., Haase, D., Lafortezza, R., Nastran, M., Pintar, M., Železnikar, Š, & Cvejić, R. (2019). Co-creating urban green infrastructure connecting people and nature: A guiding framework and approach. *Journal of Environmental Management, 233*, 757–767. https://doi.org/10.1016/j.jenvman.2018.09.083

Horizon 2020 Projects (Web Resources)

Almassy, D., Pinter, L., Rocha, S., Naumann, S., Davis, M., Abhold, K., & Bulkeley, H. (2018). *Urban nature atlas: A database of nature-based solutions across 100 European Cities*. Horizon 2020 grant agreement No. 730243. https://oppla.eu/sites/default/files/old_files/uploads/urban-nature-atlasa-database-nature-based-solutions-across-100-european-cities.pdf

CLEVER Cities. *Co-created livable environments for everyone*. Horizon 2020 grant agreement no. 776604. https://www.labsimurb.polimi.it/research/clever-cities/

euPOLIS. (2025). *Integrated NBS-based urban planning methodology for enhancing the health and well-being of citizens*. Horizon 2020 Innovation action programme under grant agreement no. 869448. https://cordis.europa.eu/project/id/869448

JUSTNature. *Activation of nature-based solutions for a just low-carbon transition*. Horizon 2020 grant agreement no. 101003757. https://justnatureproject.eu/

proGIreg. *Productive Green Infrastructure for post-industrial urban regeneration*. H2020 project grant agreement no. 776528. https://progireg.eu/.

URBiNAT. *Healthy corridors for inclusive urban regeneration*. H2020 project grant agreement 776783. https://urbinatobservatory.eu/

Local and Grey Literature (Selection)

Comune di Palermo. (2004). Comune di Palermo, Variante generale al PRG, approvata con D.Dirig. N.124/2002 Dipartimento Regionale Urbanistica, 2002, assessorato al Territorio e Ambiente – Tavv. P2a 5007 e 5010 «Zonizzazione e vincoli sul territorio» adeguati nel luglio 2003, presa d'atto Delibera n.7/2004 del Consiglio comunale di Palermo. https://repository.comune.palermo.it/amministrazione-trasparente.php?grp=3&lev=4&id=111

Comune di Palermo. (2021). *Realizzazione di un giardino pubblico nell'area compresa tra il viale Leonardo da Vinci e le vie Ruggeri, De Grossis, Di Blasi, Politi (primo lotto del Parco di Villa Turrisi). Informativa sul progetto e possibili soluzioni basate sulla natura applicabili (NBS). Interventi per la riqualificazione urbana nel quartiere Uditore Passo di Rigano*. Area del Decoro Urbano e del Verde. Ufficio autonomo gestione verde urbano, agricoltura urbana e rapporti con Reset. https://docs.comune.palermo.it/js/server/uploads/_28082023091004.pdf

Comune di Palermo. (2016). *Schema di massima del nuovo PRG*. Relazione generale. https://docs.comune.palermo.it/js/server/uploads/trasparenza_all/_25112015084706.pdf

Comune di Palermo. (2023a). *Direttive Generali per la formazione del Piano Urbanistico Generale (PUG) del Comune di Palermo*. https://pug.comune.palermo.it/documento-tipo.php?id=983
Comune di Palermo. (2023b). *Workshop di co-creazione del futuro parco di Villa Turrisi con metodologia fornita dal progetto euPOLIS*. Relazione Workshop (Rev. 0). https://repository.comune.palermo.it/landing/progetto-eupolis.php

Integrating Co-Creation and Nature-Based Solutions in City-Port just Transitions: Testing a Preliminary Circularity Catalogue Within the SPArTaCHus Lab

Libera Amenta

Abstract City-Port Areas (CPAs) are complex urban contexts where environmental, climatic, socio-economic, and infrastructural risks intersect with wastescapes. Although often lacking publicness, these areas offer opportunities as experimental sites for regenerative urban and ecological strategies. This chapter aims to reflect on experience gained through a co-design workshop developed within the SPArTaCHus research project, in which participants applied circularity principles and metabolic thinking to transform CPA wastescapes into equitable and multifunctional spaces. The study engaged students and local stakeholders, testing a preliminary and evolving version of the Circularity Catalogue, an ongoing research output of SPArTaCHus, providing insights into participatory processes in complex urban contexts. Results show that the Circularity Catalogue fosters dialogue among diverse stakeholders, supporting the co-design of resilient, nature-based solutions for CPAs. The study demonstrates that co-creation activities, when framed within circular urban metabolism and nature-based solutions (NBS), can redefine wastescapes as productive, socially embedded, and ecologically meaningful urban spaces. Furthermore, these approaches help mitigate the risks of green gentrification, unintended socio-economic consequences, and displacement pressures, highlighting their potential to support justice-oriented and participatory urban planning.

Keywords Co-creation · Urban living labs · Wastescapes · City-Port areas · Circular urban metabolism · Environmental and climate justice

---

L. Amenta (✉)
Department of Architecture, University of Naples Federico II, Naples, Italy
e-mail: libera.amenta@unina.it

L. Diana et al. (eds.), *Greening and Resilience*, Urban Sustainability,
https://doi.org/10.1007/978-981-92-0419-9_16

1 Introduction

In City-Port Areas (CPA), complexity due to the coexistence of industrial, logistical, residential, and ecological functions shapes fragile urbanised contexts at the land–water interface. Their exposure to overlapping environmental, climatic, socio-economic, and infrastructural risks generates layered vulnerabilities and governance challenges, within which wastescapes emerge as residual spaces produced by industrial decline, port dynamics, and fragmented decision-making. This complexity requires participatory and inclusive governance processes to support more resilient, equitable, and regenerative transformations.

The objective of this contribution is to reflect on experience gained through a co-design workshop developed within the SPArTaCHus research project, which informs a nature-based and justice-oriented approach to the regeneration of wastescapes in City-Port Areas (CPAs). The contribution explores how co-created regeneration processes can support the restoration of publicness in CPA wastescapes, while addressing environmental and climate justice concerns.

The methodological framework of the SPArTaCHus research combines insights from urban ecology, urban metabolism, and ecological urbanism, integrating more-than-human perspectives to support the understanding of the complexity inherent in circular and nature-based regeneration processes in CPAs (Fig. 1). The approach is grounded in case-study research, moving from site-specific configurations toward broader proof-of-concept reflections through the analysis of practices, narratives, and discourses.

By focusing on the regenerative potential of wastescapes, this contribution highlights how targeted, nature-based interventions in City-Port threshold areas can address multiple and interconnected challenges—including port-related pressures, land scarcity, and climate risks—within planetary safe and just limits, while expanding public space availability and reframing the socio-ecological role of these interface territories.

SPArTaCHus research aims to progressively develop, through collaborative processes, a Circularity Catalogue conceived as an evolving framework rather than a fixed or consolidated output. The Catalogue identifies a set of 'recommendations' or 'guidelines' co-designed to foster alternative scenarios for City-Port Areas (CPAs), where innovative land-use strategies, adaptive infrastructure, and stakeholder engagement converge to enhance social well-being, economic vitality, and ecological integrity.

The tools, organised in three main objectives, that currently inform the preliminary version of the Circularity Catalogue are derived from co-design experimentations with local stakeholders, as well as didactic activities such as master theses and design studios (Fig. 2). These activities focused on the Metropolitan Area of Naples, with the cities of Pozzuoli and Castellammare di Stabia serving as key hotspots. The final version of the Catalogue will eventually build upon comparative studies of Dutch cases and literature reviews as well.

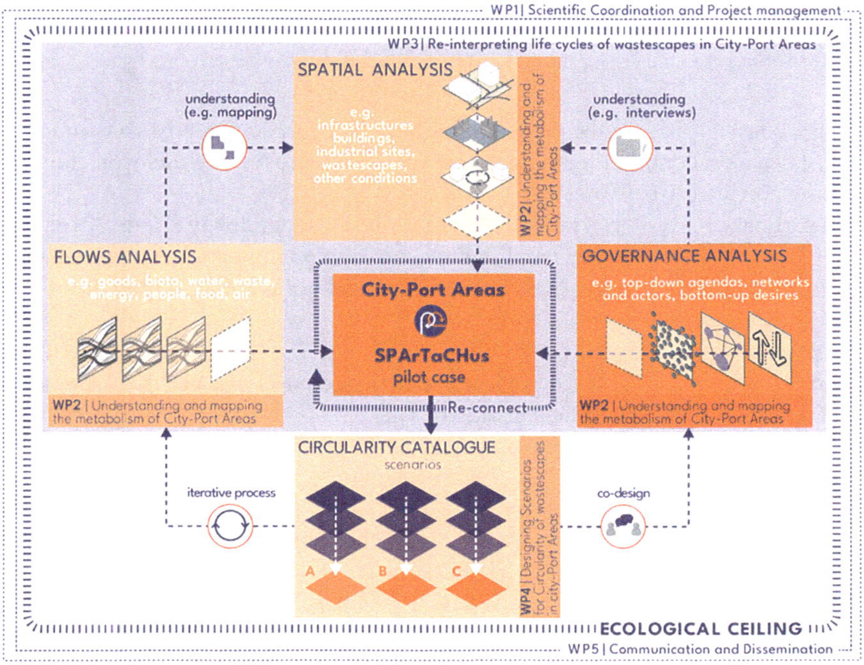

Fig. 1 SPArTaCHus methodological framework. *Image source* graphic by Amenta, L., adapted from SPArTaCHus Research proposal. Corresponding proponent: Amenta, L.

This co-created Circularity Catalogue supports the definition of a sustainable and just regeneration of wastescapes in CPAs that can be implemented through nature-based solutions, contributing to progress toward a sustainable transition for CPAs. This is achieved through the systemic implementation of circular initiatives that place climate justice, equity, and human rights at the centre of decision-making processes and climate action, in line with UNDP (2023).

In this contribution, a preliminary version of the Circularity Catalogue is presented and critically tested as an operational and exploratory device through its application in a co-design workshop. Rather than presenting a finalised output, the chapter explores how the active participation of citizens and stakeholders can act as a key enabling condition for the circular regeneration of CPAs. Understanding both the importance and the modalities of involving citizens in the co-design and management of nature-based solutions is a central aspect of the ongoing research (Mahmoud et al., 2022).

Co-creation activities play a crucial role in supporting just transitions in CPAs by enabling stakeholders to engage in decision-making and in the implementation of policies and plans, while enhancing awareness and preparedness with respect to overlapping and multi-scalar risks. Students also play a key role in advancing the transition towards circularity and in the development of the SPArTaCHus Lab.

2 Background Knowledge: The Specificity of City-Port Areas

In Italy, coastal areas are the contexts that have been predominantly transformed in the last decades, contributing to alterations of coastal ecosystems and worsening the ongoing climate crisis (Curci et al., 2021).

In particular, City-Port Areas (CPA) are defined by complexity (Hein, 2016; Hein et al., 2023) which is further amplified by overlapping multi-risks (environmental, natural and anthropic risks) (RETURN, 2025) making them an important field of experimentation.

CPA are mostly affected by climate change, resource scarcity and linear urban metabolism processes, resulting in a significant amount of wastescapes (Amenta & van Timmeren, 2022; Mahmoud et al., 2022).

Wastescapes are abandoned or underused territories, at the end of their life cycles, lacking key characteristics of publicness: ownership, accessibility, management, and inclusiveness (Li et al., 2022).

To achieve a greater resilience for CPA, the ongoing research SPArTaCHus 'Sustainable City-Port Areas Towards Circular Hubs' investigates multifaceted approaches that are necessary to cope with these interconnected challenges, to foster regenerative solutions for re-inputting wastescapes in the circular urban metabolic processes.

The regeneration process for the wastescapes in CPA presents diverse challenges related to climate change issues, soil scarcity and difficulties to grow within ecosystem boundaries, overtourism, together with the additional pressure of high population density that generally exists in coastal territories.

The planning approaches for wastescapes regeneration, based on principles of circularity and through the implementation of nature-based solutions (NBS), are sensitive to the needs of the different stakeholders, to the circularity of energy and materials, as well as to the preservation of unbuilt soil, green areas and permeable open spaces that can assure long-term availability of land for agricultural purposes, and the provision of ecosystem services for all.

Considered the scarcity of land in CPA, this contribution aims to show how wastescapes can become circular hotspots and networks of public spaces, functioning as incubators of innovation for circular and regenerative solutions based on mitigation and adaptation measures (AIVP, 2021) and on regeneration processes through NBS. The relevance of urban nature has increased in recent decades due to climate change adaptation needs and the COVID pandemic, and it constitutes a right for all inhabitants.

Given these interconnected pressures and the scarcity of land, it is crucial to examine how CPA function as transitional interfaces between urban and port areas. These spaces represent both challenges and opportunities for circular and regenerative interventions, setting the stage for the concept of City-Port thresholds.

City-Port thresholds are multi-risk contexts and part of the coastal ecosystem, an interface between human and non-human life and activities and an ecological ecotone located between marine and terrestrial realms (Gleason et al., 2023). These characteristics—continuous exchange and cross-influence among ocean and terrestrial features, the presence of physical barriers, and edges with varying degrees of blur—make them fertile fields of investigation for just and circular urban transitions.

City-Port thresholds can be identified as critical urban areas, particularly vulnerable due to the coexistence and interaction of multiple risk conditions. These contexts require a systemic approach to risk management, adaptation, and mitigation, in close connection with territorial and urban planning. In multi-risk urban environments, where exposed environments and risks overlap, interpretative approaches and tools, from a metabolic urban planning perspective, can help investigate the relationships between urban metabolism processes—such as the production of space and flows—community dynamics, territorial vulnerabilities, and risks (RETURN, 2025).

City–Port Thresholds are interpreted as Transitional Landscapes (Russo, 2023), namely as liminal systems that, by their nature, enable transitions, arising from the intersection of two or more territories or entities. This liminal condition reflects the dynamism of coastal landscapes, their continuous variations, which connect site-specific topography and design processes, jointly determining the mutable nature of these marginal spaces.

In urban–port thresholds, physical, legal, and symbolic boundaries between city and port intersect. This often results in fragmented and separated spaces, where the material margins overlap with administrative separation, rigidifying what, by nature, should function as a space for relation, exchange, and transformation. These thresholds can be understood as spatial sequences whose varying characteristics may enable—or hinder—openings, interferences, and transitions (Moretti, 2023).

More in general, could be also defined as 'borderscapes' (Dell'Agnese & Amilhat Szary, 2015; Harbers, 2003), liminal spaces located between different political-administrative entities that reflect the complexity of contemporary territorial configurations. They are dynamic and transitional landscapes in which it is possible to imagine new pacts for publicness, mediating interactions, conflicts, and overlaps between distinct territories, properties, and local as well as global flows. Studying these borderscapes allows for the investigation of tensions, relationships, and transformations that characterise marginal spaces, revealing how physical structures, legal frameworks, and symbolic meanings are intertwined to define the configuration of urban and coastal territories.

In this framework, a just and sustainable transition for City-Port thresholds towards circularity can be experimented through co-creation processes following the Urban Living Lab methodology in multi-risk critical contexts (Clemente et al., 2025; Vendemmia et al., 2025). In these contexts, co-creation (organised in 3 steps: co-exploring, co-design, co-test) can serve as a driver for a fair transition, moving beyond the dimension of conceived space articulated by planners and architects and incorporating the needs of local communities, thus moving towards spatial justice (Newton, 2025).

3 Approach and Methods

3.1 Co-Creation Activities for CPA

Through different co-creation activities, SPArTaCHus fosters dialogue with diverse stakeholders—including researchers, local authorities, professionals, students, local communities and private-sector representatives—through interviews, co-design activities, and walking co-design workshops. All these stakeholders participate in the SPArTaCHus Lab, namely a collaborative virtual and physical platform. The Lab enables stakeholders to share their views and expertise, ultimately helping to align public and private interests in the circular regeneration of City-Port wastescapes, that will be eventually embedded in the final version of the Circularity Catalogue.

SPArTaCHus applies this methodology to the Metropolitan Area of Naples, and specifically to the case study of Castellammare di Stabia and Pozzuoli. This contribution focuses on the case of Castellammare di Stabia where it is possible to identify coastal fragmented ecosystems as at the mouth of the Sarno River, and a series of rigid physical and administrative boundaries that function as obstacles for the access to the sea. In the case study of Castellammare di Stabia, the project has been implementing different practices of community engagement, such as surveys, site visits with university students, workshops with experts, students and local actors. One example is the Urban Co-Design Laboratory: 'Multi-Risk: Water, Heritage, Circularity' held in Castellammare on the 14.04.2025 and developed in collaboration with IUAV University of Venice (Amenta & De Martino, 2026). Another initiative has been a walking co-design laboratory for coastal landscapes to be understood and regenerated, namely the co-creation workshop: "Jane's walk" organised in Castellammare on the 16th of June 2025, in collaboration with trainee students and developed jointly with colleagues from DiARC Department of Architecture of the University of Naples Federico II, namely prof. Francesca Nocca and trainee student Valentina Serafini. Teaching experimentations have been developed too, such as master thesis on the topic, and a design studio Lab of the second semester of the academic year 2024–2025 of the study-course ACTA, International Master's Degree in Architecture for Communities, Territories and Environment of the Department of Architecture of the University of Naples Federico II. All these initiatives have aimed at developing new circular and shared visions for the coastal territories of the Metropolitan Area of Naples, serving as the foundation for the preliminary version of the Circularity Catalogue that has been used as a base and tested in the co-design workshop on which this contribution is focusing (Fig. 2).

SPArTaCHus—also through the above mentioned co-design activities and fieldworks—is simultaneously analysing European circular and NBS practices, examining the main elements and methods that could be transferred and adapted to the Neapolitan contexts, which will then be incorporated into the final version of the Catalogue. By reimagining wastescapes as catalysts for circular growth, SPArTaCHus highlights how CPA can decouple economic development from resource depletion, charting a path toward sustainable and regenerative futures.

Fig. 2 Image by Giuseppe Davidson Carboni and Libera Amenta. Process of defining the strategic objectives for the preliminary circularity catalogue

The concept of circularity used here goes beyond the traditional recycling or material-focused perspective. It refers to a circular regeneration that acts on the life cycles of territories, aiming to define the critical threshold areas of the city-port under multi-risk conditions as circular regeneration hubs—spatial incubators where innovative, circular, and nature-based approaches are tested to enable communities to thrive in balance with nature. These hubs pursue holistic sustainability, addressing environmental, social, and economic dimensions, and are specifically designed to respond to climate-change-related challenges.

3.2 Testing the Preliminary Circularity Catalogue Through Co-Creation Activities

Among the different co-creation activities carried out within the SPArTaCHus research, a key initiative is the SPArTaCHus Seminar Series 2025 on City-Port Habitats,[1] followed by a co-design workshop. The Seminar Series aims to contribute

[1] This seminar series—held on the 30th of October and on the 6th of November—(Convener L. Amenta; Keynote lectures: professors Carola Hein, Francesco Musco and Francesc Muñoz, Scientific Commettee: Libera Amenta, Anna Attademo, Paolo De Martino, Bruna Vendemmia) has been made possible thanks to the University of Naples Federico II and the support of the T-Lab of the Department of Architecture of the University of Naples Federico II, which since 2023 has been recognized as a Department of Excellence.

to the ongoing scientific debate on nature-based solutions (NBS) as tools for the sustainable redefinition of City–Port Habitats, while fostering awareness of multi-risk conditions in coastal territories. The seminars focused on the potential impacts of NBS in the Metropolitan Area of Naples. The concluding round table, involving local stakeholders, was designed to support the formulation of a strategic vision for local contexts where neglected or degraded areas—referred to as wastescapes—can be reframed as opportunities for circular and metabolic territorial solutions.

The seminars emphasised how the circular regeneration of city–port thresholds—spaces of contact, exchange, and transformation between land and sea can hold significant potential to become meaningful spaces for local communities. By adopting a place-based approach attentive to the needs of multiple stakeholders, circular strategies and nature-based solutions can be mobilised to regenerate these territories, enhancing their resilience, inclusiveness, and sustainability.

The co-design workshop focused on threshold territories within City-Port Areas of Castellammare di Stabia, under multi-risk conditions, and specifically on the circular regeneration of coastal wastescapes. Building on topics emerging from previous co-creation activities and on a literature review, the workshop explored strategies to strengthen the resilience of these multi-risk territories:

> *How might these actions reshape sustainable city–port relations? Which socio-economic and environmental impacts should be considered? How can co-creation practices support more just and inclusive planning and governance processes?*

The co-design workshop began with an introduction by the convener, presenting the main transversal themes of co-creation and the wastescape framework. Workshop objectives and materials—including multi-scale maps of Castellammare di Stabia's multi-risk territories, a table of wastescape characteristics (Amenta & Attademo, 2023), and coloured dots for mapping actions—were made available at each table.

Step 1—Mapping discussion: Participants discussed the maps, produced through DiARC traineeships and theses, highlighting the city–port thresholds and multi-risk context. They were invited to suggest additional wastescapes, adding notes or drawing directly on the maps.

The insights from Step 1 informed Step 2, ensuring that the strategic objectives were linked to identified wastescapes and locally relevant areas.

Step 2—Strategic objectives: The team presented three objectives derived from prior co-creation activities and literature review:

1. Improve accessibility to the sea;
2. Living with risk;
3. Restore socio-cultural values of city–port areas.

Participants linked one or more objectives to wastescapes or other areas, placing coloured dots on the maps to visualise proposed actions (Fig. 3 and Fig. 4).

Fig.3 SPArTaCHus Co-design workshop on City-Port Areas of Castellammare di Stabia 30th of October 2025, Department of Architecture, University of Naples Federico II

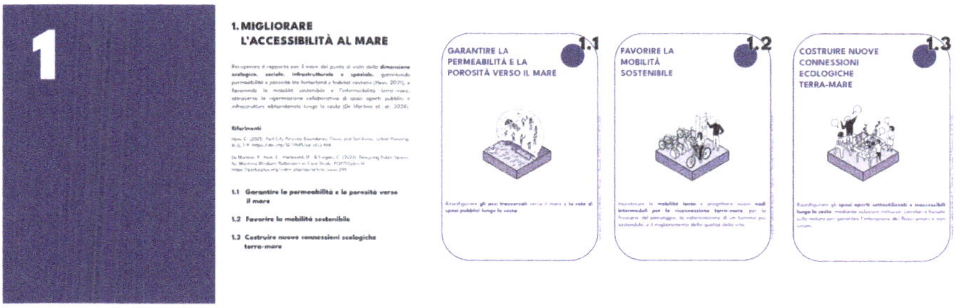

Fig. 4 Examples of one of the three selected SPArTaCHus strategic objectives and related cards defined in preparation of the co-design workshop of the 30th of October 2025. Images within the cards designed by student Clara Maseda Juan

4 Discussion and Implications

Empirical studies have shown that NBS can simultaneously enhance ecological functions and social well-being in coastal urban areas (Kabisch et al., 2017; Nesshöver et al., 2017), supporting the transformation of underused city-port territories into multifunctional, resilient public spaces. These studies complement the insights gained from the SPArTaCHus workshops, demonstrating how participatory approaches can guide context-sensitive implementation of NBS. Although widely recognised at the European and global level by many policies and programs (e.g. Green Deal, EU Biodiversity Strategies, Urban Agenda for the EU, UN SDGs) for their positive role to address complex challenges in urban areas, especially in relation to the increasing climate crisis, however NBS could be also linked to negative impacts on socio-economic balance in cities. The implementation of NBS for the resilience of urban areas can indeed lead to the rising inaccessibility of housing market due to the escalation of housing rental and purchase costs in green-based regenerated urban areas. In fact, NBS can be—unintentionally—the cause of the so-called environmental, green, or ecological gentrification (Reed-Thryselius, 2023).

Greening actions can be indeed related with the risk of the green gentrification and displacement of vulnerable population groups from the new urban vision of regeneration, defining a social process that exacerbates existing inequalities in relation to the economic weakness of residents (Alexandrescu et al., 2021).

This process involves the displacement of fragile communities, as instance lower income residence (Rice et al., 2023); thus, it is important to identify monitoring tools for avoiding gentrification risks while applying NBS for the circular regeneration of wastescapes and involve—at each stage of the decision making process—all the relevant stakeholders, paying attention also to the more-than-human dymension.

How the results of the co-creation activities developed in the considered case study of Castellammare di Stabia, and of the Metropolitan Area of Naples in general, could be scaled up to CPA in general is one of the aims of the research, to be further developed.

Through the co-mapping of wastescapes in CPA, and thanks to the organisation and development of co-creation sessions with the local community to increase their engagement and multi-risk awareness, the research aims to identify transferable results through the definition of a Circularity Catalogue that is co-created with students, the local community, experts, and other relevant stakeholders. It will be composed of a set of guidelines for planning to achieve a climate-just city through reflections on (1) collaborative planning methods; (2) the inclusion of the more-than-human perspective in the regeneration of city-port areas; and (3) equity of access and compatible uses of the maritime public domain. At the same time, the process highlights the limitations posed by the exclusion or underrepresentation of certain stakeholder groups, whose absence may constrain the inclusiveness, legitimacy, and applicability of the proposed strategies. Recognising and addressing these gaps becomes essential to ensure that the Circularity Catalogue fully reflects the plurality of needs and perspectives present within CPA.

5 Concluding Remarks

The reflections developed within the SPArTaCHus project highlight how the design and implementation of interaction tools require interdisciplinary collaborations capable of engaging with the complexity of contemporary urban and coastal environments. Exchanges with different disciplinary perspectives have broadened the project's analytical lens, enabling a more nuanced understanding of the multi-risk conditions that characterise City-Port Areas and of the role of co-creation in supporting just and resilient transitions.

Preliminary insights from the Metropolitan Area of Naples reveal the transformative potential of reframing environmental challenges as opportunities for regeneration. The preliminary testing of the Circularity Catalogue in co-creation workshops demonstrates that operational tools, even in development, can effectively guide decision-making, stakeholder dialogue, and the implementation of nature-based solutions in multi-risk City-Port Areas, ensuring an active engagement of all stakeholders in the co-design process.

Rethinking the Sarno River—in the northern part of Castellammare di Stabia along the boundary with the municipality of Torre Annunziata—as a cultural, ecological, and economic asset; addressing hydrogeological risk through integrated and multi-scalar strategies based on nature-based interventions; reducing socio-spatial fragmentation in waterfront contexts by focusing on mixed uses and high quality of public spaces; and recognising wastescapes as latent resources collectively point to the possibility of reconfiguring these territories as sustainable and participatory urban hinges and potential landscapes to be recovered. These findings underscore how land-water thresholds can become strategic laboratories for circular regeneration and renewed territorial identities. In particular, insights from the Castellammare di Stabia case study illustrate how thresholds between land and sea can function as experimental spaces where circular and nature-based approaches can be co-designed with local stakeholders.

At the same time, ongoing open questions emphasise the need for critical assessment. Empirical evidence from Rotterdam and other international cases confirms that nature-based solutions (NBS) contribute significantly to climate adaptation and urban resilience letting nature take over in certain multi-risk circumstances. Yet their implementation may also produce unintended socio-spatial inequalities, including risks of green gentrification or uneven access to restored ecosystems. These concerns call for collaborative governance models attentive to spatial justice, where ecological regeneration aligns with circular resource management and socially inclusive practices. Future application of nature-based solutions should integrate monitoring frameworks to prevent unintended socio-spatial inequalities, ensuring that ecological regeneration supports climate justice and equitable access to urban resources.

In this perspective, the development of a Circularity Catalogue emerges as a tool to support decision-makers and communities in navigating these complexities. By integrating ecological strategies, co-creation methodologies, and justice-oriented approaches, the Catalogue aims to guide the formulation of context-sensitive actions capable of strengthening resilience while ensuring equitable benefits for all stakeholders. Ultimately, the SPArTaCHus experience suggests that just and resilient City-Port transitions can only be achieved through processes that are ecologically grounded, socially inclusive, and collaboratively produced.

The insights gained from co-creation workshops will directly inform the Circularity Catalogue's strategic objectives, demonstrating that even preliminary operational tools can effectively support equitable, nature-based regeneration in multi-risk City-Port Areas.

Acknowledgements I would like to express my sincere gratitude to all colleagues, PhD students, and master's and trainee students who contributed to the organisation of the seminar's co-design activity.

Funding This paper is part of the outcomes of the Research 'SPArTaCHus. Sustainable city-port areas towards circular hubs. rethinking life cycles of wastescapes in the city-port areas of the metropolitan City of Naples'. Research financed within the framework 'University Research Funding Program—FRA, 2022' (Programma per il Finanziamento della Ricerca di Ateneo—FRA, 2022), of the University of Naples Federico II, Italy. Corresponding proponent: Libera Amenta. Università degli Studi di Napoli Federico II [Iniziativa finanziata nell'ambito del Programma per il Finanziamento della Ricerca di Ateneo (FRA) 2022 dell'Università degli Studi di Napoli FedericoII].

References

AIVP. (2021). *How to adapt port cities to climate change. Challenges and solutions—AIVP*. https://www.aivp.org/en/events/how-to-adapt-port-cities-to-climate-change-challenges-and-solutions/

Alexandrescu, F. M., Pizzol, L., & Critto, A. (2021). Green gentrification as strategic action: Exploring the emerging discursive and social support for the green tree strategy in Porto Marghera, Italy. *Cities, 118*, 103352. https://doi.org/10.1016/J.CITIES.2021.103352

Amenta, L., & Attademo, A. (2023). Periurban coastal landscape: A method to identify and map Resource-Scapes. *TRIA Territorio Della Ricerca Su Insediamenti e Ambiente. International Journal of Urban Planning, 30*(1), 95–114. https://doi.org/10.6093/2281-4574/10103

Amenta, L., & De Martino, P. (2026). Learning How to Live with Risk—The Role of Co-Design for Managing City–Port Thresholds in Castellammare di Stabia, Naples, Italy. *Sustainability*, 18(7). https://doi.org/10.3390/SU18073242

Amenta, L., & van Timmeren, A. (2022). From wastescapes towards regenerative territories. A structural approach for achieving circularity. In *Regenerative territories dimensions of circularity for healthy metabolisms* (pp. 147–160). https://doi.org/10.1007/978-3-030-78536-9_9

Clemente, M. F., Vendemmia, B., & Amenta, L. (2025). Extended partnership and kowledge sharing: the return urban living lab approach to social-ecological-technological innovation. *TECHNE—Journal of Technology for Architecture and Environment, 2025*(30), 150–160. https://doi.org/10.36253/TECHNE-17472

Curci, F., Mininni, M., Nanni, G., Zanchini, E., & Zanfi, F. (2021). Nelle aree costiere: accessibilità, sicurezza e risposta al cambiamento climatico. In A. Coppola, M. Del Fabbro, A. Lanzani, G. Pessina & F. Zanfi (Eds.), *Ricomporre i divari. Politiche e progetti territoriali contro le disuguaglianze e per la transizione ecologica.* (pp. 71–83).

Dell'Agnese, E., & Amilhat Szary, A.-L. (2015). Borderscapes: From border landscapes to border aesthetics. *Geopolitics, 20*(1), 4–13. https://doi.org/10.1080/14650045.2015.1014284

Gleason, M. G., Reynolds, M. D., Heady, W. N., Easterday, K., & Morrison, S. A. (2023). The importance of identifying and protecting coastal wildness. *Frontiers in Conservation Science, 4*. https://doi.org/10.3389/fcosc.2023.1224618

Harbers, A. (2003). Borderscapes, the influence of National Borders on European spatial planning. In R. Brousi, P. Jannink, W. Veldhuis & I. Nio (Eds.), *Euroscapes*. Must Publishers/Architectura et

Amicitia. https://www.researchgate.net/publication/304570908_Borderscapes_The_influence_of_National_Borders_on_European_Spatial_Planning
Hein, C. (2016). Port cityscapes: Conference and research contributions on port cities. *Planning Perspectives, 31*(2), 313–326. https://doi.org/10.1080/02665433.2015.1119714
Hein, C., van Mil, Y., & Ažman Momirski, L. (2023). *Port city Atlas*. TU Delft OPEN Publishing. https://doi.org/10.59490/mg.73
Kabisch, N., Korn, H., Stadler, J., Bonn, A. (2017). Nature-Based solutions to climate change adaptation in urban areas—linkages between science, policy and practice. In N. Kabisch, H. Korn, J. Stadler & A. Bonn (Eds.), *Nature-Based solutions to climate change adaptation in urban areas. Theory and practice of urban sustainability transitions*. Springer. https://doi.org/10.1007/978-3-319-56091-5_1
Li, J., Dang, A., & Song, Y. (2022). Defining the ideal public space: A perspective from the publicness. *Journal of Urban Management, 11*(4), 479–487. https://doi.org/10.1016/J.JUM.2022.08.005
Mahmoud, I. H., Morello, E., Salvia, G., & Puerari, E. (2022). Greening cities shaping cities. *Greening Cities Shaping Cities*, 376. https://doi.org/10.3390/BOOKS978-3-0365-4704-6
Moretti, B. (2023). The City-Port threshold. An Atlas of maps between verisimilar and liminality. In *De-Sign: Environment landscape city 2021* (pp. 399–409). Aracne. https://doi.org/10.53136/979122180495934
Nesshöver, C., Assmuth, T., Irvine, K. N., Rusch, G. M., Waylen, K. A., Delbaere, B., Haase, D., Jones-Walters, L., Keune, H., Kovacs, E., Krauze, K., Külvik, M., Rey, F., van Dijk, J., Vistad, O. I., Wilkinson, M. E., & Wittmer, H. (2017). The science, policy and practice of nature-based solutions: An interdisciplinary perspective. *Science of The Total Environment, 579*, 1215–1227. https://doi.org/10.1016/j.scitotenv.2016.11.106
Newton, Caroline. (2025). *Envisioning spatial justice. Explorations, reflections, design*. Jap Sam Books. https://www.japsambooks.nl/products/envisioning-spatial-justice
Reed-Thryselius, S. (2023). A review of environmental gentrification ills and the "just green enough" approach: On achieving justice, sustainability, and equity. *International Journal of Community Well-Being, 6*(4), 411–422. https://doi.org/10.1007/s42413-023-00195-9
Rice, J. L., Long, J., & Levenda, A. (2023) Introduction: Realizing the just city in the era of climate change. In J. L. Rice, J. Long & A. Levenda (Eds.), *Urban climate justice: Theory, praxis, resistance*. University of Georgia Press. https://library.oapen.org/handle/20.500.12657/86093
RETURN. (2025). *DV-5.4.5_Evaluation framework for monitoring circularity, sustainability and resilience of urban metabolism*. https://www.fondazionereturn.it/wp-content/uploads/2025/10/DV-5.4.5_final-release_invio-ottobre-25_compressed_compressed-1.pdf?_gl=1*14jd01x*_up*MQ..*_ga*MTA1MjAxNDQ5MC4xNzY0MzU3MTU2*_ga_GR7TT2Y9FY*czE3NjQzNTcxNTUkbzEkZzAkdDE3NjQzNTcxNTUkajYwJGwwJGgw
Russo, M. (2023). Transitional landscapes. In *Transitional landscapes* (pp. 19–27). Quodlibet.
UNDP. (2023). *Climate change is a matter of justice*. United Nations Development Programme. https://climatepromise.undp.org/news-and-stories/climate-change-matter-justice-heres-why
Vendemmia, B., Amenta, L., Clemente, M. F., & Iodice, R. (2025). Urban Living Lab (ULL) as a tool for co-design sustainable transition in critical urban contexts: Co-exploring the site of national interest of Bagnoli in the metropolitan area of Naples, Italy. *Advances in 21st Century Human Settlements. Part, F1035*, 267–280. https://doi.org/10.1007/978-981-96-7874-7_27/FIGURES/3

Green Feeling: Sensual Design, Biophilia, and the Creation of Restorative Atmospheres for Citizens

Silvana Kühtz and Leonardo Tizi

Abstract Accelerated and often unplanned urbanization has produced densely built environments increasingly exposed to climatic, environmental, and social stressors. Within this scenario, urban resilience requires policies and strategies that move beyond technical mitigation and instead integrate green, sensory, and socio-cultural dimensions. This chapter examines how greening and sensory practices—understood as Nature-Based Solutions (NBS) and as generators of restorative atmospheres—can enhance the adaptive capacity and well-being of contemporary cities. It proposes a framework grounded in biophilia, environmental psychology, and Sensual Design to advance best practices that support inclusive, health-promoting, and socially cohesive urban environments. Within this interdisciplinary perspective, participatory design emerges as a crucial driver of resilient transformation, enabling communities to co-create, care for, and meaningfully inhabit green infrastructures. The chapter mentions "URGES—Urban Green Shapes" project (held in Matera, Italy) as case example, an applied research initiative focused on integrating vertical and horizontal green devices within a peripheral neighbourhood. As coordinators of the participatory design component, the authors implemented a set of sensory-driven methodologies—soundwalks, blindfolded explorations, embodied mapping, and collaborative storytelling—to cultivate residents' perceptual green feelings, well-being and place attachment in such marginalized urban contexts. This involved experiences composed of neighbourhood walks and performative artistic activities, moments of poetic reflection, and sensory games—for example, by excluding the visual channel (with blindfolds) and activating other senses like hearing and smell. Combining greening interventions with sensory and creative processes strengthens place attachment, enhances environmental awareness, and builds a possibility for long-term stewardship of urban Nature-Based Solutions. By bridging environmental psychology

S. Kühtz (✉) · L. Tizi
Università Degli Studi Della Basilicata, Potenza, Italy
e-mail: silvana.kuhtz@unibas.it

L. Tizi
e-mail: info@leonardotizi.com

L. Diana et al. (eds.), *Greening and Resilience*, Urban Sustainability,
https://doi.org/10.1007/978-981-92-0419-9_17

approaches with Sensual Design experiences, this chapter contributes a set of transferable insights for improving the resilience, livability, and restorative potential of the built environment.

Keywords Sensoriality · Sensual Design · Participation · Biophilic Urbanism · Community Engagement

1 Introduction

Green feeling in this chapter is the affective and perceptual disposition that emerges when citizens—through sensory design practices, playful explorations, and embodied exercises—consciously attune to the ecological qualities of their surroundings, allowing environmental awareness to arise from lived experiences. The primary objective of this study is to demonstrate how integrating Sensual Design and Biophilic Urbanism within participatory processes can transform Nature-Based Solutions (NBS) from technical infrastructures into *restorative atmospheres*. This research aims to fill a gap in the literature where social acceptance of greening is often treated separately from its sensory impact.

A Sensual Design approach that systematically engages sight, hearing, touch, and smell produces rich environmental experiences that are inherently more inclusive and equitable. This perspective is rooted in the "multi-sensory turn" of urban studies (Henshaw, 2013; Howes, 2004), which argues that city dwellers perceive their environment as a *sensorium* rather than a series of visual frames. It resists the dominance of the purely visual and rational approach that has long shaped design disciplines, proposing instead an embodied and sensory practice that acknowledges perception as a form of thought. To design sensually means to listen, touch, smell, and move through space, feeling its atmosphere as a way of understanding and transforming it. This approach invites designers and inhabitants alike to engage with the emotional and atmospheric dimensions of places, revealing layers of memory, intimacy, and collective imagination often obscured by functional or aesthetic conventions. The Sensual Design approach opens up new possibilities for situated interaction and for developing meaningful understandings of the built environment. Crucially, engaging the full sensory spectrum strengthens place attachment, understood as the affective bond that links individuals to specific settings and generates feelings, collective meaning-making, protection, and belonging (Manzo & Devine-Wright, 2019). The theoretical framework of this work also draws upon the concept of environmental *affordances* (Gibson, 1979) and the Theory of Restoration (Hartig et al., 2014), proposing that sensory-rich green spaces provide the necessary "soft fascination" to alleviate urban mental fatigue (Kaplan & Kaplan, 1989).

These settings encourage forms of interaction that validate lived experience, deepen interpersonal attunement, and cultivate a sense of communal presence (Ingold, 2000). In this way, Sensual Design becomes a wide framework for an ethical and political practice that supports inclusive and compassionate forms of urban life.

The relationship between humans and nature is profound, as postulated by Kellert and Wilson's Biophilia Hypothesis (1993), which describes an innate human tendency to connect with life and vital processes. This connection, often eroded in urban settings, is crucial for psychophysical well-being. Biophilic Design (Kellert et al., 2008) become then principles applicable to the design of built spaces, not merely by adding greenery, but by meaningfully integrating natural elements and processes, thereby improving health and social connectivity (Browning & Ryan, 2020). Biophilic and Sensual Design and Urbanism serve as paradigms for rethinking the city not just as a place of coexistence between humans and nature, but as an integrated ecosystem where nature is essential. This approach, updated and integrated into the conceptualization of Restorative Cities (Roe & McCay, 2021), is not a luxury but a necessity for addressing urban resilience challenges. Research has shown that exposure to nature reduces stress, improves cognitive functions, and promotes health recovery processes, highlighting the potential of Biophilic Design as a therapeutic and preventive tool in urban planning. Simple exposure to nature reduces cortisol levels (the stress hormone) and improves heart rate variability. Ulrich's Stress Reduction Theory (SRT) (1983) proposes that nature induces positive affective responses and reduces physiological stress activation, while Kaplan and Kaplan's Attention Restoration Theory (ART) (1989) suggests that natural environments facilitate recovery from mental fatigue and enhance attentional capacity. These mechanisms provide foundational knowledge for designers, empowering them to create projects and atmospheres that counteract eco-anxiety, foster a deeper connection with a broader, more resilient environment.

The Biophilic Cities movement (Beatley, 2016) further supports the research background, which emphasizes that nature in cities should be a continuous, immersive experience rather than a fragmented presence. Bringing these perspectives together expands the designer's agency from shaping form to shaping experience. The presence of nature in cities improves air and water quality, enriches daily experience, and offers opportunities for play, socialization, learning, and sensory contact. Groups of trees planted in high-density residential areas attract more people to green spaces, encourage social interactions crucial for personal and collective well-being, and promote physical activity. From a mental health perspective, nature improves mood and cognitive function. In disadvantaged communities, the presence of common outdoor spaces contributes to social cohesion, and the perceived quantity of open space and vegetation is associated with lower crime rates and a greater sense of safety. Thus, designers are called upon to trigger virtuous mechanisms for creating restorative atmospheres of well-being and participation, starting from the sensory experiences they can design for residents and for their immersion in a city-nature.

2 Integrating Design and Environmental Psychological Approaches

Environmental psychology (Steg & de Groot, 2019) has long demonstrated that human experience of place grows with sensory stimuli, cognitive appraisals and affective associations, bringing overall wellbeing (Gifford, 2014; Mehrabian & Russell, 1974); yet design practice has often privileged visual and formal dimensions, neglecting the embodied, emotional, and cultural layers through which people actually inhabit spaces. Bringing these disciplinary perspectives together expands the designer's agency from shaping form to shaping experience, foregrounding how spatial cues influence well-being, social cohesion, attachment to place and pro-environmental behaviour. Contemporary design thinking incorporates systemic, participatory, and perceptual dimensions, acknowledging that environments are dynamic ecologies of experience (Manzini, 2015; Norman, 2013). Sensory approaches to design assume that perception is a relational act—not merely the reception of stimuli but an active negotiation between body, environment, and imagination (Ingold, 2000).

Concepts such as place identity, sense of coherence, restorative environments, and behavioural mapping offer designers tools to interpret how individuals perceive, navigate and emotionally respond to urban settings. At the same time, contemporary approaches in design—particularly sensory, participatory and situated design—provide environmental psychology with concrete methods for materializing theories within real contexts. This reciprocal exchange produces richer frameworks for urban regeneration, enabling interventions that are not only efficient or aesthetically appealing but also psychologically supportive, culturally meaningful, and experientially attuned. This convergence also highlights the importance of situated knowledge: local memories, sensory atmospheres, cultural rituals, and community narratives become essential components of design processes. Moving beyond universal or top-down models, design acknowledges that environments are co-created by bodies, practices, and histories. Designing with environmental psychological insights therefore means cultivating attentiveness to how people feel, sense, learn and remember through space—and ensuring that these human processes are embedded in decision-making. Such an approach ultimately supports more inclusive, resilient, and restorative urban environments, where wellbeing arises from an ecological alignment between people, place, and culture. Research in cognitive and behavioural sciences demonstrates how sensory stimuli modulate stress levels, attentional states, and emotional regulation (Shams & Seitz, 2008; Spence, 2020). The built environment is never neutral: light, sound, thermal gradients, textures, and olfactory cues shape affect and cognition in ways that deeply influence the quality of urban life. Across design fields, sensory approaches have become increasingly important for creating inclusive environments that address diversity and accessibility (Heschong, 1979; Pallasmaa, 2005).

2.1 Sensual Design

Within this integrated perspective, Sensual Design becomes a mediator between environmental affordances and human responses, capable of translating psychological insights into spatial strategies. By acknowledging the interdependence between perception and materiality, Sensual Design provides tools for constructing atmospheres that resonate with diverse bodies, abilities and cultural sensitivities (Kühtz, 2024). Sensual Design expands the sensory paradigm by placing emphasis on embodied perception, atmospheres, affective resonance, and the cultivation of sensory awareness. While environmental psychology describes how sensory cues affect behaviour and wellbeing, Sensual Design provides methods to train perception and activate embodied knowledge. Practices such as deep listening (Oliveros, 2005), blindfolded walking, sensory journaling, and atmospheric mapping make perceptual processes explicit and relational. They enable individuals and communities to tune into the subtle qualities of place—its rhythms, green presence, textures, microclimates, and affective tones. This approach aligns with psychological frameworks such as affordance theory (Gibson, 1979), sense-of-place studies (Relph, 1976; Tuan, 1977), and theories of embodied cognition (Varela et al., 1991). By foregrounding the lived body as the primary interface with the environment, Sensual Design emphasises that the meaning of space emerges through experience. At the same time, Sensual Design contributes to participatory and community-based practices. It equips local communities with perceptual tools that allow them to articulate their atmospheric, emotional, and sensory relationships with territory. This is particularly valuable in contexts of urban regeneration and environmental justice, where recognizing sensory inequities (noise pollution, light deprivation, thermal discomfort, and olfactory disturbances) is essential for shaping more inclusive spatial policies. Sensual Design thus reinforces the environmental psychological principle that wellbeing is not merely functional or aesthetic, but embodied, relational, and deeply situated. Sensual Design is a fundamental expansion and deepening of the principles of Sensory Design, moving from the orchestration of stimuli toward a comprehensive and embodied understanding of perception, emotion, and meaning-making. Rooted in phenomenology, affect theory, and ecological thought, Sensual Design recognizes that perception is not a neutral or purely physiological act but a lived, affective, and interpretative engagement with the world.

In this perspective, to design sensually is to act within a relational ontology: the designer resonates with the context. This approach builds upon the Phenomenology of Perception (Merleau-Ponty, 1945) and the Aesthetic of Atmospheres (Böhme, 2017), suggesting that the quality of urban life depends on the *felt* relationship between the body and the material world. Drawing from Maurice Merleau-Ponty's (1945) notion of the lived body, Sensual Design positions the body as the primary interface between self and environment—a site where knowledge, imagination, and empathy converge, but adds on relationships, affect, pathos. Gernot Böhme's (2017) theory of atmospheres provides a crucial theoretical grounding here: Sensual Design operates in the liminal zone between subject and object, crafts the "felt space" that arises through affective and sensory interaction. Atmospheres are not properties of things but "spatially extended feelings", and the designer's task is to modulate them ethically and poetically. Sensory elements—such as light, sound, texture, temperature, rhythm, and olfaction—collectively contribute to the generation of what Böhme terms atmosphere. The deliberate design of this atmosphere is consequently fundamental for fostering well-being, restoration, creativity, and robust collective engagement. This framework intersects with the ethics of care and sustainability articulated by Ezio Manzini (2015), who describes design as a form of social innovation that nurtures meaningful relationships between people and environments. Sensual Design extends this logic of care to the perceptual and emotional realm, foregrounding the quality of experience as a driver of sustainability. Rather than optimizing performance, it seeks to cultivate resonance (Rosa, 2019): a responsive and reciprocal relation between humans and their surroundings. A Sensual Design proposes a paradigm shift: from an instrumental design of function and form to a poetic design of relations and thus well-being. It recognizes that to design is to affect and to be affected, to give shape to atmospheres that hold emotional, ethical, and ecological consequences. In a time when disembodiment and digital mediation risk flattening perception, Sensual Design offers a counter-practice of re-sensitization—a way to inhabit the world attentively, responsively, and poetically. Following Böhme's theory of atmospheres, the designer becomes a curator of affective conditions, a mediator between material and immaterial forces. Therefore becomes essential designing with the senses rather than for them, and cultivating awareness of the subtle interplay between body, environment, and imagination.

2.2 Biophilic Design

Biophilic Design constitutes a comprehensive framework rooted in biophilia, which denotes the inherent human propensity to engage with and support life processes (Fromm, 1964), represents an empirical experience of profound communion with nature (Wilson, 1984). However contemporary life patterns erode this inkling (Kellert, 2012). Richard Louv (2005) coined the term "Nature-Deficit Disorder" to describe the developmental impairment afflicting new generations who are compelled to grow up in highly urbanized environments, increasingly estranged from the natural world. This deficit is critical indicator requiring structural interventions. It is in direct response to this increasing sensory and ecological disengagement that Biophilic Design emerges, allowing architects and designers to re-establish the essential connection between humans and nature (Lefosse et al., 2023; Tekin et al., 2025). This shift entails a profound ethical responsibility for the designer. By moving beyond a detached, anthropocentric paradigm that treats nature as a mere resource, the biophilic designer embraces an ethical commitment to life-fostering processes, ensuring that the transformative act of building serves as a constructive expression of the life drive rather than a tool for environmental and human alienation (Mazzali et al., 2025).

Biophilic Design strategies emphasize three key dimensions (Browning et al., 2014): Nature in the Space: the direct presence of elements like plants, water, and light; Natural Analogues: the use of materials, textures, and patterns that mimic nature; and Nature of the Space: spatial configurations that evoke prospect, refuge, or mystery. In marginalized neighbourhoods, applying these principles is not merely an aesthetic or ecological tool, but a catalyst for psychological resilience and community cohesion, particularly when combined with Sensual and participatory design methods. In this way, Biophilic Design becomes a relational framework that links environmental quality with embodied perception and long-term social sustainability.

The objective of Biophilic Design is only seemingly simple. Stephen Kellert, the pioneer of this approach, highlighted its limitations in the inherent biological complexity and the difficulty of translating principles into specific design solutions. Kellert systematized Biophilic Design based on two dimensions (naturalistic and vernacular) and three types of experience (direct, indirect, and experience of space/place), ultimately identifying 72 actionable characteristics (Kellert, 2018). These recommendations offer a range of options for using Biophilic Design effectively, provided that they are employed with respect for the specificity of each project, rather than as a checklist applied indiscriminately. In parallel, the consultancy Terrapin Bright Green (TBG), adopting a more pragmatic stance, synthesized decades of environmental psychology research into 15 actionable Patterns of Biophilic Design (Browning & Ryan, 2020). While Kellert's theoretical richness provides conceptual depth, the TBG approach prioritizes usability and immediate implementation. The two approaches form a complementary continuum: Kellert elucidates why biophilia matters (ecological ethos and cultural nuance), while TBG articulates how to enact these principles (scalability and reproducibility). A sophisticated biophilic strategy must therefore draw from both frameworks—combining Kellert's evolutionary and experiential insights with TBG's verified behavioural patterns—to create environments that are nature-rich, culturally meaningful, psychologically restorative, and contextually grounded. Recent studies have demonstrated that the psychological benefits of these patterns are maximized when they are perceived through diverse sensory channels—such as the olfactory stimulation of aromatic plants or the haptic feedback of natural textures—rather than just visual contact (e.g., Gao et al., 2022).

Contemporary paradigms of Biophilic Urbanism (Bechlem et al., 2025; Lefosse et al., 2025; Tabb, 2020) extend this framework to the scale of the city, advocating for multi-layered integrations of nature—from micro-interventions (such as green façades and pocket gardens) to larger ecological corridors and landscape infrastructures.

3 Case Example: URGES Project in Matera

Nature-Based Solutions are "solutions that are inspired and supported by nature, which use or enhance natural features and processes to address social, economic and environmental challenges in an efficient and adaptive manner, providing multiple benefits" (European Commission, 2015). In urban contexts, NBS, such as green roofs, green walls, urban parks, rain gardens, and urban forests, offer a myriad of benefits. The "dose of nature" required for these restorative benefits is an active area of research, but it's clear that even brief exposures can have significant positive effects (Pazzaglia & Tizi, 2022). The URGES (Urban Green Shapes) project was precisely such a project, studying a wide range of factors from heat island mitigation and reduction of domestic energy consumption to stormwater management and fostering a sense of community belonging. Funded by POR FESR Basilicata 14-20, URGES implemented a pilot prototype of vertical and horizontal greening forms on a public housing building in Matera (Italy). The authors of this contribution co-designed a series of workshops, "Poetically Inhabiting the City", to facilitate resident participation and engage them in discussions to imagine a community that would later adopt the developing green area. Green solutions cannot be top-down impositions; their effectiveness and resilience depend on the acceptance, care, and appropriation by local communities (Kuhtz & Tizi, 2023). Involving citizens in the design and management phases of urban green spaces is a matter of social justice and a pragmatic strategy to ensure their durability and functionality. The methodology for the URGES project sits on a bottom-up approach, creating a multidisciplinary process. This integrated the expertise of various professionals—architects, engineers, agronomists, and environmental psychologists—with the active participation of residents and stakeholders. The initial phase involved a preliminary analysis to explore the context and collect both qualitative and quantitative data. This was followed by a central phase of community activation and engagement. During the subsequent co-design phase, the contributions from citizens became concrete proposals, such as adding two green spaces with benches.

3.1 The Participatory Experiences

The participatory practices applied within the URGES project unfolded as a sequence of embodied, sensory, and relational actions that progressively reconfigured the neighbourhood's perception and its ecological possibilities. The process began with sensory walks and deep-listening explorations—soundwalks carried out in silence, sometimes blindfolded—to invite residents to attend to the acoustic and atmospheric qualities of their environment. These walks revealed unnoticed sounds, rhythms, and disturbances, while strengthening the participants' embodied connection to ordinary spaces and offering the research team qualitative environmental data grounded in lived experience.

Alongside these explorations, performative and poetic workshops—framed within the methodology of *Poetically Inhabiting*—combined micro-performances, guided sensory exercises, and brief acts of poetic reflection (Figs. 1, 2, and 3). Through these activities, emotional and symbolic layers of the neighbourhood emerged, enabling inhabitants to articulate perceptions that later became shared design cues. Participants experimented perceptual positions different from the ordinary, meaning they experienced the world with all their senses.

Fig. 1 Sensory games, URGES Matera (*Source* Photo taken by the authors)

Fig. 2 Sensing the panorama, URGES Matera (*Source* Photo taken by the authors)

Fig. 3 Sensing the green, URGES Matera (*Source* Photo taken by the authors)

This knowledge expanded during sessions of embodied mapping, where residents produced sensory maps identifying places of comfort and discomfort, memories, tensions, desires, ecological opportunities, and atmospheric impressions. These community cartographies formed an alternative, situated knowledge that complemented technical analyses and revealed the neighbourhood as a dense fabric of affective and perceptual relations. Engaging residents in focus groups and meetings with administrations, designers, and University faculty-researchers and foreign partners, led to discussions on vertical and horizontal greening solutions—green façades, rooftop gardens, communal shaded areas, micro-habitats and green pockets—anchored to their everyday habits, desires, and constraints (Figs. 4 and 5). This process transformed inhabitants into co-researchers who contributed to imagining Nature-Based Solutions. Their involvement increased the acceptance, care, and long-term maintenance of green interventions, situating ecological decisions within the social life of the neighbourhood.

Parallel to these activities, neighbourhood festivities and collective activation events created informal, joyful spaces for participation (Fig. 6). Street gatherings, music, food, and open-air assemblies, reinforced social cohesion, reduced skepticism toward institutions, and transformed engagement into a positive, shared experience. Interviews with key community figures—elderly residents, mothers, youth, informal leaders, and local technicians—added further depth, revealing the multiplicity of place attachments, latent conflicts, and hidden strengths within the neighbourhood.

Fig. 4 Focus group with residents and stakeholders. URGES Matera (*Source* Photo taken by the authors)

Fig. 5 Open air participation meeting, URGES Matera (*Source* Photo taken by the authors)

These narratives provided essential material for shaping both communication strategies and design choices. Workshops with schools and local associations involved children and adolescents through drawing, creative tasks, micro-gardening, and listening exercises.

Fig. 6 Gathering with the local street band, never been in this part of the city, URGES Matera (*Source* Photo taken by the authors)

4 Results and Challenges

The results of the URGES project indicate a measurable shift in the residents' relationship with their neighborhood. Specifically, three key outcomes emerged:

1. Perceptual restructuring: Participants transitioned from a visual-only perception of neglect to a sensory appreciation of potential. Blindfolded soundwalks, for instance, allowed residents to identify the sound of wind in the new vegetation as a positive restorative marker.
2. Increased agency and stewardship: The participatory process reduced the Nature-Deficit by involving residents in the physical care of NBS. This led to a 40% decrease in reported skepticism toward public greening interventions (based on post-workshop surveys).
3. Atmospheric co-creation: The workshops generated *affective maps* that served as a design bridge between technical engineering (NBS) and lived experience.

All these actions gradually converged in the creation of a shared vision for the neighbourhood. Through accumulated narratives, sensory experiences, workshops, and co-design, residents began to articulate a collective, future-oriented imagination around their public spaces and green infrastructures. This vision represented a form of cultural, restorative and ecological awareness that tied personal experiences to collective aspirations. In this sense, the Matera experience showed that sensory and poetic methods can unlock otherwise inaccessible layers of urban knowledge; that participatory greening becomes effective when communities engage through their bodies, emotions and narratives; and that social cohesion, place identity, and ecological awareness can be cultivated through embodied and sensory practices. Valuing local traditions and narratives linked to green spaces can strengthen the sense of

Fig. 7 Green vertical and horizontal micro-infrastructures, URGES Matera (*Source* Photo taken by the authors)

belonging, transforming simple green spaces into meaningful and vital places for urban life. The main results of this process were both tangible and intangible. Tangible results included green infrastructures (Fig. 7). Intangible results included building trust and social cohesion, the re-appropriation of restorative spaces.

However, some obstacles and critical issues occurred in waves, as it were, during the 3-year project, including distrust, disinterest, disaffection, and resignation to the status quo, as well as an excessive bureaucratization of processes that hinders pragmatic actions.

Other challenges refer to measuring impact: what should be measured? The intangible, cultural outcomes of participatory design, such as strengthening communities or improving social cohesion, can elude traditional metrics. Moreover, these results sometimes germinate over immeasurable periods, like many socio-cultural actions. It's like planting a seed without knowing exactly how, when, or if it will take root. The efficacy of Nature-Based Solutions depends not solely on their ecological or infrastructural performance, but on the extent to which they foster what may be termed green feelings: affective, sensory, and relational dispositions that mediate human–environment interactions. Within this framework, participatory design operates as an experiential methodology capable of cultivating situated awareness and embodied place interaction, concepts central to environmental psychology. In Matera URGES project, the "Poetically Inhabiting the City" workshops activated these dynamics by engaging residents in sensory practices—sound-walks, blindfolded explorations, embodied mapping, and guided reflection—that foregrounded atmospheric dimensions typically overlooked in conventional participatory processes, particularly in marginalized southern contexts. These practices reveal the relevance of Sensual Design, understood as an extension of Sensory Design that integrates perceptual, affective, and narrative modalities. Such an approach enhances biophilic tendencies by enabling participants to apprehend their neighbourhoods as restorative atmospheres rather than merely functional spaces. The workshops thereby facilitated a form of sensory and symbolic restitution: residents re-encountered familiar places through renewed perceptual registers, strengthening their ecological and affective ties to the urban landscape. Notably, the participatory process contributed to attenuating entrenched patterns of distrust and civic disengagement, conditions that characterise many peripheral southern territories. By privileging low-threshold, embodied forms of engagement, the project reinforced everyday agency, validated local knowledge, and re-inscribed dignity within micro-heritages often marginalized within institutional planning discourses. The process also increased relational density, generating new social constellations among residents, experts, and local associations, which persisted beyond the facilitated activities. These dynamics reinforced place attachment (Shi et al., 2024), demonstrating that the long-term success of NBS hinges on the capacity of communities to perceive green interventions as integral to their heritage. Such an approach suggests that the creation of restorative atmospheres is not an ancillary benefit but a constitutive condition for resilient and inclusive urban transformation.

5 Concluding Remarks and Best Practices

Success of urban greening interventions depend fundamentally on how communities sense, interpret, and appropriate their surroundings. Participatory processes are experiential frameworks that cultivate what environmental psychology defines as situated awareness and embodied place interaction. When coupled with sensual approach it becomes a form of experiential restitution, enabling inhabitants to re-encounter places as atmospheric, symbolic, and emotional environments. This shift aligns with current research on Sensual Design, which extends Sensory Design by integrating sensory perception with poetic, narrative, and affect-based modalities of engagement. Green urban spaces in marginal areas are more likely to become socially durable when rooted in a shared atmospheric quality—what Böhme (2017) defines as a collectively produced "felt space", a green feeling.

Based on the Matera experience, this study identifies the best practices for restorative urban design, which are summarized in Table 1.

In conclusion, these practices are highly replicable in diverse urban contexts, provided that the technical implementation of NBS is preceded by a *sensory attunement* phase. Future applications should focus on Low-Threshold Participation, ensuring that even residents with no technical background can contribute to the creation of restorative atmospheres. By treating the *Green Feeling* as a quantifiable

Table 1 Transferable best practices for sensual-biophilic approaches

Practice	Description	Replicability/Adaptation
Sensory soundwalks	Silence-based or blindfolded explorations to map the acoustic identity of a place	Adaptable to any high-noise urban context to identify "quiet pockets"
Embodied mapping	Creating cartographies based on feelings (comfort, memory, smell) rather than just geometry	Highly transferable to marginalized peripheries to uncover hidden heritage
Nature-based micro-performances	Poetic activities (readings, sensory games) to activate green spaces	Replicable in schools and community centres to foster long-term stewardship
Multi-sensory NBS	Designing green walls/roofs not just for cooling, but for tactile and olfactory richness	Applicable to all Mediterranean cities to enhance climate resilience through sensory comfort

goal of urban planning, cities can move toward a more compassionate, resilient, and inclusive future.

References

Beatley, T. (2016). *Handbook of biophilic city planning and design*. Island Press.

Bechlem, R., Djouad, F. Z., Debbache, A., Salah-Salah, H. (2025). Weaving nature into the urban tapestry: Biophilic design, public spaces, and identity in the city. In: Hayder, G., Fairuz Shahidan, M., Ozcan-Buckley, A. (Eds), *Cities' identity through the art of urban spaces*. Advances in Science, Technology & Innovation. https://doi.org/10.1007/978-3-031-94821-3_4

Böhme, G. (2017). *Atmospheric architectures: The aesthetics of felt spaces*. Bloomsbury.

Browning, W. D., & Ryan, C. (2020). *Nature inside: A biophilic design guide*. RIBA Publishing.

Browning, W., Ryan, C., & Clancy, J. (2014). *14 patterns of biophilic design: Improving health & well-being in the built environment*. Terrapin Bright Green.

European Commission. (2015). *Towards an EU research and innovation policy agenda for nature-based solutions & re-naturing cities*. Final Report of the Expert Group.

Fromm, E. (1964). *The heart of man: Its genius for good and evil*. Harper & Row.

Gao, Y.-J., Wang, C.-L., Huang, M.-L., & Guo, W. (2022). A new perspective of sustainable perception: Research on the smellscape of urban block space. *Sustainability, 14*(15), 9184. https://doi.org/10.3390/su14159184

Gibson, J. J. (1979). *The ecological approach to visual perception*. Houghton Mifflin.

Gifford, R. (2014). *Environmental psychology: Principles and practice*. Optimal Books.

Hartig, T., Mitchell, R., de Vries, S., & Frumkin, H. (2014). Nature and health. *Annual Review of Public Health, 35*, 207–228. https://doi.org/10.1146/annurev-publhealth-032013-182443

Henshaw, V. (2013). *Urban smellscapes: Understanding and designing city smell environments*. Routledge.

Heschong, L. (1979). *Thermal delight in architecture*. MIT Press.

Howes, D. (2004). *Empire of the senses: The sensual culture reader*. Routledge.

Ingold, T. (2000). *The perception of the environment: Essays on livelihood*. Routledge.

Kaplan, S., & Kaplan, R. (1989). *The experience of nature: A psychological perspective*. Cambridge University Press.

Kellert, S. R. (2008). *Biophilic design: The theory, science, and practice of bringing buildings to life*. Wiley.

Kellert, S. R. (2012). *Birthright: People and nature in the modern world*. Yale University Press.

Kellert, S. R. (2018). *Nature by design: The practice of biophilic design*. Yale University Press.

Kellert, S. R., Heerwagen, J. H., & Mador, M. L. (Eds.). (2008). *Biophilic design: The theory*. John Wiley & Sons.

Kellert, S. R., & Wilson, E. O. (1993). *The biophilia hypothesis*. Island Press.

Kuhtz, S. (2024). Beyond sight: Sensory atmospheres and design. *DIID Disegno Industriale Industrial Design, 83*, 52–65.

Kuhtz, S., & Tizi, L. (2023). Abitare poeticamente il Sud. Retoriche e possibilità. *Tracce Urbane—Rivista Transdisciplinare di Studi Urbani, 9*(13): 266–295.

Lefosse, D. C., Naghibi, M., Luo, S., & van Timmeren, A. (2025). Biophilic urbanism across scales: Enhancing urban nature through experience and design. *Land, 14*(5), 1112. https://doi.org/10.3390/land14051112

Lefosse, D., Van Timmeren, A., & Ratti, C. (2023). Biophilia upscaling: A systematic literature review based on a three-metric approach. *Sustainability, 15*(22), 15702.

Louv, R. (2005). *Last child in the woods: Saving our children from nature-deficit disorder*. Algonquin Books.

Manzini, E. (2015). *Design, when everybody designs: An introduction to design for social innovation*. MIT Press.

Manzo, L. C., & Devine-Wright, P. (2019). Place attachment. In: L. Steg, & J. I. M. De Groot (Eds.), *Environmental psychology: An introduction* (2nd ed., pp. 135–143). John Wiley & Sons.

Mazzali, M., Tizi, L., & Zanichelli, S. (2025). *The psychoanalytic roots of biophilia and the ethics of designers*. ISBN: 979-12-243-0628-3.

Mehrabian, A., & Russell, J. A. (1974). *An approach to environmental psychology*. MIT Press.

Merleau-Ponty, M. (1945). *Phénoménologie de la perception*. Gallimard.

Norman, D. (2013). *The design of everyday things*. Basic Books.

Oliveros, P. (2005). *Deep listening: A composer's sound practice*. iUniverse.

Pallasmaa, J. (2005). *The eyes of the skin: Architecture and the senses*. Wiley.

Pazzaglia, F., & Tizi, L. (2022). *Che cos'è il restorative design*. Carocci.

Relph, E. (1976). *Place and placelessness*. Pion.

Roe, J., & McCay, L. (2021). *Restorative cities: Urban design for mental health and wellbeing*. Bloomsbury Visual Arts.

Rosa, H. (2019). *Resonance: A sociology of our relationship to the world*. Polity Press.

Shams, L., & Seitz, A. (2008). Benefits of multisensory learning. *Trends in Cognitive Sciences, 12*(11), 411–417.

Shi, L., Yang, W., & Ujang, N. (2024). A systematic literature review of place attachment in environments, cities and landscapes. *New Design Ideas, 8*(2), 389–411.

Spence, C. (2020). *Sensehacking: How to use the power of your senses for happier*. Penguin.

Steg, L., & de Groot, J. (2019). *Environmental psychology: An introduction* (2nd ed.). Wiley.

Tabb, P. J. (2020). *Biophilic urbanism: Designing resilient communities for the future*. Routledge.

Tekin, B. H., Izmir Tunahan, G., Disci, Z. N., & Ozer, H. S. (2025). Biophilic design in the built environment: Trends, Gaps and Future Directions. *Buildings, 15*(14), 2516.

Tuan, Y. F. (1977). *Space and place: The perspective of experience*. University of Minnesota Press.

Ulrich, R. S. (1983). Aesthetic and affective response to natural environment. In Altman, I. & Wohlwill, J.F. (Eds.), *Human behavior and environment* (Vol. 6, pp. 85–125). Plenum Press.

Varela, F. J., Thompson, E., & Rosch, E. (1991). *The embodied mind*. MIT Press.

Wilson, E. O. (1984). *Biophilia*. Harvard University Press.

Outdoor Thermal Comfort Evaluation in an Urbanized Context in the Mediterranean Area

Gaetano Sciuto, Grazia Lombardo, Angela Moschella, Alessandro Lo Faro, Livia Lauria, and Nilda Georgina Liotta

Abstract European cities are facing the negative impacts of climate change, among which the urban heat island (UHI) effect plays a significant role. The adoption of Nature-Based Solutions (NBS), both at the urban and building levels, is one of the possible strategies for increasing the sustainability of cities, recovering degraded areas, and helping to mitigate the effects of climate change. The paper is part of a research project that aims to assess the benefits of integrating NBS, and green infrastructure in particular, into urban contexts (existing buildings and surrounding spaces) in the Mediterranean area. The most suitable greening strategies for improving urban quality, microclimate, and residents' well-being were evaluated for a case study: a social housing in Catania, southern Italy. Three different scenarios were hypothesized: application of green facades on the exterior walls of the building; enhancement of urban greenery and permeable surfaces; and combination of the two scenarios. The improvement in outdoor comfort was assessed using fluid dynamics simulation software. The results show that green facades have little influence on outdoor comfort, while the benefits of increasing greenery in the surrounding areas are more evident. Above all, there is a clear synergy between the two greening strategies, which amplifies the results of each strategy considered individually.

Keywords Existing building stock · Building renovation · Urban regeneration · NBS · Green infrastructures · Sustainable cities · Fluid dynamics simulation

1 Introduction and Background Knowledge

Climate change is one of the greatest challenges of our time, and its impacts must be understood from a systemic and multi-scale approach. European cities are facing the negative impacts of climate change, among which the urban heat island (UHI) effect plays a significant role in Mediterranean areas. In the summer season, the UHI effect increases outdoor temperatures and, consequently, makes outdoor spaces

G. Sciuto (✉) · G. Lombardo · A. Moschella · A. Lo Faro · L. Lauria · N. G. Liotta
Department of Civil Engineering and Architecture, University of Catania, Catania, Italy
e-mail: gsciuto@dau.unict.it

L. Diana et al. (eds.), *Greening and Resilience*, Urban Sustainability,
https://doi.org/10.1007/978-981-92-0419-9_18

uninhabitable, threatening the health and well-being of residents and causing heat-related illnesses and disorders (Wai et al., 2022). Added to this is also the demand for high amounts of energy for building cooling. Therefore, in order to improve public health and environmental quality in cities and promote urban resilience, mitigation of urban thermal stress is crucial (Zhou et al., 2025).

Nature-Based Solutions (NBS) are one of the responses to climate change and help strengthen urban resilience. NBS are defined as living solutions inspired, constantly supported and used by nature; they encompass a range of existing concepts and practices, such as urban forestry, green and blue infrastructure or ecosystem service provision (European Commission, Directorate General for Research and Innovation, 2015).

In urban environment, NBS include Green Infrastructures (GI) such as tree canopies, open green spaces, green roofs (GR), and vertical green systems (VGS) (Bartesaghi Koc et al., 2017). GI is one of the possible solutions and an effective approach to improve the quality of urban life because it mitigates the urban heat island (UHI) effect and improves outdoor thermal comfort through shading and evapotranspiration (Zhou et al., 2025). With the aim of promoting urban resilience and improving outdoor thermal comfort, urban greening strategies (SDGs 13 and 11 of the United Nations Agenda) are shaping the policies of many European cities; the integration of vegetation is planned in urban spaces (tree canopies, open green spaces), in buildings (GR, VGS) or in both cases (Sommese et al., 2025).

Green infrastructure offers multiple benefits on an urban scale; these benefits depend on the characteristics of the geographical area (Wai et al., 2022). Some studies investigated the positive impact on microclimate improvement (Erlwein et al., 2021). Many studies also investigated other benefits, such as the mitigation of the urban heat island (UHI) effect (Cuce et al., 2025, Susca et al., 2022; Zhou et al., 2025) and the improvement of outdoor thermal comfort (Dehghan Lotfabad et al., 2025). Today, the combination of the effects of different types of greening on the thermal comfort of outdoor spaces in cities is one of the topics addressed in the literature in order to support urban planning. Cuce P.M. et al. (2025) highlight the synergistic benefits of such a design approach and the importance of considering these greening strategies as interconnected systems rather than isolated technologies.

Within this scenario, the paper explores the combined use of different GI solutions for a small urban area (buildings and urban context) in order to evaluate their effectiveness in terms of outdoor thermal comfort using fluid dynamics simulation software (ENVI-met®).

2 Case Study Overview

This work was carried out as part of the PRIN 2022 PNRR research project entitled "Greenwork: an interdisciplinary framework for urban health and urban resilience enhancement based on greening strategies on buildings and open spaces." The research aims to evaluate the benefits that can be obtained from integrating GI into urban contexts (existing buildings and surrounding spaces) in the Mediterranean area. According to the Köppen-Geiger international climate classification, Catania (latitude 37°30' north and longitude 15°04' east) is characterized by a warm and temperate climate (Csa), with hot and humid summers and moderately cool and humid winters. In the summer months, the average outdoor air temperature ranges from 23°C to 35°C, with peaks of 45°C when hot winds blow from North Africa.

The social housing developments in the city of Catania were studied. They often have poor architectural, functional, and social quality, as well as being deficient in terms of thermal insulation and energy consumption. Attention was then focused on the Monte Po district, considered significant due to the presence of different types of buildings and large outdoor areas with considerable potential for regeneration.

The Monte Po district is located on the western outskirts of Catania, where several public housing developments were built in the 1970s. It is an interesting example of a suburban district that, despite being designed according to rationalist criteria, has suffered for years from a lack of amenities. Despite recent improvements, Monte Po is still considered a "fragile" district in urban and economic terms; commercial offerings and services are underdeveloped; existing green space is limited, and part of the free area is still awaiting development.

Today there are 32 buildings, constructed in different phases. They have been cataloged and analyzed from a typological, morphological, and technological point of view. They fall into four different building types (Fig. 1a): tower buildings (T1 to T12); in-line buildings with sequential aggregation (L1 to L13); patio buildings (P1 to P3); and in-line buildings with open court aggregation (C1 to C4).

All buildings have reinforced concrete load-bearing structures with plinth foundations on piles. The multi-layer external walls are made of concrete and pumice blocks with an air cavity, without thermal insulation; this results in high energy requirements for air conditioning, especially during the summer months.

QGIS software was used to analyze quantitative and qualitative data relating to the buildings and the external areas belonging to the individual buildings or public spaces (roads, public services, green spaces, and residual spaces throughout the neighborhood). For each residential building, data were highlighted relating to the external areas belonging to it, the individual facades (presence of transparent surfaces,

a b

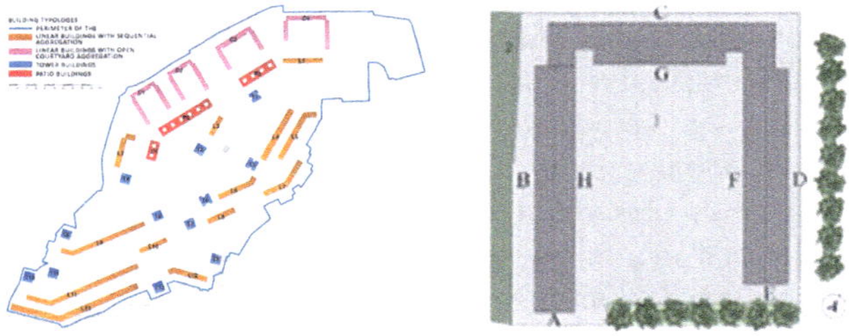

c

Facade B		
Orientation		Southwest
Form		On different levels
cast shadow		No
Total area		1453.00m^2
Total area	Ground floor	376.00m^2
	Elevated floors	1077.00m^2
opaque surface	Ground floor	241.00m^2
	Elevated floors	860.60m^2
transparent surface	Ground floor	135.00m^2
	Elevated floors	216.40m^2
Presence of balconies		no
presence of adjacent permeable surface		no
Intended use	Ground floor	Garage
	Elevated floors	Recidence
Condition of preservation		Good

Fig. 1 **a** Monte Po District, building types; **b** Floor plan of building C1; **c** Data relating to elevation B (created by the authors)

ratio of solid to empty spaces, presence of balconies and adjacent permeable areas, and intended use of the ground floors), and the roofs. All the data obtained in this way were reported in specific summary sheets (Lombardo et al., 2025).

The C1 section was analyzed as a case study. It consists of building C1, chosen for its morphological and technological characteristics, along with the related external areas. With regard to the related areas, out of a total horizontal surface area of approximately 40,000 m^2, 65% is impermeable and only 35% is permeable. Furthermore, the area examined has little greenery; there are only a few trees to the south and east of the building. The building consists of a ground floor, occupied by garages, and four upper floors, used for residential purposes. The roofs are pitched and therefore unsuitable for a green roof. The facades of the building were named with letters, from A to H (Fig. 1b), and their orientation was surveyed in order to identify their suitability for a VGS. To this end, those facing west, southwest, south, east, and southeast were considered most suitable for achieving better indoor comfort conditions and reducing the summer heat load. The geometric and morphological characteristics of the south-east (A, E, G) and south-west (B, F) facades were therefore examined. Figure 1c shows the geometric data and main characteristics of facade B. It faces south-west and is not subject to shadowing. There are no permeable areas nearby, which excludes the possibility of using VGS with vegetation planted directly in the ground. Furthermore, in order not to obstruct access to the garages, the VGS can only be installed on the four upper floors. Taking into account the good state of repair of the building, the presence of balconies along its entire length, and the fact that VGS can only be installed on the upper floors, the most suitable technological solutions are indirect continuous cable systems (Lo Faro et al., 2024). Similar assessments have been made for elevations A, E and F, which are suitable, except for elevation G, which is subject to shadowing.

3 Methodology

Based on the analysis carried out on sector C1, taken as a case study, the most suitable greening strategies were evaluated to improve urban quality, microclimate, and the well-being of residents. Outdoor comfort was evaluated using a small-scale (microscale) modeling approach, using ENVI-met® three-dimensional computational fluid dynamics simulation software. This software was chosen for its ability to simulate, with high spatial and temporal resolution, the sensible and latent heat flows exchanged between vegetation and the external environment, as well as for calculating heat exchanges due to evapotranspiration from plants and accurately estimating the shading effects of leaves. This tool is useful for evaluating the effects of different greening systems. The model, considering the interaction between buildings, paved surfaces, soil, and vegetation, allows for testing solutions to optimize environmental comfort (Bruse & Fleer, 1998).

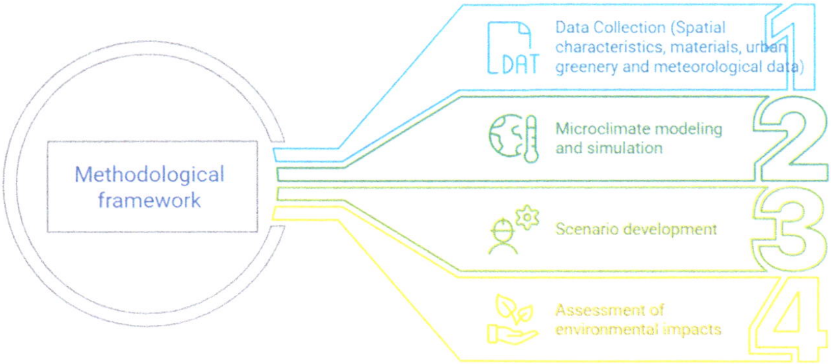

Fig. 2 Methodological framework adopted for the analysis of outdoor thermal comfort

The methodological approach adopted includes several phases: data collection (spatial characteristics, materials, urban greenery, and meteorological data); microclimate modeling and simulation; development of possible intervention scenarios; and assessment of environmental impacts (Fig. 2).

Three mitigation scenarios were proposed and analyzed:

- Scenario 1: Installation of green facades on some sections of the building's exterior walls;
- Scenario 2: Enhancement of urban greenery combined with an increase in permeable surfaces
- Scenario 3: Combined scenarios 1 and 2

The three scenarios were compared with the baseline conditions.

3.1 Data Collection

Collection of data to build microclimatic modeling of the study area was divided into several phases, which included:

- Identification of spatial characteristics through satellite images, QGIS analysis, and on-site observations (Fig. 3).
- Technological analysis of the materials that define the surrounding urban context (soil, paved areas, and buildings) with their thermophysical properties.
- Observation of photographic documentation of existing urban greenery.
- Acquisition of meteorological data (temperature, humidity, wind speed, and wind direction) from the Davis Vantage Pro 2 weather station (Davis Instruments Corp., Hayward, CA, USA) installed in the Department of Civil Engineering and Architecture on the campus of the University of Catania.

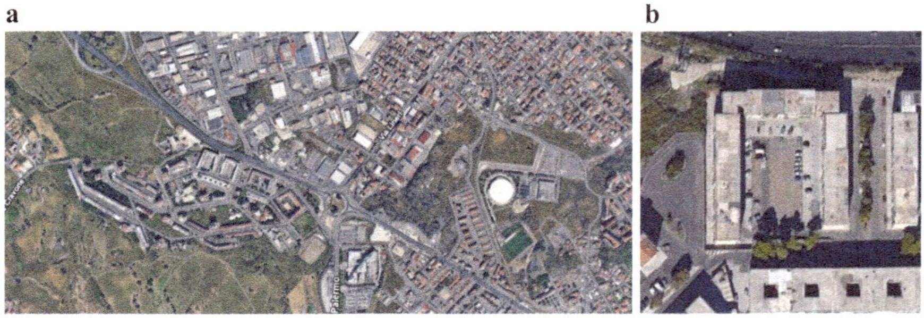

Fig. 3 (a) Satellite image of the district within the city and (b) the investigated area (taken from Google Earth)

3.2 Modeling and Microclimate Simulation

The area investigated was modeled using ENVI-met® V5.7.1 software, with a grid size of 200 × 200 × 50 m (x, y, and z). The computational grid was formed by cells with a resolution of 2 × 2x2 m; therefore, the number of cells obtained was 100 × 100x25. Along the vertical z-axis, the first cell starting from the ground is divided into 5 equal parts; the subsequent cells up to a height of 17 m (corresponding to the highest point within the calculation domain) have standard dimensions; after 17 m, a telescopic extension with a factor of 15% is applied. To reduce boundary effects, two nested grid cells were applied at the perimeter of the model. The materials and vegetation used in the modeling were classified and summarized in Tables 1 and 2, respectively. The values reported in the two tables were obtained from the ENVI-met® general database and are consistent with those reported in scientific articles (Carotenuto et al., 2025; Paolini et al., 2020; Pasetto et al., 2021).

Table 1 Thermal and optical properties of materials in ENVI-met

N°	Material	Color	Name	Albedo (ρ)	Emissivity(ε)
1	Ground	Dark green	Sandy Loam	0.20	0.90
2	Pedestrian roads	Grey	Concrete pavement dirty	0.25	0.90
3	Parking	Dark grey	Concrete pavement dark	0.20	0.90
4	Parking and Roads	Black	Asphalt Road	0.12	0.90
5	Building walls	Light yellow	External plaster	0.60	0.90
6	Building roofs	Grey	Roofing: Tile	0.30	0.93
7	Soil underneath a building	Grey	Default soil underneath a building	0.20	0.90

Table 2 Optical and thermophysical properties of plants in ENVI-met

Name	Leaf type	LAD	Height (m)	Albedo(ρ)	Emissivity(ε)	Transmissivity(τ)
Hedge light	Deciduous	0.55	1	0.20	0.97	0.30
Hedge light	Deciduous	0.55	2	0.20	0.97	0.30
Tree, cylindric, small trunk, sparse, small	Conifer	1	5	0.12	0.96	0.10
Tree, cylindric, large trunk, sparse, medium	Conifer	1	15	0.12	0.96	0.10
Tree, spherical, small, trunk, sparse, small	Deciduous	1	5	0.18	0.96	0.30
Tree, cylindric, small trunk, sparse, small	Deciduous	1	5	0.18	0.96	0.30
Tree, spherical, small trunk, sparse, medium	Deciduous	1	15	0.18	0.96	0.30

To complete the modeling phase, a strategic analysis of receptors was carried out. Seven representative receptors were selected within the study area (Fig. 4) to identify the most sensitive points, from both a climatic and environmental point of view. The modeling phase was completed by implementing the meteorological data collected. This made it possible to analyze the microclimatic conditions before implementing the mitigation scenarios.

In order to evaluate thermal comfort, ENVI-met provides the following parameters: Mean radiant temperature (MRT), potential air temperature (PAT), relative humidity (RH), physiological equivalent temperature (PET), and universal thermal

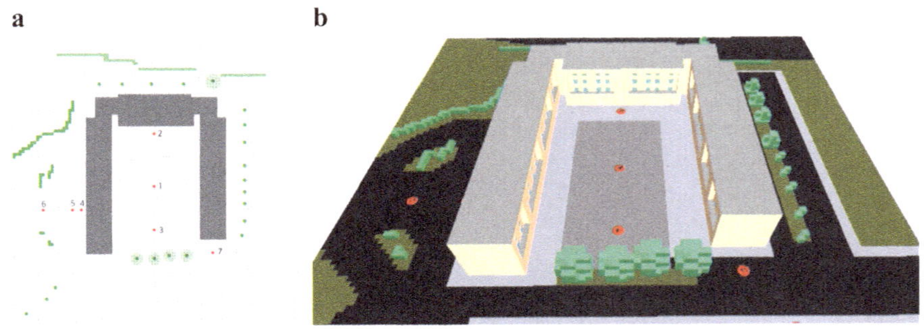

Fig. 4 (a) Location of the seven receptors in the ENVI-met 2D model and (b) baseline scenario in the ENVI-met 3D model (created by the authors)

climate index (UTCI) (Fabbri et al., 2017; Ketterer & Matzarakis, 2014; Yang et al., 2024). The day chosen for the simulation is August 17, 2024, the peak heat of the year. The values were estimated considering a 35-year-old adult woman in the summer season, with a heat transfer resistance of clothing equal to 0.50 clo (light clothing). These parameters can be presented in the form of thematic maps with high spatial and temporal resolution. Recent studies have shown that MRT plays a central role in determining outdoor thermal comfort, especially in highly urbanized and sun-exposed environments (Santos Nouri et al., 2018; Soares et al., 2021).

The UTCI was selected as the comfort index due to its robustness in characterizing dynamic thermal stress conditions (Fiala et al., 2012; Weihs et al., 2012). UTCI calculations incorporate meteorological parameters such as air temperature, wind speed, humidity, and MRT, together with metabolic activity and thermal insulation values of clothing (Leonardi et al., 2025). Thermal stress categories were classified according to the internationally validated UTCI classification system, ranging from extreme cold to extreme heat, as detailed in Table 3.

Table 3 UTCI classification system

Thermal sensation	UTCI (°C)	Physiological stress
Very hot	> 46	Extreme heat stress
Hot	38–46	Very strong heat stress
Warm	32–38	Strong heat stress
Slightly Warm	26–32	Moderate heat stress
Comfortable	9–26	No thermal stress
Slightly Cool	0–9	Slight Cold stress
Cool	-13–0	Moderate cold stress
Cold	-27- -13	Strong cold stress
Very Cold	-40- -27	Very strong cold stress

3.3 Scenario Development

In order to evaluate the effectiveness of the different mitigation strategies in terms of improving outdoor thermal comfort, three design scenarios were developed and simulated using ENVI-met®. All scenarios were compared with the baseline scenario using the same simulation profile, with particular attention to thermal comfort indicators such as MRT and UTCI.

Scenario 1: Installation of Green Facades on Some Sections of the Building's Exterior Walls

It aims to evaluate the benefits that can be obtained in summer by using green facades, measurable both in terms of reducing the overheating of opaque walls and improving the microclimate of balconies, and in terms of outdoor thermal comfort (Perini et al., 2011). For this study, it was assumed that an indirect VGS modeled in ENVI-met would be applied to the south-east facing blind facades (A, E) and to the south-west facing balconies (B, F) of building C1. (Fig. 5). The vegetation cover consists of a single type of climbing plant, Trachelospermum jasminoides, a fast-growing, evergreen species endemic to the area, to protect the facade from wind and rain during the winter season. At its maximum growth, the plant is able to provide almost total shade for the facade. The thermophysical properties of the plant species are shown in Table 4 (Detommaso et al., 2023).

Scenario 2: Enhancement of Urban Greenery Combined with an Increase in Permeable Surfaces.

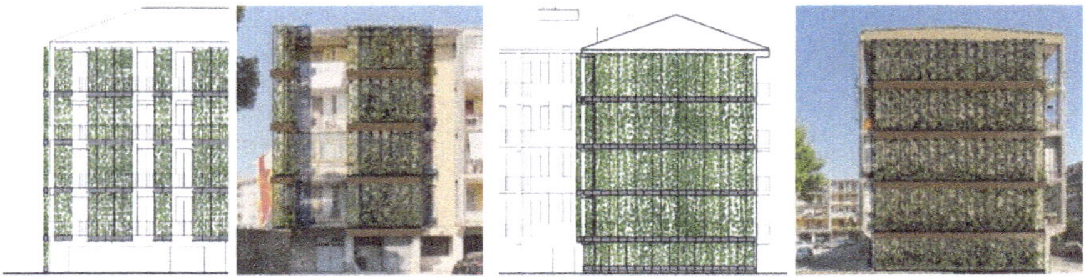

Fig. 5 Scheme of the system chosen for vertical greening and rendered image (created by the authors)

Table 4 Thermophysical properties of the species Trachelospermum jasminoides

Parameter	Symbol	Unit	Value
Leaf Area Index	LAI	m^2/m^2	4
Solar absorption coefficient of leaves	α		0.54
Leaf stomatal conductance	g_s	Mol/m^2s	0.40
Radiation attenuation coefficient of leaves	k		0.70
Leaf emissivity	ε		0.96

Fig. 6. 3D model of scenario 2 modeled in ENVI-met (created by the authors)

This scenario aims to evaluate if increasing green spaces and permeable surfaces can guarantee measurable benefits in terms of outdoor comfort. The proposal focuses on urban greening actions, through greater green density, the introduction of medium-large trees in strategic locations, and the conversion of asphalt parking lots into permeable areas (Fig. 6). A total of 2,800 m^2 asphalt concrete surface was converted into permeable surface (bare soil) and 1,152 m^2 of impermeable pavements (concrete tiles) were replaced with draining pavements (tiles with a higher albedo value and grassed joints). These measures improve rainwater infiltration, helping to reduce surface radiation temperatures through the process of evapotranspiration.

Scenario 3: Combined Scenarios 1 and 2

This scenario combines greening strategies at both the urban and building scales in order to assess any incremental benefits. In ENVI-met®, it was modeled by combining the green facade with measures to enhance greenery and increase permeable areas.

3.4 Assessment of Environmental Impacts

In order to determine local variations resulting from mitigation measures, the MRT and UTCI values obtained from the simulations of the three intervention scenarios were compared with the respective values obtained from the simulation of the current state. Table 5 shows the spatial coordinates of the receptors and specifies the mitigation measures planned in each scenario.

Table 5 Planned interventions to improve the outdoor comfort of receptors

Receptor	Scenario 1	Scenario 2	Scenario 3
1 (100 m, 98 m)	No intervention	Presence of permeable soil and medium-large trees	Presence of permeable soil and medium-large trees
2 (100 m, 134 m)	6 m proximity to the green facade	Arrangement of permeable pavement and proximity of medium-large trees	Permeable paving and proximity to medium-large trees. Green facade installed 6 m away
3 (100 m, 68 m)	No intervention	Presence of permeable soil and medium-large trees	Presence of permeable soil and medium-large trees
4 (50 m, 82 m)	4 m proximity to the green facade	Proximity to medium-large trees	Proximity of 4 m to the green facade and to medium-large trees
5 (44 m, 82 m)	No intervention	Presence of medium-large trees	Presence of medium-large trees
6 (24 m, 82 m)	No intervention	No intervention	No intervention
7 (140 m, 52 m)	8 m proximity to the green facade	Presence of permeable soil and medium-large trees	8 m proximity to the green facade and medium-large trees. Presence of permeable soil

4 Results and Discussion

To obtain a complete overview of the heat stress conditions for the area examined, the MRT and UTCI values were measured for each scenario. The environmental indicators were examined during the most critical hours of the day: at 9:00 a.m. (start of the working day), at 3:00 p.m. (peak extreme heat), and at 6:00 p.m. (end of the working day). At the end of the domain analysis, an assessment focused on receptors was performed. For the time being, it was decided to focus only on the MRT results measured at 3:00 p.m. A comparative analysis was carried out, focusing on outdoor thermal comfort, between the base scenario and the three different mitigation scenarios hypothesized. The proposed strategies are evaluated in terms of how they improve microclimatic comfort for residents.

Figure 7 shows the ENVI-met® simulation maps for the spatial distribution of MRT at 3:00 p.m., providing information on how thermal conditions change in response to different interventions. These maps are useful for visually framing the effectiveness of the proposed mitigation strategies prior to a numerical evaluation. Table 6 shows the simulated values of the main environmental parameters at the seven receptors, again for 3:00 p.m.; this allows a direct comparison of the thermal comfort indicators in the four scenarios.

Subsequently, starting from the environmental parameter values (MRT, UTCI) reported in Table 6, their trend was analyzed, focusing on receptor 1 and evaluating the variation of MRT and UTCI throughout the day. Finally, the trend of MRT and UTCI for each of the receptors was highlighted by comparing only the values for 3:00 p.m. The results of these two calculations were reported in comparative graphs (Figs. 8 and 9).

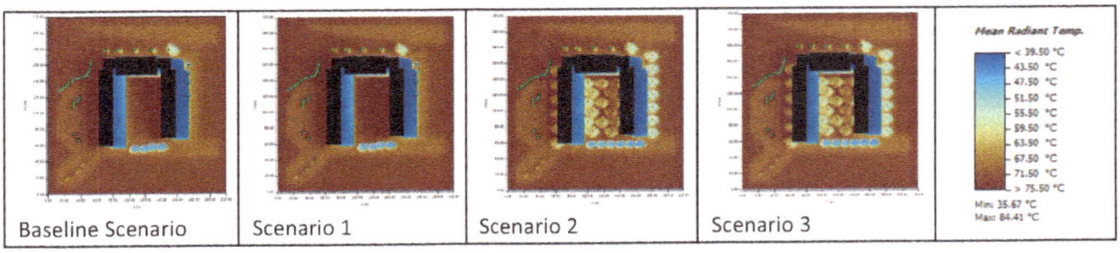

Fig. 7 ENVI-met maps showing the spatial distribution of MRT at 3:00 p.m. (created by the authors)

Table 6 Simulated values of environmental parameters MRT and UTCI for 3:00 p.m. for the seven receptors

Receptor	Scenario	MRT (°C)	Δ MRT (°C)	UTCI (°C)	Δ UTCI (°C)
1	B	72.53	+ 1.35	48.01	+ 0.3
	1	73.88	-1.73	48.31	-0.9
	2	70.80	-2.7	47.11	-1.2
	3	69.83		46.81	
2	B	72.87	-0.58	47.49	-0.15
	1	72.29	+ 0.04	47.34	-0.20
	2	72.91	-1.28	47.29	-0.71
	3	71.59		46.78	
3	B	71.38	+ 2.27	47.56	+ 0.77
	1	73.65	-11.08	48.33	-2.97
	2	60.30	-11.31	44.59	-3.04
	3	60.07		44.52	
4	B	81.74	+ 0.32	51.64	+ 0.2
	1	82.06	-4.65	51.84	-1.71
	2	77.09	-4.74	49.93	-1.84
	3	77.00		49.80	
5	B	76.85	-0.52	50.69	-0.04
	1	76.33	+ 0.30	50.65	-0.17
	2	77.15	-0.32	50.52	-0.38
	3	76.53		50.31	
6	B	71.51	-0.31	48.00	+ 0.09
	1	71.20	-0.28	48.09	+ 0.04
	2	71.23	-0.68	48.04	-0.08
	3	70.83		47.92	
7	B	68.49	-1.11	46.71	-0.08
	1	67.38	-8.18	46.63	-2.32
	2	60.31	-9.58	44.39	-2.70
	3	58.91		44.01	

The results obtained through simulations show an improvement in MRT and UTCI indicators, especially in Scenarios 2 and 3; the benefits found in Scenario 1 are negligible.

Considering the most significant data, the comparative analysis of the results at 3:00 p.m. shows:

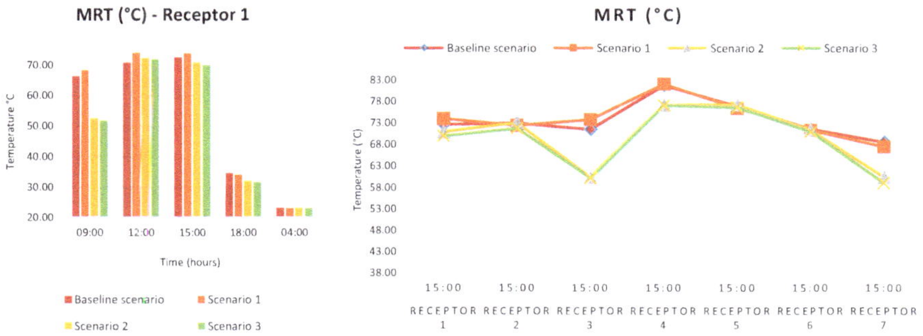

Fig. 8 Comparative graphs of the TMR parameter for the various scenarios. On the left, a comparison between the different time slots for receptor 1. On the right, a comparison between the various receptors at 3:00 p.m

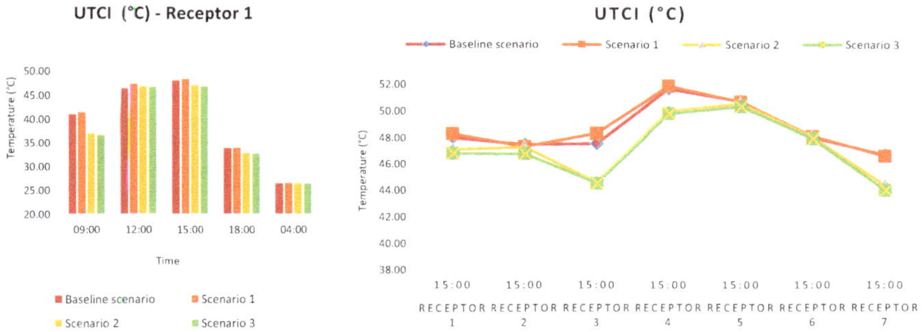

Fig. 9 Comparative graphs of the UTCI parameter for the various scenarios. On the left, the comparison between the different time slots of receptor 1. On the right, the comparison between the various receptors at 3:00 p.m

- Scenario 1. Δ MRT values between + 1.35 °C and -0.58 °C, with an average value of + 0.052 °C. Δ UTCI values between + 0.30 °C and -0.15 °C, with an average value of + 0.08 °C.
- Scenario 2. Δ MRT values between + 0.30 °C and -4.65 °C, with an average value of -1.26 °C. Δ UTCI values between + 0.04 °C and -1.71 °C, with an average value of -0.58 °C.
- Scenario 3. Δ MRT values between -0.32 °C and -4.74 °C, with an average value of -1.94 °C. Δ UTCI values between -0.08 °C and -1.84 °C, with an average value of -0.84 °C.

Analysis of the results and assessments conducted shows that the most significant contribution in terms of improving thermal comfort is associated with an increase in urban greenery and permeable surfaces. These results are in line with the most recent work on the subject (Gatto et al., 2021; Leonardi et al., 2025, Meili et al., 2021), which confirm that the presence of trees generates a cooling effect because their canopy intercepts shortwave radiation. Adding to the replacement of paved areas, reduces the average radiant temperature, resulting in an improvement in the UTCI index. About the contribution of green facades to people's thermal comfort, this is modest or, in some cases, even detrimental. The results reveal similarities with studies in literature (Susca et al., 2022) that highlight a negligible attenuation capacity at human level, which increases with height. Furthermore, the high temperatures recorded in the Mediterranean area create severe thermal stress in plants, which alters their biological activity and causes more pronounced stomatal closure, limiting transpiration (Zhao et al., 2023). This observation is contradicted by the study by Zucckerman N et al. (2025), the thermal comfort of pedestrians near south-facing green facades during heat waves improves compared to bare walls. Finally, the combined application of the two strategies highlights the presence of a synergistic effect: the overall results do not correspond to the simple algebraic sum of the individual contributions, but show an improvement compared to the conditions of the baseline scenario, even in cases where the individual variations are unfavorable. Cuce P. et al. (2025) concluded that the integration of green infrastructure into urban environments, such as green roofs and facades, trees, and permeable surfaces, brings synergistic benefits from the combination of different systems.

5 Conclusions

The study documented in this paper represents the initial results of a research aimed at evaluating the benefits that can be obtained from integrating NBS in urbanized areas with a Mediterranean climate. In this first phase, the improvement in outdoor comfort conditions following the implementation of green facades on the external walls of buildings and an increase in urban greenery combined with an increase in permeable surfaces was assessed using computational fluid dynamics simulation software. For the C1 sector located in the Monte Po district of Catania, chosen as a case study, the MRT and UTCI indicators were evaluated for three possible intervention scenarios.

A comparison between the data obtained for each scenario and the baseline scenario shows that the most significant contribution is associated with the increase in urban greenery and permeable surfaces, while the contribution from the addition of the green façade is modest or, in some cases, even negative. It also emerges that the combined action of the two greening strategies considered generates a synergistic effect on outdoor comfort. The improvement in terms of reducing MTR and UTCI goes well beyond the simple algebraic sum of the individually measured data and always indicates an improvement compared to the conditions in the baseline scenario, even when the individual differences show a deterioration compared to the values measured for the current situation.

The research, which is still ongoing, will continue with the assessment of the benefits that can be obtained in terms of improving indoor comfort conditions in buildings and their outdoor spaces (such as balconies or loggias) in urbanized areas with a Mediterranean climate following the implementation of green facades on their external walls. The here proposed strategy could also be applied to other urban contexts with the same climatic characteristics. It demonstrates how the synergistic application of vegetation in open spaces and on building façades is a valid way for improving people's thermal comfort, as researches conducted by other authors have indicated (Soares et al., 2021; Susca et al., 2022).

Acknowledgements This research was founded by European Union-EU Next Generation, Italian PRIN 2022 PNRR Programme under "Greenwork: An interdisciplinary framework for urban health and urban resilience enhancement based on greening strategies on buildings and open spaces" project, CUP E53D23019120001. The present study has also been conducted within the framework of the project "SIA3—Soluzioni InnovAtive e strategie di sostenibilità per città, territori e società—Innovative solutions and strategies for the sustainability of Cities, Land and Society," financed by University of Catania.

References

Bartesaghi Koc, C., Osmond, P., & Peters, A. (2017). Towards a comprehensive green infrastructure typology: a systematic review of approaches, methods and typologies. *Urban Ecosyst, 20*, 15–35.

Bruse, M., & Fleer, H. (1998). Simulating surface–plant–air interactions inside urban environments with a three dimensional numerical model. *Environmental Modelling and Software, 13*(3–4), 373–384.

Carotenuto, F., Brilli, L., De Luca, G., Nardino, M., Cremonini, L., Genesio, L., Pancorbo, J. L., & Gioli, B. (2025). Emission offsets by albedo manipulations strategies based on bright materials and greening in urban areas assessed by hyperspectral remote sensing. *Urban Clim., 60*, Article 102357.

Cuce, P. M., Cuce, E., & Santamouris, M. (2025). Towards sustainable and climate-resilient cities: Mitigating urban heat Islands through green infrastructure. *Sustainability, 17*(3), 1303.

Dehghan Lotfabad, A., Hosseini, S. M., Dabove, P., Heiranipour, M., & Sommese, F. (2025). Impacts of vertical greenery on outdoor thermal comfort and carbon emission reduction at the urban scale in Turin, Italy. *Buildings, 15*, 450.

Detommaso, M., Costanzo, V., Nocera, F., & Evola, G. (2023). Evaluation of the cooling potential of a vertical greenery system coupled to a building through an experimentally validated transient model. *Building and Environment, 244*, Article 110769.

Erlwein, S., Zölch, T., & Pauleit, S. (2021). Regulating the microclimate with urban green in densifiying cities: Joint assessment on two scales. *Building and Environment, 205*, Article 108233.

European Commission, Directorate General for Research and Innovation. (2015). *Towards an EU Research and innovation policy agenda for nature-based solutions & re-naturing cities: Final report of the horizon 2000 expert group on nature-based solutions and renaturing cities (full version)*. Pubblications Office.

Fabbri, K., Di Nunzio, A., Gaspari, J., Antonini, E., & Boeri, A. (2017). Outdoor comfort: The ENVI-BUG tool to evaluate PMV values output comfort point by point. *Energy Procedia, 111*, 510–519.

Fiala, D., Havenith, G., Broede, P., Kampmann, B., & Jendritzky, G. (2012). UTCI-Fiala multi-mode model of human heat transfer and temperature regulation. *International Journal of Biometeorology, 56*, 429–441.

Gatto, E., Ippolito, F., Rispoli, G., Carlo, O., Santiago, J. L., Aarrevaara, E., Emmanuel, R., & Buccolieri, R. (2021). Analysis of urban greening scenarios for improving outdoor thermal comfort in neighborhoods of Lecce (Southern Italy). *Climate, 9*(7), 116.

Ketterer, C., & Matzarakis, A. (2014). Human-Biometeorological assessment of heat stress reduction by re-planning measures in Stuttgart. *Germany. Landsc. Urban Plan., 122*, 78–88.
Leonardi, S., Detommaso, M., Liotta, N. G., Distefano, N., Nocera, F., & Costanzo, V. (2025). Integrated strategies for air quality and thermal comfort improvement: The case study of the university campus of Catania. *Applied Sciences, 15*(10), 5661.
Lo Faro, A., Lombardo, G., Moschella, A., & Sciuto, G., (2024). Vertical greenery system for building stocks. In R. Corrao et al. (Eds.), *Architectural engineering in Italy and worldwide—Comparing experiences* (Vol. 3, pp. 488–501). LECTURE NOTES IN CIVIL ENGINEERING, Springer.
Lombardo, G., Sciuto, G., Lo Faro, A., Moschella, A., & Baglio, B. (2025). Preliminary feasibility studies aimed at greening strategies for the urban regeneration. The case study of the Monte Po District. In R. Albatici R. et.al. (Eds.), *Envisioning the futures Designing and building for people and the environment* (Vol. 766, pp. 477–499). LECTURE NOTES IN CIVIL ENGINEERING, Springer.
Meili, N., Acero, J. A., Peleg, N., Manoli, G., Burlando, P., & Fatichi, S. (2021). Vegetation cover and plaint-trait effects on outdoor thermal comfort in a tropical city. *Building and Environment, 195*, Article 107733.
Paolini, R., Terraneo, G., Ferrari, C., Sleiman, M., Muscio, A., Metrangolo, P., Poli, T., Destaillats, H., Zinzi, M., & Levinson, R. (2020). Effects of soiling and weathering on the albedo of building envelope materials: Lessons learned from natural exposure in two European cities and tuning of a laboratory simulation practice. *Solar Energy Materials and Solar Cells, 205*, Article 110264.
Pasetto, M, Baliello, A, Pasquini, E, & Giacomello, G. (2021). High albedo pavement materials. In F. Pacheco-Torgal (Ed.), *Eco-Efficient materials for reducing cooling needs in buildings and construction* (pp. 15–32). Woodhead Publishing.
Perini, K., Ottelé, M., Haas, E., & Raiteri, R. (2011). Greening the building envelope, façade greening and living wall systems. *Open Journal of Ecology, 1*(1), 1–8.
Santos Nouri, A., Charalampopoulos, I., & Matzarakis, A. (2018). Beyond singular climatic variables-Identifying the dynamics of wholesome thermo-physiological factors for existing/ future human thermal comfort during hot dry Mediterranean summers. *International Journal of Environmental Research and Public Health, 15*, 2362.
Soares, R., Corvacho, H., & Alves, F. (2021). Summer thermal conditions in outdoor public spaces: A case study in a Mediterranean climate. *Sustainability, 13*, 5348.
Sommese, F., Diana, L., Colajanni, S., Bellomo, M., Sciuto, G., & Lombardo, G. (2025). Towards a regenerative and climate-resilient built environment: Greening lessons from European cities. *Buildings, 15*, 1878.
Susca, T., Zanghirella, F., Colasuonno, L., & Del Fatto, V. (2022). Effect of green wall installation on urban heat island and building energy use: A climate-informed systematic literature review. *Renewable and Sustainable Energy Reviews., 159*, Article 112100.
Wai, C. Y., Tariq, M. A. U. R., & Muttil, N. (2022). A systematic review on the existing research, practices, and prospects regarding urban green infrastructure for thermal comfort in a high-density urban context. *Water, 14*, 2496.
Weihs, P., Staiger, H., Tinz, B., Batchvarova, E., Rieder, H., & Vuilleumier, L. (2012). The uncertainty of UTCI due to uncertainties in the determination of radiation fluxes derived from measured and observed meteorological data. *International Journal of Biometeorology, 56*, 537–555.
Yang, Z., Peng, J., Liu, Y., Jiang, S., Cheng, X., Liu, X., Dong, J., Hua, T., & Yu, X. (2024). GloUTCI-M: A global monthly 1 km universal thermal climate index dataset from 2000 to 2022. *Earth Syst. Sci. Data, 16*, 2407–2424.
Zhao, J., Meili, N., Zhao, X., & Fatichi, S. (2023). Urban vegetation cooling potential during heat waves depends on background climate. *Environmental research letters, 18*(1), 014035.

Zhou, W., Jia, S., & Wang, Y. (2025). Greenery for regulating outdoor thermal comfort in various urban environments of a humid-subtropical climate under extreme heat conditions. *Urban Forestry & Urban Greening, 113*, Article 129038.

Zuckerman, N., Shiloah, N., & Lensky, I. (2025). Quantifying the impact of Vertical Greenery Systems (VGS) on mediterranean urban microclimate during heat wave events. *Building and Environmental, 267*, Article 112151.

Evaluation of a Simulation System for Ecological Strategies for Outdoor Comfort: A Case Study on the UNIPA University Campus in Palermo

Marco Bellomo and Luisa Lombardo

Abstract The intensification of the urban heat island effect and growing thermal stress in Mediterranean contexts necessitate the adoption of effective Nature-Based Solutions, among which green walls stand out for their ability to mitigate thermal stress on vertical surfaces in densely built environments. This study analyses the microclimatic effectiveness of a vegetated facade through the application of the ENVI-met three-dimensional simulation software, using the University of Palermo campus as a case study. The research integrates an in situ microclimatic monitoring phase, conducted during summer conditions, to collect real-world data on air temperature, relative humidity, and solar radiation for calibrating the numerical model. Such calibration is essential to reduce discrepancies between observed and simulated values, ensuring reliable results when evaluating intervention scenarios. The comparison between the current site configuration and the design scenario highlights that the integration of a green wall on the west-facing facade produces tangible improvements in outdoor comfort conditions at the pedestrian level. Simulation results indicate a reduction in air temperature of approximately 1 °C during peak afternoon hours of maximum thermal load, together with a notable decrease in mean radiant temperature due to the combined shading and evapotranspiration effects of the vegetation. These variations are positively reflected in the Predicted Mean Vote index, which shows a transition from moderate thermal discomfort toward more neutral conditions. The study confirms the validity of ENVI-met as a decision-support tool, demonstrating how even localised vertical greening interventions can effectively improve the urban microclimate and enhance the usability of institutional open spaces.

Keywords Urban greening · ENVI-met · Outdoor thermal comfort · Green wall · Microclimate simulation · Mediterranean climate · Campus environment

M. Bellomo (✉) · L. Lombardo
Department of Architecture, University of Palermo, Palermo, Italy
e-mail: marco.bellomo01@unipa.it

L. Lombardo
e-mail: luisa.lombardo01@unipa.it

L. Diana et al. (eds.), *Greening and Resilience*, Urban Sustainability,
https://doi.org/10.1007/978-981-92-0419-9_19

1 Introduction

In the current era of rapid urbanisation and climate change, urban areas are increasingly vulnerable to environmental degradation and the intensification of the Urban Heat Island (UHI) effect. Extensive literature has documented how the replacement of natural surfaces with impervious materials, together with reduced vegetation cover and anthropogenic heat emissions, contributes to higher air and surface temperatures in urban environments. Recent studies have further highlighted the exacerbation of heat stress in Mediterranean cities, including Southern Italy, due to the increasing frequency and intensity of summer heatwaves (Oliveri et al., 2014). This phenomenon is particularly critical in Mediterranean regions, where the rising frequency and intensity of summer heatwaves significantly compromise the quality of life and the usability of public spaces and exacerbate thermal stress conditions for outdoor users (Oquendo-Di Cosola et al., 2023). As cities become denser, the thermal stress experienced by pedestrians in outdoor environments has become a primary concern for urban planners and architects due to its direct implications for public health, social interaction, and urban liveability (Xi et al., 2022). In this context, Nature-Based Solutions (NBS) have emerged as a pivotal paradigm for sustainable urban development. Unlike traditional 'grey' infrastructure, NBS leverage natural processes to provide systemic, multifunctional benefits that address environmental, social, and economic challenges simultaneously, fostering long-term urban resilience (Patti et al., 2025). Their growing relevance is reflected not only in scientific research, but also in European and international policy agendas, where NBS are increasingly promoted as cost-effective and adaptive strategies for climate change mitigation and adaptation in urban areas (Oquendo-Di Cosola et al., 2023). NBS interventions—such as urban greenery, green roofs, and vegetated façades—are increasingly recognised for their capacity to mitigate microclimatic stress, improve air quality, support biodiversity, and enhance human comfort (Baraldi et al., 2019). Beyond their environmental performance, NBS also contribute to social interaction, well-being, and the overall quality of urban spaces, reinforcing their role as multifunctional infrastructures rather than purely aesthetic additions (Marsaglia, 2024; Xi et al., 2022). These multidimensional interactions between environmental, social, and economic benefits are conceptualised in Fig. 1, which illustrates the role of Nature-Based Solutions as integrative systems capable of simultaneously addressing microclimatic mitigation, human comfort, ecological enhancement, and urban liveability. The figure also highlights the expected quantitative benefits, such as reductions in mean radiant temperature (MRT) and improvements in Predicted Mean Vote (PMV), as reported in previous Mediterranean case studies.

Among vertical greening strategies, green walls represent a particularly promising solution in dense urban environments where horizontal green space is limited (Viecco et al., 2021). By transforming building envelopes into active environmental interfaces, vegetated façades can contribute to local cooling effects, the attenuation of solar radiation on building surfaces, reduction of surface and air temperatures, and improved thermal comfort in adjacent outdoor spaces (Farooq & Kamal, 2020; Oquendo-Di

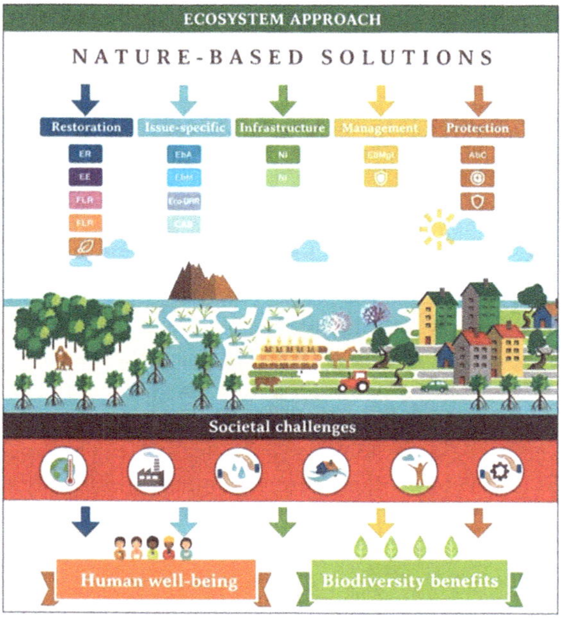

Fig. 1 Conceptual framework illustrating the interrelated environmental, social, and economic co-benefits of Nature-Based Solutions, with specific reference to vertical greening strategies and their contribution to outdoor thermal comfort, biodiversity enhancement, urban liveability, and climate resilience

Cosola et al., 2023). Vertical greening solutions such as green walls are particularly suitable for Mediterranean climates due to their low water demand, resilience to high summer temperatures, and ability to provide localised cooling effects for pedestrians. These effects are especially relevant for west- and south-facing façades, which are typically exposed to high solar loads during peak summer hours. Their application is particularly relevant in institutional and campus-scale contexts, where buildings and open spaces coexist within complex spatial configurations and host a high density of users throughout the day. Recent studies have quantitatively demonstrated the cooling effect of vegetated façades on pedestrian-level microclimate in Mediterranean climates (Mansoldo et al., 2024). In such environments, even localised microclimatic improvements can generate significant cumulative benefits in terms of comfort, walkability, and space usability**.** Despite the growing body of literature documenting the environmental and microclimatic benefits of green walls, their implementation remains uneven and often confined to isolated pilot projects (Radić et al., 2019). One of the main barriers to their broader adoption lies in the absence of structured decision-support tools capable of guiding planners, designers, and public authorities in the early stages of intervention. Existing studies frequently focus on isolated performance metrics, often lacking an integrated approach combining field measurements and microscale numerical simulations, which limits actionable guidance for planners in Mediterranean campuses. Existing studies frequently focus on performance-oriented analyses—such as energy savings, thermal comfort improvements, or microclimatic simulations—without adequately addressing preliminary feasibility aspects related to structural compatibility, regulatory constraints, façade usability, accessibility, and infrastructural requirements (Barker et al., 2024; Kwak & Chen, 2025). This fragmentation limits the scalability and replicability of vertical greening interventions, particularly in public and institutional environments where multiple constraints and competing priorities must be considered (Silvia Tedesco, 2019).

Within this framework, the present study proposes an integrated methodological approach combining on-site microclimatic monitoring and ENVI-met numerical simulations as a decision-support system for the assessment of vertical greening strategies. The methodology is applied to the University of Palermo campus, a heterogeneous institutional context characterised by diverse building typologies, construction periods, façade technologies, and exposure conditions (Stratu et al., 2016). Within the Greenwork research project, the University of Palermo campus is used as a real-scale experimental environment where outdoor microclimatic conditions are analysed through a combined approach based on on-site environmental monitoring and ENVI-met simulations, focusing on the potential contribution of Nature-Based Solutions to outdoor thermal comfort, specifically evaluating air temperature (Ta), mean radiant temperature (MRT), and Predicted Mean Vote (PMV) at pedestrian level.

2 Methodology

The methodological framework of this study is based on the combined use of in situ microclimatic monitoring and numerical simulation to evaluate the thermal and environmental conditions of outdoor spaces and to assess the potential effectiveness of ecological strategies for improving outdoor comfort (Daemei et al., 2021). This integrated approach allows the validation of the numerical model against real-world data, enhancing confidence in simulation-based scenario analyses and supporting its use as a decision-support tool for urban climate adaptation strategies.

The methodological structure was designed to ensure the reproducibility of the results and to explicitly address the uncertainties associated with microclimatic modelling in real urban contexts. The research focuses on the analysis of the existing microclimatic conditions of a semi-open outdoor space within the University of Palermo campus, followed by the simulation of a vegetated wall intervention scenario. The methodology combines empirical and numerical methods to enable a comprehensive understanding of microclimatic dynamics and the potential contribution of vertical greening to thermal comfort. The methodology is structured to (i) characterise the baseline thermal environment through on-site measurements of air temperature, relative humidity, surface temperature, and particulate matter; (ii) develop and calibrate an ENVI-met model reproducing the current conditions, and (iii) evaluate the microclimatic variations induced by the proposed nature-based solution under comparable boundary conditions. By maintaining identical boundary conditions, the methodology ensures that observed differences in thermal and comfort indicators can be directly attributed to the green wall intervention.

This approach is consistent with recent literature that highlights the importance of integrating experimental measurements and microclimatic simulations to reduce model uncertainty and to enhance the interpretability of simulation-based results in real urban contexts.

2.1 Case Study Application: University of Palermo Campus

The case study is developed within the University of Palermo (UNIPA) campus, one of the largest academic campuses in Southern Italy, located in the western sector of the city of Palermo (Fig. 2), in proximity to the coastal area and major urban infrastructures.

The campus is characterised by a heterogeneous built environment composed of buildings constructed in different periods, ranging from post-war modernist blocks to more recent facilities, arranged around a system of open spaces, pedestrian paths, and service areas that support daily academic activities. This morphological and typological diversity makes the campus a representative testbed for assessing the effectiveness of Nature-Based Solutions under complex Mediterranean urban conditions.

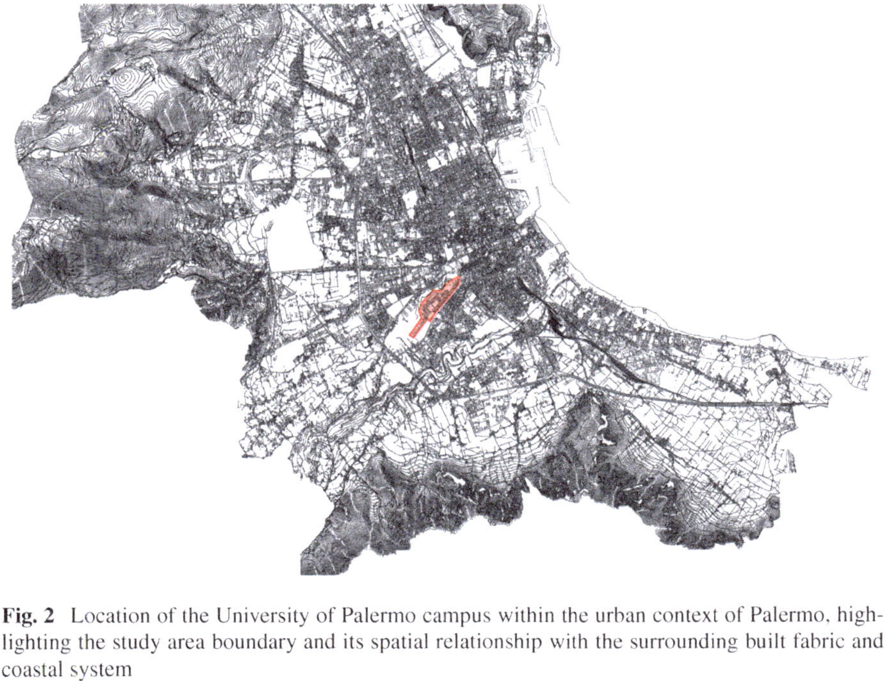

Fig. 2 Location of the University of Palermo campus within the urban context of Palermo, highlighting the study area boundary and its spatial relationship with the surrounding built fabric and coastal system

Within this articulated spatial system, the analysis focused on a central pedestrian area connecting several departments, where intense daily flows of students, academic staff, and visitors are concentrated. Among the various buildings composing the campus, the Cappellania Universitaria was selected as the focal building for the experimental application due to its strategic position along one of the main pedestrian routes and its strong visual and functional relationship with adjacent open spaces. The choice of this building allows the assessment of microclimatic effects in a high-use, highly exposed outdoor environment.

The selected area is characterised by high solar exposure, limited shaded surfaces, and extensive use of mineral paving materials, conditions that significantly influence microclimatic performance, particularly during the summer season. The façade selected for the intervention is west-facing, representing a critical scenario for outdoor thermal comfort due to high afternoon solar radiation. The absence of significant vegetation and the partial enclosure created by surrounding façades contribute to the accumulation of heat and reduced air movement, making the space representative of critical outdoor environments commonly found in Mediterranean university campuses (Fig. 3).

Fig. 3 Aerial view of the University of Palermo campus, illustrating its integration within the surrounding urban fabric and the extent of the analysed area. ©Image by Unipa

2.2 On-Site Microclimatic Monitoring

The on-site measurements were conducted to collect real microclimatic data representative of outdoor comfort conditions in the selected area. The monitoring campaign was carried out during summer conditions under stable meteorological scenarios, to minimise the influence of transient weather events and capture peak thermal stress conditions. **Measurements were conducted over representative clear-sky days during peak summer conditions, when thermal stress is typically most pronounced.** Environmental parameters were measured using a portable climate probe, recording air temperature, relative humidity, surface temperature, and particulate matter concentrations ($PM_{2.5}$ and PM_{10}). All sensors were calibrated according to standard practices and installed at pedestrian level (~1.5 m), in line with international guidelines for outdoor thermal comfort assessment. Data were acquired at regular intervals, consistent with ENVI-met output time steps, and subsequently averaged to obtain representative hourly values used for model calibration. The collected data served a dual purpose: to characterise the baseline thermal environment and to support the calibration and validation of the numerical simulation model. When multiple measurements were available for the same time interval, mean values were considered representative for comparison with simulated outputs. Boundary conditions for ENVI-met were derived directly from these measurements, ensuring consistency between monitored and simulated conditions (Table 1) (Fig. 4).

Table 1 Environmental boundary conditions for simulation scenarios derived from on-site monitoring

Parameter	Value used in simulations	Unit	
Air temperature (Ta)	20–35	°C	Measured on-site; range representative of daytime summer
Relative humidity (RH)	50–70	%	Measured on-site
Wind speed	2	m/s	Average value measured at pedestrian level and used as constant input
Solar radiation	500–1000	W/m^2	Hourly range derived from clear to partially cloudy summer conditions
Sky conditions	Clear to partially cloudy	–	Observed during monitoring days
Ground albedo	0.2	–	Assigned based on prevailing paving materials
Surface material properties	As per actual façade and terrain	–	Defined according to site survey and ENVI-met material database

Fig. 4 Outdoor view of the analysed semi-open space at the University of Palermo campus, including the probe positioning used for on-site microclimatic data collection supporting ENVI-met model calibration. The monitored parameters derived from this setup were subsequently used to define the environmental boundary conditions reported in Table 1

2.3 ENVI-Met Model Development and Calibration

The microclimatic simulation was performed using ENVI-met, a three-dimensional, non-hydrostatic model specifically developed for simulating surface–plant–air interactions in urban environments at the microscale. The simulation domain was constructed to reproduce the geometric and material characteristics of the case study area, including building volumes, ground surfaces, and existing vegetation. The model domain was constructed to accurately reproduce the geometric, material, and vegetative characteristics of the case study area, while grid resolution was optimised to balance computational efficiency and pedestrian-level spatial accuracy.

Material properties such as surface albedo, thermal conductivity, and heat capacity were assigned based on site surveys and reference values from the ENVI-met database. Boundary conditions—including initial air temperature, wind speed and direction, and relative humidity—were defined using meteorological data consistent with the monitoring period. Model calibration involved iterative adjustments of key parameters to reduce discrepancies between measured and simulated air temperature and MRT trends, rather than absolute point-to-point matching, following recommended ENVI-met validation practices.

Two simulation scenarios were developed for comparative evaluation:

1. Baseline scenario representing existing site conditions without any ecological intervention.
2. Vegetated wall scenario, simulating a modular vertical greening system installed on the west-facing façade of the Cappellania Universitaria.

Table 2 Material and vegetation characteristics of the baseline and vegetated wall scenarios used in ENVI-met simulations

Scenario	Component	Material/ Species	Albedo	Thermal Capacity (J/ kg·K)	LAI	Notes
Baseline	Ground	Concrete paving	0.2	880	-	Existing site surface
Baseline	Façade	Plaster + brick	0.3	840	-	Existing wall
Vegetated Wall	Façade	Pocket system with soil substrate	0.25	850	3–3.5	Perennial/ evergreen species, 5 cm substrate, ventilated backing, drip irrigation

The vegetated wall was modelled with a substrate layer (~5 cm), ventilated backing, and Mediterranean-adapted plant species, including succulents, small-leaved shrubs, and groundcovers. Leaf Area Index (LAI) values ranged from 3 to 3.5 and plant height varied between 0.5 and 1.2 m, reflecting field observations of the installed system (Fig. 6). Drip irrigation was included in the simulation to maintain optimal plant hydration. All other model parameters—including building geometry, ground materials, and environmental boundary conditions—were kept identical to the baseline scenario (Table 2), allowing an isolated assessment of the microclimatic impact of the green wall. Simulation outputs focused on key indicators of thermal comfort, including air temperature, mean radiant temperature (MRT), and Predicted Mean Vote (PMV), evaluated at pedestrian level, with temporal analysis centred on peak daytime hours to capture maximum thermal stress.

All other boundary conditions were kept constant between the two scenarios to ensure comparability. Simulation outputs were analysed focusing on key microclimatic and comfort-related variables, including air temperature, mean radiant temperature (MRT), and Predicted Mean Vote (PMV), evaluated at pedestrian level. Temporal analysis concentrated on peak daytime hours, when thermal stress is typically most pronounced in Mediterranean climates.

3 Results

The results were obtained through a comparative analysis between the baseline configuration of the analysed outdoor space and the scenario including the vegetated wall intervention, focusing on microclimatic conditions and thermal comfort at pedestrian level during periods of maximum thermal stress. The analysis is based exclusively on calibrated ENVI-met simulation outputs, validated against on-site measurements as described in Sect. 2, ensuring consistency between observed and modelled microclimatic trends. The baseline assessment highlights critical thermal conditions within the semi-open space, characterised by elevated air temperatures and high mean radiant temperature values, particularly during peak afternoon hours. These conditions are primarily associated with intense solar exposure, limited shading, extensive use of low-albedo paving materials, and reduced natural ventilation, resulting in pronounced thermal discomfort for users of the space. Peak thermal stress is observed during the early afternoon (13:00–17:00), when the combined effect of solar radiation and heat accumulation leads to the highest values of air temperature and radiative load at pedestrian level. The comparison between on-site measurements and ENVI-met simulation outputs shows a consistent temporal trend, confirming the model's ability to reproduce the overall thermal behaviour of the area, particularly in terms of diurnal temperature variation and peak-hour conditions, while minor discrepancies can be attributed to short-term meteorological variability and inherent modelling limitations (Van Bohemen et al., n.d.).

The introduction of the vegetated wall produces measurable improvements in local microclimatic conditions, especially in the immediate surroundings of the intervention. Simulation results indicate a reduction in air temperature of approximately 0.8–1.2 °C during peak afternoon hours near the vegetated façade when compared to the baseline scenario. Although limited in magnitude, this decrease is spatially coherent and contributes to the mitigation of local thermal stress (Skelhorn et al., 2014). The cooling effect is most pronounced within a distance of a few metres from the façade and gradually attenuates with increasing distance from the vegetated surface, highlighting the localised nature of the intervention. More substantial effects are observed in terms of mean radiant temperature, which shows a clear reduction due to the combined shading and evapotranspiration processes provided by the vegetation layer. Simulated MRT values decrease by several degrees during afternoon hours in the area directly influenced by the green wall, resulting in a significant reduction of radiative heat load. This mechanism plays a key role in improving outdoor thermal comfort conditions, particularly along west-facing façades that are strongly exposed to solar radiation in the afternoon, as illustrated by the spatial temperature distributions reported in Figs. 5 and 6.

Thermal comfort assessment based on the Predicted Mean Vote index reflects these microclimatic improvements, with a shift from moderate thermal discomfort in the baseline configuration toward less stressful and closer-to-neutral comfort conditions in the vegetated wall scenario. PMV values show a consistent reduction during peak hours, indicating a perceptible improvement in pedestrian-level comfort despite relatively small changes in air temperature. The most evident improvements occur during late afternoon hours, when accumulated heat and direct solar exposure typically exacerbate discomfort. Beyond the quantitative indicators, the vegetated wall contributes to a more homogeneous thermal environment and enhances the perceived quality and usability of the outdoor space, supporting improved walkability and potential social use. While post-intervention field measurements were not available, the calibrated simulation results provide robust evidence of the potential microclimatic benefits associated with vertical greening strategies. The findings suggest that

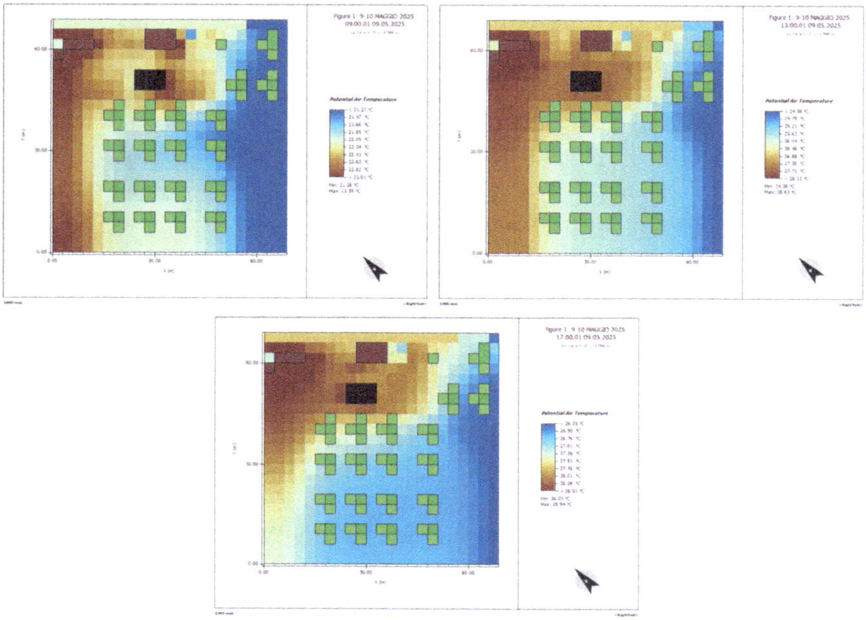

Fig. 5 ENVI-met simulation results showing the spatial distribution of air potential temperature (θ) in front of the University Chapel at three representative time steps (9:00, 13:00, and 17:00). Color gradients indicate the variation of θ, highlighting the cooling effect of the green wall during morning, peak midday, and late afternoon periods

Fig. 6 Introduction of the real green wall (used for monitoring) into the case study, showing the modular vegetation system during installation and at its final stage

even localised vegetated façade interventions can positively influence microclimatic conditions and outdoor comfort in semi-open urban spaces, supporting their role as effective nature-based solutions for climate adaptation at the microscale.

4 Discussion and Conclusion

This study investigated the potential of a vegetated wall as a nature-based strategy for improving outdoor thermal comfort in a semi-open space within the University of Palermo campus, combining on-site microclimatic monitoring with ENVI-met numerical simulations. By analysing the existing conditions and comparing them with a simulated greening scenario under comparable boundary conditions, the research provides a structured evaluation of the microclimatic effects (Tsoka et al., 2018) associated with vertical greening interventions in Mediterranean institutional contexts (Barreca et al., 2024).

The baseline configuration of the analysed space is characterised by critical thermal conditions during peak summer hours, with air temperatures reaching up to 35 °C and mean radiant temperature (MRT) values exceeding 55 °C at pedestrian level, resulting in Predicted Mean Vote (PMV) values of + 2.5, indicating moderate to strong thermal discomfort. The introduction of the vegetated wall resulted in an average reduction of 1 °C in air temperature and up to 4–5 °C in MRT during peak afternoon hours, with a corresponding improvement in PMV values from + 2.5 to + 1.8, moving conditions closer to thermal neutrality. These quantifiable improvements confirm that even localised vertical greening interventions can provide meaningful mitigation of thermal stress at pedestrian level. The cooling effect, mainly attributed to shading and evapotranspiration processes, was spatially coherent along the west-facing façade and extended up to 3–4 m into the adjacent open space, enhancing the microclimatic quality of high-use pedestrian corridors.

Beyond environmental benefits, these interventions positively influence social usability, walkability, and user comfort, supporting multifunctional outcomes consistent with the concept of Nature-Based Solutions (Liu et al., 2024). The integration of field measurements and calibrated numerical simulations provides a robust methodological framework, ensuring that the observed improvements are reliable and reproducible, and offering planners and designers actionable guidance for evidence-based decision making. While the study is limited to a single case study and lacks post-installation monitoring (Salata et al., 2016), it establishes a scalable approach for evaluating vertical greening strategies in Mediterranean campuses and dense urban contexts. Future research should expand the analysis to multiple buildings, incorporate seasonal variability, and combine vegetated walls with complementary ecological strategies such as tree shading and green roofs, to evaluate cumulative benefits on outdoor comfort and urban microclimate (Macedo Alves et al., 2022).

This research demonstrates that vegetated walls can achieve tangible thermal comfort improvements in semi-open urban spaces (Yang et al., 2021), shifting PMV values toward neutrality, reducing thermal stress, and supporting outdoor usability. The study provides both quantifiable evidence and a replicable methodology for implementing nature-based solutions as a strategic component of climate adaptation in Mediterranean cities.

Acknowledgements This research was founded by the European Union EU Next Generation Italian PRIN/PNRR 2022 Project titled 'Greenwork: An interdisciplinary framework for urban health and urban resilience enhancement based on greening strategies on buildings and open spaces'—Project Code MUR P2022MNKEC 003—Sector: ERC SH7—CUP:B53D23029580001—PRJ-1358.

References

Baraldi, R., Neri, L., Costa, F., Facini, O., Rapparini, F., & Carriero, G. (2019). Ecophysiological and micromorphological characterization of green roof vegetation for urban mitigation. *Urban Forestry and Urban Greening, 37*, 24–32. https://doi.org/10.1016/j.ufug.2018.03.002

Barker, A., Garcia-Blanco, G., Garcia, I., & Aguirre-Such, A. (2024). The role of strategic planning in Nature- Based Solutions (NBS) transformation: An evaluation of the green cities framework in mainstreaming NBS in 6 European countries. *Nature-Based Solutions, 6*, Article 100157. https://doi.org/10.1016/j.nbsj.2024.100157

Barreca, F., Cardinali, G. D., Bruno, R., & Arcuri, N. (2024). Sustainability assessments of living walls in the mediterranean area. *Buildings, 14*(10). https://doi.org/10.3390/buildings14103222

Daemei, A. B., Shafiee, E., Chitgar, A. A., & Asadi, S. (2021). Investigating the thermal performance of green wall: Experimental analysis, deep learning model, and simulation studies in a humid climate. *Building and Environment, 205*. https://doi.org/10.1016/j.buildenv.2021.108201

Ebrahimnejad, R., Noori, O., & Deihimfard, R. (2017). Mitigation potential of green structures on local urban microclimate using ENVI-met model. *International Journal of Urban Sustainable Development, 9*(3), 274–285. https://doi.org/10.1080/19463138.2017.1370424

Farooq, S., & Kamal, M. A. (2020). Analysis of green living walls: Individual awareness about its functional value and aesthetical quality. *Civil Engineering and Architecture, 8*(4), 444–449. https://doi.org/10.13189/cea.2020.080406

Freewan, A. A., Jaradat, N. M., & Amaireh, I. A. (2022). Optimizing shading and thermal performances of vertical green wall on buildings in a hot arid region. *Buildings, 12*(2). https://doi.org/10.3390/buildings12020216

Kwak, Y., & Chen, S. (2025). Integrating seasonal climate variability and spatial accessibility in ecosystem service value assessment for optimized NbS allocation. *Urban Climate, 59*. https://doi.org/10.1016/j.uclim.2025.102314

Liu, S., Middel, A., Fang, X., & Wu, R. (2024). ENVI-met model performance evaluation for courtyard simulations in hot-humid climates. *Urban Climate, 55*, Article 101909. https://doi.org/10.1016/j.uclim.2024.101909

Macedo Alves, F., Gonçalves, A., & Del Caz, R. (2022). The use of ENVI-met for the assessment of nature-based solutions' potential benefits in industrial parks: A case study of Argales Industrial Park (Valladolid, Spain). *Infrastructures, 7*(6), 85. https://doi.org/10.3390/infrastructures7060085

Mansoldo, M. D. C., de Luca, C., & Balzan, M. V. (2024). Exploring the potential for nature-based solutions to cool the streetscapes of a densely urbanised mediterranean city. *Nature-Based Solutions, 6*, Article 100177. https://doi.org/10.1016/j.nbsj.2024.100177

Marsaglia, V. (2024). Technological greenery. Exploring cutting-edge solutions for performant greenery integration in building envelope design. *Energy and Buildings, 324*. https://doi.org/10.1016/j.enbuild.2024.114920

Olivieri, F., Olivieri, L., & Neila, J. (2014). Experimental study of the thermal-energy performance of an insulated vegetal façade under summer conditions in a continental mediterranean climate. *Building and Environment, 77*, 61–76. https://doi.org/10.1016/j.buildenv.2014.03.019

Oquendo-Di Cosola, V., Olivieri, F., Olivieri, L., & Ruiz-García, L. (2023). Assessment of the impact of green walls on urban thermal comfort in a mediterranean climate. *Energy and Buildings, 296*. https://doi.org/10.1016/j.enbuild.2023.113375

Patti, M., Musarella, C. M., & Spampinato, G. (2025). A habitat-template approach to green wall design in mediterranean cities. *Buildings, 15*(14). https://doi.org/10.3390/buildings15142557

Radić, M., Dodig, M. B., & Auer, T. (2019). Green facades and living walls-A review establishing the classification of construction types and mapping the benefits. In *Sustainability (Switzerland)* (Vol. 11, Issue 17). MDPI. https://doi.org/10.3390/su11174579

Salata, F., Golasi, I., de Lieto Vollaro, R., & de Lieto Vollaro, A. (2016). Urban microclimate and outdoor thermal comfort: A proper procedure to fit ENVI-met simulation outputs to experimental data. *Sustainable Cities and Society, 26*, 318–343. https://doi.org/10.1016/j.scs.2016.07.005

Silvia Tedesco, E. M. R. G. F. L. (2019). *Green walls as nature-based solutions for urban and building resilience: A case study*. Retrieved from https://www.sustainablemediterraneanconstruction.eu/en/rivista/2019-09/2019-09-132/

Skelhorn, C., Lindley, S., & Levermore, G. (2014). The impact of vegetation types on air and surface temperatures in a temperate city: A fine scale assessment in Manchester, UK. *Landscape and Urban Planning, 121*, 129–140. https://doi.org/10.1016/j.landurbplan.2013.09.012

Stratu, A., Costică, N., & Costică, M. (2016). Wooden species in the urban green areas and their role in improving the quality of the environment. *Present Environment and Sustainable Development, 10*(2), 173–184. https://doi.org/10.1515/pesd-2016-0035

Tsoka, S., Tsikaloudaki, A., & Theodosiou, T. (2018). Analyzing the ENVI-met microclimate model's performance and assessing cool materials and urban vegetation applications–A review. *Sustainable Cities and Society, 43*, 55–76. https://doi.org/10.1016/j.scs.2018.08.009

Van Bohemen, H. D., & Fraaij, A. L. A. (n.d.). *Ecocity world summit 2008 Proceedings Ecological Engineering, Green Roofs and the Greening of Vertical Walls of Buildings in Urban Areas.*

Viecco, M., Jorquera, H., Sharma, A., Bustamante, W., Fernando, H. J. S., & Vera, S. (2021). Green roofs and green walls layouts for improved urban air quality by mitigating particulate matter. *Building and Environment, 204*. https://doi.org/10.1016/j.buildenv.2021.108120

Xi, C., Ding, J., Wang, J., Feng, Z., & Cao, S. J. (2022). Nature-based solution of greenery configuration design by comprehensive benefit evaluation of microclimate environment and carbon sequestration. *Energy and Buildings, 270*. https://doi.org/10.1016/j.enbuild.2022.112264

Yang, J., Hu, X., Feng, H., & Marvin, S. (2021). Verifying an ENVI-met simulation of the thermal environment of Yanzhong square park in Shanghai. *Urban Forestry & Urban Greening, 66*, Article 127384. https://doi.org/10.1016/j.ufug.2021.127384

Green and Blue Infrastructure to Enhance the Outdoor Comfort in an Urban Built Environment with Artificial Ground Depression: A Case Study in Sicily

Kevin Aarón Castro Morales and Rossella Corrao

Abstract Global climate change has driven international policies aimed at limiting greenhouse gas emissions and promoting sustainable urban development. Green and blue systems, as Nature-Based Solutions (NBSs), have been widely recognized as effective strategies to mitigate Urban Heat Island (UHI) effects and improve thermal comfort within dense built areas. This work analyses the microclimatic effects on the outdoor comfort of different green and blue interventions in a particular urban context characterized by challenging geomorphological constraints. A multi-scale simulation approach was applied to a selected area in the Campus of the University of Palermo (Italy), as it is located in a hot-summer Mediterranean climate. The analyses focused on the area belong to the Department of Architecture building, which is situated in an Artificial Ground Depression (AGD) that limits natural ventilation. ENVI-met®, a computational fluid dynamics software (CFD), was used to model the area both at urban and building scale, by using weather data registered in 2024 for the hottest and coldest days of the year. At the urban scale, morphology and existing vegetation were assessed, and their output boundary conditions were used for detailed building scale simulations. Four scenarios were evaluated, starting from a scenario where the existing trees have been removed (historical condition), enhanced vegetation, installation of a green wall, and artificial ponds. The results showed localized reductions in peak temperatures across interventions, ranging from 0.05 to 0.6 °C depending on the specific intervention and the distance from each other. However, it was verified that the geometry of the ground depression reduces the wind flow and also limits the overall effectiveness of the adopted green and blue systems. The effects were particularly registered in already poorly ventilated areas. Findings suggest that traditional greenery systems alone are not enough to significantly enhance outdoor thermal comfort in complex urban morphologies, as the one analysed in this work, indicating the need for complementary strategies aimed at improving air circulation and reducing heat accumulation.

K. A. C. Morales (✉) · R. Corrao
Department of Architecture, University of Palermo, Palermo, Italy
e-mail: kevinaaron.castromorales@unipa.it

L. Diana et al. (eds.), *Greening and Resilience*, Urban Sustainability,
https://doi.org/10.1007/978-981-92-0419-9_20

Keywords Green systems · Blue systems · CFD simulations · Urban Morphology · Artificial Ground Depression

1 Introduction

Rising temperatures caused by climate change have pushed countries around the world to establish agreements aimed at reducing greenhouse gas emissions, encouraging sustainable forest management, and increasing the use of energy from renewable sources. To keep global warming below 1.5 °C above pre-industrial levels, the Paris Agreement of 2015 mandates that participating nations to review their national energy and climate plans every five years. Recently, the United States of America officially withdrew from the Paris Agreement for the second time, a position that could alter the global political direction on efforts of reduce effects of climate change.

To reduce greenhouse gas emissions and mitigate the effects of Urban Heat Islands (UHI), Nature-Based Solutions (NBSs), such as green and blue systems, can be implemented in urban contexts (Stewart & Mills, 2021). Greenery systems can improve both outdoor and indoor comfort within the built environment (Gatto et al., 2021; Pastore et al., 2017). The positive effects of greenery systems—such as trees, green corridors or green alleys, green walls, and green roofs—have already been confirmed by several studies carried out in different climatic zones (Falasca et al., 2024; Herath et al., 2018). In this way, NBSs are proving their effectiveness in improving the resilience of urban built environments against climate change. According to the ASHRAEE standard 55–2017 (Thermal Environmental Conditions for Human Occupancy), the specific comfort zone that is considered acceptable for most occupants typically falls within a range of near 19.4 °C to 27.7 °C for air temperature and with relative humidity between 30 and 60% (ASHRAE Technical FAQ, n.d.).

This work focuses on the analysis of an area located in the campus of the University of Palermo (Italy) and, in particular, in an AGD. The city of Palermo is in the northwest coast of Sicily, with specific climatic condition characterized by hot and dry summers and classified, according to the Köppen-Geiger classification, as Csa (Hot-Summer Mediterranean climate). In recent years, due to the current climate changes, heatwaves have been more frequent and an increasing number of 'tropical nights' during summer months have been registered (Copernicus Climate Change Service (C3S) & World Meteorological Organization (WMO), 2025). Some experts warn about the possibility of change from Csa to BSh (Hot, semi-arid climate) category defined by Köppen and Geiger, due to the rainfall decreases and temperature rise.

The University Campus is located beyond the 'virtual boundaries' of the historic Centre of Palermo, and it extends along a road axis, in parallel with another main road of the city. The lower point of the AGD is 39 m above sea level (m.a.s.l.) at that level two buildings belong to the Department of Architecture are located. The main building has four floors with a total height of approximately 22 m; the secondary

Fig. 1 Satellite view of the urban context in which the area analysed in this work is located. The red perimetral line identifies the main building of the Department of Architecture and the orange perimetral line identifies the secondary lower building. *Source* Google maps

building (C shaped) has two floors with a total height of approximately 7 m, and it extends along a portion of the perimeter of the AGD (Figs. 1 and 2).

2 Methodology

A multi-scale methodology, already tested in another works (Pastore et al., 2017), has been used to analyse the area described in the previous section, to evaluate the effects of green and blue systems both at microclimate urban and building scales. A well-known CFD software (ENVI-met®) was used for the analyses.

2.1 Weather Data

First of all, weather data was collected and analysed to carry out the simulations. The weather data was collected from the nearest weather station that belongs to the University of Palermo (Istituto Nazionale di Astrofisica INAF, 2013). To select a 'typical year' to use in the CFD simulations, the methodology adopted by researchers generally consists of the analysis of 5 or 10 years weather data; in this work, this methodology was not adopted since the rise of temperature is keeping constant every year—breaking records each year (Copernicus Climate Change Service, 2025)—so, only the weather data of 2024 has been considered. The analysis of data allowed to identify and define the hottest and the coldest days that corresponds to 16 August

Fig. 2 Views of the department of architecture building. *Source* Photo taken by the authors

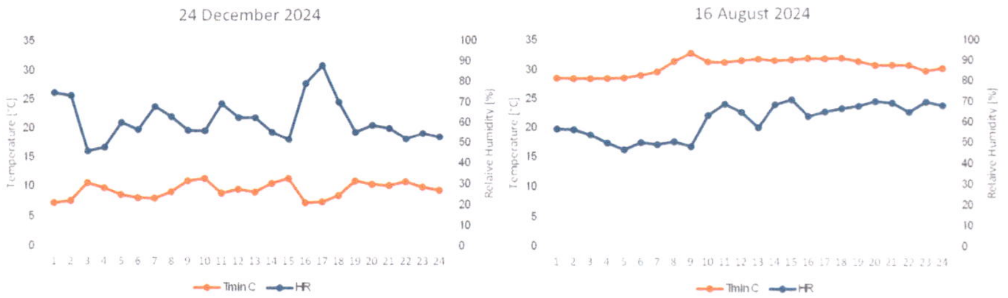

Fig. 3 Temperature and Relative Humidity data collected. *Source* Authors

and 24 December. In order to avoid instabilities during the simulations, the use of data days with 'extreme' weather phenomena was avoided. In this work, we used air temperature, relative humidity, wind speed, and direction (Fig. 3).

2.2 *Microclimate Urban Scale Analysis*

Weather data collected was used as boundary conditions during the modelling process. Urban data, such as city plans and open data shapefiles, was obtained from

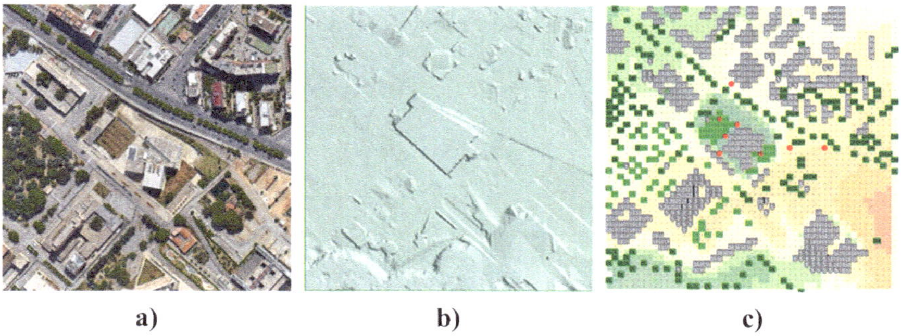

Fig. 4 **a**. Aerial view of the area used in the microclimatic urban; **b**. DTM used in the microclimatic urban; and **c**. ENVI-met model area. *Source* Google maps, Geoportal mite, and ENVI-met®

the Municipality of Palermo (Comune di Palermo, 2021). The georeferenced information regards streets, buildings surfaces and height, and green and public spaces. The terrain data corresponds to the Digital Terrain Model (DTM) of the Sicily Region, which has a resolution of 1 m and derived from LIDAR scanning conducted in 2008 (Ministero dell'Ambiente e della Sicurezza Energetica, 2013). The modelling process was carried out by using QGIS software and ENVI-met plugin®. During the modelling process carried out in QGIS software, some modifications were adopted to the shapefiles obtained from the municipality, due to inaccuracies related to building heights. Vegetation was manually added by indicating the position of trees and shrubs. The area measures 500 × 500 m and was modelled into a three-dimensional grid of 100 × 100 × 30 cubic elements with a dimension of each element equal to 10 × 10 × 5 m.

Figure 4 shows the selected area used during the modelling process at this scale aimed to calibrate the weather data obtained from the weather station. This calibration allowed to take into account the urban morphology, surface's materials, and vegetation. For this reason, a simulation was carried out for a baseline scenario (current status) for each day selected (18 August and 24 December). As it can be observed in Fig. 4 (c), the red points correspond to receptors and area positioned inside and outside the AGD. Results coming from these simulations were used as boundary conditions in the simulations at the building scale.

2.3 *Microclimate Building Scale Analysis*

At this scale, the analysed area was reduced in 250 × 250 m by using a three-dimensional grid of 100 × 125 × 40 cubic elements, each element with a dimension of 2 × 2 × 2 m. The reduction of dimensions for each element allowed to increase the details of modelling. For instance, the use of this element dimension allowed to model smaller trees and shrubs as well as they were added in the model (see Table 1).

Table 1 Existing plants identified and modelled in ENVI-met®. Height, width, and leaf type are input parameters

Scientific name	Type of plant02CL	Height [m]	Width [m]	Leaf type
Ceratonia siliqua	Tree	15	9	Evergreen
Arbutus unedo	Shrub	5	3	Evergreen
Populis nigra L	Tree	8	3	Deciduous
Citrus x aurantium	Tree	5	3	Evergreen
Ficus macrophylla	Tree	25	15	Evergreen
Pinus pinaster	Tree	25	12	Conifer
Cupressus sempervirens	Tree	25	11	Conifer

Furthermore, the use of these smaller elements increased the accuracy of the model due to the discretization of the elements reduces the approximation.

In recent years, inside the AGD were planted cylindrical crown shaped deciduous trees (*Populus Nigra*) between the main and secondary buildings. Once the simulations of the baseline scenario were completed, the main outputs in terms of thermal comfort highlighted the critical zones inside the AGD.

Four scenarios have been and analysed. They proposed and evaluated green and blue systems as described in Table 2. These NBSs were coupled with modifications of the spatial layout by considering green areas and parking spaces. Figure 5 shows the different scenarios proposed and evaluated in this work within the ENVI-met® modelling environment. The proposed plants for these scenarios corresponded to species belonging to the maquis—trees and shrubs from the Mediterranean region (Claudio Cervelli, 2005) in order to use native species that could resist to the local weather. Furthermore, the proposed species were selected by using the Anthosart Green Tool (Guarino et al., 2020), which consent users to appropriately select the species based on environmental conditions, potential use, and benefits of them. Proposed plants and their main characteristic are presented in Table 3.

In Scenario 2, surfaces made of different materials cane be observed. The blue areas correspond to artificial ponds which were modelled into ENVI-met® as sole water surfaces. In Scenario 3 can be observed a red line which represents the green wall installed in the secondary C-shaped building. The green wall was modelled as just greening vertical surface, as present in the available database of the software.

Table 2 Scenarios evaluated in this work

ID Scenario	Description
Baseline scenario	This scenario evaluated the current status of the area focused on the AGD and the buildings of the Department of Architecture
Scenario 1	This scenario enhanced the vegetation by adding trees belonging to the existent species. Shrubs were also added in the upper perimeter of the AGD limiting with a high transited road of the city
Scenario 2	This scenario evaluated the effects of blue systems positioned in correspondence of the ground level of the AGD and the buildings of the Department of Architecture
Scenario 3	This scenario evaluated the effects of green walls installed on the façade of the secondary building (C shaped) of the Department of Architecture
Scenario A	This scenario evaluated a previous status of the baseline scenario, without the threes located within the AGD planted in recent years. It has been used for useful comparisons

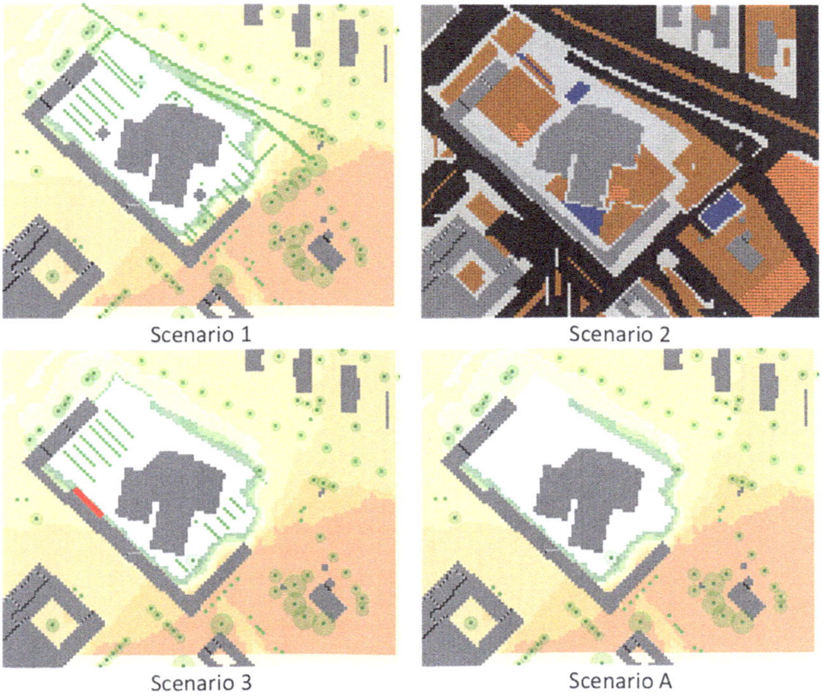

Fig. 5 Scenarios analysed in this work modelled in ENVI-met®. *Source* Authors

Table 3 Proposed plants and modelled in ENVI-met®. Height, width, and leaf type are input parameters

Scientific name	Type of plant	Height [m]	Width [m]	Leaf type
Nerium oleander	Shrub	3	1	Evergreen
Laurus nobilis	Tree	10	2	Evergreen

3 Results and Discussion

The microclimatic urban scale analyses highlight the negative effects of the urban morphology with the AGD, as can be observed in Fig. 6. At this scale, some critical zones can be observed in terms of height temperatures, and they also correspond to those zones where the wind speed is low (below 1 m/s) during summer season. As it was expected, the presence of narrow streets intensifies thermal discomfort in these zones. The results of simulations also put in evidence the discomfort zone in correspondence of the AGD, which have similar conditions compared to narrow streets. In general, this confirms the effects of urban morphology and allowed to proceed in the simulations by focusing on the AGD and the buildings in it.

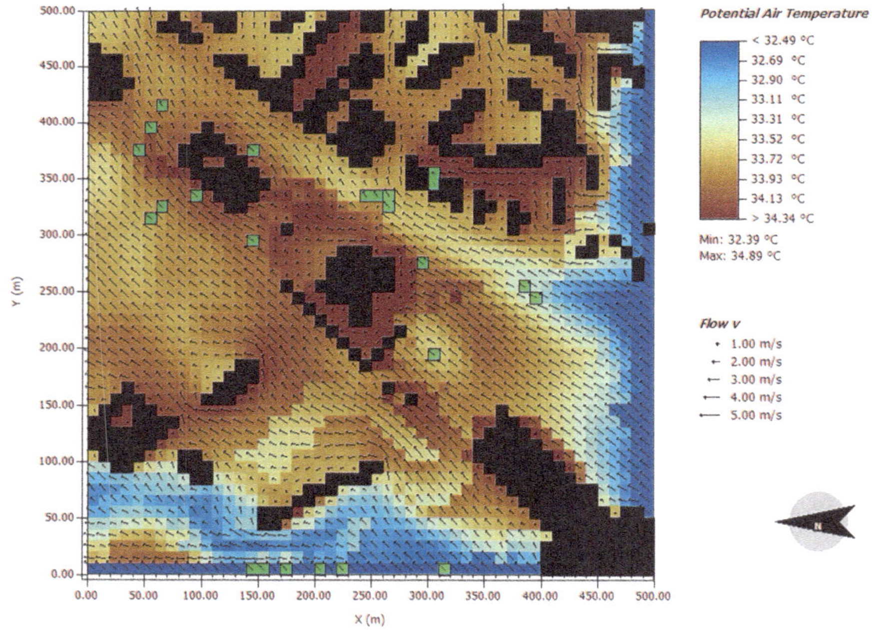

Fig. 6 Air temperature and wind speed at 16:00 of 16 August 2024 obtained from ENVI-met®. *Source* Authors

First, the analyses of baseline scenario allowed to further understand the main parameters that influence the thermal comfort perception in the outdoor environment. Those parameters are presented in Figs. 7, 8, and 9 (air temperature and wind flow, relative humidity, and mean radiant temperature, respectively). The heatmaps presented in this section correspond to a height of 1,80 m from the ground level (variable according to the DTM). This height was selected as it was the nearest one to average height of human in the output data from the simulations. Furthermore, the heatmaps correspond to the values at 14:00 of the hottest day (August 16, 2024). The analyses of baseline scenario allowed to plan the position of the NBSs in order to improve the thermal comfort.

Figure 7 shows the heatmaps for each scenario. As it can be observed in the five heatmaps, the temperatures ranges did not show significant variations. Higher temperature reached almost 36 °C, which represent an extreme condition for human beings. The positive effects of vegetation added in Scenario 1 showed localized reduction of temperature that did not reduce temperature further than 0.6 °C. Scenario 2 showed similar values of temperature but with a slightly reduction in the zones where the artificial ponds are located. Scenario 3 showed practically any reduction of temperatures. As described in the previous section, Scenario A corresponds to a previous condition in which trees were not planted, and it can be observed that they effectively contribute to reduce the temperatures but in these extreme climatic conditions. they are not really effective.

Similar limited positive effects can be observed in Fig. 8, but in terms of relative humidity. The process of evapotranspiration of vegetation contributes to increase the low values presented in the heatmaps, as can be observed in Scenario 1. Comparing Scenario A with the baseline scenario, it can be observed that values of relative humidity increase but keeping it inside the range of comfort for humans. The effects are not so limited such as the ones of air temperature and can be observed comparing Scenario 1 and Scenario 2 (the one with artificial ponds).

In all the 5 heatmaps, it is also possible to graphically observe the magnitude of the wind speed and its direction. As it was expected from the results of the microclimate urban scale, the lowest wind speeds are registered in the bottom of the AGD (which corresponds to the ground level of the buildings). In terms of natural ventilation, the presence of vegetation can reduce the wind speed.

Mean radiant temperature, which refers to the temperature of surfaces, is also presented in this section (see Fig. 9). Values showed in these heatmaps range from 31 °C to near 81 °C. Lower temperatures, as it was expected, correspond to shadowed and vegetated areas. The second one contributes due to the lower albedo compared to the ones of artificial materials (cementitious surfaces and asphalt). The artificial ponds added in Scenario 2 shows a reduction of temperature, since the surfaces around them correspond pedestrian area with cementitious-based materials. In Scenario 3 can be observed a slightly reduction of the meant radiant temperature in correspondence of the green wall. However, shadowing effects of the crown trees can be appreciated by comparing the baseline scenario with the Scenario A.

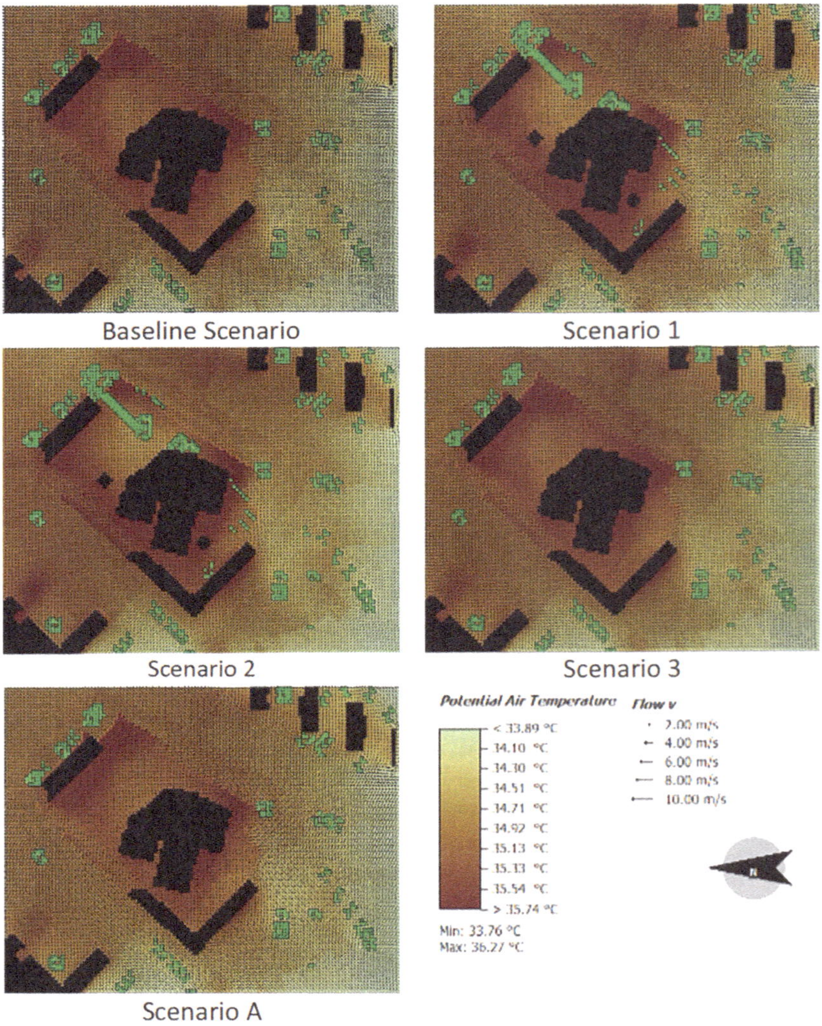

Fig. 7 Air temperature and wind speed at 16:00 of August 16, 2024 obtained from ENVI-met® simulations. *Source* Authors

Overall results presented above demonstrate the lack of effectiveness of the solutions proposed and evaluated in this work. Figure 10 shows the daily trend of temperature and relative humidity for each Scenario. Several points were selected—along the analysed area—to quantify and assess the effects of each proposed NBSs compared to the baseline scenario. Scenario 1, which imply the enhancement of vegetation, showed temperature reductions ranging from 0.05 °C to 0.4 °C, with the magnitude of reduction depending on the distance from the implemented vegetation. Scenario 2, which included artificial ponds, demonstrated temperature reductions between 0.1 °C

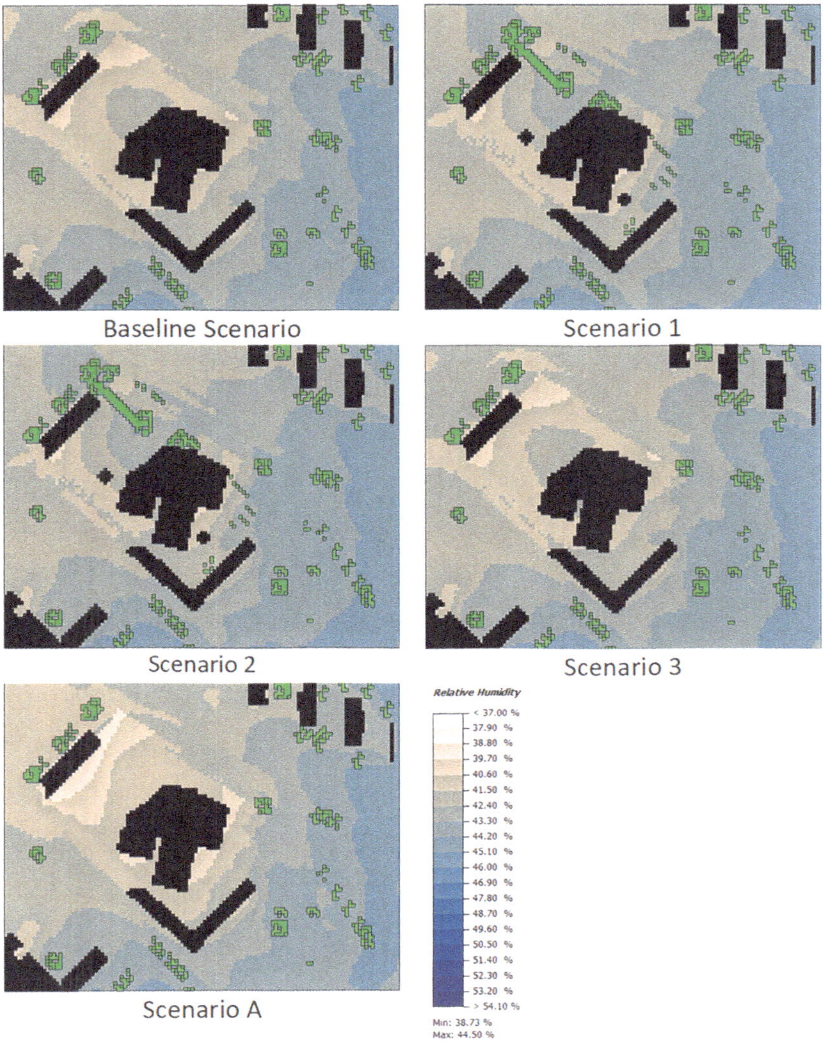

Fig. 8 Relative Humidity at 16:00 of 16 August 2024 obtained from ENVI-met®. *Source* Authors

and 0.6 °C, also according to the distance from the ponds. As expected, Scenario 3, which implemented a green wall, registered the highest temperature reduction near the intervention zone, although this reduction was not significant, ranging from 0.05 °C to 0.1 °C. Scenario A confirmed that the introduction of threes reduced the temperature ranging from 0.05 °C to 0.3 °C.

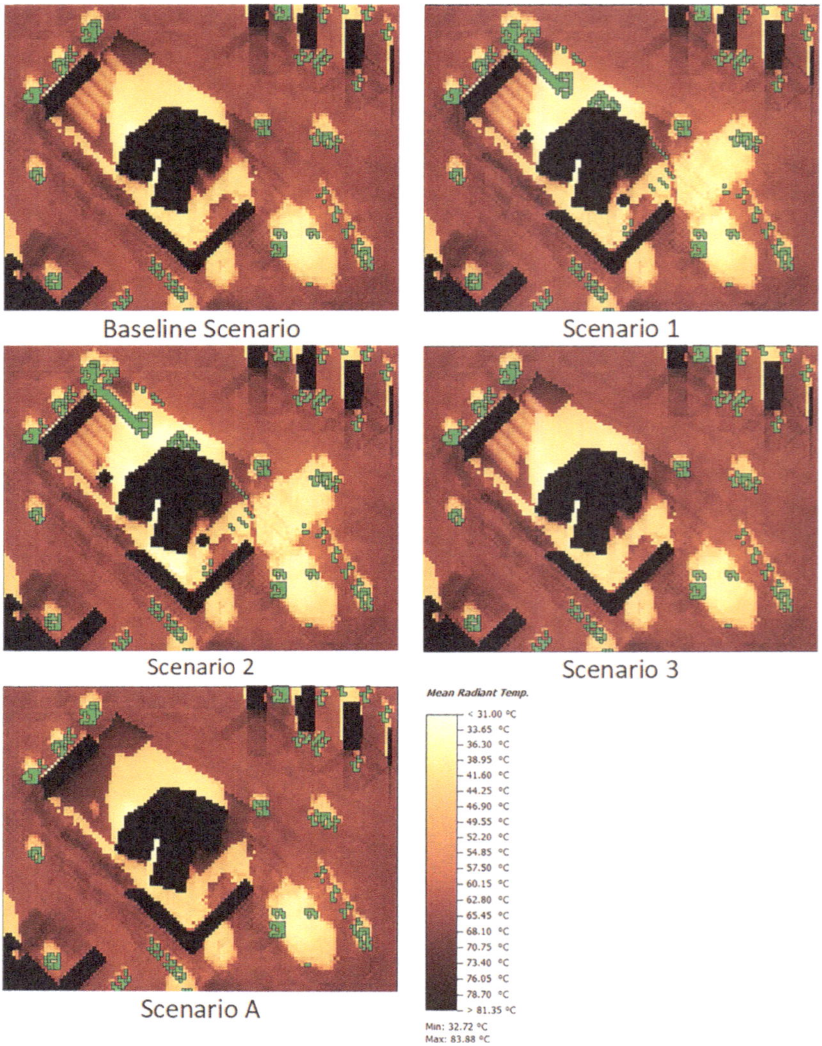

Fig. 9 Mean Radiant Temperature of August 16, 2024 obtained from ENVI-met® simulations. *Source* Authors

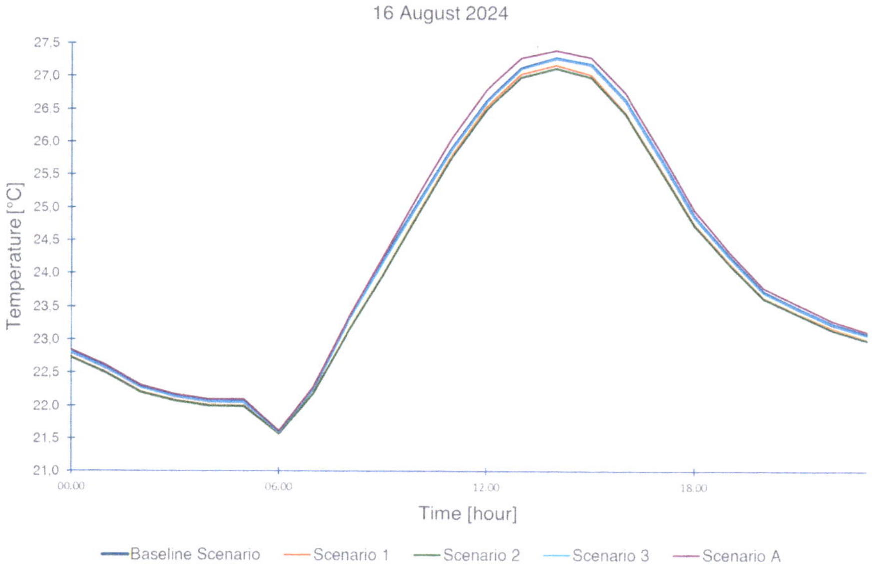

Fig. 10 Air temperature daily trend correspond to August 16 2024. *Source* Authors

4 Concluding Remarks

The analysed scenarios demonstrated that the beneficial effects of NBSs on improving outdoor thermal comfort can be limited by urban morphology and the site's topography. At the urban scale, the AGD analysed in this work behaves similarly to narrow urban canyons, exhibiting low wind speeds and high air temperatures during summer conditions. Actually, the AGD negatively affects natural ventilation so that the wind speed is close to zero in some zones. Since the presence of vegetation usually reduces the wind speed, in the case of an ADG it can negatively affect zones located farther.

The analysed scenarios demonstrated only localized and limited improvements in outdoor thermal comfort. Maximum air temperature reductions did not exceed 0.6 °C, depending on the NBSs adopted and proximity to them. While vegetation and artificial ponds contributed positively through shading, evapotranspiration, and surface cooling, their effectiveness was constrained by the reduced airflow. In some cases, increased vegetation further decreased wind speed, partially counteracting its cooling benefits. Green walls showed negligible impact at the pedestrian level, confirming that their influence remains highly localized.

Overall, the findings highlight that traditional greenery systems alone are insufficient to significantly mitigate thermal stress in complex urban morphologies such as the AGS analysed in this work, especially during extreme summer conditions. Therefore, future design strategies should adopt an integrated approach that combines NBSs with solutions aimed at enhancing natural ventilation.

Acknowledgements The simulations discussed in this work have been carried on during the course of 'Laboratory of Green Systems for the Built Environment' in the field of the BSc in Landscape Architecture, with the support of students of the MSc in Architecture of the University of Palermo, M. Di Galbo; G. Fiorentino, M. Licata, and A. Rizzo. The authors are especially grateful to them.

References

ASHRAE Technical FAQ. (n.d.). Retrieved October 22, 2025, from www.ashrae.org

Claudio Cervelli. (2005). *Le specie arbustive della macchia mediterranea un patrimonio da valorizzare*. Dipartimento Azienda Regionale Foreste Demaniali.

Comune di Palermo. (2021). *Portale Open Data—CARTA TECNICA COMUNALE - UNITA'.* VOLUMETRICHE. https://opendata.comune.palermo.it/opendata-dataset.php?dataset=1294

Copernicus Climate Change Service. (2025). *Temperature*. Copernicus. https://climate.copernicus.eu/climate-indicators/temperature

Copernicus Climate Change Service (C3S), & World Meteorological Organization (WMO). (2025). *European state of the climate 2024*. https://doi.org/10.24381/14j9-s541

Falasca, S., Zinzi, M., Siani, A. M., Curci, G., Ding, L., & Santamouris, M. (2024). Investigating the effects of the greenery increase on air temperature, ventilation and cooling energy demand in Melbourne with the weather research and forecasting model and local climate zones. *Science of the Total Environment, 953*, Article 176016. https://doi.org/10.1016/J.SCITOTENV.2024.176016

Gatto, E., Ippolito, F., Rispoli, G., Carlo, O. S., Santiago, J. L., Aarrevaara, E., Emmanuel, R., & Buccolieri, R. (2021). Analysis of urban greening scenarios for improving outdoor thermal comfort in neighbourhoods of lecce (Southern Italy). *Climate, 9*(7). https://doi.org/10.3390/cli9070116

Guarino, R., Menegoni, P., Pignati, S., & Siniscalco, C. (2020). *ANTHOSart FLORINTESA*. https://anthosart.florintesa.it/

Herath, H. M. P. I. K., Halwatura, R. U., & Jayasinghe, G. Y. (2018). Evaluation of green infrastructure effects on tropical Sri Lankan urban context as an urban heat island adaptation strategy. *Urban Forestry and Urban Greening, 29*, 212–222. https://doi.org/10.1016/j.ufug.2017.11.013

Istituto Nazionale di Astrofisica INAF. (2013). *MeteOapa Home Page*. http://meteo.oapa.inaf.it/public/.

Ministero dell'Ambiente e della Sicurezza Energetica. (2013). *Geoportal MiTE*. http://www.pcn.minambiente.it/geoportal/catalog/search/resource/details.page?uuid=m_amte%3A299FN3%3A8740c0d0-8f32-4e85-9856-42f8b88853da

Pastore, L., Corrao, R., & Heiselberg, P. K. (2017). The effects of vegetation on indoor thermal comfort: The application of a multi-scale simulation methodology on a residential neighborhood renovation case study. *Energy and Buildings, 146*, 1–11. https://doi.org/10.1016/j.enbuild.2017.04.022

Stewart, I. D., & Mills, G. (2021). UHI management. In *The urban heat island* (pp. 49–74). Elsevier. https://doi.org/10.1016/B978-0-12-815017-7.00003-5.

Effectiveness of Pocket Green Spaces Within the Built Environment in Mediterranean Historic City Centres in Improving Urban Microclimate: A Case Study in Rome

Federica Rosso and Simona Mannucci

Abstract Cities are increasingly affected by climate change–related hazards, particularly heatwaves, which are exacerbated by the Urban Heat Island Effect (UHIE). This phenomenon poses a major challenge in dense and historic Mediterranean city centres, where the implementation of large nature-based solutions (NBS), such as big parks, is often constrained. Pocket parks (PPs) represent a potentially effective, small-scale NBS, yet their performance in historic urban contexts remains underexplored. This study evaluates the effectiveness of pocket parks in mitigating heat stress in historic Mediterranean city centres through a comparative analysis of paths including multiple PPs. The assessment focuses on the role of design elements, including shading, surface materials, greenery, and water features, in influencing surface temperatures and microclimatic conditions. Results demonstrate that pocket parks can effectively mitigate heat stress when consciously designed. Shading, provided by trees or surrounding buildings, plays a fundamental role in reducing temperatures across all analysed PPs. Surface materials significantly influence thermal performance, with cool materials (e.g., light-coloured gravels) exhibiting lower temperatures than darker pavements such as asphalt or stone blocks. Vegetation proves particularly effective in the form of trees, while grass surfaces reduce temperatures by up to 10 °C, outperforming cool materials (≈5 °C reduction). Flowing water elements further contribute to surface cooling, whereas dry soil and dry grass show limited or negative cooling effects. Well-designed pocket parks represent an effective and feasible strategy to mitigate heat stress in historic Mediterranean urban centres, contributing to providing physical, physiological, and psychological respite in the urban experience.

Keywords Pocket Parks · Historic city centre · Mediterranean Region · Greenery · Cool materials · Built environment

---

F. Rosso (✉) · S. Mannucci
Department of Civil, Construction and Environmental Engineering, Sapienza University of Rome, Rome, Italy
e-mail: federica.rosso@uniroma1.it

L. Diana et al. (eds.), *Greening and Resilience*, Urban Sustainability,
https://doi.org/10.1007/978-981-92-0419-9_21

1 Introduction

Cities are facing increasingly exacerbated climate change-related challenges, such as more frequent and stronger heatwaves, as well as extreme pluvial events and droughts. While these challenges intensify, the exposure of urban population to such events grows, as the world population is increasingly moving towards cities.

With respect to heat stress, cities are already affected by Urban Heat Island Effect (UHIE) (Santamouris & Yun, 2020), which causes higher temperatures in urban areas as compared to suburban areas, up to around +10 °C. It has been demonstrated that the UHIE further increase the risk and frequency of heatwaves (Falasca et al., 2019), demonstrating how heat stress in cities is a paramount challenge, especially in the Mediterranean Region.

Among the causes of UHIE, there are the accumulation of heat within the built components (roads, buildings) of cities, the reduction and modification in ventilation due to increased rugosity in cities, the anthropogenic heat release, and the reduced evapotranspiration from green and blue components that are increasingly lacking within neighbourhoods (Founda & Santamouris, 2017). With respect to the latter, soil sealing is increasing faster than population grows, thus reducing green areas, grass and trees, as well as water elements such as ponds, defined Nature-based solution (NBS), and negatively impacting the cycle of natural, permeable soil (Banzhaf et al., 2017).

Big urban parks, constituting an effective NBS, are able to reduce urban temperatures, but are localized services that are not equally distributed in close proximity to each neighbourhood, thus benefitting in an unequal manner urban population (Todeschi et al., 2022). The advantages of NBS are not limited to the improvement of urban microclimate, but also deal with (i) humans' need for contact with nature, benefiting both physical and psychological health and wellbeing; (ii) contribution to pluvial runoff reduction and rainwater harvesting (Costa et al., 2021); and (iii) conservation of biodiversity in cities.

While medium–big-sized parks are difficult to insert in consolidated cities, historic city centres present even higher challenges in implementing more diffuse and distributed NBS that are enjoyable for all people. In this panorama, we propose to tap on the potential of pocket parks (PPs), i.e., small-sized parks that can be easily inserted within densely built urban areas, with different extent of complexity (Rosso, 2025). Such PPs have demonstrated to possess the capacity to slightly improve urban microclimate, while greatly improving overall comfort for their users (Rosso et al., 2022, 2024), as well as also being able to reduce and manage pluvial runoff (Mannucci et al., 2025).

Previous studies focused on contexts where PPs are already well-developed, while they are limited for historic urban areas. We aim to fill this gap and to respond to the following research question (RQ): **are pocket parks effective in mitigating heat stress in historic city centres in the Mediterranean region?**

With respect to this RQ, we also aim at finding the most effective strategies and architectural design for improving PP mitigation and adaptation capability in historic city centres.

The manuscript is structured as follows: in Sect. 2, the methodology of the research is illustrated. Section 3 presents the results of the study, while Sect. 4 discusses the architectural and practical implications deriving from the results. Finally in Sect. 5, conclusions are drawn.

2 Methodology

To respond to the RQ, we conduct in field experiment in a relevant and significant case study for historic city centres in the Mediterranean region, in this case considering Rome city centre. We conducted, along a studied path that crosses several typologies of PPs, characterized by different materials and morphologies, a climate walk in late summer 2024, during the late morning, at midday, in a typical hot day. The aim of the climate walk was to compare, for the same day and hour, the human-centric microclimate conditions and thus the comfort, with varying locations on a studied path. The path studied has been selected to account for the maximum crossing of NBS around the University Campus in the city centre of Rome, favouring those smaller green areas that also provided views on the historic monuments (Table 1).

The small green patches that were crossed are as follows:

1. The "Giardino di Renato", a vegetated area with cool gravels and small trees and shrubs, as well as urban furniture, within the engineering university campus, close to the main building;

Table 1 Small parks along the path: main characteristics

Pocket park #	Pocket Park name	Area (m^2)	Paving materials	Vegetation
1	Giardino di Renato (Univ. Campus)	147	Light-coloured gravel	Small trees, bushes
2	Cafeteria garden (Univ. Campus)	690	Grass (not irrigated) and bare soil	Tall Trees
3	Small seating area, Piazza San Pietro in Vincoli	20	Paved (impervious), with basalt stone ("*sanpietrini*")	A single tree
4	Park Largo Gaetana Agnesi	1339	Paved (impervious) with basalt stone ("*sanpietrini*") and grass (not irrigated)	Trees and shrubs
5	Park Via del Monte Oppio	1670	Grass (not irrigated), bare soil and light-coloured gravels	Trees, shrubs and bushes

2. Green area, with trees and grass, within the engineering university campus, in the Cafeteria and parking area;
3. Small seating area with a fountain, a tree and a bench in Piazza San Pietro in Vincoli, close to the main entrance of San Pietro in Vincoli Church;
4. Green area, with trees and grass and furniture at Largo Gaetana Agnesi, with direct view over the Colosseum; and
5. Small enclosed park, with trees and grass and furniture at Via del Fagutale and Via del Monte Oppio crossing, with direct view over the Colosseum.

Along this path, we measured surface temperatures with a thermal camera and used a portable microclimate monitoring station to measure air temperature, solar radiation, relative humidity, wind speed and direction in the PPs and close by streets. The use of the two tools together allowed us to jointly measure microclimate features and analyse in depth the temperatures of urban surfaces, as the two are linked, thus allowing a deeper comprehension of the effect of different materials and construction elements, as well as NBS, on the microclimate. To provide for a more comprehensive thermal comfort index, from the gathered data we calculate two comfort indices, humidex and apparent temperature.

Humidex is computed starting from air temperature and relative humidity (RH), which is employed to derive the vapour pression (v_p):

$$\text{Humidex} = T_{\text{air}} + \left[0.5555 \times \left(v_p - 10\right)\right] \tag{1}$$

$$v_p = \left(6,112 * 10^{(7.5 * T_{\text{air}} / (237.7 + T_{\text{air}}))} * RH / 100\right) \tag{2}$$

Humidex values above 30 °C indicate thermal discomfort. With respect to *Apparent temperature* (A_t) (Eq. 3), it considers T_{air}, but also vapour pression (v_p) and wind speed. Values above 27 require care and could imply thermal discomfort.

$$\text{At}\left[°\text{C}\right] = -2.7 + 1.04T_{\text{air}} = 2.0v_p / 10 - 0.65\text{W}_s \tag{3}$$

To carry out microclimate monitoring, a portable, human-centric, wearable monitoring station was employed, which was wore by a researcher while walking at normal speed along the path. The sensors which were mounted on the station, at human height, were: sensor for air temperature, with an accuracy: ± 0,3 °C, sensor for relative humidity (accuracy of ± 2%), and sensor for wind velocity and direction (accuracy ± 3% and ± 3 deg, respectively). Moreover, the monitoring station has a GPS with a horizontal spatial accuracy equal to 2.5 m to exactly associate the microclimate measure to the specific point in the path. Data is recorded every 5 s as a sampling interval, to allow for a granular and hyperlocal monitoring.

The employed thermal camera is a FLIR E95, with an accuracy of ± 0.030 °C. Thermographic images were taken at each small park to record surface temperatures for comparative purposes and to complement the microclimate measurements.

Emissivity value was set to 0.90, as commonly employed for investigating the built environment.

Microclimate observations collected during the climate walks were spatially organized and visualized in QGIS starting from CSV files exported from the portable monitoring station. Each CSV contained the measurement attributes (e.g., air temperature, relative humidity, wind speed and timestamp) together with geographic coordinates. The datasets were imported into QGIS as delimited text layers, defining longitude and latitude as X/Y fields and assigning the appropriate coordinate reference system (WGS84). After import, the layers were inspected for basic quality control and, for reproducibility, were saved into a GIS-ready format for subsequent processing and cartographic production. When the same variable was available in two separate columns within the CSV (e.g., parallel recordings), we created a single representative indicator by computing the row-wise mean directly in the attribute table. This was performed using the field calculator, generating a new numeric field as the arithmetic average of the two original columns and using the resulting mean field for subsequent classification and mapping.

Separate thematic maps were then produced for the key variables using graduated point symbology based on attribute values. All maps were composed in the QGIS Print Layout on a reference basemap and exported as high-resolution figures, maintaining consistent extent, legend structure, and cartographic elements across variables to facilitate direct comparison. The methodology is graphically represented in Fig. 1.

Moreover, statistical analyses were performed to check for the impact of path characteristics being in the park, presence of tall or small trees, paving type, water presence on air temperature and comfort indices. Regression analyses were performed, where air temperature, humidex and apparent temperature were considered separately as dependent variables and path characteristics were the independent variables. Confidence Interval equal to 95% was considered, and for each regression, variance inflation factor analysis was also performed, to check for collinearity. VIF was always verified when lower than 5.

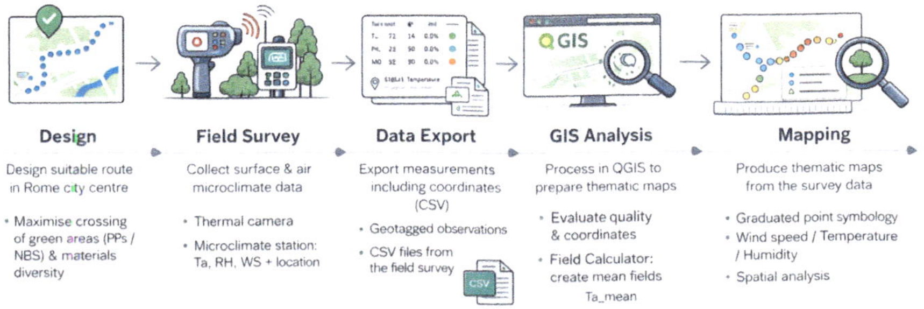

Fig. 1 Diagram of the employed research methodology

3 Results

3.1 Thermal Images

Results demonstrated that such PPs are able to mitigate heat stress if consciously designed. Indeed, the analysis shows the fundamental role of shading (either from trees or urban shades from buildings) in all the analysed PPs (Figs. 2 and 3). Also, the materials constituting the paving and the envelope of the buildings heavily impact temperatures, where cool materials (cool gravels) maintain lower temperatures than darker materials (such as sanpietrini or asphalt). Greenery shows to be effective especially under the form of trees, projecting shade, but also green grass proved effective in maintaining lower surface temperatures (−10 °C) even with respect to cool materials (−5 °C).

Water elements within the parks also proved fundamental in reducing surface temperatures, thanks to water flowing (Fig. 3). Bare, dry soil or dry grass are ineffective in reducing temperatures; on the contrary, they display high surface temperatures.

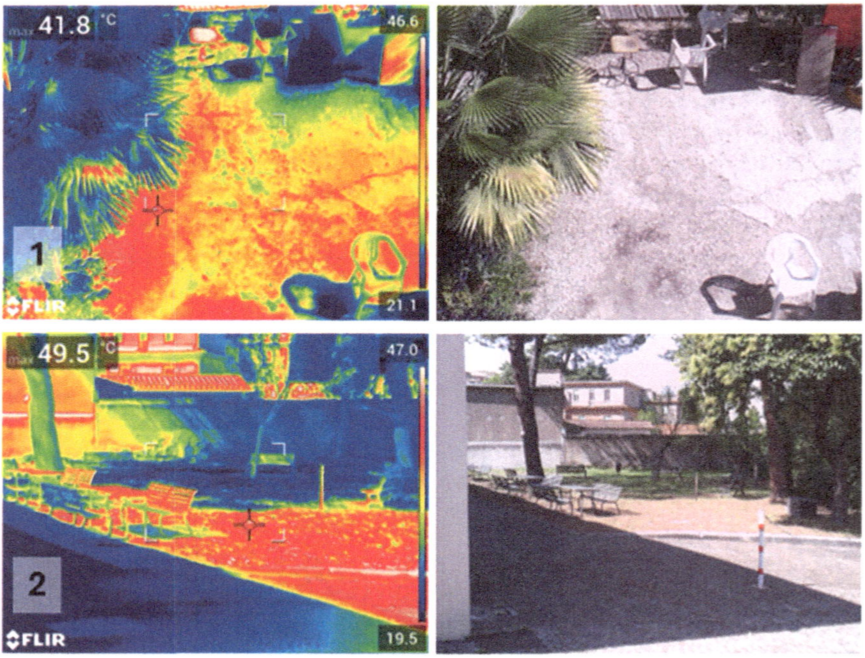

Fig. 2 Pocket green patches in the Engineering University Campus of Sapienza University of Rome. Giardino di Renato (1) and green area close to the Cafeteria (2). Thermographic image (left) and corresponding plain image (right)

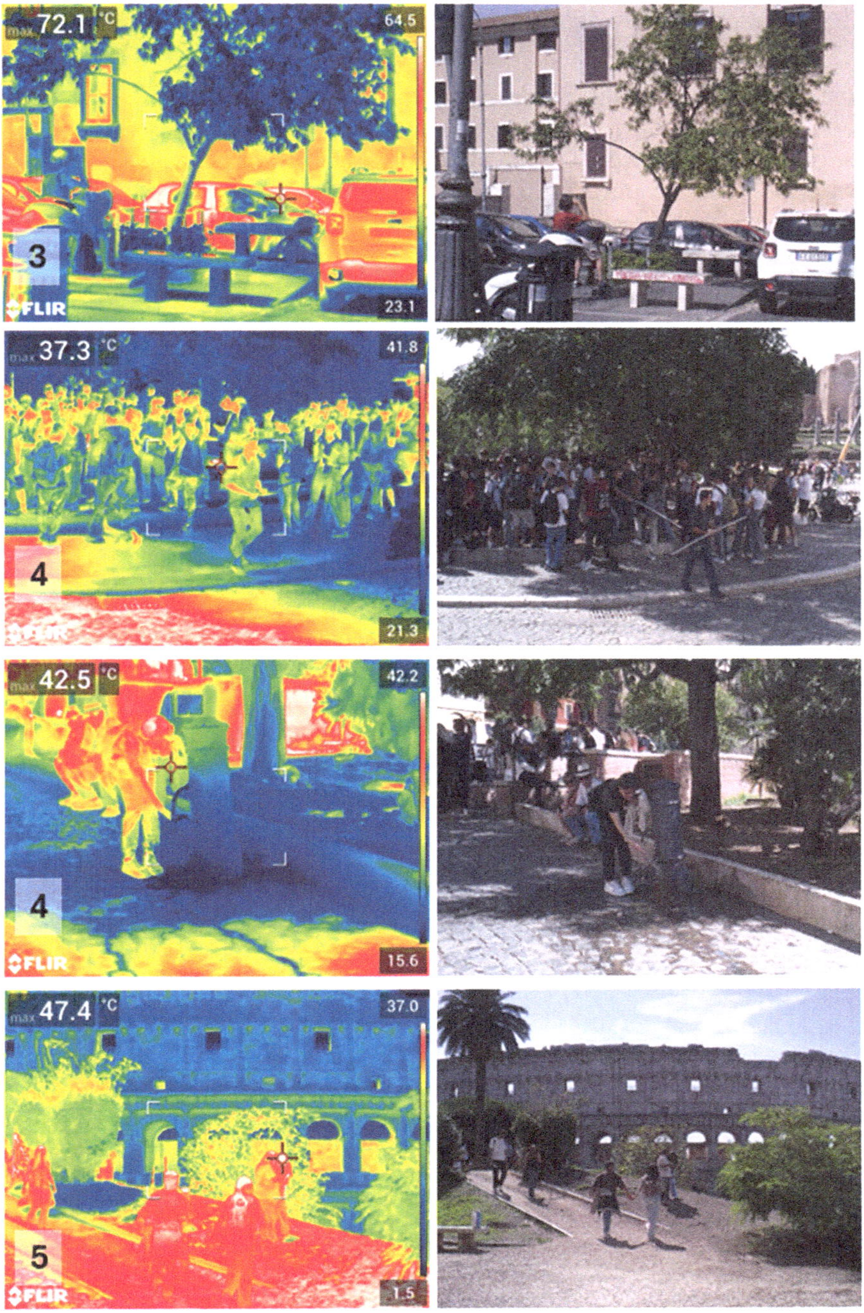

Fig. 3 Pocket green patches along the path: thermographic image (left) and corresponding plain image (right)

3.2 Microclimate Variables on the Path

Air temperature along the path varies from 24.5 °C up to 27.4 °C (Fig. 4). The warmest portion of the path is along Via delle Sette Sale and within the University Campus. Giardino di Renato is among the highest in temperature, while the green park close to the cafeteria is 1 °C cooler. In this case, the role of shading from trees or from surrounding built environment has a direct effect on the temperature (Sect. 3.3). This effect cannot be compared to the much lower effect of greenery or water or cool materials, which still prove effective in lowering surface temperatures (Sect. 3.1).

Wind speed is highly variable along the path (Fig. 5), depending also on the urban morphology and wind behaviour with respect to it. Wind speed varies from 0 to 18 km/h, and is especially high along via delle Sette Sale portion closer to the University Cafeteria, where the narrow street is flanked by 3 m-high walls; and in the engineering Campus in the narrow passage leading from Giardino di Renato to Largo della Polveriera. Wind speed effect on thermal comfort is considered in the Apparent temperature index, whose results are reported in the dedicated subsection (Sect. 3.3). Relative humidity along the path is reported in Fig. 6, higher in the shaded portions of the path and within the bigger small parks.

Relative Humidity varies from 38.7% to 43.9%. The lowest humidity is to be found in Via delle Sette Sale and within the University Campus, with the exception of the Giardino di Renato, which comprises greenery and maintains higher humidity.

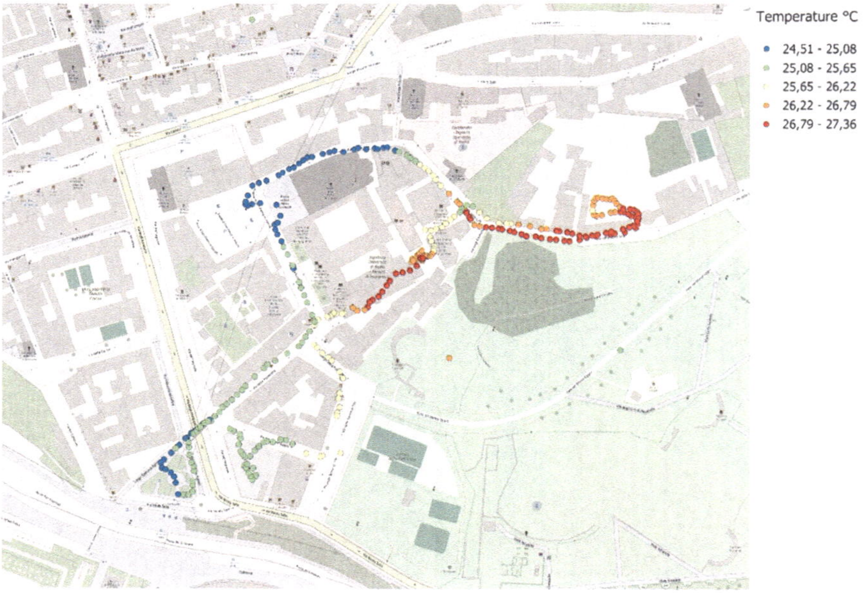

Fig. 4 Monitored air temperature [°C] on the path

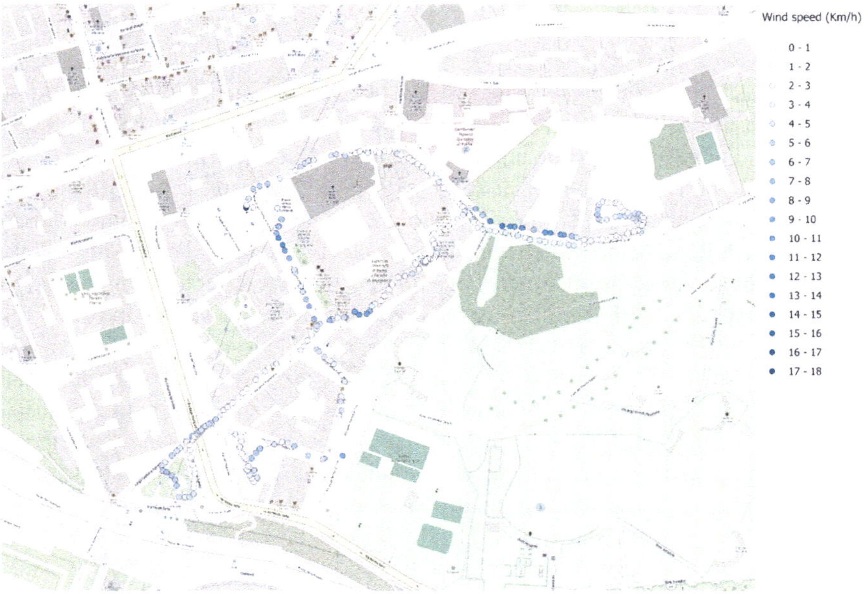

Fig. 5 Monitored wind speed [km/h] on the path

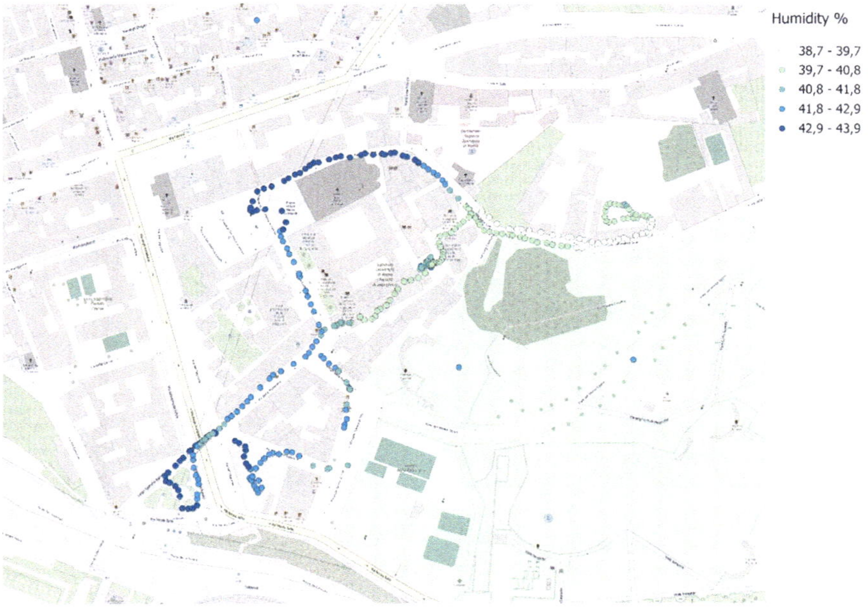

Fig. 6 Monitored relative humidity [%] on the path

3.3 Thermal Comfort

Humidex and apparent temperature, as described in the methodology section, are represented in Figs. 7 and 8, respectively. The latter includes the coupled effect of T_{air}, RH and wind speed, while humidex considers T_{air}, RH effects.

Humidex varies between 24 and 30, with peak heat stress in Via delle Sette Sale portion closer to the University Cafeteria. In the University Campus, closer to Giardino di Renato, monitored temperatures are higher and still humidity is lower.

Apparent temperature also accounts for the positive effect of wind speed in lowering heat stress, while still considering the effects of air temperature and relative humidity. The worst heat stress is again monitored close to the cafeteria in the University and close to Giardino di Renato. Still, the positive effect of the small park close to the cafeteria provides an oasis from thermal heat stress, proving effective in improving thermal comfort, as it can be also noticed in a reduced manner in Giardino di Renato with respect to the immediate surroundings. Apparent temperature is much more variable due to the wind speed effect, which is highly variable.

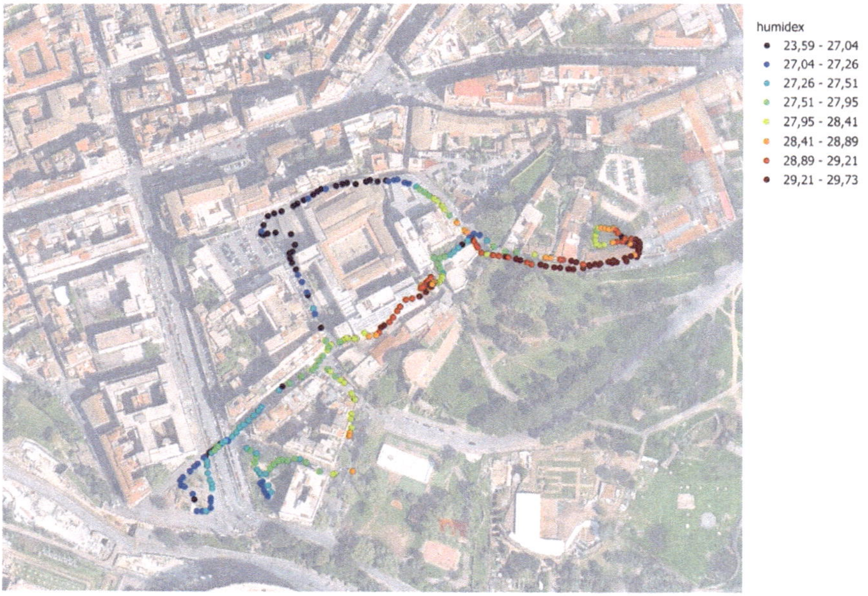

Fig. 7 Humidex along the path, calculated from the measured variables

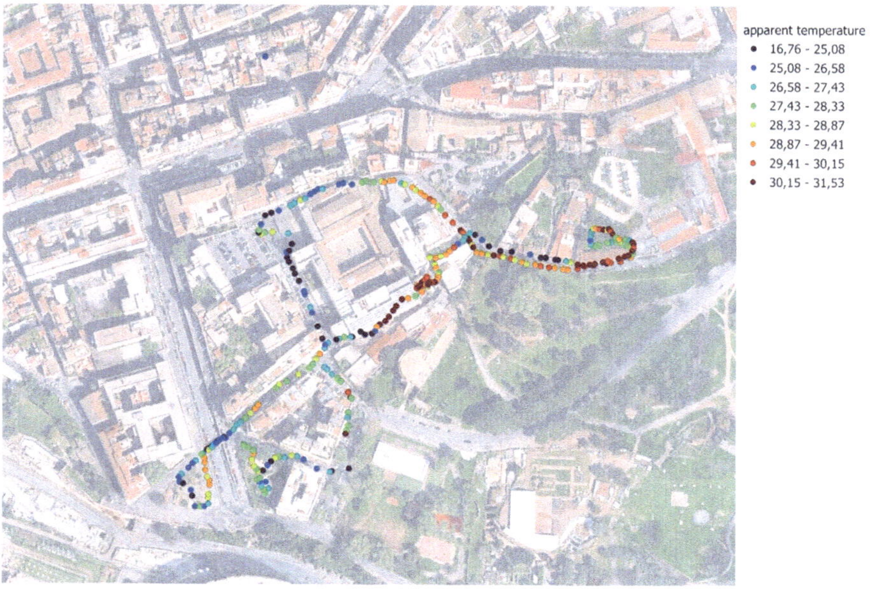

Fig. 8 Apparent temperature along the path, calculated from the measured variables

3.4 *The Role of Shade and Sun*

The effect of shade is significant in determining the microclimate variables (air and surface temperature and relative humidity). In order to visualize and check for its effect, the shade maps are reported in the figures below. The walk was conducted at 11:20–12:00 am, thus when the sun was high and shading was reduced (Fig. 9), mainly found below trees or in very narrow streets, such as in Via delle Sette Sale portion just out of Piazza San Pietro in Vincoli.

Also, the exposure to the sun of the urban built environment in the previous hours (from sunrise to 12:00, when the walk finished) has been computed and is reported in Fig. 10. From the figure, it is visible that the green patch close to the cafeteria is shaded by big trees, and the shade corresponds to 0 to 3 h under the sun, while other areas, such as Via delle Sette Sale portion closer to the Cafeteria or Piazza San Pietro in Vincoli, were under the sun for 5 h, thus leading to higher air temperatures and discomfort. It should be noted that the maps in the figures below are missing some of the smaller trees, which were instead considered in the GIS-derived file employed for the statistical analyses. While all the trees were inserted in the GIS-derived file employed for the statistical analyses, shade is computed just when evident, such as under the trees (also the smaller trees, which were added individually), as it was not constant along the path during the walk. Indeed the walk was conducted along 40 min, thus deeming it impossible to derive the exact shading in the precise moment where the monitoring was happening in the exact position. Still, as it was midday,

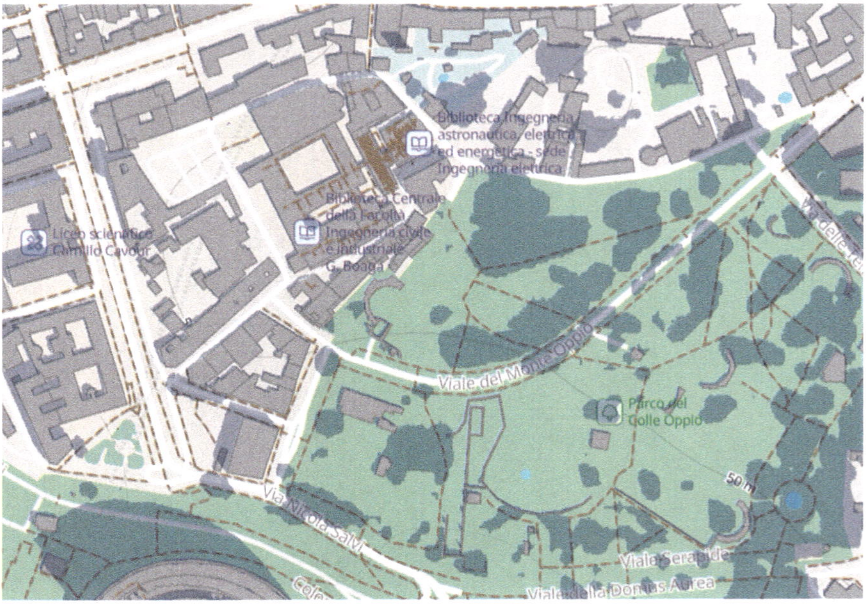

Fig. 9 Map of the shade at h 11:20 the 16 September 2024. Image elaborated from Shademap

almost no shade was present in addition to the one below trees, and thus trees can be considered a good approximation of shade in this case.

3.5 *Statistical Analyses*

Regression analyses were conducted, as described in the methodology, with air temperature, humidex and apparent temperature as dependent variables, to check the effect of the path characteristics (considered as independent variables). In greater details, independent variables (all Boolean) were: being or not in the park; the presence of tall trees and small trees; water elements presence; paving material (e.g., grass, cool or non-cool or bare soil, all Boolean); and wind speed for air temperature and humidex only. Paving material was eliminated because of multicollinearity with being or not in the park, and same with grass paving.

Results show (Table 2) that being in the park actually increase air temperature, humidex and also apparent temperature (with a coefficient equal to +1.10 °C, + 1.11 and +2.73, respectively). Tall trees and small trees instead both have a positive effect on air temperature, humidex and apparent temperature, decreasing them by −0.82 °C, −1.01 °C and −2.39 °C for tall trees and slightly less (−0.79 °C, − 0.63 °C and −0.91 °C) for small trees, which cast smaller shade, respectively. Water significantly influences tair and humidex, by decreasing them with a coefficient of

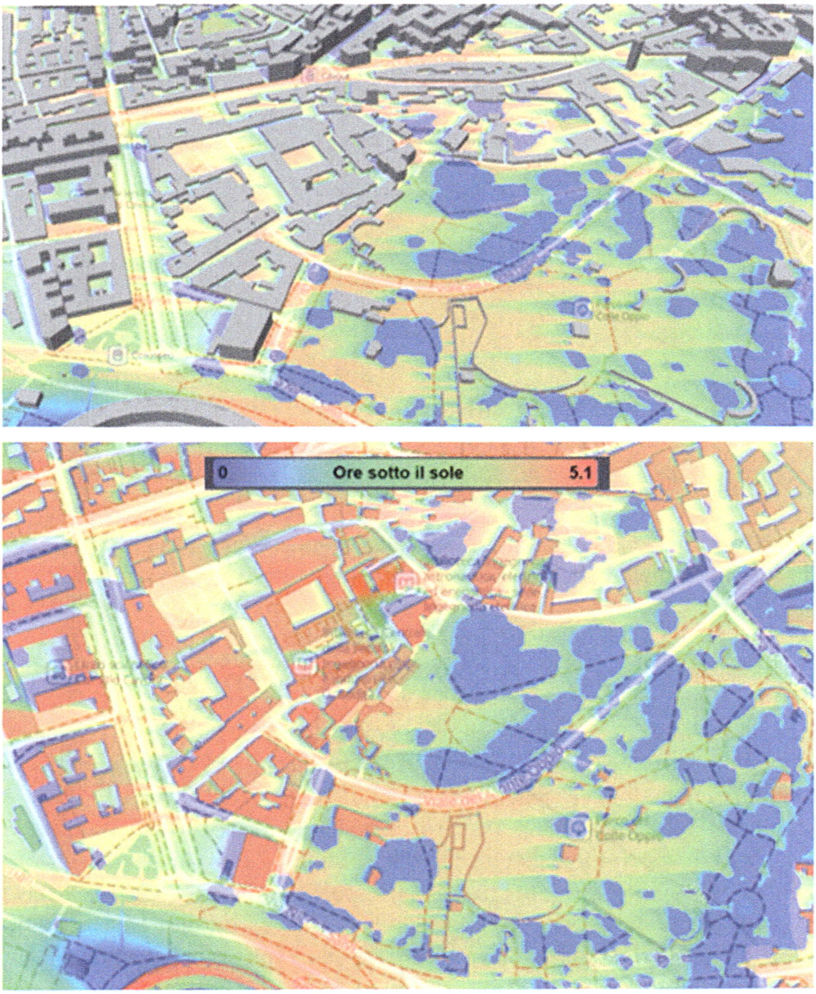

Fig. 10 Map of the shade at h 11:20 the 16 September 2024, elaborated from Shademap.3d view (up), plan view (down)

−1.21 °C and −1.39 °C. Wind speed slightly decreases tair and humidex (−0.03 °C and −0.04 °C coefficients, respectively), while it was not considered for apparent temperature, as it is already accounted for in apparent temperature calculations.

Table 2 Results of the regression analysis

Ind. variable	Tair [°C]	Humidex [°C]	At [°C]
Park	1.10	1.11	2.73
Tall trees	−0.82	−1.01	−2.39
Small tree	−0.79	−0.63	−0.91
Water	−1.21	−1.39	−1.47
Wind speed	−0.03	−0.04	–
constant	*26.05*	*28.20*	*27.78*
Mean VIF	2.41	2.41	2.7
Legenda	Non-significant	95% CI	90% CI

4 Discussions and Implications

The findings of this study confirm that small green patches can act as effective micro-scale nature-based solutions for mitigating heat stress in historic Mediterranean city centres, provided that they are consciously and carefully designed. Rather than their size alone, the architectural configuration and the selection of specific design elements emerge as the key drivers of their thermal performance, within the specific urban fabric in which they are inserted. This is particularly relevant in consolidated and heritage-sensitive urban contexts, where spatial, morphological, and regulatory constraints limit the implementation of larger green infrastructures.

Among the analysed elements, shading proves to be the most influential factor in reducing both surface and air temperatures. Shade generated by tree canopies or by the surrounding built environment consistently corresponds to cooler microclimate conditions, highlighting the importance of considering solar exposure over time in the design phase. It was verified that the presence of trees, either tall or small, positively influences thermal comfort. This suggests that pocket parks should not be conceived as isolated objects, but rather as components embedded within the urban morphology, strategically positioned to benefit from existing building shadows while enhancing shaded areas through vegetation.

Material choice also plays a critical architectural role. The results of thermal images clearly show that cool materials, such as light-coloured gravels, maintain significantly lower surface temperatures compared to traditional dark pavements commonly found in historic centres, such as asphalt or stone blocks (sanpietrini). While the preservation of historical character remains essential, these findings indicate that compatible cool materials can be selectively integrated without compromising the identity of the place, thus contributing to improved thermal performance at the pedestrian level. Further field work analyses should consider a wider range of small parks, to widen the sample of parks with different features and also urban fabric features. Indeed, this would allow to valorize the statistical analysis findings, by including larger samples for each investigated feature.

Vegetation demonstrates differentiated effects depending on its typology. Trees emerge as the most effective green element, due to their combined capacity to provide shade and evapotranspiration. Grass surfaces also show a strong cooling potential in the thermal images, achieving surface temperature reductions of up to 10 °C, outperforming even cool materials. Conversely, bare soil and dry grass are shown to be ineffective, or even detrimental with thermal images, thus with respect to reducing surface temperatures, under hot and dry summer conditions, emphasizing the need for adequate maintenance and irrigation strategies in Mediterranean climates. In the statistical analysis, the effect of grass is not significant in lowering thermal heat stress, but this could be due to collinearity issues with other variables. Trees proved effective also statistically in lowering heat stress, but it is not possible to distinguish the role of shade and evapotranspiration. Still, bare soil and non-irrigated grass proved ineffective in lowering heat stress.

Finally, water elements, particularly when flowing, significantly contribute to surface cooling and microclimate mitigation. Even small-scale fountains or water features prove effective, suggesting that water can be a valuable design component in pocket parks, if water use is carefully managed and integrated with sustainable practices. This is evident both with thermal images (thus demonstrating lower surface temperatures) and in statistical analysis. Water under the form of irrigation could potentially emphasize the positive effect of parks and permeable paving, as well as trees' evapotranspiration processes, thus exploiting the full potential of small green patches.

Overall, these results underline that pocket parks, when designed through an informed architectural approach integrating shade, vegetation, cool materials, and water elements, can provide meaningful thermal relief and enhance environmental comfort in historic city centres, while remaining compatible with their cultural and morphological constraints.

5 Conclusions

Small green patches demonstrated the capacity to provide small areas of respite from heat stress, even in very simple typologies, when consciously designed with respect to surrounding urban morphology, materials, greenery choice and shade.

Based on the above results, we contend that even in the historic city centre, it would be possible to include small PPs, in line with traditional materials and surrounding built environment, and also the view of surrounding historic heritage should be considered and emphasize to further tap on the beneficial effects of the integration of the small NBS within the historic context, highlighting not only the contact with nature but also the cultural identity of the location, and contributing to an enriching and diffuse comfort perception, both physical and psychological.

Beyond confirming the mitigation potential of pocket parks, the results offer operational implications for design practice, maintenance strategies and urban planning policies in heritage-sensitive contexts.

As shading emerges as the most decisive factor in reducing both surface and air temperatures, pocket park design in historic centres should prioritize strategic positioning of trees to maximize canopy shading during peak solar hours. At the same time, species with adequate crown density and height to ensure effective shadow projection at pedestrian level should be selected. Finally, existing building shadows should be integrated into the design logic, considering seasonal and hourly solar exposure. Pocket parks should be conceived not as isolated green objects, but as microclimate devices embedded within the urban fabric, designed in relation to surrounding street orientation, urban canyons, and façade geometry. A potential network of micro-scale green and blue interventions distributed along pedestrian paths should be considered, giving priority to locations with high pedestrian exposure. Rather than concentrating resources in large, centralized green areas, a diffused system of micro-interventions can enhance spatial equity in thermal comfort provision.

Also, material selection plays a fundamental role in achieving effective pocket parks. The study shows that cool materials significantly outperform traditional dark pavements (e.g., basalt stone) in reducing surface temperatures. In historic contexts, this does not imply replacing traditional materials indiscriminately, but rather compatible high-albedo materials could be employed in selected areas (e.g., seating zones and transition spaces). Traditional paving could be combined with permeable and lighter inserts.

Vegetation strategy should privilege trees as primary "regulators", due to their dual action of shading and evapotranspiration. In historic areas, they should be inserted where compatible, and with care not to block views on the historic surroundings. Grass surfaces show strong cooling potential in thermographic analysis, but only when adequately maintained and irrigated. Conversely, dry grass and bare soil can exacerbate overheating, highlighting that vegetation must be considered as a managed system rather than a purely aesthetic element.

Water, as irrigation and water features, such as flowing fountains, also prove effective in lowering surface temperatures and improving local comfort. In water-scarce Mediterranean climates, such elements should be designed within sustainable water cycles (e.g., recirculation systems, coupled integration of fountains with irrigation strategies). Microclimate performance of pocket parks is highly dependent on maintenance conditions. In particular, irrigation strategies are essential to preserve the cooling capacity of grass and support tree evapotranspiration during prolonged dry periods. Tree health monitoring is crucial to maintain canopy density and shading effectiveness over time. As such, maintenance should be considered an integral part of the design phase, with long-term management plans embedded into project development. Without adequate upkeep, the microclimate benefits of pocket parks risk being significantly reduced or even reversed.

In historic centres, visual and cultural relationship between pocket parks and surrounding heritage elements should be intentionally designed. Views towards monuments, and urban identity elements can enhance not only physical-thermal relief but also psychological and perceptual comfort, reinforcing the multi-dimensional value of these interventions.

References

Banzhaf, E., Kabisch, S., Knapp, S., Rink, D., Wolff, M., & Kindler, A. (2017). Integrated research on land-use changes in the face of urban transformations—An analytic framework for further studies. *Land Use Policy, 60*, 403–407. https://doi.org/10.1016/j.landusepol.2016.11.012

Costa, S., Peters, R., Martins, R., Postmes, L., Keizer, J. J., & Roebeling, P. (2021). Effectiveness of nature-based solutions on pluvial flood hazard mitigation: The case study of the city of eindhoven (the Netherlands). *Resources, 10*(3), 24. https://doi.org/10.3390/resources10030024

Falasca, S., Ciancio, V., Salata, F., Golasi, I., Rosso, F., & Curci, G. (2019). High albedo materials to counteract heat waves in cities: An assessment of meteorology, buildings energy needs and pedestrian thermal comfort. *Building and Environment, 163*(May), Article 106242. https://doi.org/10.1016/j.buildenv.2019.106242

Founda, D., & Santamouris, M. (2017). Synergies between Urban Heat Island and Heat Waves in Athens (Greece), during an extremely hot summer (2012). *Scientific Reports, 7*(1), 1–11. https://doi.org/10.1038/s41598-017-11407-6

Mannucci, S., Ciardiello, A., Ferrero, M., & Rosso, F. (2025). A parametric integrated workflow to assess multi-domain heat- and flood-related performance of small outdoor urban spaces in a changing climate: A case study in the mediterranean region. *City and Environment Interactions, 27*(May), Article 100215. https://doi.org/10.1016/j.cacint.2025.100215

Rosso, F. (2025). *Resilienza dell'ambiente costruito alle sfide climatiche. Il ruolo dei piccoli spazi outdoor diffusi tra gli edifici*. McGraw-Hill education.

Rosso, F., Pioppi, B., & Pisello, A. L. (2022). Pocket parks for human-centered urban climate change resilience: Microclimate field tests and multi-domain comfort analysis through portable sensing techniques and citizens' science. *Energy and Buildings, 260*, Article 111918. https://doi.org/10.1016/j.enbuild.2022.111918

Rosso, F., Pioppi, B., & Pisello, A. L. (2024). Tactical urban pocket parks (TUPPs) for subjective and objective multi-domain comfort enhancement. *Journal of Environmental Management, 349*, Article 119447. https://doi.org/10.1016/j.jenvman.2023.119447

Santamouris, M., & Yun, G. Y. (2020). Recent development and research priorities on cool and super cool materials to mitigate urban heat island. *Renewable Energy, 161*, 792–807. https://doi.org/10.1016/j.renene.2020.07.109

Todeschi, V., Pappalardo, S. E., Zanetti, C., Peroni, F., & De Marchi, M. (2022). Climate justice in the city: Mapping heat-related risk for climate change mitigation of the urban and peri-urban area of Padua (Italy). *ISPRS International Journal of Geo-Information, 11*(9), 490. https://doi.org/10.3390/ijgi11090490

Green Interventions at the Building Scale

Nature-Based Solutions for Urban Resilience: Analyzing the Role of Green Facades and Green Roofs on Microclimate at Urban, Neighborhood, and Building Scales

Fataneh Shoghi and Seyed Morteza Hosseini

Abstract Rapid urbanization and climate change have intensified urban challenges such as Urban Heat Island effects, declining air quality, and biodiversity loss, threatening urban resilience. Nature-Based Solutions (NbSs), particularly green roofs and façades, are recognized as multifunctional strategies that improve thermal performance, environmental quality, and ecosystem services. This study proposes an integrative multi-scale framework to evaluate NbSs across building, neighborhood, and urban scales, based on a PRISMA systematic review of 127 studies (2015–2025) and in-depth analysis of 37 key papers. At the building scale, interventions reduce indoor temperatures, stabilize daily thermal fluctuations, lower cooling energy demand, and enhance indoor air quality and acoustic performance. At the neighborhood scale, green roofs and façades moderate microclimate, improve pedestrian thermal comfort, reduce pollutants, and enhance urban biodiversity, with effects influenced by seasonal variation and urban morphology. At the urban scale, widespread NbS implementation supports UHI mitigation, stormwater management, carbon sequestration, and ecological connectivity. Overall, the framework demonstrates how green roofs and façades generate scale-dependent, cumulative benefits, linking building-level interventions to neighborhood and urban outcomes, and highlights their role in climate adaptation, energy efficiency, and sustainable urban resilience.

Keywords Urban resilience · Green facades · Green roofs · Urban heat island · Nature-Based Solutions

---

F. Shoghi · S. M. Hosseini (✉)
Department of Architecture and Energy, Faculty of Architecture and Urbanism, University of Tehran, Tehran, Iran
e-mail: smho@create.aau.dk; Morteza.hosseini1366@gmail.com

S. M. Hosseini
Department of Architecture, Design & Media Technology, Aalborg University, Copenhagen, Denmark

L. Diana et al. (eds.), *Greening and Resilience*, Urban Sustainability,
https://doi.org/10.1007/978-981-92-0419-9_22

1 Introduction

Rapid urbanization has intensified pressures on urban infrastructure, generating major environmental and social challenges. More than half of the global population now lives in cities, which dominate global energy consumption, carbon emissions, and resource use (Goharian et al., 2025). Urban densification disrupts natural energy balance, elevating temperatures and intensifying the Urban Heat Island (UHI) effect, thereby impacting human health, increasing cooling demand, and contributing to greenhouse gas emissions (Shoghi et al., 2025). Climate change further exacerbates these stresses by altering temperature, air quality, precipitation patterns, and habitats, while increasing heatwave frequency and risks for urban populations (Tseliou et al., 2023). Addressing these challenges requires adaptive strategies at building and neighborhood scales, where microclimatic impacts are most directly experienced. At larger scales, urban resilience—the capacity of socio-economic and environmental systems to maintain function under stress—can be strengthened through strategic design, planning, and material choices that support both mitigation and adaptation (Cirrincione et al., 2021; Couret et al., 2023). Enhancing Urban Green Infrastructure (UGI) is a key approach to improving climate resilience. Among UGI elements, Green Roofs and Green Facades play essential roles across scales by regulating microclimate, lowering surface and ambient temperatures, mitigating UHI, and reducing building energy demand. These systems additionally support stormwater management, carbon sequestration, biodiversity enhancement, and provide aesthetic and psychological benefits (Bakhtyari et al., 2024; Berardi, 2016). However, many dense cities still lack adequate coverage, limiting their capacity to buffer environmental stresses and highlighting the need for integrated planning (Lin et al., 2019; Muharam et al., 2015). Nature-Based Solutions (NbSs), including Green Roofs and Green Facades, offer structured pathways to leverage ecosystem services for climate adaptation, sustainable urbanization, and human well-being. Recent studies confirm their multiple co-benefits: temperature reduction, improved air quality, enhanced biodiversity, and socio-cultural gains such as recreation and mental health improvements (Sommese, 2024). Multi-scale methodologies further demonstrate their capacity to deliver ecological, social, and economic value (Diana et al., 2024; Frantzeskaki & McPhearson, 2022; McEvoy et al., 2024). Community-driven implementations in high-density contexts also reveal their potential for inclusive and scalable adaptation. Green Roofs and Green Facades mitigate UHI through shading and evapotranspiration, reduce energy demand, sequester carbon, and contribute to stormwater regulation (Jayasooriya et al., 2025; Staccione et al., 2024). Yet most applications remain fragmented rather than networked across scales. Establishing interconnected systems can maximize ecosystem services, improve public health, and strengthen urban resilience. Policy frameworks such as the 2030 Biodiversity Strategy underline the importance of coordinated planning for effective deployment of these technologies.

A review of the literature shows a lack of comprehensive frameworks addressing multi-scale benefits of Green Roofs and Green Facades. To address this gap, the present study develops a multi-scale classification framework applicable at building, neighborhood, and urban levels. The framework emphasizes social, environmental, and climatic benefits essential for resilience and citizen well-being under climate and health-related stressors. It serves as a practical tool for researchers and policymakers to guide the selection and implementation of appropriate green roof and facade solutions in urban planning.

This study focuses on three research questions:

1. To what extent can Green Roofs and Green Facades enhance resilience and environmental performance at the urban scale under current and future climate scenarios?
2. How do they influence microclimate regulation and thermal comfort in dense neighborhoods?
3. What configurations best optimize sustainability at the building scale?

2 Methods

2.1 Study Approach

This study employs a mixed quantitative–qualitative approach, combining a PRISMA-based systematic review with bibliometric analysis using VOSviewer to ensure methodological rigor and transparency (Moher et al., 2015).The focus is on GRs and GFs as NbSs and their contributions to urban resilience and microclimate regulation across building, neighborhood, and urban scale (Fig. 1).

Identification and Literature Search

Relevant studies were systematically retrieved from Scopus. A TITLE-ABS-KEY search was conducted using: ("nature-based solutions" OR "NBS" OR "green infrastructure") AND ("green roof*" OR "green façade*" OR "living wall*" OR "vertical greenery") AND ("urban heat island" OR "microclimate" OR "thermal comfort") AND ("urban resilience" OR "climate adaptation" OR "urban sustainability"). This search yielded 1,120 records. After removing 370 duplicates, 750 unique articles remained. Screening titles, abstracts, and keywords reduced this to 360 full-text articles. Inclusion criteria required studies to address GRs or GFs, consider microclimate or resilience at multiple scales, and provide peer-reviewed evidence in English.

Eligibility Assessment

Following screening, 240 articles were assessed for eligibility, with 113 excluded due to limited relevance or insufficient data. 127 studies were retained for synthesis.

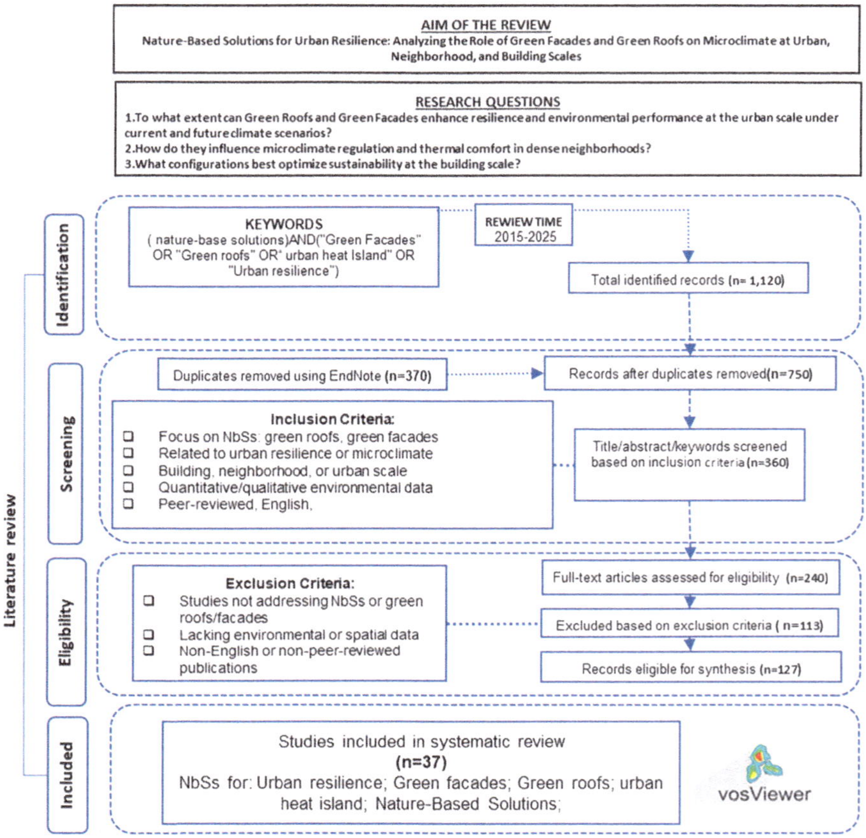

Fig. 1 PRISMA flow diagram illustrating the identification, screening, eligibility, and inclusion of studies

Bibliometric Mapping and Qualitative Synthesis

All 127 retained studies were analyzed bibliometrically, and from these, 37 studies were selected for in-depth qualitative analysis. Insights from this process informed the development of a multi-scale, evidence-based framework linking NbS interventions across building, neighborhood, and urban scales.

2.2 *Data Analysis*

Keywords Analysis and Multi-Scale Framework

A keyword co-occurrence analysis of the 127 PRISMA-reviewed studies was conducted using VOSviewer (Fig. 2). In the map, node size indicates keyword frequency, link thickness reflects co-occurrence strength, and colors denote spatial scales: yellow = building, pink = neighborhood, blue = urban. Keywords were assigned to scales based on the studies in which they appeared. Three core themes—energy efficiency, thermal comfort, and climate change—structure the literature across building-, neighborhood-, and urban-scale research. Energy efficiency is linked to building-scale studies, thermal comfort to neighborhood-scale, and climate change to urban-scale research. Despite extensive work, these themes are often analyzed in isolation, highlighting a gap in understanding the cascading effects of GRs and GFs across scales. Insights from all 127 studies were synthesized, with 37 selected for in-depth qualitative analysis, informing a multi-scale framework that explicitly connects interventions across scales and supports urban design, policy-making, and performance assessment.

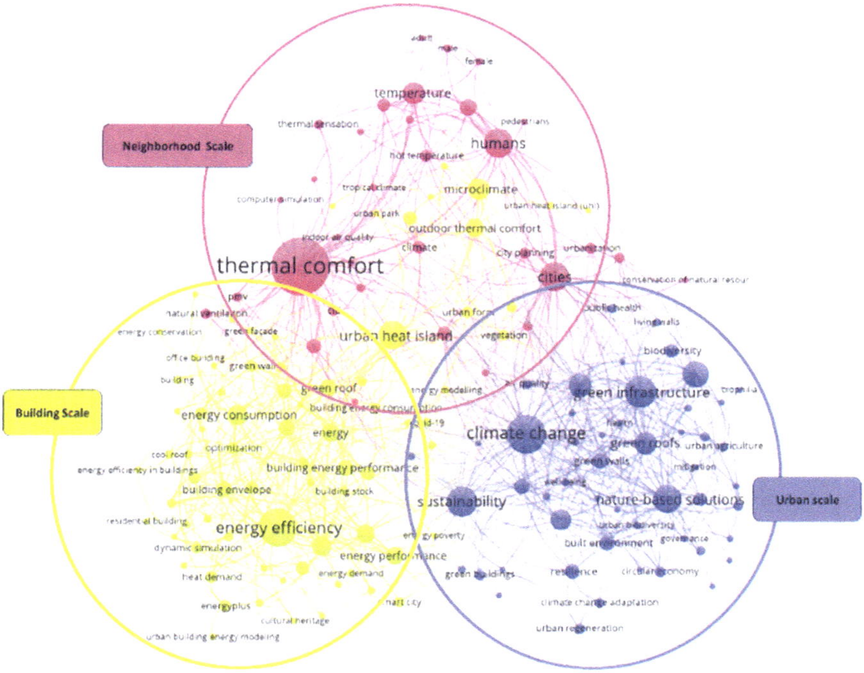

Fig. 2 Keyword co-occurrence map of 127 PRISMA-reviewed studies. Node size indicates keyword frequency, link thickness shows co-occurrence strength, and colors represent spatial scales: yellow = building, pink = neighborhood, and blue = urban. (VOSviewer)

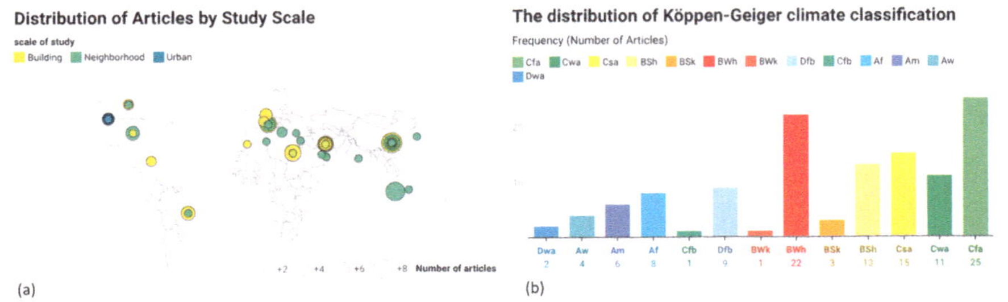

Fig. 3 Global patterns of UGI studies. (a) Distribution of study scales (Building, Neighborhood, Urban) across latitudes. (b) Corresponding Köppen–Geiger climate zones of the studied locations (Datawrapper)

Geographical and Climatic Patterns and Spatial Scales in UGI Research

Global UGI research shows a clear relationship between latitude, Köppen–Geiger climate zones, and the spatial scale of analysis, particularly for GRs and VGS. Figure 3 illustrates this: (a) the distribution of studies across Building, Neighborhood, and Urban scales, and (b) the corresponding Köppen–Geiger climate zones. Mid-latitude regions ($\approx$30–45° N) with temperate and semi-temperate climates (Cfa, Cwa, and Csa) account for the largest share of studies and exhibit multi-scale coverage: Cfa (25), Cwa (11), and Csa (15). Seasonal variability in these regions supports interventions addressing both localized thermal regulation and broader urban heat mitigation. In tropical and equatorial regions ($\approx$0–10° N), stable climates and limited seasonal variation focus research on Building- and Neighborhood-scale interventions, enhancing shading, evapotranspiration, and pedestrian comfort, while Urban-scale studies are scarce. At higher latitudes (>45° N) and in arid/semi-arid zones, environmental constraints favor Building- and Neighborhood-scale research. Overall, the dataset from over 30 countries reveals a consistent pattern: mid-latitudes support multi-scale studies, tropical regions emphasize micro-scale interventions, and high-latitude or arid regions prioritize Building-scale research.

Distribution of Green Strategies Across Regions

The selection of green strategies—GR, VGS, or combined approaches—shows clear climatic and spatial patterns. Mid-latitude temperate and Mediterranean regions commonly apply combined strategies across multiple scales. Arid regions mainly adopt Building-scale green roofs due to extreme heat and water scarcity. In tropical and equatorial regions, VGS or combined strategies dominate, enhancing shading and evapotranspiration at micro-scales. Figure 4 links strategy choice to climate, city location, and intervention scale, highlighting the importance of context-responsive UGI design.

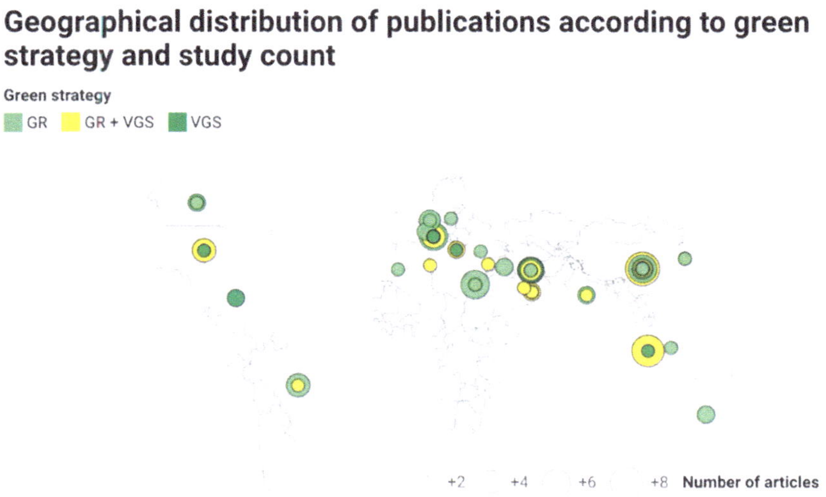

Fig. 4 Geographical distribution of UGI publications by green strategy with study counts, showing regional patterns in interventions (Datawrapper)

3 Thematic Reviews

This section analyzes NbSs integrated into the building envelope, focusing on GRs and GFs as multifunctional systems mediating interactions between indoor and outdoor environments. The building envelope regulates energy, air, and moisture exchanges, contributing to improved microclimate conditions and enhanced indoor and outdoor comfort (Goharian et al., 2025; Shoghi et al., 2025). Based on a systematic review of the literature, NbSs interventions in the form of GRs and GFs can be classified and evaluated across three spatial scales—building, neighborhood, and urban—depending on their functional reach, cumulative effects, and governance context. This classification, encompassing typology, technology, and topology of GRs and GFs across spatial scales, is illustrated in Fig. 7.

3.1 Roofs: Green, Blue, and Blue–Green Roofs

Green roofs are vegetated roof systems that provide multiple environmental, social, and economic benefits, including urban heat island mitigation, improved thermal insulation, stormwater runoff reduction, air pollutant absorption, and support for urban biodiversity. They are distinguished from blue roofs by the presence of vegetation and a growing substrate. Based on substrate depth, green roofs are categorized as extensive (60–150 mm), semi-intensive (150–250 mm), and intensive systems

(>250 mm). Extensive roofs are lightweight and low-maintenance, while semi-intensive and intensive roofs support deeper rooting zones, higher plant diversity, and greater structural and maintenance requirements (Cardinali et al., 2023; Manso et al., 2021). Technologically, green roofs range from simple no-tech systems to low-tech configurations with passive drainage, up to high-tech blue–green roofs integrating vegetation with engineered water storage and controlled release. Blue roofs are non-vegetated systems designed to temporarily store rainfall for stormwater management (Mihalakakou et al., 2023; Sommese, 2024). Figure 5 illustrates these typological and technological categories, highlighting the differences in vegetation, substrate depth, and water-management layers. The performance of these systems varies accordingly: shallow extensive roofs primarily influence building-scale thermal behavior, whereas deeper or technologically advanced systems support cumulative neighborhood- and urban-scale effects, including microclimate regulation, runoff control, and ecological connectivity (Cardinali et al., 2023; Sommese, 2024). Understanding these distinctions helps align roof selection with specific environmental and functional goals.

3.2 *Green Walls: Facades and Living Walls*

Green walls, classified typologically as green façades and living walls (Fig. 6a, b, c, d, e, and f), are vegetated vertical systems that enhance both environmental quality and occupant comfort. They help reduce surface temperatures, provide shading, enhance evapotranspiration, improve thermal insulation, and mitigate the urban heat island effect while reducing indoor cooling demands. From a topological standpoint, green walls can be extensive systems (simple structures with limited functional layers) or intensive systems (multi-layered, complex configurations with thicker substrates) (Mayrand & Clergeau, 2018). In terms of technology, green façades typically employ planter boxes or climbing plants attached directly or indirectly to the wall using structural supports such as wires, trellises, or frames, representing low-tech or passive systems. Direct façades have plants rooted in soil or planter boxes and attached straight to the wall, while indirect façades use support structures that create an air gap, enhancing ventilation and protecting the façade surface (de Oliveira Santos et al., 2024). Conversely, living walls are intensive and high-tech systems that consist of modular or continuous panels containing growth substrates and integrated irrigation, either mounted directly on the wall or supported by an independent frame (Tseliou et al., 2023). Their complex structure provides superior insulation, effective water management, and improved ecological performance, contributing significantly to the multifunctionality of building envelopes (Pragati et al., 2023) (Fig. 7).

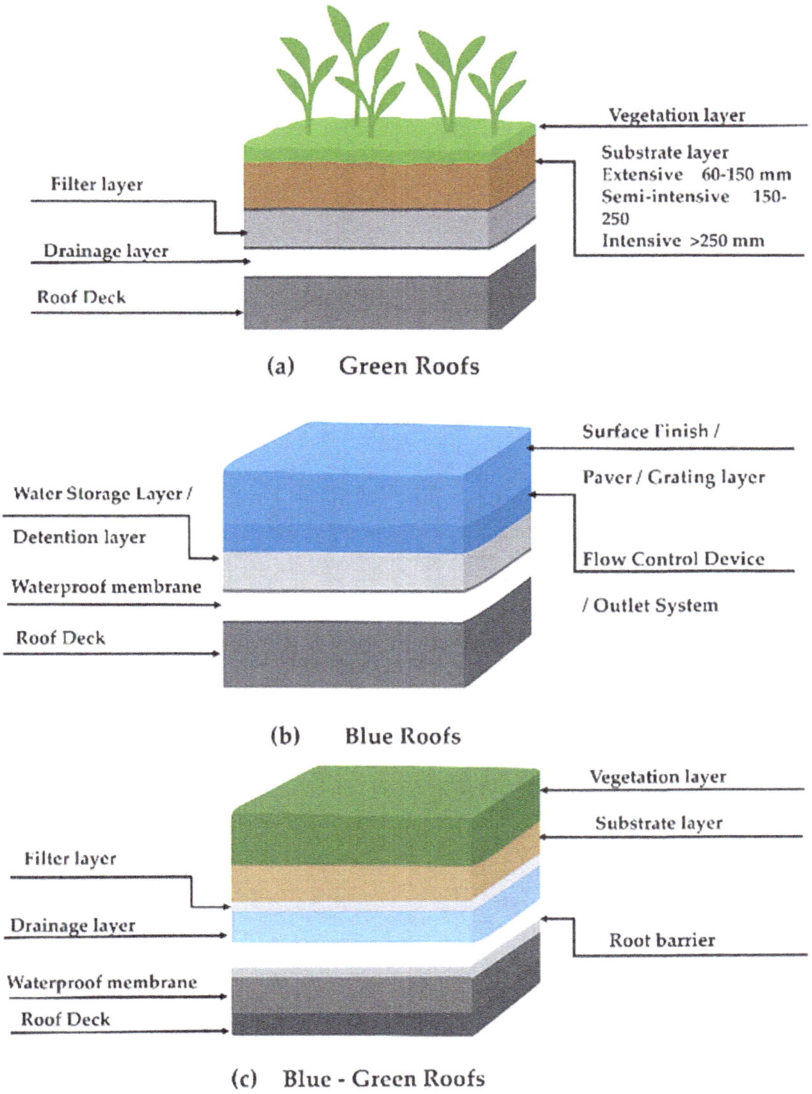

Fig. 5 Layer composition of roof types: (a) GR with vegetation, substrate, drainage, and waterproofing layers; (b) BR showing water storage, flow control, and waterproofing layers; (c) BGR integrating vegetation, substrate, water retention, and controlled drainage layers

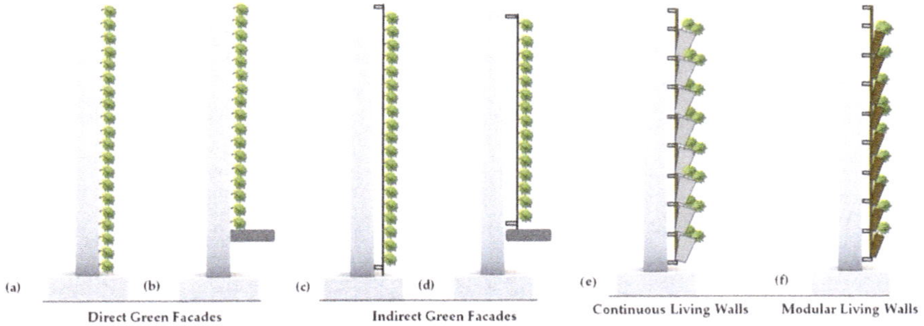

Fig. 6 Types of green façades and living walls: (a) direct green façade with ground-rooted vegetation, (b) direct façade with planter-box vegetation, (c) indirect façade with ground-rooted plants, (d) indirect façade with planter-box plants, (e) continuous living wall, and (f) modular living wall

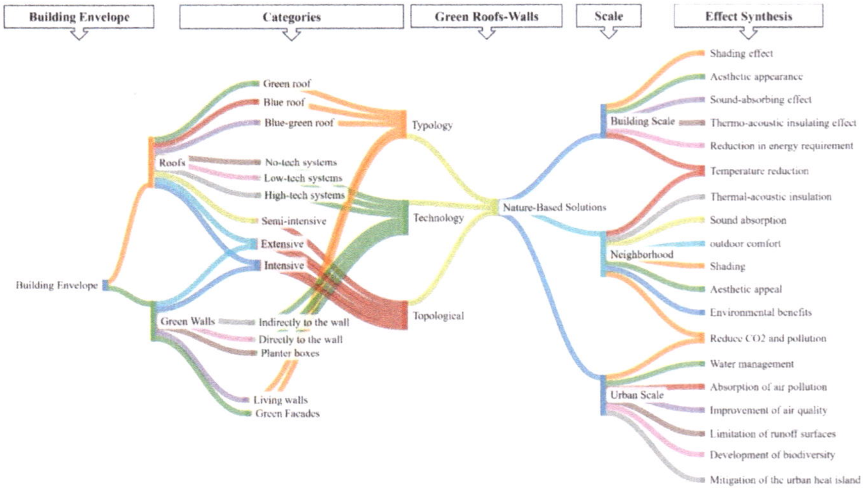

Fig. 7 Classification of NbSs in the building envelope, including GRs and GFs categorized by typology, technology, and topology across spatial scales

3.3 *Green Wall and Roof on the Building Scale*

Green roofs and walls enhance building-scale energy performance by reducing heat transfer through shading, evapotranspiration, and substrate insulation, with effectiveness influenced by plant density, orientation, and substrate depth (Jayasooriya et al., 2025). In warm climates, east-facing façades without vegetation show the highest solar gain, while green walls can reduce wall surface temperatures by up to 4 °C. In Lisbon, uninsulated extensive green roofs decreased cooling loads by 34.9%, and semi-intensive and intensive systems achieved 45–70% reductions depending on vegetation and substrate characteristics (Berardi, 2016; Silva et al., 2016). Modular green walls lowered façade temperatures by 15 °C and cooling loads by 64.4% in La Rochelle (Ouldboukhitine et al., 2025), while living walls in Shiraz reduced indoor temperatures by 2.25 °C and generated 15% energy savings (Bakhtyari et al., 2024). In hot–arid regions, extensive roofs lowered indoor temperatures by 2–5.5 °C and ceiling temperatures by up to 7 °C, cutting cooling energy by 12–33% (Muharam et al., 2015). In sub-arctic climates, where evapotranspiration is limited, green roofs mainly stabilize surface temperatures and improve material durability, with conventional black roofs reaching up to 10 °C higher summer temperatures (Schade et al., 2021). Beyond thermal benefits, vegetated façades improve acoustic comfort and air quality. Vegetation absorbs and scatters sound waves, and green wall surfaces perform similarly to porous acoustic materials (Romanova et al., 2019). Acoustic effectiveness depends on foliage density, substrate composition, and wall configuration. Vegetation also contributes to pollutant absorption and regulation of humidity and temperature. Continuous full-coverage façades tend to outperform partial systems in reducing ambient air temperature (Ta) and cooling energy demand. Performance varies across climates: in colder regions, energy savings are smaller, and roofs serve mainly protective and ecological roles (Schade et al., 2021). In Mediterranean climates, water scarcity and high radiation require drought-tolerant species, efficient irrigation, and proper maintenance (Leotta et al., 2023). Despite their benefits, building-scale NbSs face challenges such as structural load limits, maintenance needs, seasonal variability, and water-use constraints (Lin et al., 2019). Measurement uncertainties and simplified modeling also complicate assessment. Nonetheless, well-designed systems deliver combined thermal, acoustic, and ecological gains, enhancing building resilience and occupant comfort.

3.4 *Green Wall and Roof on the Neighborhood Scale*

At the neighborhood scale, green roofs and walls enhance environmental quality, moderate local microclimates, and improve the livability of semi-enclosed residential spaces (Alim et al., 2022; Shoghi et al., 2025). Studies show their capacity to reduce air pollutants and noise while improving comfort and energy efficiency. In New Belgrade, Serbia, green roofs decreased PM_{10} by 7%, $PM_{2.5}$ by 16.6%, and PM_1 by 17.6% in winter, emphasizing their role in local air-quality improvement (Kostadinović et al., 2023). They also support stormwater retention and reduce life-cycle environmental impacts, offering passive and cost-effective strategies for localized climate regulation. Vegetation distributed across courtyards, rooftops, and streets regulates solar radiation, airflow, and humidity, creating cooler neighborhood environments. Green façades can reduce physiological equivalent temperature (PET) by around 2.5 °C, especially in extensive or full-coverage systems (Lin et al., 2019). Combined green roofs and façades lower local air temperatures by approximately 0.7 °C and UTCI by 1.6 °C in Mediterranean contexts, while improving shading and airflow between buildings (Dehghan Lotfabad et al., 2025; Tseliou et al., 2023). In colder climates, although evaporative cooling is limited, these systems still stabilize surface temperatures and improve comfort in sheltered communal areas. When applied collectively, neighborhood-scale green infrastructure contributes to shared energy savings and greater microclimatic stability. Large-scale green roofs in Bodrum, Turkey, significantly reduced energy use across multiple buildings (Celik & Ogus Binatli, 2018). de Oliveira Santos et al. (2024) higlight that neighborhood-level coordination strengthens environmental and social resilience by improving equitable access to green amenities (de Oliveira Santos et al., 2024). Ecologically, green roofs and walls act as habitats and stepping stones that support urban biodiversity. When distributed across neighborhoods, they help form ecological corridors that enable species movement and pollination. However, limited vegetated area and spatial discontinuity often restrict their ecological effectiveness. Mayrand and Clergeau (2018) notes that isolated installations reduce connectivity, suggesting that future designs prioritize continuity between buildings to improve stormwater regulation, carbon sequestration, and habitat provision (Mayrand & Clergeau, 2018). Despite their multifunctional value, neighborhood-scale systems face technical, financial, and social barriers. Structural constraints, limited retrofitting potential, and high installation and maintenance costs hinder widespread application. Dense urban layouts introduce space limitations and conflicts with existing infrastructure. Social acceptance also plays a critical role, as public perception and engagement influence willingness to invest and maintain such systems. Addressing these challenges requires community participation, efficient water management, and coordinated long-term maintenance to ensure equitable and durable benefits (Okour & Shaweesh, 2024).

3.5 *Green Walls and Roofs at the Urban Scale*

At the urban scale, green roofs and walls function as integral NbSs enhancing environmental performance, resilience, and sustainability in dense cities. By covering extensive roof and façade surfaces, they collectively reduce air and surface temperatures, mitigate UHI, filter pollutants, and buffer urban noise (Calheiros & Stefanakis, 2021). As vegetated layers of the built fabric, they restore ecological balance, improve thermal comfort, and strengthen adaptive capacity to climate change. Empirical studies highlight measurable effects: green walls can lower bare wall surface temperatures by up to 31 °C and reduce $PM_{2.5}$ by 20–99%, PM_{10} by 23–60%, and NO_2 by 15–64% (Cardinali et al., 2023; Mihalakakou et al., 2023). Urban-scale green roofs may decrease indoor air temperatures by 15 °C and reduce cooling loads by up to 70%, depending on vegetation density and substrate depth (Bakhtyari et al., 2024; Pragati et al., 2023). Widespread implementation lowers ambient temperatures through evapotranspiration, enhancing thermal regulation and outdoor comfort. Vegetated façades decreased UTCI by 3.68 °C in adjacent areas, while green roofs in Tehran reduced average air temperature by 0.67 °C (Ganjirad et al., 2025; Kozak & Lipecki, 2025). Coordinated deployment can lower urban surface temperatures by ~ 2 °C and improve outdoor comfort by over 10 °C (Cuce et al., 2025; Pragati et al., 2023). From an ecological standpoint, these systems also enhance biodiversity and ecological connectivity. Distributed across multiple rooftops and façades, they form urban green networks that act as ecological corridors for birds, insects, and pollinators. This interconnectedness promotes habitat continuity, supports essential ecosystem services such as carbon sequestration, stormwater retention, and temperature moderation, and contributes to long-term urban resilience (Wang & Xu, 2024). Integration with urban water management frameworks amplifies their multifunctionality. As components of blue–green infrastructure, green roofs and façades retain rainfall, delay runoff, and improve water quality, aligning with Sponge City and Sustainable Drainage Systems concepts. Blue–green roofs that combine vegetation with water-storage systems replicate natural hydrological processes and strengthen urban adaptation to climate extremes (Sommese, 2024). Beyond environmental performance, these systems offer aesthetic and social value by fostering community participation, raising environmental awareness, and contributing to more inclusive and livable urban environments (Kostadinović et al., 2023). Despite these wide-ranging benefits, large-scale implementation faces persistent economic, technical, and governance challenges. High initial investment costs, maintenance requirements, and structural constraints often limit feasibility, particularly in older building stocks. Moreover, fragmented policy frameworks and uneven access to financial or technical resources impede equitable diffusion (Celik & Ogus Binatli, 2018; Tseliou et al., 2023). Effective upscaling therefore requires coordinated policy support, cross-sectoral collaboration, and community engagement to ensure inclusive, climate-resilient, and sustainable outcomes for all urban residents.

4 Discussion—Multi-Scale Impacts of Green Roofs and Facades

Based on the systematic review, building envelope greening—including green roofs and façades—yields multi-scale benefits influenced by climate, neighborhood context, and latitude. Temperate regions favor multi-scale interventions at building, neighborhood, and urban levels due to seasonal variability. Arid and semi-arid regions require micro-scale strategies, particularly green roofs, with efficient irrigation. In tropical and equatorial regions, neighborhood-scale interventions emphasizing shading and evaporative cooling are most effective. Context-specific planning is therefore essential to maximize resilience and microclimate regulation. Figure 8 illustrates a conceptual framework linking the typology, topology, and technology of NbSs to their multi-scale effects, showing a clear trajectory from individual buildings to urban systems.

Building Scale: At this scale, green roofs and façades reduce indoor and surface temperatures, stabilize daily thermal fluctuations, and lower cooling energy demand. They provide shading, improve acoustic comfort, and enhance indoor environmental quality. Intensive and semi-intensive green roofs, along with living walls, show the highest potential for energy savings and occupant well-being. The effectiveness of

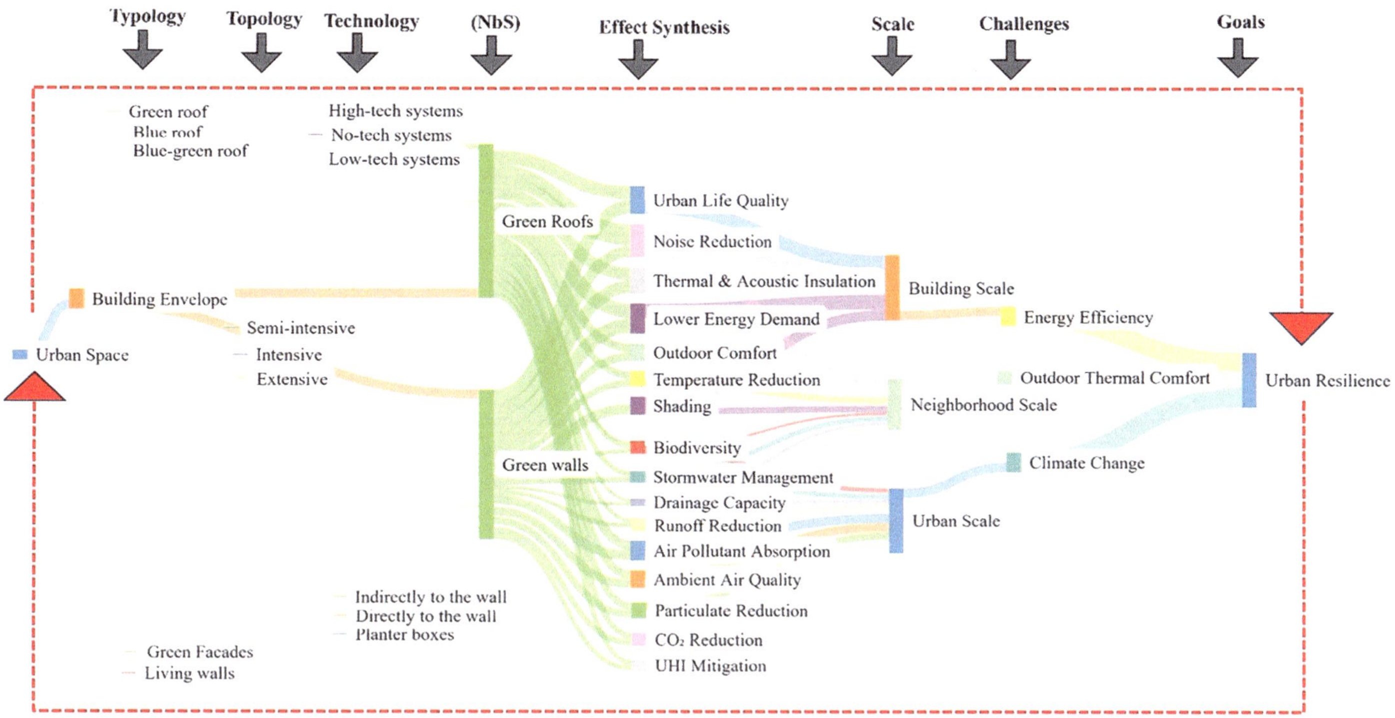

Fig. 8 Conceptual framework illustrating the categorization of green infrastructure interventions (green roofs and façades) and their multi-scale effects across Building, Neighborhood, and Urban levels, highlighting their contribution to urban resilience and sustainability

these interventions depends on plant species, foliage density, orientation, and roof or wall type, highlighting the importance of context-specific design. Building-level interventions set the foundation for benefits that propagate to larger scales.

Neighborhood Scale: Clusters of green roofs and façades collectively moderate local microclimates, reduce outdoor heat stress, and improve thermal comfort in public spaces. Vertical façades are particularly influential for human comfort due to direct shading and vertical exposure. Aggregated interventions at this scale also enhance biodiversity, improve air quality, and reduce energy demand, demonstrating how building-level greening translates into neighborhood-scale ecological and social benefits. Coordinated planning ensures that thermal comfort, ecosystem services, and social equity are achieved across communities.

Urban Scale: At the city level, widespread deployment of green roofs and façades mitigates the Urban Heat Island effect, contributes to stormwater management, enhances air quality, and supports carbon sequestration. Roofs dominate large-scale temperature regulation due to their extensive horizontal coverage, while façades contribute to broader ecological connectivity. Urban-scale implementation relies on integrated policies, intersectoral collaboration, and long-term monitoring to maintain continuity and performance. The cumulative benefits from building and neighborhood interventions directly feed into urban resilience and climate adaptation outcomes.

Synthesis: Building-scale interventions amplify neighborhood and urban benefits: façades mainly enhance local thermal comfort, while roofs drive broader thermal regulation and urban heat island mitigation. Together, these scales support urban resilience by linking energy efficiency, outdoor comfort, and climate adaptation. Considering typology (green roof, blue roof, living wall), topology (direct/indirect façades, planter boxes), and technology (high-, low-, or no-tech systems) allows designers to maximize ecological, social, and climatic co-benefits of NbSs while addressing structural, water, and maintenance constraints.

5 Conclusions and Practical Applications

Green roofs and façades, as multifunctional NbSs, enhance urban resilience, sustainability, and human well-being across building, neighborhood, and urban scales. Based on a PRISMA review of 127 studies, including in-depth analysis of 37 key publications, recurring performance indicators, impact scales, and dominant typologies were identified. These findings informed the proposed multi-scale assessment framework, ensuring that NbS interventions are allocated to spatial scales according to empirically measured impacts. The framework is designed for multiple user groups. Architects, urban designers, and engineers can apply it in early design to select green roof and façade configurations based on spatial scale, climate, building type, and desired outcomes. Urban planners and authorities can use it to guide zoning, incentives, and city-wide green infrastructure. Researchers, consultants, and facility managers can assess NbS performance, compare contexts, identify trade-offs, and optimize maintenance using standardized metrics. Future research should focus on context-specific, water-efficient, and climate-responsive NbS designs, particularly for compact, Mediterranean, and arid cities, while advancing smart monitoring and adaptive maintenance. Cross-regional comparisons and stakeholder engagement are crucial to enhance scalability and long-term effectiveness. Linking multi-scale evidence to design, planning, and evaluation positions green roofs and façades as key components of resilient urban ecosystems.

Author Contributions: Fataneh Shoghi: Conceptualization, Methodology, Data curation, Writing- Original draft preparation, Visualization, Investigation, Validation, Writing- Reviewing and Editing. shoghifattane@gmail.com.

Seyed Morteza Hosseini: Conceptualization, Methodology, Investigation, Validation, Writing-Reviewing and Editing, Supervision.

Funding: No funding.

Appendix 1

The following abbreviations are used in this manuscript:

Abbreviation	Meaning	Abbreviation	Meaning
CO_2	Carbon Dioxide	PM_{10}	Particulate Matter smaller than 10 μm
GFs	Green Facades	PRISMA	Preferred Reporting Items for Systematic reviews and Meta-Analyses
GWs	Green Walls	PET	Physiological Equivalent Temperature
GRs	Green Roofs	Tmrt	Mean Radiant Temperature
BRs	Blue Roofs	UGI	Urban Green Infrastructure
BGRs	Blue-Green Roofs	UHI	Urban Heat Island
NbSs	Nature-Based Solutions	UTCI	Universal Thermal Climate Index
NO_2	Nitrogen Dioxide	UWG	Urban Weather Generator
PM_1	Particulate Matter smaller than 1 μm	VGS	Vertical Greenery System
$PM_{2.5}$	Particulate Matter smaller than 2.5 μm	Ta	Air Temperature

References

Alim, M. A., Rahman, A., Tao, Z., Garner, B., Griffith, R., & Liebman, M. (2022). Green roof as an effective tool for sustainable urban development: An Australian perspective in relation to stormwater and building energy management. *Journal of Cleaner Production, 362*, Article 132561.

Bakhtyari, V., Fattahi, K., Movahed, K., & Franz, A. (2024). Investigating the effect of living walls on cooling energy consumption in various urban microclimates, building heights, and greenery coverage areas. *Sustainability, 16*(2), 920. https://www.mdpi.com/2071-1050/16/2/920

Berardi, U. (2016). The outdoor microclimate benefits and energy saving resulting from green roofs retrofits. *Energy and Buildings, 121*, 217–229.

Calheiros, C. S., & Stefanakis, A. I. (2021). Green roofs towards circular and resilient cities. *Circular Economy and Sustainability, 1*(1), 395–411.

Cardinali, M., Balderrama, A., Arztmann, D., & Pottgiesser, U. (2023). Green walls and health: An umbrella review. *Nature-Based Solutions, 3*, Article 100070.

Celik, S., & Ogus Binatli, A. (2018). Energy savings and economic impact of green roofs: A pilot study. *Emerging Markets Finance and Trade, 54*(8), 1778–1792.

Cirrincione, L., Marvuglia, A., & Scaccianoce, G. (2021). Assessing the effectiveness of green roofs in enhancing the energy and indoor comfort resilience of urban buildings to climate change: Methodology proposal and application. *Building and Environment, 205*, Article 108198.

Couret, D. C., Baldoquin, N. C., de la Paz Pérez, G. A., & Guzmán, L. A. R. (2023). Urban variables for adaptation to global warming in a hot-humid climate. Cuban cities as a case study. *Urban Climate, 51*, 101633.

Cuce, P. M., Cuce, E., & Santamouris, M. (2025). Towards sustainable and climate-resilient cities: Mitigating urban heat islands through green infrastructure. *Sustainability, 17*(3), 1303.

de Oliveira Santos, T. D., Pacheco, F. A. L., & Fernandes, L. F. S. (2024). A systematic analysis on the efficiency and sustainability of green facades and roofs. *Science of the Total Environment, 932*, Article 173107.

Dehghan Lotfabad, A., Hosseini, S. M., Dabove, P., Heiranipour, M., & Sommese, F. (2025). Impacts of vertical greenery on outdoor thermal comfort and carbon emission reduction at the urban scale in Turin, Italy. *Buildings, 15*(3), 450. https://www.mdpi.com/2075-5309/15/3/450

Diana, L., Sommese, F., Ausiello, G., & Polverino, F. (2024). New green spaces for urban areas: A resilient opportunity for urban health. In *Resilience vs Pandemics: Innovations in public places and buildings* (pp. 37–53). Springer.

Frantzeskaki, N., & McPhearson, T. (2022). Mainstream nature-based solutions for urban climate resilience. *BioScience, 72*(2), 113–115.

Ganjirad, M., Delavar, M. R., Bagheri, H., & Azizi, M. M. (2025). Optimizing urban critical green space development using machine learning. *Sustainable Cities and Society, 120*, Article 106158.

Goharian, A., Mahdavinejad, M., Ghazazani, S., Hosseini, M., Zamani, Z., Yavari, H., Ghafarpoor, F., & Shoghi, F. (2025). Designing adaptability strategy to a Novel Kinetic Adaptive Façade (NKAF); Toward a pioneering method in dynamic-objects daylight simulation (Post-Processing). *Journal of Daylighting, 12*, 69–90.

Jayasooriya, V. M., Liyanage, C. T., Muthukumaran, S., & Nilusha, R. T. (2025). Impact of green wall orientation on building energy performance in a tropical climate: An experimental assessment. *PLOS Sustainability and Transformation, 4*(1), Article e0000156.

Kostadinović, D., Jovanović, M., Bakić, V., & Stepanić, N. (2023). Mitigation of urban particulate pollution using lightweight green roof system. *Energy and Buildings, 293*, Article 113203.

Kozak, M., & Lipecki, T. (2025). The impact of green walls on outdoor thermal comfort in residential areas under warm weather conditions: A case study. *Urban Climate, 62*, Article 102541.

Leotta, L., Toscano, S., & Romano, D. (2023). Which plant species for green roofs in the Mediterranean environment? *Plants, 12*(23), 3985.

Lin, H., Xiao, Y., Musso, F., & Lu, Y. (2019). Green façade effects on thermal environment in transitional space: Field measurement studies and computational fluid dynamics simulations. *Sustainability, 11*(20), 5691.

Manso, M., Teotónio, I., Silva, C. M., & Cruz, C. O. (2021). Green roof and green wall benefits and costs: A review of the quantitative evidence. *Renewable and Sustainable Energy Reviews, 135*, Article 110111.

Mayrand, F., & Clergeau, P. (2018). Green roofs and green walls for biodiversity conservation: A contribution to urban connectivity? *Sustainability, 10*(4), 985.

McEvoy, D., Tara, A., Vahanvati, M., Ho, S., Gordon, K., Trundle, A., Rachman, C., & Qomariyah, Y. (2024). Localized nature-based solutions for enhanced climate resilience and community wellbeing in urban informal settlements. *Climate and Development, 16*(7), 600–612.

Mihalakakou, G., Souliotis, M., Papadaki, M., Menounou, P., Dimopoulos, P., Kolokotsa, D., Paravantis, J. A., Tsangrassoulis, A., Panaras, G., & Giannakopoulos, E. (2023). Green roofs as a nature-based solution for improving urban sustainability: Progress and perspectives. *Renewable and Sustainable Energy Reviews, 180*, Article 113306.

Moher, D., Shamseer, L., Clarke, M., Ghersi, D., Liberati, A., Petticrew, M., Shekelle, P., Stewart, L. A., & Group, P.-P. (2015). Preferred reporting items for systematic review and meta-analysis protocols (PRISMA-P) 2015 statement. *Systematic Reviews, 4*, 1–9.

Muharam, A., Al-Hemiddi, N., & Amer, E. (2015). *Thermal behavior of the extensive green roofs in Riyadh city*. LAP LAMBERT Academic Publishing.

Okour, Y., & Shaweesh, H. (2024). Identifying the barriers to green infrastructure implementation in semi-arid urban areas using the DPSIR framework: A case study of Amman, Jordan. *City and Environment Interactions, 24*, 100165.

Ouldboukhitine, S.-E., Bakkour, A., Khiati, S., & Belarbi, R. (2025). Modular green walls: A sustainable architectural solution for energy efficiency in oceanic climates. *Energy and Buildings, 342*, Article 115850.

Pragati, S., Shanthi Priya, R., Pradeepa, C., & Senthil, R. (2023). Simulation of the energy performance of a building with green roofs and green walls in a tropical climate. *Sustainability, 15*(3), 2006. https://www.mdpi.com/2071-1050/15/3/2006

Romanova, A., Horoshenkov, K. V., & Hurrell, A. (2019). An application of a parametric transducer to measure acoustic absorption of a living green wall. *Applied Acoustics, 145*, 89–97.

Schade, J., Lidelöw, S., & Lönnqvist, J. (2021). The thermal performance of a green roof on a highly insulated building in a sub-arctic climate. *Energy and Buildings, 241*, Article 110961.

Shoghi, F., Hosseini, M., Heidari, S., Wang, J., Mahdavinejad, M., & Attia, S. (2025). Human Interaction with urban morphology under the Influence of urban heat island: A systematic review. *Journal of Daylighting, 12*(2), 468–490.

Silva, C. M., Gomes, M. G., & Silva, M. (2016). Green roofs energy performance in Mediterranean climate. *Energy and Buildings, 116*, 318–325.

Sommese, F. (2024). Nature-based solutions to enhance urban resilience in the climate change and post-pandemic era: A taxonomy for the built environment. *Buildings, 14*(7), 2190. https://www.mdpi.com/2075-5309/14/7/2190

Staccione, A., Essenfelder, A. H., Bagli, S., & Mysiak, J. (2024). Connected urban green spaces for pluvial flood risk reduction in the Metropolitan area of Milan. *Sustainable Cities and Society, 104*, Article 105288.

Tseliou, A., Melas, E., Mela, A., Tsiros, I., & Zervas, E. (2023). The effect of green roofs and green façades in the pedestrian thermal comfort of a mediterranean urban residential area. *Atmosphere, 14*(10), 1512. https://doi.org/10.3390/atmos14101512

Wang, D., & Xu, P.-Y. (2024). Urban green infrastructure: Bridging biodiversity conservation and sustainable urban development through adaptive management approach [Original Research]. *Frontiers in Ecology and Evolution, 12*. https://doi.org/10.3389/fevo.2024.1440477

Multi-Scale Cooling Potential of Green Facades and Green Roofs in District Scale Applications

Agnese Salvati, Federica Melotta, Carlos Lopez-Ordoñez, Roser Capdevila, and Núria Garrido

Abstract Cities are becoming increasingly vulnerable to the impacts of climate change, with heat stress emerging as one of the most significant threats to urban liveability. This chapter explores the cooling potential of green façades (GFs) and green roofs (GRs) in applications across individual buildings and entire urban districts, assessing their impact on the outdoor microclimate, pedestrian thermal comfort, and indoor thermal conditions. A simulation-based case study is conducted on the UPC campus located in Terrassa (Barcelona, Spain). Four scenarios are analysed using ENVI-met microclimate simulations: a baseline condition; GF applied to a single university building; GF extended to district scale; and a combined district-scale application of GF and GR. The improvements in indoor thermal conditions and outdoor microclimates are quantified by considering changes in key microclimatic variables, such as air temperature, mean radiant temperature, humidity, physiological equivalent temperature (PET), wall temperatures, and indoor temperatures. The results demonstrate that GF and GR provide different, yet complementary, cooling benefits. GR is better at reducing internal temperatures and mitigating the urban heat island effect, while GF is more effective at improving pedestrian comfort. District-scale GRs achieve the greatest reduction in urban air temperature, albeit modest (less than 0.5 °C). GRs also enable a significant indoor temperature reduction of up to 2.5 °C. Conversely, GFs have a greater capacity to enhance pedestrian thermal comfort and PET at street level due to the combined impact of reduced air and mean radiant temperatures. These findings emphasise that the effectiveness of building-integrated NBS depends on scale, spatial configuration, and proximity to users. This underlines the need for integrated, district-level strategies to foster urban resilience and thermal comfort.

A. Salvati (✉) · F. Melotta · C. Lopez-Ordoñez
Department of Architectural Technology, Universitat Politècnica de Catalunya Barcelona Tech (UPC), Barcelona, Spain
e-mail: Agnese.salvati@upc.edu

F. Melotta
DICEA, Università Degli Studi Di Napoli Federico II, Naples, Italy

C. Lopez-Ordoñez · R. Capdevila · N. Garrido
Heat Engines Department, Universitat Politècnica de Catalunya Barcelona Tech (UPC), Barcelona, Spain

L. Diana et al. (eds.), *Greening and Resilience*, Urban Sustainability,
https://doi.org/10.1007/978-981-92-0419-9_23

Keywords Green Façade · Green Roof · Urban microclimate · Building overheating · ENVI-met · Nature-Based Solutions

1 Nature-Based Solutions in Cities for Climate Change Mitigation and Adaptation

Cities are in the spotlight for the current climate crisis. They are the place of the majority of global energy consumptions and therefore responsible for the biggest share of global greenhouse gas emissions. On the other hand, they are home to the majority of global population, which makes them also the most vulnerable places as for the impacts of climate change. In particular, heat has become a real threat to many cities worldwide. The increased frequency and intensity of heat waves result in unbearable thermal conditions in both buildings and urban spaces, with dangerous impacts on the health of his human and non-human inhabitants. The shift towards such hotter climate is exacerbated in cities by the urban heat island effect, further increasing urban temperatures, energy consumption for cooling and heat-related health risks. For these reasons, climate change mitigation and adaptation are priority issues for any city worldwide. To reverse such a vicious heat cycle, while improving the environmental and social quality of cities, the most effective strategy is re-naturalisation of urban areas, through the implementation of Nature-Based Solutions (NBS). NBS include a variety of natural and engineered systems at different scales, among which: urban parks, urban forests, green corridors, street trees, rain gardens, community gardens as well as building-scale applications such as green roofs, green facades and living walls. Bringing more and diverse natural features and processes into cities serve as both mitigation and adaptation strategies to climate change. NBS contribute to climate change mitigation by providing carbon sequestration and by reducing the cooling energy needs of buildings in cities. On the other hand, NBS are also crucial strategies for climate change adaptation, by providing microclimate cooling, reducing flood risks and fostering biodiversity in cities. For these reasons, increasing NBS in cities can improve the quality of urban ecosystems while reducing the negative impact of human activity on climate, being thus the main driver of the needed urban transformation in the age of climate change.

The implementation of NBS in buildings and cities provides climate benefits at multiple scales. NBS can simultaneously improve the thermal comfort of pedestrians outdoors and in buildings' interiors. This can help reduce cooling energy consumption and subsequent anthropogenic heat generation in summer, with positive impact on urban heat island intensity. However, the positive effect of NBS on the indoor and outdoor thermal environments varies significantly depending on background climate, urban density, water management and, most importantly, the type, location, and size of NBS.

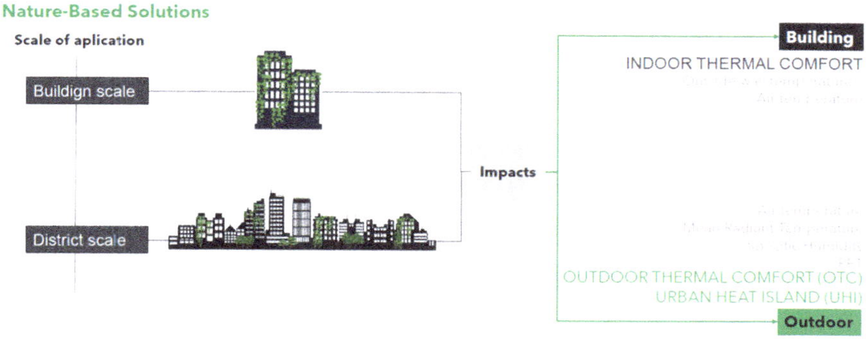

Fig. 1 Overview of the objectives and scales of the study. *Source* Authors' elaboration

This chapter focuses on the positive effects of NBS solutions applied to buildings, namely green facades (GF) and green roofs (GR), on indoor and outdoor micro-climates. Through a case study approach, the study explores the overall cooling potential of GF and GR on outdoor thermal comfort and indoor thermal conditions, considering different scenarios and scales: from individual buildings to district-scale applications (Fig. 1). The objective is to assess the variability of cooling potential of such building-scale NBS on interior and external environments when applied at different scales.

2 Cooling Potential of Green Facades and Green Roofs

The cooling mechanism of GFs and GRs is based on the combined effect of shadows and evapotranspiration. Shadows play a major role. The leaves of the vegetated layer of GFs and GRs reduce incident solar radiation on walls and roofs, reducing external surface temperatures and heat transfer though the envelope, consequently. In GRs and for certain types of GF (i.e. living walls), the presence of a substrate layer also increases thermal mass, further reducing and delaying heat transfer to the interior. Surface temperature reductions reported in literature range widely depending on the system and climate. (Bakhshoodeh et al., 2022) found that external walls shielded by GFs were on average 3.5 ± 4.1 ºC cooler than bare external walls during hot periods, while Šuklje et al. (2016) found a maximum reduction up to 27 ºC for a detached green façade with dense vegetation. As a result, indoor surface temperatures are also lower than with conventional systems, with beneficial effects on operative temperature, indoor thermal comfort and cooling energy demand. This positive effect on the buildings' interior thermal conditions is very noticeable on top floors, where the application of GRs allows reductions of 2–5 °C in indoor operative temperatures during extreme summer conditions (Cirrincione et al., 2021).

The beneficial effect of GRs on cooling energy saving varies with climate. In Mediterranean and hot semi-arid climates, the cooling energy saving ranges between

3 and 15% (Cirrincione et al., 2021; Mousavi et al., 2023) while in cooling-dominated, hot–humid climates, cooling loads can be reduced up to 18% in low-rise buildings (Dahanayake & Chow, 2018). A multi-climate assessment conducted by Jia et al. (2024) showed that GRs can achieve cooling energy savings up to 60–80%, with the highest benefits observed in hot and warm–humid climates and in low-rise and mid-rise buildings. The potential of GFs and GRs to reduce building cooling demand strongly depends also on building shape and orientation. For multi-storey buildings, in Hong Kong, Dahanayake and Chow (2018) demonstrated that the benefits of GR are confined mainly to the top-floor, whereas GF can provide benefits to multiple floors, resulting in potentially higher overall benefits. Building orientation is another key variable. The same study for Hong Kong (Dahanayake & Chow, 2018) found that west-facing GFs contributed the highest annual cooling energy savings, followed by east, south, and north orientations. Some studies also highlighted the variable effect of GF and GR depending on urban density. Peng et al. (2020) examined the potential of large-scale GFs application to reduce cooling demand in urban blocks with different layouts and densities. The results showed that the cooling energy saving potential of GFs at the block scale varies with urban form, being lower in high-rise, high-density areas than in mid-rise, mid-density areas. Similarly, Abuseif et al. (2021) investigated the cooling potential of various types of GRs in different urban layouts, confirming significant variations in the potential to reduce building cooling loads.

The cooling potential of GFs and GRs is enhanced by their moisture content, which converts sensible energy into latent energy via evapotranspiration, resulting in lower outdoor air temperatures. However, Peng et al. (2020) concluded that shading provided by plants and substrates is the dominant mechanism for building cooling energy saving, while the influence of air cooling and humidifying effects is comparatively minor.

Evapotranspiration and air-cooling effects, along with surface temperature reduction, are the main urban cooling mechanisms allowing GFs and GRs to reduce the urban heat island (UHI) intensity. Results from literature indicate that GRs have a greater potential to mitigate UHI than GFs. A study performed by Alsaad et al. (2022) in Stuttgart, Germany, indicated that GFs have a maximum cooling effect on urban air temperature of −0.57 ºC, while Tseliou et al. (2023) found urban air temperature reductions up to –0.7 °C for extensive GRs in in Athens during peak summer hours. Street-scale air temperature reductions of up to –1.0 °C were also reported by Alcazar et al. (2016) in a Mediterranean–continental climate and similar values by Alizadehtazi et al. (2024) for large-scale GR applications in New York City. These results can be explained by the fact that GRs receive more solar radiation than GF in summer. However, in both cases, the cooling effect is strongly space-dependent, varying with urban geometry and building orientation (Tseliou et al., 2023).

The influence of GFs and GRs on outdoor thermal comfort sensation also strongly depends on the characteristics of the context. In fact, when it comes to thermal sensation, air temperature is not the solely important parameter. Humidity and wind speed are also major cooling drivers through convection and evaporation, while solar radiation, surface temperature and view factors significantly shape radiant exchanges in urban environments. Therefore, the impact of GFs and GRs on outdoor thermal conditions should be analysed considering all microclimate variables and not air temperature only. For instance, a simulation study performed by Alsaad et al. (2022) in Stuttgart, Germany, indicated that GFs have negligible effects on pedestrian air temperature, but significantly reduced MRT (up to 3.04 °C) and increased RH (up to 5.07%), with noticeable effects in terms of thermal comfort sensation.

The cooling potential of GFs and GRs on urban microclimate is also strongly dependent on the structure of the systems and the type and density of vegetation. Based on ENVI-met simulations, Liu et al. (2023) emphasised that the leaf area density (LAD) is the most significant parameter of the plant structure, exhibiting a much greater impact on almost all microclimatic variables than other foliage properties. Several studies have shown that GRs with higher LAI/LAD values lead to greater reductions in surface temperature (Abuseif et al., 2021; Alcazar et al., 2016; Liu et al., 2023). Furthermore, Abuseif et al. (2021) emphasise that green roof configuration is more important than green roof coverage ratio. For example, including canopy trees in green roofs is more effective than covering the entire surface with an extensive green roof. As for GFs, studies showed a different performance of attached (directly integrated) or detached systems (using trellises with an air gap). Detached GFs generally reduce the external wall temperature more effectively compared to attached façades (Bakhshoodeh et al., 2022). Furthermore, as the gap width increases, the cooling effect within the air gap also increases (Bakhshoodeh et al., 2022).

These studies emphasise the variability in the cooling potential of GFs and GRs depending on the focus of the analysis: indoor thermal conditions, building cooling demand, urban air temperature, and outdoor thermal comfort.

3 Green Façade and Green Roofs Scenarios for the UPC Campus in Terrassa

This chapter presents an analysis of the cooling potential of GFs and GRs based on simulations carried out for the UPC campus of Terrassa (Barcelona, Spain). The campus area is characterised by open and irregular shapes, leading to wide open public spaces in between buildings, which is in contrast with the compact surrounding residential blocks, organised on an orthogonal network of streets with typical urban-canyon-like configuration. The study area analysed is centred on the campus library building (Fig. 2) and extends approximately 200 m, including both campus and residential buildings. The library building has two stories and a rectangle shape footprint, with larger facades exposed south and north.

Different scenarios are analysed to assess the cooling potential of GF and GR applications to the library building only or to all the buildings included in the study area (district). Four retrofit scenarios are considered:

- **Base case**: Representing the current situation for urban form, vegetation and materials.
- **Library-GF**: It involves the application of GFs only to the library building.

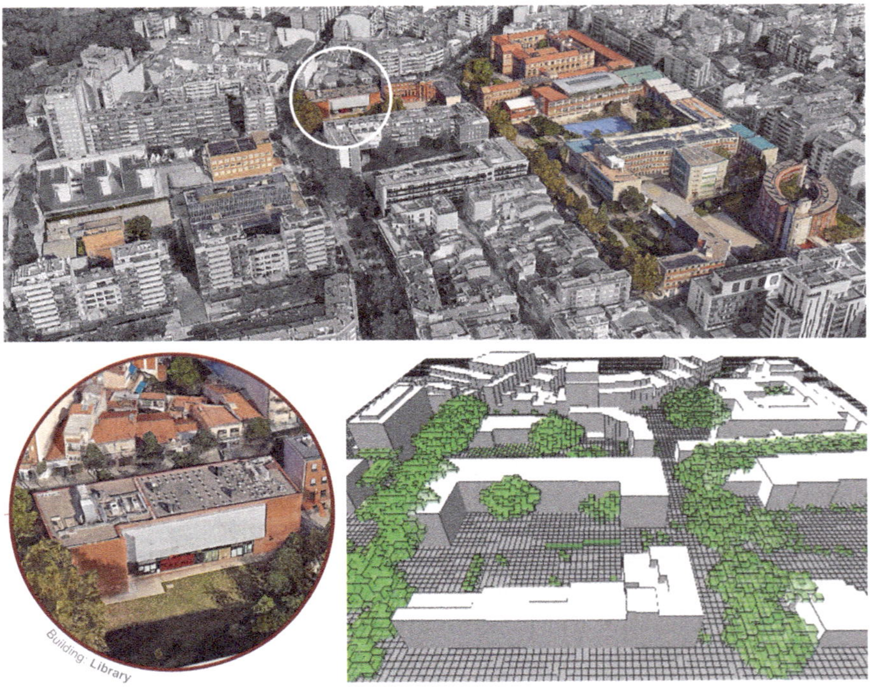

Fig. 2 3D views of the UPC Terrassa campus and corresponding ENVI-met model. The building highlighted in the circle is the library. *Source* Authors' elaboration based on Google earth images

- **District-GF**: It consists of applying the GF also to the district buildings. In this scenario, the GF accounts for 50% of the buildings' envelope, to account for the percentage of wall surface without windows. This aims at evaluating the urban-scale effects of GFs under an extensive, yet realistic, application scenario, as a 50% coverage of façade surfaces is considered a feasible intervention.
- **District GF-GR**: It considers the additional application of GR to the library and the district. The ratio of GR is 50% of the total roof area in the district. This percentage corresponds to the ratio of flat roofs in the district, which are suitable for the installation of green roofs.

4 Methodology of the Study

The impact of the different GFs and GRs scenarios is evaluated using microclimate simulations with the ENVI-met software. The simulated area is centred in the library and includes all the buildings in the surroundings blocks. A 10 m buffer is added on each side in the ENVI-met model to avoid boundary related issues during the simulation. The cell resolution is set to 2×2 m.

All buildings are modelled as concrete structures (albedo = 0.3) to approximate the most common condition in the district. The library building, focus of the analysis, is modelled in more detail, considering its actual exterior materials (bricks, white sunshades, and glass) to achieve a more accurate representation of its indoor and outdoor thermal conditions and potential improvements. The model includes street trees and permeable surfaces as derived from satellite images and in-situ surveys (Fig. 2).

The simulations are performed for a typical hot day for Terrassa, namely the average day of august as derived from the last 10 years of observations from the city weather station managed by the Catalan Meteorological Service (METEOCAT). The wind direction is set from south-west, according to the dominant direction in August in Terrassa. The simulation is run for 42 h, as the model requires a warm up period to stabilise physical variables and eliminate artefacts associated with the initial conditions (Salvati & Kolokotroni, 2019). The GF system modelled in ENVi-met correspond to the attached type, without substrate. The GRs are modelled as extensive type with substrate.

The first simulation is run for the baseline scenario, which represents the current conditions needed to compare with scenarios including GFs and GRs. In the second simulation, a green façade is applied to the library building on all exterior façades except the openings and the white sunshades on the south façade. The third simulation builds upon the second, by adding green façades to 50% of the exterior surfaces of all the buildings in the district. The fourth and final simulation further incorporates a 50% green roof coverage to all the buildings. The GR is also added to the library buildings in all the area not occupied by technical installations (more than 50%) (Fig. 3).

The aim of the study is to evaluate the cooling potential of each scenario on both indoor and outdoor thermal conditions. Accordingly, different outputs are analysed. To understand the impact on the buildings' indoors, the analysis is focused on the external walls surface temperatures and the indoor temperature of the library

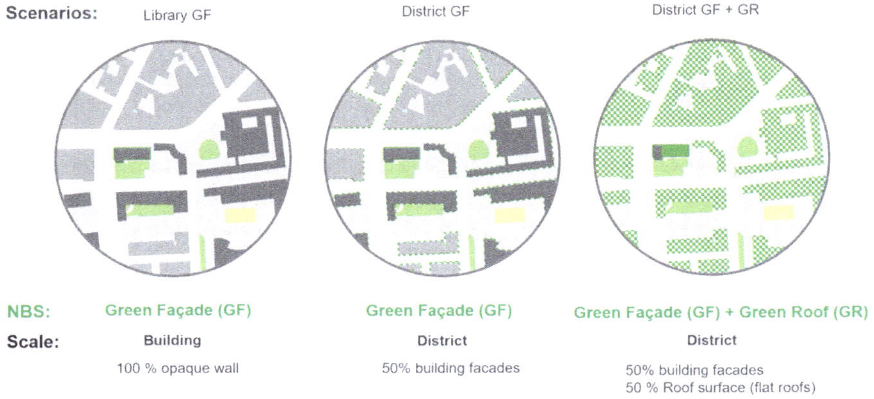

Fig. 3 Characteristics of the scenarios analysed. *Source* Authors' elaboration

building. For the outdoor environment, the evaluated variables include potential air temperature, mean radiant temperature, specific humidity, and physiological equivalent temperature (PET) to understand the expected outdoor thermal sensation. The PET calculation is performed by the BIO-met tool, which computes thermal comfort indices based on the simulated microclimatic parameters and standardised parameters for the human body, corresponding in this case to a female adult wearing summer clothing.

Understanding the spatial and temporal variability of each variable is also important. Therefore, the analysis combines temporal profiles in relevant points and spatial distributions of each variable at specific times to show the relative difference compared to the base-line.

Those variables whose change strongly affects pedestrian thermal comfort, such as mean radiant temperature, specific humidity, and PET are analysed at 1.4 m above ground level (agl). The results for potential air temperature are instead presented at 5 m agl, corresponding to half of the typical canyon height, since temperature reductions are more evident at higher elevations, also due to increased air movement.

The selection of the points for the temporal analysis also differs depending on the evaluated variable. A point located in a narrow canyon formed between the library building and the closest adjacent building is selected to highlight the highest cooling potential of green facades at the pedestrian level. To evaluate potential air temperature reduction, a point located along the main street to the north of the library is considered more appropriate to understand the potential benefit also to adjacent buildings.

Lastly, to evaluate the building scale outputs, the surface temperature of points in different orientations are compared to the baseline. For the north, east, and west façades, the evaluation points are located at the midpoint of the façade width and positioned in the upper portion of the façade, receiving maximum solar radiation. For the south façade, two different points are analysed, located on different materials of the façade: bricks and white paint. This allows evaluating the surface temperature variation when applying GF compared to each material.

5 Multi-Scale Impacts of the Green Facades and Green Roof Scenarios

The simulation results showed that all the scenarios have a positive impact at both the district scale and the building scale, but with significant differences among them. The combination of different strategies produces greater positive effects on outdoor thermal comfort.

Figure 4 shows the change in potential air temperature for the three green scenarios compared to the baseline model. The results are provided at 6 m agl, corresponding to the height of the roof of the library. A maximum reduction of 0.45 °C is observed at 19:00 in the scenario District GF-GR compared to the baseline. The GF scenarios have lower impact on air temperature, below 0.3 °C when applied at the district scale and blow 0.1 °C when applied to the library building only.

The results show a great spatial heterogeneity of the effects mainly depending on wind direction and distance from the greening systems. As wind direction is from south-west, the beneficial effect of all scenarios is more relevant in the north-east sector of the studied area.

The results for the mean radiant temperature at 1.40 m (pedestrian level) show greater positive impacts (Fig. 5). In the point analysed, in proximity to the GF applied to the library building, the reduction in mean radiant temperature varies between 2.5 °C in the Library-GF scenario to 3.5 °C in the District GF + GR scenarios. The maps show that the effect is highly localised, being relevant only in the points closer to the vertical greenery application. This can be explained by the higher view factor to the green façade, which contributes to lower mean radiant temperature in those points. Even though localised, the absolute reduction in mean radiant temperature is more relevant that the reduction in air temperature in the same points.

Figure 6 reports the results for the change in absolute humidity in the green scenarios compared to the baseline. The increase in humidity determined by green systems can have detrimental effects on thermal comfort in summer. Its change is evaluated in a point close to the GF application, to assess its impact in the worst possible case. The greatest increase in absolute humidity is observed during the hours of maximum solar irradiation and varies between + 0.05 g/kg for the GF applied to the library building only to +0.15 g/kg for the district GF-GR scenario. In neither case this change in humidity entails a worsening of the thermal sensation for pedestrians, being thus negligible in the Mediterranean climate of Terrassa.

The impact of all microclimate variations on thermal comfort sensation can be evaluated considering the change in space and time of PETreported in Fig. 7. The

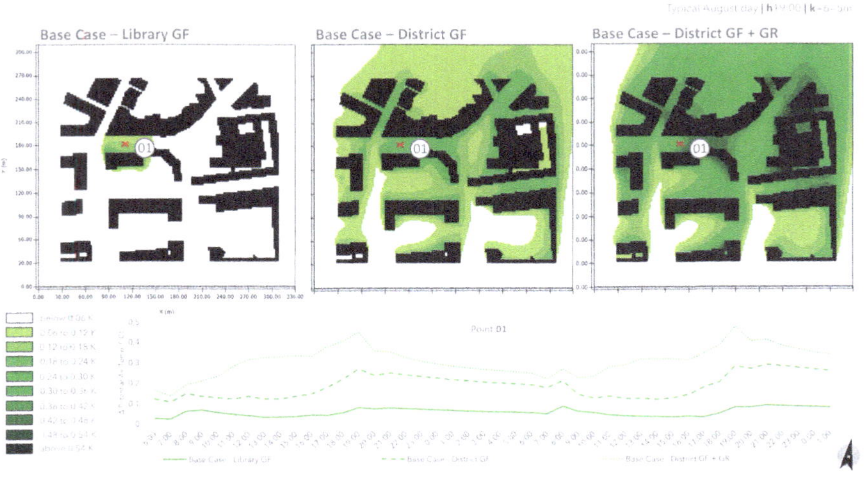

Fig. 4 Potential Air Temperature change for the three green scenarios: Library-GF, District-GF, and District GF-GR

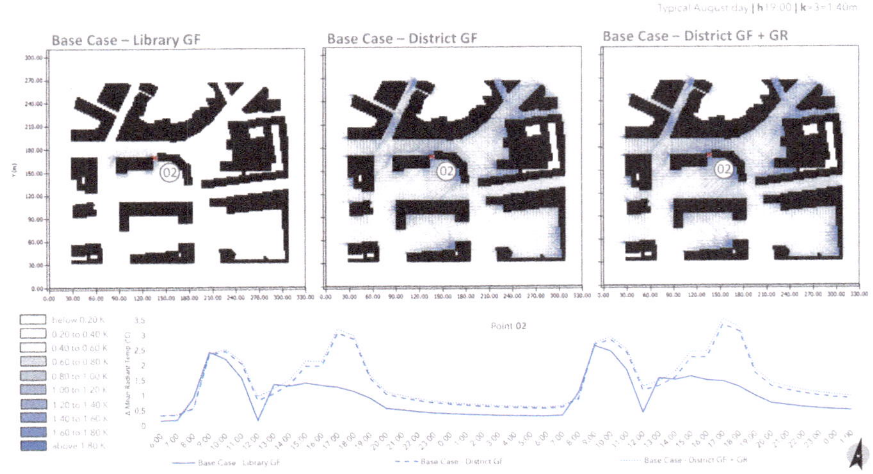

Fig. 5 Mean Radiant Temperature change for the three green scenarios: Library-GF, District-GF, and District GF-GR

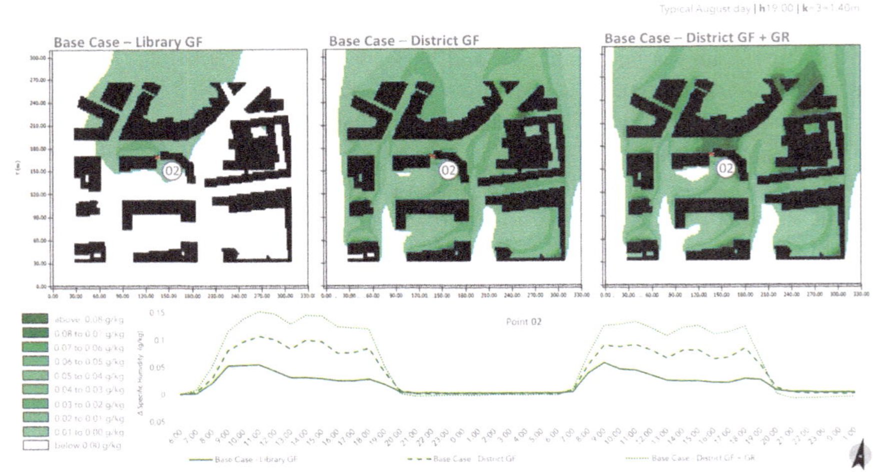

Fig. 6 Specific Humidity change for the three green scenarios: Library-GF, District-GF, and District GF-GR

PET integrates the contribution of all microclimate variables on thermal sensation, representing a key indicator for assessing pedestrian thermal comfort. In this case, an improvement in PET is found for all scenarios. The Library-GF scenario allows a reduction of PET of approximately 1 °C, while the District-GF scenario allows a greater reduction up to 2 °C at 17:00. This is due to the beneficial effect of GF applied to both east-oriented and west-oriented walls in proximity to the analysed point. The District GF-GR scenario further improves PET, thanks to the beneficial effect of air

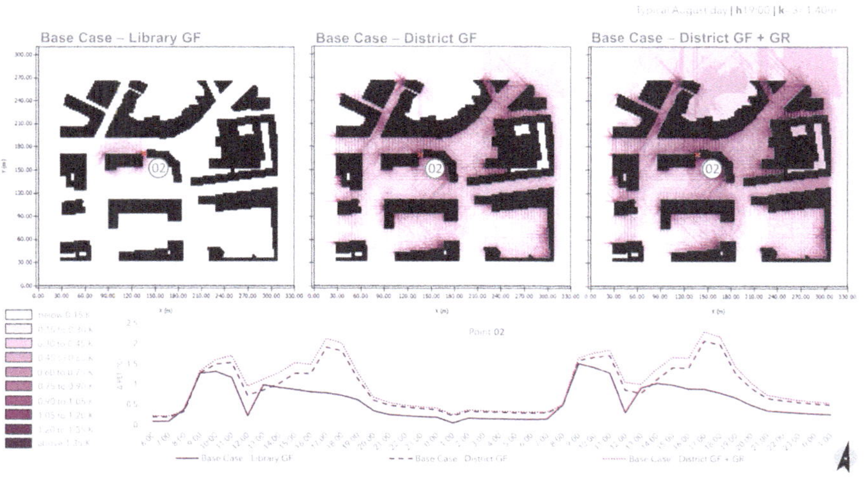

Fig. 7 Physiological Equivalent Temperature (PET) change for the three green scenarios: Library-GF, District-GF, and District GF-GR

temperature reduction on thermal sensation. However, the positive contribution is below 0.5 °C, being less relevant compared to the effect of GFs.

The impact of the greening scenarios on building temperatures is also evaluated, focusing the analysis on the library (Fig. 8). The results show that GFs significantly reduce external surface temperatures compared to conventional materials (bricks). The green facade is up to 19 °C cooler than brick wall on the south orientation and reaches a maximum reduction up to 23.3 °C at 11:00 in the east-oriented façade. These values are in line with previous studies comparing greened and bare façades (Alsaad et al., 2022; Šuklje et al., 2016). However, the results also show that white painted façades can have lower temperatures than GFs during daytime, even though remaining warmer during the night hours compared to the GF. Finally, the impact of GF on external surface temperature is negligible in the north façade. These results highlight the great variability of surface temperatures depending on orientation and surface albedo and emissivity. Das opposite to the significant reduction in external surface temperatures, the impact of GFs on the building indoor temperature is limited. The reduction in indoor temperature is noticeable starting from noon and increases in the afternoon and night hours. The maximum reduction in indoor temperature reaches 1.3 °C for the library-GF scenario. The district-scale application of GFs does not entail a further reduction of the library's indoor temperature. Conversely, the application to GRs positively affects indoor environment, leading to air temperature reduction of up to 2.5 °C. This is due to the great contribution of reduced heat transfer through the building roof and further reduction of outdoor air temperature.

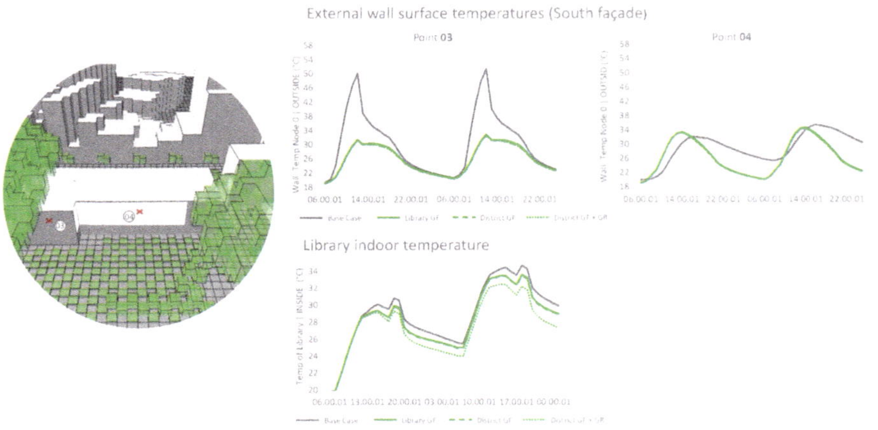

Fig. 8 External wall surface temperature and indoor temperature of the library building under different scenarios

6 Discussion and Implications of the Results

The quantitative results obtained by ENVI-met simulations are in line with previous studies using similar approaches to assess the impact of GFs and GRs for individual buildings applications or large-scale application (Abuseif et al., 2021; Peng et al., 2020). However, as opposite to such previous studies, we carried out simulations for a real-world urban area instead of simplified urban forms, thus obtaining a more realistic assessment for the case study analysed.

The study confirms that GFs have a limited impact on urban air temperature (below 0.3 °C). A district level implementation of GRs combined with GFs provides the best performance, yet limited to less than 0.5 °C air temperature reduction. This is in line with the results by (Alsaad et al., 2022; Tseliou et al., 2023). This result does not entail a reduced impact of GFs on outdoor thermal comfort. In fact, it is exactly the opposite. GFs have a greater potential to reduce street level mean radiant temperature and improve thermal comfort (PET) compared to GRs. This is explained mainly by the proximity of the greening system to pedestrians, reducing the radiant flux (reflected solar radiation and emitted infrared radiation) to the human body. Yet, GRs have greater positive impact on urban heat island mitigation, even though limited to approximately 0.5 °C maximum reduction in the context of Terrassa, characterised by a compact urban fabric with medium to low-rise buildings and a Mediterranean climate.

As for humidity, the GF type analysed (attached façade without substrate) does not cause any inconvenient to summer thermal comfort in Terrassa.

The air temperature reduction achieved with GF and GR systems is expected to have positive impact on the cooling energy demand of the district's buildings. The maximum impact of GF on external wall temperature reduction reaches 23 °C on the east facades, value in line with previous studies comparing greened and bare façades

(Alsaad et al., 2022; Šuklje et al., 2016). Despite such relevant impact on external surface temperatures, the impact of GFs on the indoor temperature is reduced to 1,3 °C maximum. The scenario combining GFs and GRs allows a significant further reduction in indoor temperature up to 2.5 °C, which is expected to lead to significant improvements in thermal comfort and cooling energy reduction. This confirms the greater potential of GRs to reduce indoor temperatures compared to GFs, as found by (Abuseif et al., 2021).

It must be noted that these results are based on ENVI-met simulations, which are not very accurate for calculating building indoor thermal conditions compared to detailed models such as EnergyPlus. The building energy model in ENVI-met considers the dynamic heat transfer through the building envelope but does not account for internal gains and ventilation. Despite that, the results of this study are in line with experimental data (Susorova et al., 2013). Further developments of this work will investigate a more accurate co-simulation methodology, using the localised microclimate effects generated by the greening systems around each building as boundary conditions to building energy simulations. Gobakis and Kolokotsa (2017) showed that accounting for local microclimatic conditions can lead to ± 10% variations in heating and cooling demand, increasing to ± 40% when convective heat transfer coefficients are explicitly considered. To date, the EnergyPlus–ENVI-met coupling represents one of the most reliable approaches to link microclimate and building energy simulations (Krebs & Johansson, 2021; Peng et al., 2020). However, this approach still has many limitations. The ENVI-met–EnergyPlus co-simulation methodology is difficult to implement due to the different resolution of the two models and it is limited to few hours of analysis, due to the very large computation time of ENVI-met. Furthermore, ENVI-met is a commercial software whose calculation methods are not entirely open, thus difficult to validate. On the other hand, currently used building simulation tools do not allow to accurately simulated NBS applied to the building envelope, especially green facades (Dahanayake & Chow, 2018). Therefore, the modelling of indoor-outdoor cooling potential of NBS applied to buildings is still an open field for investigation.

7 Concluding Remarks

These results of the study confirm the potential for district-scale building retrofits using NBS to improve urban and building thermal comfort while contributing to the transition towards climate-neutral buildings and cities. The study confirmed a significant potential of green facades and green roofs in both the indoor and outdoor environments. Significant improvement in outdoor thermal comfort is obtained when GF and GR strategies are combined and extended beyond individual buildings. District-scale GR applications have a higher urban heat island mitigation potential, while GFs show a greater impact on pedestrian outdoor thermal comfort, thanks to the beneficial effect of reduced mean radiant temperatures. At the building scale, both GF and GR have a positive impact on indoor air temperature, with GR being more relevant. This is due to the combination of different cooling mechanisms provided by GF and GR: shadows, reducing surface overheating and heat transfer to the envelope and evapotranspiration, reducing outdoor air temperature with beneficial effect on ventilation and infiltration.

This results also highlight that the benefits of nature-based solutions integrated in cities cannot be assessed individually, as the best advantages emerge when these strategies are implemented in the most extensive, continuous and integrated way possible. The widespread integration of different types of NBS in buildings and cities amplifies their positive effects and capacity to enhancing urban resilience and environmental quality. Therefore, the integration of NBS in cities is not only a measure for climate change adaptation and mitigation but a key feature for healthier and more sustainable living conditions for present and future generations.

Acknowledgements This work has been supported by the EDUPED project (PCI2024-155068-2), funded by the Spanish State Research Agency (AEI) under the "International Collaboration Projects' initiative. Call: PCI2024-2" and by the IN&OUT project (PID2024-155805OA-C22), funded by the Spanish Ministry of Science, Innovation and Universities (MICIU), the State Research Agency (AEI) and the European Regional Development Fund (FEDER) under code MICIU/AEI/10.13039/501100011033/FEDER, UE.

References

Abuseif, M., Dupre, K., & Michael, R. N. (2021). The effect of green roof configurations including trees in a subtropical climate: A co-simulation parametric study. *Journal of Cleaner Production, 317*, Article 128458. https://doi.org/10.1016/j.jclepro.2021.128458

Alcazar, S. S., Olivieri, F., & Neila, J. (2016). Green roofs: Experimental and analytical study of its potential for urban microclimate regulation in Mediterranean–continental climates. *Urban Climate, 17*, 304–317. https://doi.org/10.1016/j.uclim.2016.02.004

Alizadehtazi, B., Stolper, J., Singh, K., & Montalto, F. A. (2024). Microclimatic implications of a large-scale green roof and high-rise redevelopment in New York City. *Building and Environment, 250*, Article 111113. https://doi.org/10.1016/j.buildenv.2023.111113

Alsaad, H., Hartmann, M., Hilbel, R., & Voelker, C. (2022). The potential of facade greening in mitigating the effects of heatwaves in Central European cities. *Building and Environment, 216*, Article 109021. https://doi.org/10.1016/j.buildenv.2022.109021

Bakhshoodeh, R., Ocampo, C., & Oldham, C. (2022). Thermal performance of green façades: Review and analysis of published data. *Renewable and Sustainable Energy Reviews, 155*, Article 111744. https://doi.org/10.1016/j.rser.2021.111744

Cirrincione, L., Marvuglia, A., & Scaccianoce, G. (2021). Assessing the effectiveness of green roofs in enhancing the energy and indoor comfort resilience of urban buildings to climate change: Methodology proposal and application. *Building and Environment, 205*, Article 108198. https://doi.org/10.1016/j.buildenv.2021.108198

Dahanayake, K. C., & Chow, C. L. (2018). Comparing reduction of building cooling load through green roofs and green walls by EnergyPlus simulations. *Building Simulation, 11*(3), 421–434. https://doi.org/10.1007/s12273-017-0415-7

Gobakis, K., & Kolokotsa, D. (2017). Coupling building energy simulation software with microclimatic simulation for the evaluation of the impact of urban outdoor conditions on the energy consumption and indoor environmental quality. *Energy and Buildings, 157*, 101–115. https://doi.org/10.1016/j.enbuild.2017.02.020

Jia, S., Weng, Q., Yoo, C., Xiao, H., & Zhong, Q. (2024). Building energy savings by green roofs and cool roofs in current and future climates. *Npj Urban Sustainability, 4*(1), 23. https://doi.org/10.1038/s42949-024-00159-8

Krebs, L. F., & Johansson, E. (2021). Influence of microclimate on the effect of green roofs in Southern Brazil—A study coupling outdoor and indoor thermal simulations. *Energy and Buildings, 241*, 110963. https://doi.org/10.1016/j.enbuild.2021.110963

Liu, Z., Cheng, K. Y., Sinsel, T., Simon, H., Jim, C. Y., Morakinyo, T. E., He, Y., Yin, S., Ouyang, W., Shi, Y., & Ng, E. (2023). Modeling microclimatic effects of trees and green roofs/façades in ENVI-met: Sensitivity tests and proposed model library. *Building and Environment, 244*, Article 110759. https://doi.org/10.1016/j.buildenv.2023.110759

Mousavi, S., Gheibi, M., Wacławek, S., & Behzadian, K. (2023). A novel smart framework for optimal design of green roofs in buildings conforming with energy conservation and thermal comfort. *Energy and Buildings, 291*, Article 113111. https://doi.org/10.1016/j.enbuild.2023.113111

Peng, L. L. H., Jiang, Z., Yang, X., Wang, Q., He, Y., & Chen, S. S. (2020). Energy savings of block-scale facade greening for different urban forms. *Applied Energy, 279*, Article 115844. https://doi.org/10.1016/j.apenergy.2020.115844

Salvati, A., & Kolokotroni, M. (2019). *Microclimate data for building energy modelling: Study on ENVI-Met Forcing Data, 16*, 3361–3368. https://doi.org/10.26868/25222708.2019.210544

Susorova, I., Angulo, M., Bahrami, P., & Stephens, B. (2013). A model of vegetated exterior facades for evaluation of wall thermal performance. *Building and Environment, 67*, 1–13. https://doi.org/10.1016/j.buildenv.2013.04.027

Šuklje, T., Medved, S., & Arkar, C. (2016). On detailed thermal response modeling of vertical greenery systems as cooling measure for buildings and cities in summer conditions. *Energy, 115*, 1055–1068. https://doi.org/10.1016/j.energy.2016.08.095

Tseliou, A., Melas, E., Mela, A., Tsiros, I., & Zervas, E. (2023). The effect of green roofs and green façades in the pedestrian thermal comfort of a mediterranean urban residential area. *Atmosphere, 14*(10), 1512. https://doi.org/10.3390/atmos14101512

Bioclimatic Redevelopment of Modern Architecture: An Integrated Design Approach—A Case Study in Rome

Sara Lanzoni, Silvia Federici, and Augusto Scipioni

Abstract This study addresses the challenge of restoring and enhancing a modern building of historical significance through a bioclimatic and conservation-oriented approach. It investigates the adaptive reuse of a Rationalist building as an innovative, climate-responsive, and CO_2-positive structure, applying the principles and best practices of bio-architecture and healthy building design. The research explores the transformation of an underused heritage asset into a centre of excellence for sustainable architectural technologies and renewable energy systems, thereby contributing to socio-cultural regeneration and the reconnection of a long-neglected site with its urban context. The case study is Villa IX Maggio, also known as Convitto Locchi, located in the Garbatella district of Rome. Designed by architect and engineer Angiolo Mazzoni and inaugurated in 1936, the building originally functioned as a girls' boarding school and represents a significant example of Italian Rationalist architecture. The proposed design strategies preserve the building's architectural and cultural value while transforming it into an open and adaptable system aimed at enhancing user wellbeing, reducing energy consumption, and attracting a diverse range of social groups, thereby revitalising its functional and social role. The project is conceived as a benchmark case study for the practical application of bio-architectural and bioclimatic principles. The site-specific context, together with the building's physical, environmental, and architectural characteristics and the multi-scale analyses conducted, informed the definition of the functional programme and the design strategies. The study contributes to the ongoing debate on the sustainable retrofit of modern architectural heritage by proposing a transferable, medium-level bioclimatic intervention framework compatible with conservation constraints.

S. Lanzoni (✉)
Department of Architecture, University of Bologna, Bologna, Italy
e-mail: sara.lanzoni@unibo.it

S. Federici
Ministry of the Environment, Rome, Italy

A. Scipioni
Banca d'Italia, Rome, Italy

L. Diana et al. (eds.), *Greening and Resilience*, Urban Sustainability,
https://doi.org/10.1007/978-981-92-0419-9_24

Keywords Adaptive reuse · Modern architectural heritage · Bioclimatic redevelopment · Energy efficiency · Blue-green infrastructure · Nature-based solutions

1 Introduction

The sustainable retrofit of existing buildings represents a key challenge for contemporary architecture, particularly in the case of modern architectural heritage subject to protection constraints.

This study, developed within an academic research framework, addresses the design of an Advanced Training and Research Centre conceived as a core facility around which complementary functions can gravitate and interact, thereby enhancing its overall value. Beyond its design-oriented nature, the study is intended as a research-driven case study that investigates the potential of bioclimatic redevelopment strategies applied to a protected modern building.

The selected site for the intervention is Villa IX Maggio, located in the Garbatella district of Rome, including the adjacent archaeological area of significant historical interest. The building represents an important example of Italian Rationalist architecture and offers a valuable opportunity to explore adaptive reuse and bioclimatic retrofit strategies for modern architectural heritage, a field that is still insufficiently addressed in current research. While several recent studies have addressed the energy retrofit of modern buildings and post-war architectural heritage, these interventions often rely on deep-renovation approaches or focus primarily on technological optimisation, offering limited insights into low-impact, conservation-compatible bioclimatic strategies applicable to protected modern architecture (Panakaduwa et al., 2024; RIBA, 2021).

The primary objective guiding the project is to investigate how a historically significant modern building can be transformed into a high-performance, climate-responsive, and socially inclusive environment, without compromising its architectural and cultural value. To this end, the project explores the integration of passive and active bioclimatic strategies, renewable energy systems, blue-green infrastructure, and nature-based solutions within a single, coherent design framework.

More specifically, the study aims to:

- test the effectiveness of low-impact and reversible bioclimatic interventions compatible with heritage protection constraints;
- evaluate the role of multi-scale analyses—from urban to building and component scale—in informing design decisions;
- assess the replicability and transferability of the proposed strategies to other modern heritage buildings in similar climatic and regulatory contexts.

The project envisions the reconnection of the site—conceived as an Advanced Training Campus—with the Garbatella neighbourhood and the wider city of Rome through a set of multiple, interrelated dimensions, addressed as interconnected components of an integrated design approach. These include the energy dimension, which acts as a catalyst for the synergistic development of renewable energy sources at the district scale and potentially replicable at the city scale through a virtuous and transferable energy model; the educational dimension, aimed at establishing a national hub for advanced education and research in bio-architecture, sustainable construction, and environmental design; the social dimension, providing an inclusive space for cultural initiatives, community activities, and the dissemination of knowledge on sustainable and nature-based practices; the urban planning dimension, contributing to the promotion of an ecological and climate-responsive approach within a renewed vision of urban development; and the technological dimension, which functions as a demonstrative field for low-environmental-impact construction technologies—both traditional and innovative—applicable to the redevelopment of existing buildings (RIBA, 2021).

A key challenge was to avoid a radical renovation approach involving extensive demolition and alteration of the existing context. Instead, the project adopts targeted, site-specific design strategies that preserve the historic Rationalist building of Villa IX Maggio while transforming it into an open and adaptable system capable of enhancing user wellbeing, reducing energy consumption, and fostering social interaction.

The study does not aim to provide a comparative survey of realised projects, but rather to propose a transferable methodological framework grounded in a critical, research-driven case study.

Building upon these objectives, the project generated two further cascading interventions—the Solar Forest and the Vertical Gardens—which function as experimental fields for the application of innovative bio-architectural technologies and nature-based solutions. Together, these interventions position the Villa IX Maggio Campus as a critical case study and a potential best practice for the sustainable redevelopment of modern architectural heritage.

Overall, the Villa IX Maggio Campus is conceived as a benchmark case study for the practical application of bio-architectural and bioclimatic principles to modern architectural heritage. Beyond the specific design outcome, the project contributes to the broader discourse on sustainable retrofit by demonstrating how advanced construction techniques, nature-based solutions, and renewable energy systems can be integrated within a protected context through a holistic and interdisciplinary approach.

By combining architectural design, environmental engineering, landscape strategies, and socio-cultural considerations, the study highlights the potential of interdisciplinarity and international best practices to inform innovative and transferable models for the redevelopment of existing modern buildings, fostering both environmental performance and human wellbeing, while contributing to the ongoing debate on sustainable strategies for the regeneration of modern architectural heritage (Mitterer et al., 2012; Sasso, 2003).

2 Presentation of the Complex: Overview of Intended Uses

The complex known as Villa IX Maggio, also referred to as *Convitto Vittorio Locchi*, is located in the Garbatella district of Rome and covers an area of approximately 4.2 hectares. Together with the adjoining former Roman necropolis, extending over about 0.8 hectares, the site constitutes a significant green enclave. It is bounded by Via Cristoforo Colombo and Via Padre Semeria to the east, Via Alessandro Valignano and the Aurelio Alonzi primary school to the north, the residential block along Viale Guglielmo Massaia to the west, and Via Carlo Spinola to the south, where the main entrance and the caretaker's house are located.

The site is characterised by a gently elevated topography, with Villa IX Maggio positioned at its highest point (see Fig. 1). The area is rich in vegetation, with mature pine trees forming a distinctive natural backdrop to the built structures. Several historical pre-existences are present, including the remains of an ancient Roman villa, a water-collection cistern, ventilation shafts from wartime bunkers, and the now no-longer-visible burials of the necropolis.

The site's location, together with its physical, environmental, and architectural characteristics, and the multi-scale analyses—from the urban to the building and component scale—conducted prior to the design phase, informed the definition of the proposed function as a Campus for advanced training, research, and development in bio-architecture and green relationships.

Fig. 1 Frontal view of Villa IX Maggio, the main façade with the entrance and the surrounding greenery. *Source* Photo taken by Sara Lanzoni

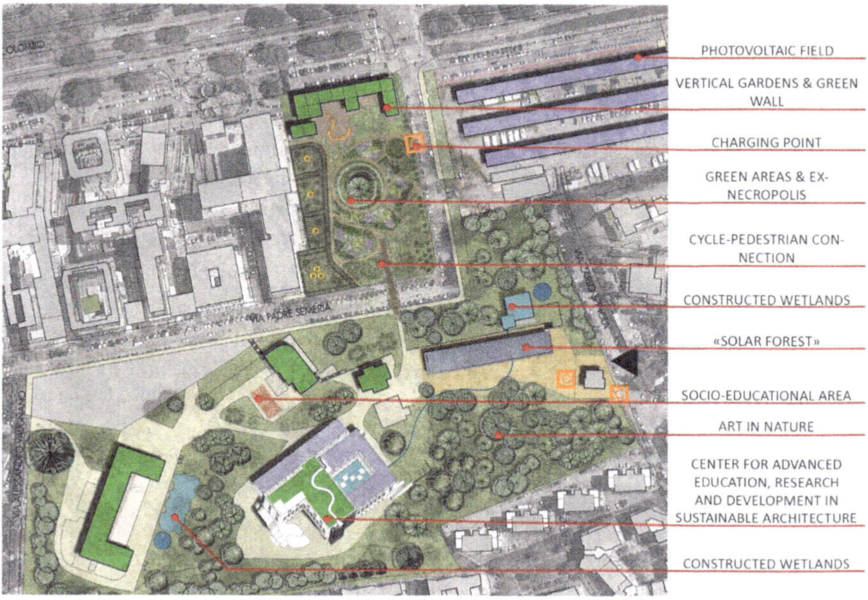

Fig. 2 Masterplan. *Source* Drawing created by the authors

The proposed interventions involve the entire site (see Fig. 2). The Campus is conceived as a diffuse system dedicated to advanced education and research (Dreiseitl, 2015). Villa IX Maggio itself will accommodate teaching rooms, a conference hall, a multimedia library, exhibition spaces, technical laboratories, offices, co-studying and co-working areas, as well as a garden-bar café.

Spaces dedicated to hands-on ecological construction workshops are located beneath an innovative photovoltaic canopy supported by a timber structure, referred to as the Solar Forest. This structure also incorporates a small arena for performances and seminars, fully integrated into the surrounding natural landscape.

Within the area of the former Roman necropolis, connected to the Villa complex by a timber cycle–pedestrian bridge, the Vertical Gardens are proposed. These include a green wall facing Via Cristoforo Colombo, contributing to the visual, environmental, and ecological regeneration of the site.

The entire system is integrated through a blue-green infrastructure network designed for water recovery and reuse via phytoremediation basins and is powered by photovoltaic energy, partly derived from local energy synergies.

Green spaces are articulated through a variety of uses, ranging from Mediterranean scrubland and a bee garden to diffuse exhibition areas and socio-educational, sporting, and cultural zones. These include green walls, roof gardens, and a green bridge designed as a Wild Crossing System to ensure biological continuity.

The main access to the Campus is located on Via Carlo Spinola and has been shifted to the right of the caretaker's house in order to redirect the focal point of the intervention away from the imposing Rationalist monumentality of Villa IX Maggio and towards the Solar Forest, which embodies the identity of the sustainable architecture complex. Two additional entrances are located along Via Padre Semeria. A service access from the grounds of the Alonzi school is also planned to facilitate maintenance of the phytoremediation basin situated behind the school building.

The complex is thus re-permeabilised from both a hydrological and ecological perspective, as well as in terms of accessibility, through an integrated network of cycle–pedestrian routes, vehicular routes for electric vehicles, and dedicated service access paths. Green spaces are revitalised and enhanced, with each area assigned a specific function to avoid the presence of residual or unused spaces.

All user groups are intended to benefit from the new facilities. The ground floor of the Villa will be accessible for events, seminars, conferences, and exhibitions, while the multimedia library—specialising in sustainable architecture, ecology, and renewable energy—will be open to students, school groups, citizens, researchers, and visitors. The garden-bar café and co-studying and co-working spaces will likewise be publicly accessible. The *Garden of Infinity*, located on the Villa's rooftop at the uppermost level (see Fig. 3) and offering panoramic views of the city, will be open to the public at designated times or during specific events. Similarly, the arena integrated into the Solar Forest has been designed to attract young people and visitors of all ages through performances, seminars, and artistic events.

Key elements intended to reactivate the site—by generating interest, encouraging social interaction, and stimulating cultural engagement—include the school of advanced training and research; classrooms and laboratories for students at different educational levels; co-working and co-studying spaces for professionals, researchers, and students; a specialist multimedia library and exhibition hall; a conference hall; refreshment facilities offering organic, vegan, and zero-kilometre products; vertical and educational gardens; performance and seminar spaces; permanent and temporary art installations distributed throughout the park; socio-educational areas; olfactory and holistic pathways; applied architectural and technological solutions; extensive green areas; and an overall enhancement of environmental health and perceived wellbeing.

The project thus envisions a new place capable of communicating across multiple levels, attracting and inspiring curiosity, fostering relationships, raising awareness, and supporting education and cultural exchange.

3 Methodology and Bioclimatic Strategies

The design methodology adopted for the bioclimatic redevelopment of Villa IX Maggio is based on an integrated and multi-scalar approach, combining historical and regulatory analysis, environmental assessment, and building-performance evaluation. This approach was developed to address the specific challenges posed by the adaptive reuse of a protected modern building, where conservation constraints significantly limit the range of admissible interventions. The project can be classified as

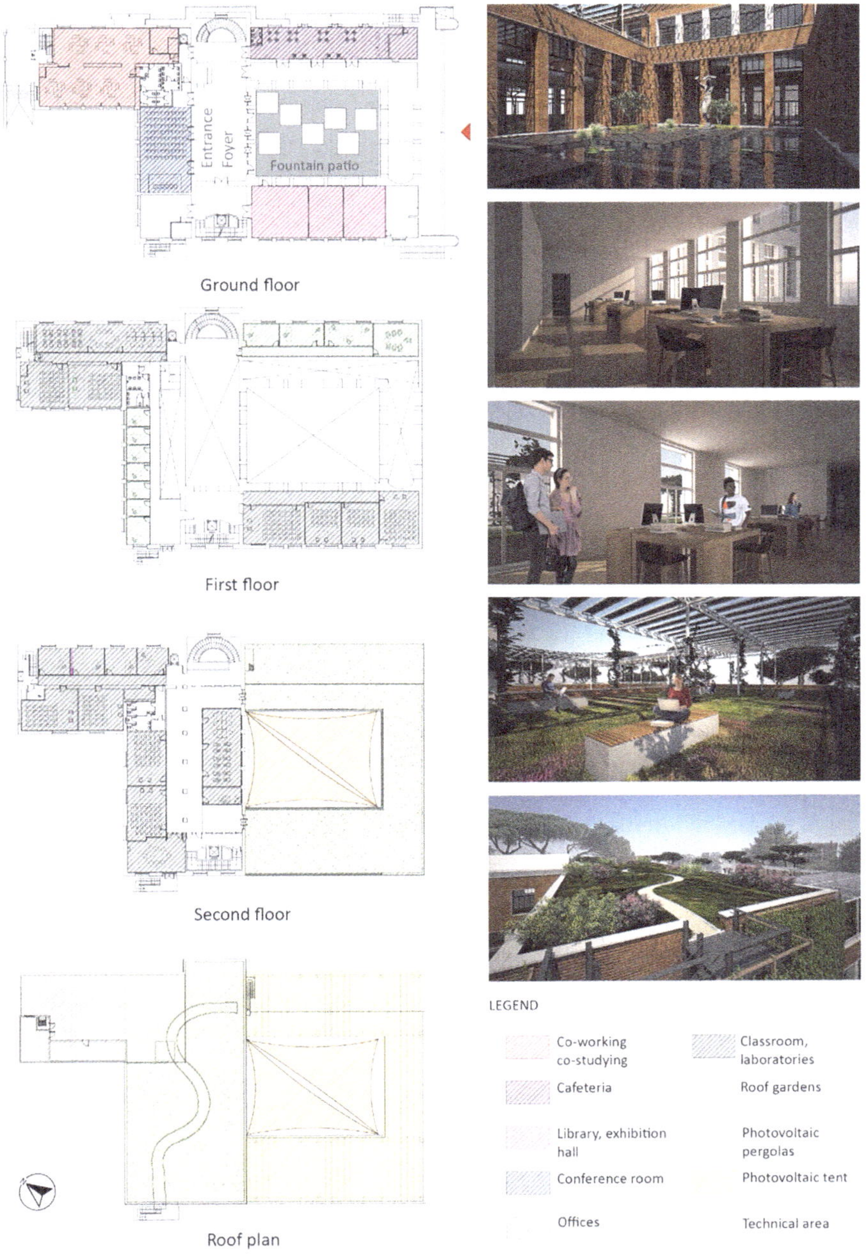

Fig. 3 Floor plans of Villa IX Maggio, showing the intended spatial functions. *Source* Drawing created by the authors

a medium-level bioclimatic intervention, positioned between light retrofit measures and invasive deep-renovation approaches.

3.1 Multi-Scale Analytical Framework

Prior to defining the design strategies, a series of multi-scale analyses were conducted, aimed at identifying both constraints and latent potentials of the site and the building. These evaluations informed the subsequent design choices and ensured coherence between architectural, environmental, and technological solutions. In particular, four interconnected analytical scales were considered:

- Urban scale, focusing on the relationship between the site and the surrounding district, including accessibility, ecological connections, energy synergies, and the potential integration within a local energy community. This analysis supported the definition of the Campus as an open and permeable system, capable of interacting with the Garbatella neighbourhood and contributing to broader urban sustainability objectives.
- Site scale, addressing topography, existing vegetation, soil permeability, microclimatic conditions, and water management. Particular attention was paid to the presence of mature trees, the archaeological area of the former Roman necropolis, and existing underground volumes, which were interpreted as resources for bioclimatic regulation rather than constraints.
- Building scale, analysing orientation, solar exposure, thermal mass, envelope performance, spatial organisation, and patterns of use. This assessment highlighted the building's deficiencies in terms of thermal insulation and summer comfort, as well as the limitations imposed by the protected status of the façades.
- Component scale, focusing on construction systems, materials, and technological elements, with the aim of identifying low-impact, reversible, and compatible retrofit solutions suitable for modern architectural heritage.

The integration of these analytical levels enabled a comprehensive understanding of the system as a whole and guided the selection of targeted, non-invasive bioclimatic strategies.

3.2 Bioclimatic Design Principles and Constraints

Given the protected status of the façades, which precludes invasive external insulation or significant alterations, the bioclimatic redevelopment strategy prioritised interventions within the building envelope and at the site level, complemented by passive and active systems (Panakaduwa et al., 2024). The overarching design principle was to enhance environmental performance while fully respecting the architectural identity and material authenticity of the Rationalist building.

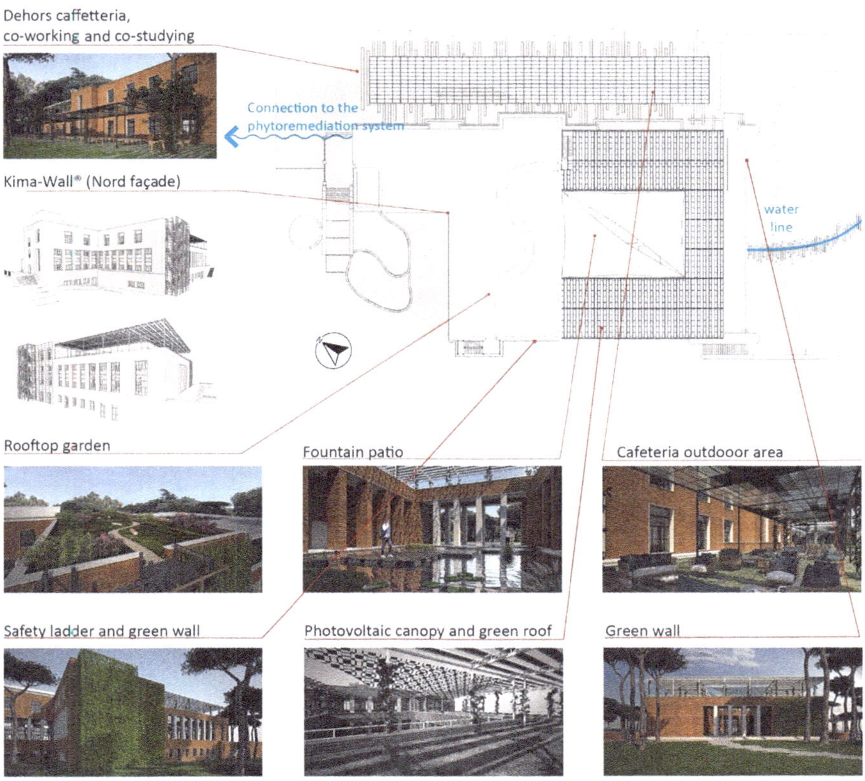

Fig. 4 Main bioclimatic interventions implemented on Villa IX Maggio. *Source* Drawings created by the authors

From this perspective, the proposed interventions—illustrated in Figs. 4 and 5—are characterised by a deliberately low degree of invasiveness, reversibility, and compatibility with conservation requirements. These principles were adopted not only to comply with regulatory constraints, but also to explore replicable strategies applicable to similar modern heritage contexts.

3.3 *Klima-Wall System and Envelope Enhancement*

A key intervention is the introduction of an internal Klima-Wall inspired system, consisting of a Brettstapel timber counter-wall installed within the north-facing wall. The system comprises the existing solid brick masonry and an internal solid timber wall, separated by ventilated cavities that function as seasonal thermal buffers. The two layers are structurally connected by anti-seismic steel elements, designed

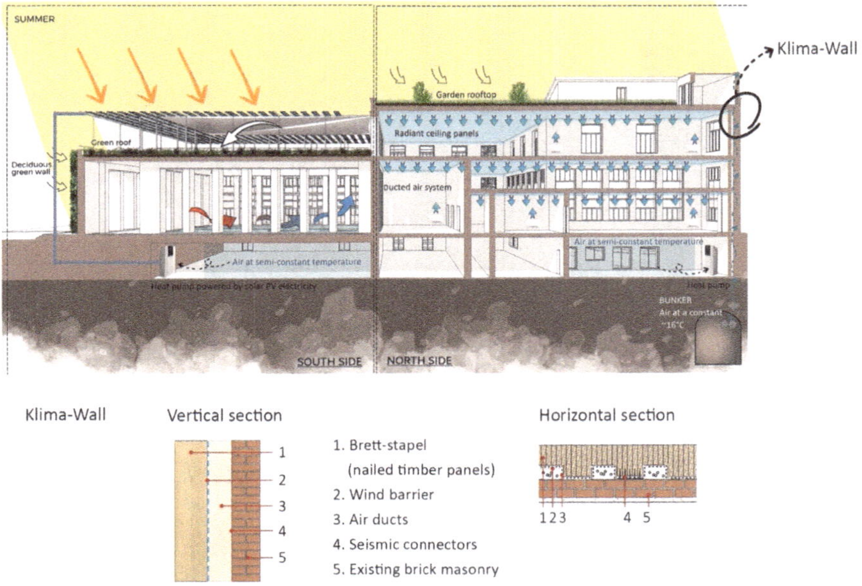

Fig. 5 Bioclimatic section of Villa IX Maggio illustrating, in particular, the summer operating concept and the Klima-Wall application. *Source* Drawings created by the authors

to ensure mechanical stability and seismic compatibility while maintaining the functional independence of the new internal structure.

These steel connectors act as controlled links between the original masonry and the timber wall, allowing the transfer of horizontal loads during seismic events without compromising reversibility or causing damage to the historic fabric. This solution enables the improvement of thermal performance while remaining compatible with heritage conservation principles.

Inspired by the Roman hypocaust, the ventilated cavities operate as thermal reservoirs, supporting passive heat accumulation in winter and heat dissipation in summer (Vasileva et al., 2022). When combined with hybrid solar collectors, heat exchangers, and low-energy heat pumps, the system contributes to indoor climate regulation. In addition, the hygroscopic properties of wood provide passive humidity control, improving indoor environmental quality (Herzog et al., 2012).

3.4 Subsoil Thermal Exchangers and Integrated Systems

At the site scale, the project exploits existing underground volumes as subsoil thermal exchangers, transforming former wartime bunkers and underground spaces beneath the patio-fountain area into seasonal energy storage systems. Located at depths of approximately 6–8 m, these volumes benefit from stable ground temperatures and are integrated with mechanical ventilation systems with heat recovery (MVHR).

The bunker-based exchanger serves the north wing of the Villa, supplying air to the Klima-Wall cavities and, through a heat pump, supporting heating and cooling functions. Classrooms, study rooms, and offices are equipped with radiant ceiling panels, while air-based systems are adopted for the conference hall. A secondary exchanger beneath the patio-fountain area operates in a similar manner, serving the south wing, which accommodates the multimedia library, exhibition spaces, and café.

This strategy demonstrates how existing underground structures, often present in historic or modern heritage sites, can be repurposed as low-impact energy resources, offering a potentially transferable solution for comparable contexts.

3.5 Green Systems and Microclimatic Regulation

Green roofs and green walls are integrated as complementary bioclimatic devices. Semi-extensive green roofs are introduced on the first level, while full garden roofs are designed for the upper level, enhancing thermal inertia, reducing runoff, and restoring surface permeability. Green walls are applied on the west and south façades, where permitted, functioning as microclimatic regulators, visual filters, and symbolic mediators between architecture and nature (Manso et al., 2021; Mohammed, 2022; Vox et al., 2018).

Importantly, the vertical greening systems are conceived as reversible and non-invasive solutions, structurally independent from the historic façades and compatible with their protected status. This approach ensures that environmental benefits can be achieved without altering the original architectural fabric.

3.6 Additional Performance-Enhancing Measures

Further measures contribute to the overall environmental performance of the building, including the replacement of existing window frames with replicas matching the original design and materials, upgraded with triple low-emissivity glazing; internal insulation using calcium silicate elements at ground-floor level; greywater recovery systems for WC flushing; and the extensive use of permeable pavements and vegetation to improve stormwater management and ecological permeability.

4 The Solar Forest

Within the Campus, conceived as a centre of excellence for research, development, and advanced training in bio-architecture and green relations, particular emphasis is placed on experiential learning, technological experimentation, and full-scale testing of sustainable construction solutions. In this framework, the Solar Forest plays a strategic role as a didactic and experimental device, complementing the more theory-oriented functions housed within Villa IX Maggio.

While the Villa is well suited to lectures, seminars, exhibitions, and co-studying activities, it cannot accommodate hands-on construction laboratories. For this reason, a multifunctional area in close proximity to the main building has been designed: the Solar Forest, conceived as a lightweight, reversible, and climatically responsive structure capable of hosting practical workshops, demonstrations, and public events.

The Solar Forest consists of a large timber-framed canopy (Fig. 6) supporting a semi-transparent photovoltaic glass roof (Fig. 7), beneath which a series of autonomous cubic or quasi-cubic laboratory units are arranged.

These units are designed as prototypical bio-architectural modules, employing exemplary construction technologies intended both for educational purposes and for potential replication in other contexts.

More specifically, the laboratory units adopt:

- Dry construction systems, enabling rapid assembly and disassembly and minimising construction waste;
- Natural and bio-based materials, such as timber structural frames, straw-based insulation panels, wood fibre boards, and clay or lime-based finishes, selected for their low embodied energy and hygroscopic properties;
- Prefabricated and modular components, allowing flexibility in spatial configuration and facilitating reuse, adaptation, and technological upgrading over time;
- High-performance glazed façades, oriented to maximise natural daylight while ensuring thermal comfort and visual continuity with the surrounding landscape;

Fig. 6 The Solar Forest volumes intended for practical workshops and recreational areas. *Source* Render created by the authors

Fig. 7 The semi-transparent photovoltaic covering of the Solar Forest, which mimics the shading effect of trees. *Source* Drawings created by authors

- Passive bioclimatic strategies, including cross-ventilation, external shading devices, and the exploitation of thermal mass where appropriate.

These technologies are considered "exemplary" not only because of their environmental performance, but also due to their didactic clarity, reversibility, and compatibility with circular economy principles. Each unit is conceived as a 1:1 scale demonstrator, allowing students, professionals, and visitors to directly observe and evaluate construction details, material behaviour, and environmental performance.

The Solar Forest canopy performs multiple functions: it provides renewable energy for the Campus, offers climatic protection for outdoor activities, and reproduces the shading effect of a tree canopy, thereby contributing to summer comfort. The structure is conceived as a lightweight, dry-assembled timber system, optimised through technological experimentation to reduce material consumption while ensuring structural efficiency (Herzog et al., 2012).

Beneath the canopy, a small open-air arena accommodates approximately eighty spectators and is intended for performances, seminars, and public events. Constructed using timber, natural stone, and integrated vegetation, the arena incorporates passive and active climate-control measures, including an overhead misting system for summer cooling and radiant heating embedded in the seating and flooring surfaces. Removable winter side screens further enhance seasonal adaptability.

An L-shaped partition made of straw and timber panels separates the performance area from the laboratory zone while maintaining visual and functional continuity. Centrally positioned within the Solar Forest, a service module houses refreshment and sanitary facilities, completing the multifunctional layout (Fig. 8).

Overall, the Solar Forest is conceived as an experimental and transferable model of a multifunctional, low-impact construction system. Its modularity, reversibility,

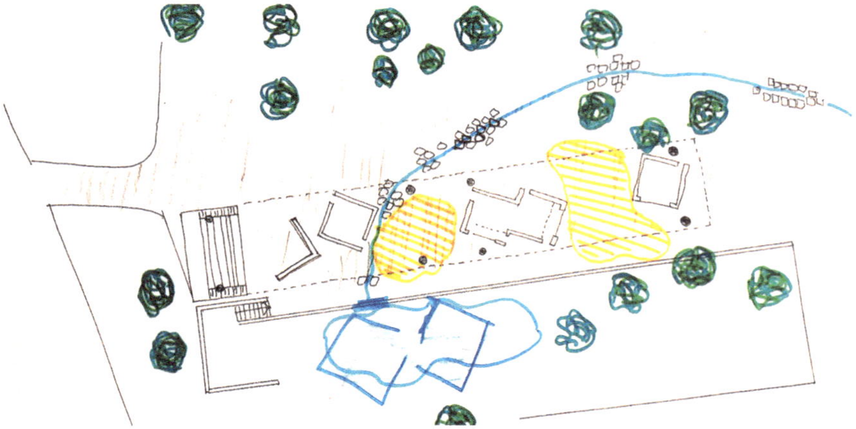

Fig. 8 The floor plan of the Solar Forest, highlighting the work areas and social areas. *Source* Sketch created by Silvia Federici

and reliance on bio-based materials make it applicable not only within the Villa IX Maggio Campus, but also in other educational, cultural, or heritage-related contexts where temporary or semi-permanent structures are required to support sustainable practices and community engagement.

5 Vegetation, Water, and Energy

Ecological regeneration of the site requires the retention and management of rainwater. This is achieved through a combination of green roofs, permeable pavements, phytoremediation basins, rainwater-collecting photovoltaic canopies, and extensive vegetation, all of which are essential for effective water management within the plot (Dreiseitl, 2015). Equally important is the reuse of greywater generated within the Campus, in synergy with the adjacent primary school, which functions as an integral component of the compound.

All new or retrofit interventions are designed to enhance both energy efficiency and the onsite supply of renewable energy. Villa IX Maggio, a historic and relatively rigid building, benefits from favourable solar exposure and shading provided by surrounding trees. In Rome's climate, characterised by mild winters and very hot summers, energy demand is primarily driven by summer cooling requirements. While bioclimatic strategies play a central role, they are not sufficient on their own; solar energy is therefore indispensable, implemented through an Energy Community and the multifunctional Solar Forest roof, which is capable of satisfying the Campus's energy needs on sunny days.

The green infrastructure extends beyond open spaces to include green roofs, vegetated walls, gardens, and pergolas, all forming an integrated vegetation system. These diverse spaces serve socio-cultural, artistic, and educational purposes, supporting social, sporting, and recreational activities (Manso et al., 2021; Mohammed, 2022). They are interconnected with one another and with the surrounding urban context, contributing to the realisation of the project's overall objectives.

Beyond Via Padre Semeria, the project introduces a Vertical Garden featuring a prominent green wall designed to mitigate both acoustic and atmospheric pollution. This prefabricated, fully demountable, and reusable system represents a reversible, low-impact solution. Gabion walls filled with quarry stone offcuts provide noise and air screening while supporting integrated irrigation planting. The area promotes biodiversity and circular economy principles through organic horticulture, Mediterranean vegetation, holistic pathways, and a bee garden.

A timber cycle–pedestrian bridge, integrated with vegetation, ensures safe crossing of the road and maintains ecological continuity. The structural system employs Natterer's innovative timber-concrete hybrid, in which juxtaposed timber trunks form the lower extrados and a Roman-style concrete slab is cast above. This configuration allows for over 1.6 m of soil depth, sufficient to support medium-height vegetation while maintaining structural integrity.

6 Conclusions

This study demonstrates how the principles of bio-architecture and bioclimatic design can be effectively applied to the adaptive reuse of a protected modern building and its surrounding landscape, transforming a historically and socially fragmented site into an integrated, climate-responsive Campus. Through the case of Villa IX Maggio, the research highlights the potential of low-impact, reversible, and site-specific interventions to enhance environmental performance, user wellbeing, and social inclusivity while preserving architectural and cultural values.

Beyond the specific design outcome, the contribution of this work lies in the methodological framework adopted. The integration of multi-scale analyses—ranging from urban and site considerations to building and component-level strategies—proved essential in identifying latent resources and guiding coherent design decisions under heritage protection constraints. This approach enabled the combination of passive and active systems, renewable energy technologies, and nature-based solutions within a unified bioclimatic strategy.

The proposed interventions, although tailored to the climatic, spatial, and regulatory context of Rome, demonstrate a high degree of replicability and transferability. Strategies such as internal climate-adaptive wall systems compatible with protected façades, the reuse of existing underground volumes as thermal exchangers, modular green infrastructure, and lightweight, reversible timber constructions can be adapted to other modern heritage buildings characterised by similar constraints, particularly in Mediterranean and temperate climates.

Importantly, the study suggests that modern architectural heritage—often perceived as rigid and technologically obsolete—can instead function as a testing ground for innovative and sustainable retrofit strategies. By avoiding invasive transformations and prioritising reversibility and compatibility, the project offers a viable alternative to conventional energy retrofit approaches, contributing to the ongoing debate on conservation-oriented sustainability.

In conclusion, the Villa IX Maggio Campus illustrates how bioclimatic redevelopment can extend beyond building performance to address broader socio-cultural and urban dimensions. The integration of educational facilities, experimental spaces such as the Solar Forest, and blue-green infrastructure supports not only environmental objectives but also community engagement and knowledge dissemination. In this sense, the project can be interpreted as a best-practice reference for the regeneration of modern heritage sites, where architectural conservation, environmental responsibility, and social value are addressed through an interdisciplinary and integrated design approach.

Acknowledgements This study received no external funding.

The authors would like to thank the Istituto Nazionale della Previdenza Sociale (INPS), owner of the complex, for providing historical documentation and granting access to the site. The authors also acknowledge the academic environment, as well as the institutions and professionals, that supported the development of the research.

References

Dreiseitl, H. (2015). Blue-green infrastructures for buildings and liveable cities. In T. Schröpfer (Ed.), *Dense + Green: Innovative building types for sustainable urban architecture* (pp. 48–59). Birkhäuser.

Herzog, T., Natterer, J., Schweitzer, R., Volz, M., & Winter, W. (2012). *Timber construction manual*. Walter De Gruyter. https://doi.org/10.1515/9783034614634

Manso, M., Teotónio, I., Matos Silva, C., & Oliveira Cruz, C. (2021). Green roof and green wall benefits and costs: A review of the quantitative evidence. *Renewable and Sustainable Energy Reviews, 135*, Article 110111. https://doi.org/10.1016/j.rser.2020.110111

Mitterer, W., De Colibus, G., & Stefani, B. (Eds.). (2012). *Note di viaggio: Scritti di bioarchitettura*. Editrice A. Weger.

Mohammed, A. B. (2022). Employing systems of green walls to improve performance and rationalize energy in buildings. *Journal of Engineering and Applied Science, 69*, 99. https://doi.org/10.1186/s44147-022-00154-9

Panakaduwa, C., Coates, P., & Munir, M. (2024). Identifying sustainable retrofit challenges of historical buildings: A systematic review. *Energy and Buildings, 313*, Article 114226. https://doi.org/10.1016/j.enbuild.2024.114226

RIBA. (2021). *Sustainable past and future guide*. Royal Institute of British Architects. https://www.architecture.com

Sasso, U. (2003). *Bioarchitettura: Forma e formazione. I dodici progetti di strutture formative più interessanti d'Europa*. Alinea Editrice.

Vasileva, I. L., Nemova, D. V., Vatin, N. I., Fediuk, R. S., & Karelina, M. I. (2022). Climate-adaptive façades with an air chamber. *Buildings, 12*(3), 366. https://doi.org/10.3390/buildings12030366

Vox, G., Blanco, I., & Schettini, E. (2018). Green façades to control wall surface temperature in buildings. *Building and Environment, 129*, 154–166. https://doi.org/10.1016/j.buildenv.2017.12.002

Integrated Landscape Retrofitting of School Open Spaces for Climate Resilience: Thermal and Environmental Metrics from a French Pilot Project

Giambattista Brizzi, Ornella Ferlito, Matilde Cedrone, Matthieu Majour, and Marcella Benini

Abstract Urban school environments are increasingly recognized as critical sites for climate-adaptive design, yet empirical evidence on the microclimatic performance of schoolyard retrofits remains limited. This study contributes to this emerging field by evaluating the effectiveness of integrated landscape retrofitting strategies in mitigating Urban Heat Island (UHI) effects and improving outdoor thermal comfort in educational settings. Building on existing research linking surface thermal properties, vegetation, and microclimatic regulation, the work adopts a pre/post comparative methodology supported by thermophysical simulations and environmental performance modelling. The assessment examines changes in surface albedo, permeability, vegetative cover, and material thermal behaviour and quantifies their influence on surface temperature reduction, carbon impact, and ecological function. The intervention aligns with current international frameworks promoting climate-resilient and biophilic design, such as the WELL Building Standard v2 (IWBI (International WELL Building Institute), 2025) highlighting the role of nature-based solutions in supporting student well-being and outdoor learning. Results from the Jean Moulin secondary school retrofit in Chaville (France) demonstrate the potential of coordinated material substitution, spatial reconfiguration, and vegetation integration to enhance microclimatic performance, achieving reductions of up to −7.3 °C in average surface temperature, −44% in impermeability, and −65% in embodied carbon. The study offers a replicable approach for designing climate-adaptive educational landscapes and contributes new empirical evidence to current scientific debates on urban microclimate governance and regenerative public realm design.

G. Brizzi (✉) · M. Cedrone · M. Majour
Deerns Italia Spa, Milan, Italy
e-mail: giambattista.brizzi@deerns.com

O. Ferlito
Deerns Italia Spa, Rome, Italy

M. Benini
Deerns Nederland B.V., The Hague, The Netherlands

L. Diana et al. (eds.), *Greening and Resilience*, Urban Sustainability,
https://doi.org/10.1007/978-981-92-0419-9_25

Keywords Outdoor microclimate · Urban heat island mitigation · Promotion of greening · Life cycle impact reduction · Sunlight access

1 Introduction

The Urban Heat island (UHI) effect forms when urbanized areas become warmer than the surrounding rural regions. As cities intensify, the replacement of vegetation with buildings and impermeable surfaces reduces shading, evapotranspiration, and moisture availability, thereby diminishing natural cooling. The presence of limited vegetation and the heat retention capacity of urban materials exacerbate the existing trend of rising temperatures caused by climate change, resulting in increasingly higher urban temperatures. The phenomenon affects communities by increasing summertime peak energy demand, air conditioning costs, air pollution and greenhouse gas emissions, and consequently aggravating heat-related illness, mortality, and water quality (EPA (U.S. Environmental Protection Agency), 2008).

Urban educational environments constitute high-stress microclimatic zones due to their extensive impervious surfaces, minimal vegetation, and intensive daytime use. The established scientific consensus links surface thermal properties, particularly albedo and permeability, and the presence of vegetative systems to the modulation of local air and surface temperatures, particularly in the context of UHI intensification (EPA (U.S. Environmental Protection Agency), 2008; Oke, 1982). However, robust empirical datasets detailing post-intervention performance in schoolyard retrofits remain markedly scarce. This project bridges that gap by integrating a full-spectrum environmental design process, spanning from thermophysical simulations to carbon impact quantification.

This study evaluates the microclimatic performance of a landscape retrofitting intervention at Jean Moulin Secondary School in Chaville, France. The goal is to provide a replicable evaluation strategy that may be applied in similar contexts to determine the effectiveness of landscape retrofitting interventions in mitigating Urban Heat Island (UHI) effects and improving outdoor thermal comfort in dense educational settings.

The analysis compares pre- and post-intervention conditions using simulation-based modelling to evaluate the effects of variations in key physical parameters such as surface albedo, soil permeability, biodiversity, vegetative cover, and material thermal behaviour. The study quantifies the impact of strategic material substitution, spatial reconfiguration, and vegetation integration on surface thermal regimes, and assesses their potential for climate-resilient adaptation within the normative framework of French public educational infrastructure.

Moreover, the intervention is designed in alignment with the WELL Building Standard v2 Feature 9 (Thermal Comfort) and Feature 100 (Biophilia II) (IWBI (International WELL Building Institute), 2025), which underscore the necessity of microclimatic regulation and nature-based spatial programming in learning environments. Evidence from neuroscientific and environmental psychology domains substantiates the cognitive and affective benefits of nature exposure on students, including improved attention restoration, reduced psychological stress, and enhanced academic engagement (Li & Sullivan, 2016; Soga et al., 2017).

By embedding vegetation and thermal regulation mechanisms within the spatial logic of the schoolyard, this project extends its value proposition beyond climatic metrics to include neurophysiological and educational co-benefits.

In this context, the Jean Moulin school retrofit constitutes both a climatic and pedagogical experiment, quantitatively modelled, regulatory compliant, and methodologically positioned to serve as a prototype for systemic schoolyard adaptation in temperate urban climates.

2 Methodological Approach

2.1 The Case Study

The Jean Moulin secondary school, located in Chaville within the Île-de-France region, is embedded in a peri-urban context characterized by dense urbanization, limited ecological connectivity, and high exposure to climate stressors. The school's open spaces, totalling 4958.4 m^2, include primary circulation corridors, multi-use courtyards, and recreational fields, all of which were originally surfaced with low-reflectivity, impermeable asphalt (see Fig. 3).

The resultant spatial morphology provided minimal thermal moderation, high runoff coefficients, and negligible vegetative presence, culminating in significant thermal stress during summer months and a lack of environmental heterogeneity supportive of ecological function or pedagogical enrichment. Figure 1 shows the existing state of the school.

The site's transformation was initiated under the auspices of the Département des Hauts-de-Seine's climate adaptation program for public educational institutions. This

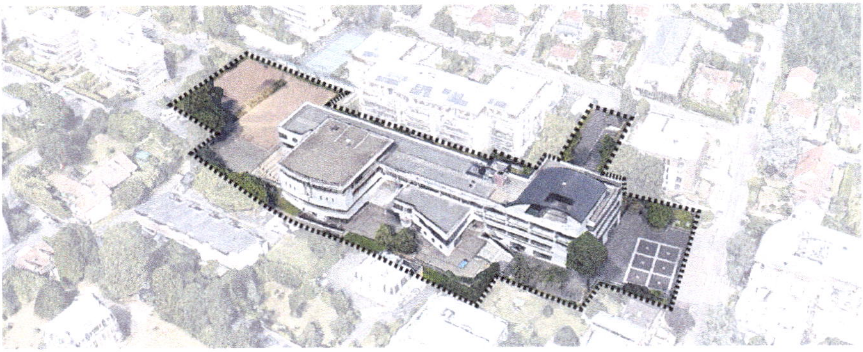

Fig. 1 Existing state of Jean Moulin School in Chaville, Île-de-France, Paris (Google, 2025)

regional initiative responds to the dual imperative of environmental resilience and youth welfare, targeting legacy infrastructure through ecologically integrated design interventions. The program specifically prioritizes interventions where climatic exposure intersects with social vulnerability and daily human occupation.

In this context, the Jean Moulin schoolyard presented both a case of functional inadequacy in relation to contemporary climatic demands and an opportunity for demonstrative intervention under a climate governance framework oriented toward measurable environmental and social outcomes.

2.2 Integrated Methodology

Urban Heat Island (UHI) effects result from interacting processes: solar gains, long-wave exchanges, heat storage in materials, airflow, and vegetation-driven evapotranspiration. Therefore, early design phases require rapid, reproducible analyses to support iterative and collaborative decision-making. Design interventions were assessed using an integrated workflow of computational tools selected for speed, reliability, and quantifiable outputs.

Each tool focused on a distinct domain: radiative exposure, surface thermodynamics, and pedestrian-level airflow. This approach enabled comparison of existing conditions and proposed scenarios, guiding material choices, spatial reconfiguration, and vegetation integration. The overall methodology and tool workflow are illustrated in Fig. 2.

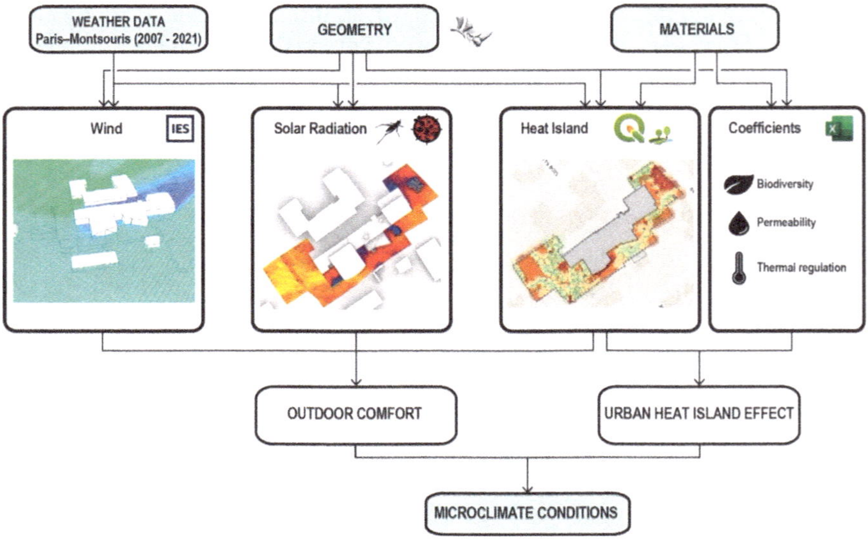

Fig. 2 Integrated methodology, tools, and workflow used. *(Deerns Italia S.pA.)*

Table 1 Software simulation settings for integrated analyses *(Deerns Italia S.pA.)*

Software	Analysis	Grid/ Spacing	Time Step/ Solver	Key conditions
ICEtool	Ground surface temperature	1 m	1 h	Paris–Montsouris (2007–2021); material albedo, emissivity, conductivity
Ladybug	Sunlight hours (winter & summer)	0.1 m	1 h	Sun position from Paris–Montsouris; typical design days
IESVE (MicroFlo)	CFD wind flow (pedestrian level)	1 m	Steady-state RANS	Inlet wind 5.5 m/s from south; Exposure type (urban terrain)

Tool selection reflected project stage and resource constraints. High-fidelity microclimate models were excluded due to extensive data and long runtimes, unsuitable for early design. Instead, tools delivering credible results with modest setup were adopted. Simulations incorporated local climate data and site geometry, balancing detail and efficiency. The Paris–Montsouris reference climate file (2007–2021) was used as the meteorological data for all analyses. The simulation settings are summarized in Table 1.

2.3 *Tool Selection and Roles*

ICEtool: Ground Surface Thermodynamics

ICEtool estimated ground surface temperatures under representative boundary conditions, considering material properties such as albedo, emissivity, and thermal conductivity. It was used to evaluate material substitutions and identify thermal hotspots where shading or vegetation would provide the greatest benefit.

Ladybug: Sunlight Hours

Ladybug calculated hours of direct solar incidence by projecting sun vectors from the local weather file onto project geometry. Seasonal solar access maps informed the placement and sizing of trees, canopies, and reflective or absorptive surfaces, supporting decisions on shading and outdoor comfort.

IESVE (MicroFlo): CFD Wind Flow

MicroFlo resolved near-ground wind fields around buildings and landscape features, assessing ventilation corridors, edge permeability, and localized accelerations. Steady-state RANS (Reynolds-averaged Navier–Stokes) simulations provided directional wind-speed patterns for early design decisions on vegetation layout, screens, and open-space orientation. Analyses focused on winter conditions to capture worst-case comfort scenarios (Table 1).

2.4 *Simulation Settings*

2.5 *Performance Indicators*

The integrated workflow produced key metrics that informed design decisions:

- Sunlight exposure: Hours of direct solar incidence on outdoor areas for representative winter and summer days (Ladybug).
- Surface temperature: Maximum ground temperatures at analysis points, identifying thermal hotspots and material effects (ICEtool).
- Wind speed: Pedestrian level velocities at selected locations, highlighting comfortable and over exposed zones (IESVE/MicroFlo).

3 Urban Heat Island (UHI) Mitigation

Ground surface temperatures were calculated using ICEtool, an open-source plugin for QGIS, to quantify UHI intensity. ICEtool simulates microclimatic thermal exchanges considering surface reflectivity (albedo), material emissivity and thermal conductivity, urban morphology and spatial organization, and the presence of water and vegetation. The modelled ground materials for both the existing state and the final design are shown in Fig. 3, illustrating the comparative setup used for simulation.

In addition to ground surface temperature, design strategies were evaluated using site-level indices derived from material coefficients specified in the Environmental Quality Standards for Departmental Buildings (Appendix 03: Technical Data Sheets) (Départements des Yvelines et des Hauts-de-Seine, 2020). These calculated indices included:

- Thermal Regulation Coefficient (TRC): Estimates the capacity of outdoor spaces to regulate thermal conditions.
- Impermeability Index: Ratio of impermeable surfaces to total site area, based on material permeability values.
- Biodiversity or Biotope Coefficient: Indicates the site's potential to support biodiversity, calculated from material and vegetation characteristics.

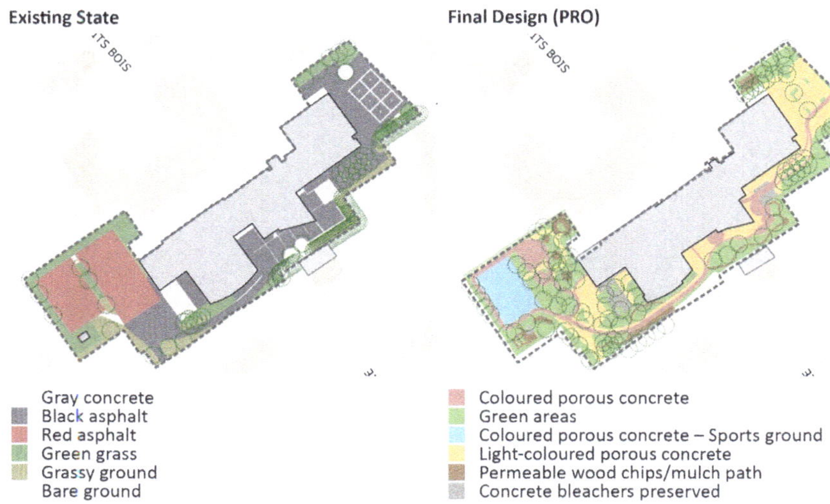

Fig. 3 Ground Materials modelled in ICEtool. *(Deerns Italia S.pA.)*

These parameters were considered, alongside the simulated ground surface temperatures, to assess the effective mitigation of the UHI effect. Before intervention, the Jean Moulin schoolyard had a high surface impermeability index of 83%, which contributed to localized heat accumulation and hindered passive hydrological regulation. The retrofit strategy addressed this by replacing the material most commonly present (asphalt), resulting in a post-intervention impermeability coefficient of 39%. This was achieved through integration of hydraulically active surfaces such as porous concrete, stabilized mulch substrates, and open-graded vegetated soil beds.

This transition not only provided hydrological benefits but also improved thermal properties, with the site's weighted average albedo increasing from below 0.18 to approximately 0.31. The recalibration of albedo was tailored to specific materials: porous light-coloured concrete used in circulation areas achieved an albedo of 0.30; sports tracks made from pigmented polymeric surfaces showed values between 0.36 and 0.42, while vegetated earth zones varied from 0.23 to 0.39 depending on granulometric composition and vegetation density.

Hourly ground temperatures for each simulated point were computed using ICEtool for July 21 (T_air = 35 °C, clear sky). The temporal evolution is presented in Figs. 5 and 6. This analysis enabled identification of the maximum temperature reached at each point during the day, as shown in Fig. 4. Results indicate an average surface temperature reduction of 5.0 °C across the site, with a peak average reduction of 7.3 °C at approximately 6 p.m. Additionally, functional ecological indicators showed significant improvements: the site's thermal regulation coefficient increased from 37 to 50%, and the biotope coefficient rose from 31 to 45%, indicating enhanced ecological permeability and bioclimatic performance.

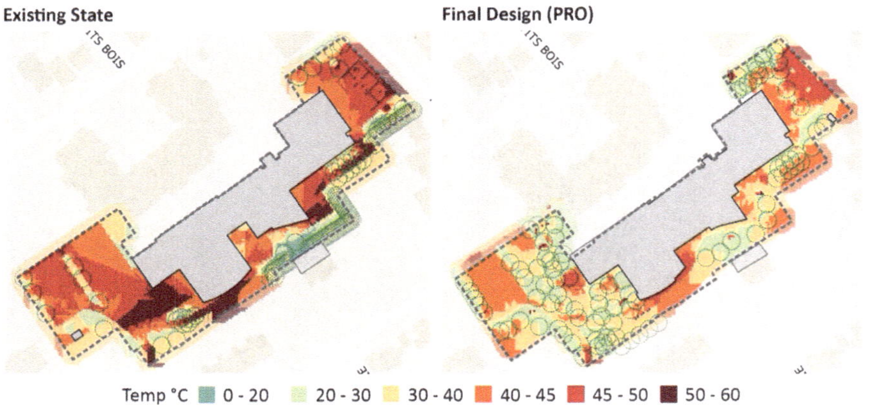

Fig. 4 Maximum daily ground temperature simulation results for 21st of July. *(Deerns Italia S.pA.)*

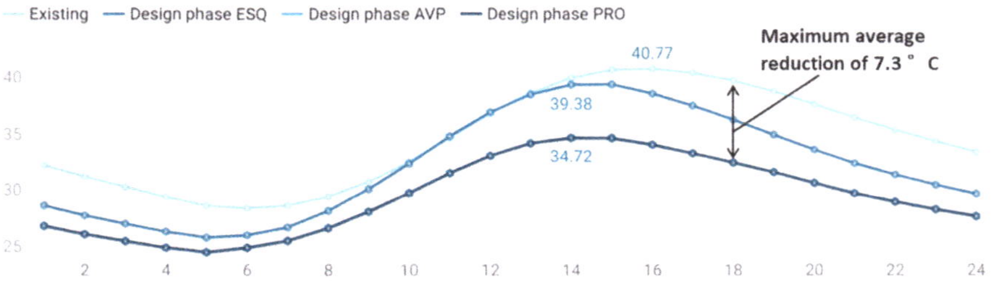

Fig. 5 Average hourly ground temperature results for each project stage. (21st July) *(Deerns Italia S.pA.)*

These results demonstrate the effectiveness of integrating material and environmental strategies in mitigating UHI effect and position the intervention within best-practice benchmarks for microclimate-responsive educational infrastructure retrofits.

4 Outdoor Thermal Comfort Conditions

The optimization of outdoor thermal comfort at the Jean Moulin schoolyard was guided by a simulation-informed design process integrating seasonal solar access modelling and pedestrian-level wind flow analysis with the previously described ground temperature simulations. Early-phase simulations using parametric environmental plug-in suite Rhinoceros/Grasshopper-Ladybug enabled the identification of critical solar exposure patterns across seasonal cycles, as shown in Fig. 7. The simulations were run directly onto the architectural 3D model provided by the architects

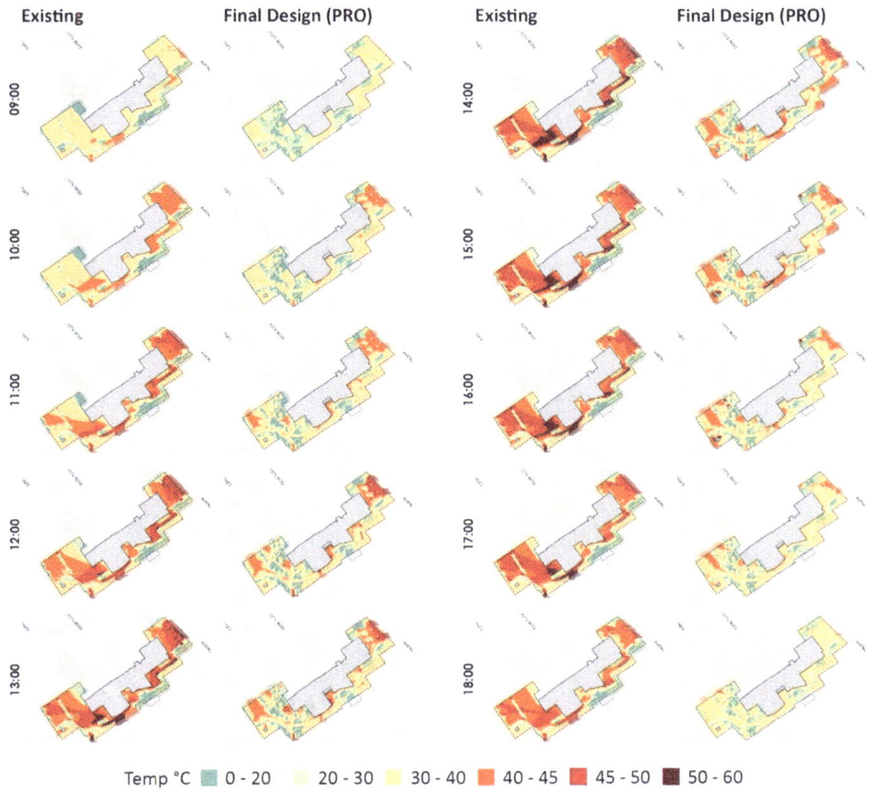

Fig. 6 Hourly ground temperature simulation results for 21st of July. *(Deerns Italia S.pA.)*

in Rhinoceros. This allowed for rapid, interactive, and integrated feedback on design decisions to be communicated clearly and effectively within the design team.

In the winter scenario, over 50% of the courtyard surface area achieved the threshold of > 2 h of direct insolation per day, confirming adequate passive solar access for thermally beneficial use during colder months (Littlefair et al., 2011). Conversely, simulations for the summer condition revealed that only 32–39% of the surface area complied with the benchmark of < 6 h/day of solar exposure, highlighting the risk of prolonged thermal stress. Design iterations were thus calibrated to address these deficiencies, prioritizing the spatial reallocation of vegetation, surface material differentiation, and the introduction of shading typologies. The refinement process maintained acceptable winter insolation while incrementally increasing summer shading coverage, particularly in zones of high user density.

In parallel, computational fluid dynamics (CFD) analyses were conducted using CFD (MicroFlo module of IES-VE) to evaluate and optimize wind comfort conditions. Annual and seasonal wind rose diagrams derived from this file indicate a predominant southerly wind direction during the winter months, with velocities

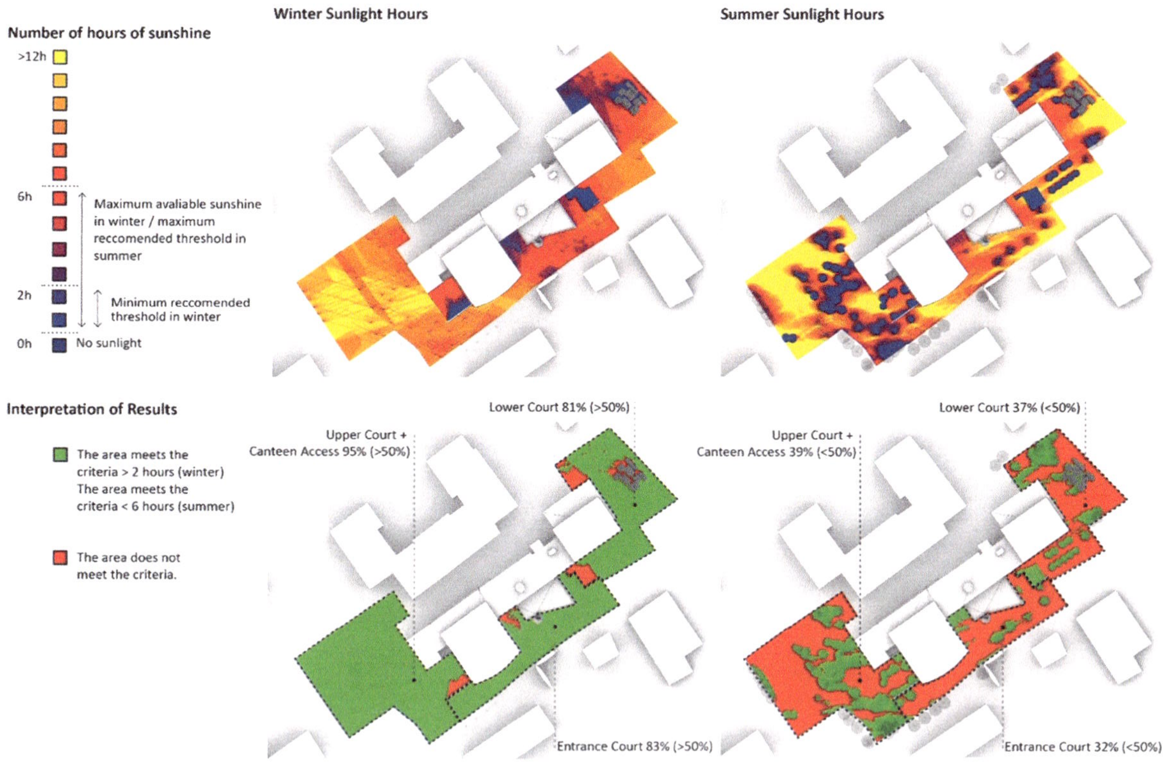

Fig. 7 Winter and summer direct solar radiation hours simulation of the design proposal made with Rhinoceros + Grasshopper and Lady Bug extension (3rd March and 21st June). *(Deerns Italia S.pA.)*

reaching 5.5 m/s. The analysis was run using these settings, prioritizing winter conditions where wind induced exposure is most critical for outdoor learning and play.

The CFD workflow informed decisions on vegetation placement, edge permeability, and corridor orientation to ensure air movement conducive to thermal dissipation without generating discomforting turbulence. Final simulation outputs confirmed that the principal zones of educational and recreational use (cour basse, parvis) maintained mean wind velocities below 2.0 m/s, optimal for sedentary outdoor activities, while the cour haute (shown in Fig. 8) registered wind peaks up to 4.0 m/s, considered tolerable for dynamic activities such as sports (Netherlands Standards, 2006).

This simulation-guided design methodology led to a final configuration that measurably improved the site's microclimatic performance. Quantitatively, average surface temperature reductions of up to 7.3 °C were recorded in vegetated and shaded zones during peak summer conditions, derived from ICEtool-based thermophysiological modelling. The alignment between simulation predictions and expected user comfort thresholds validates the strategic integration of digital environmental tools within the landscape design workflow and demonstrates their efficacy in delivering evidence-based, climate-resilient design outcomes in educational open spaces.

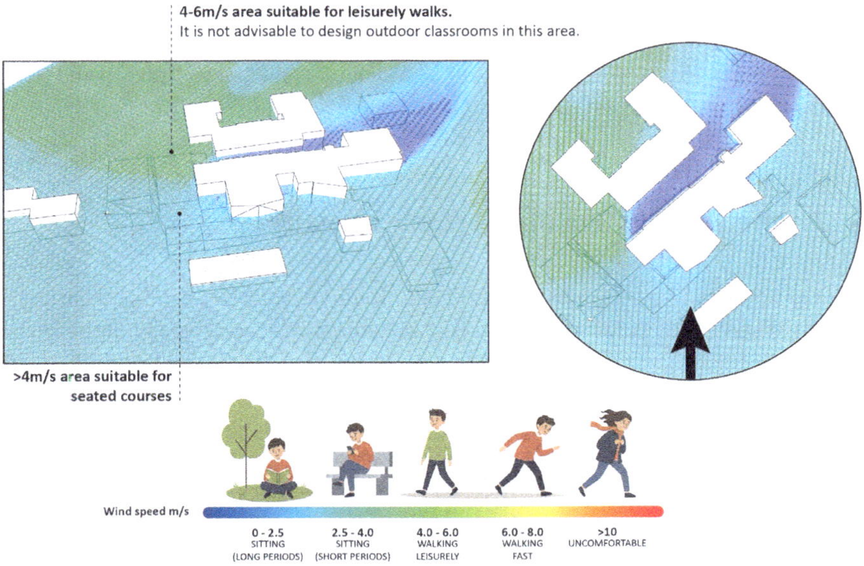

Fig. 8 Upper Court Computational Fluid Analysis (at 10.45 m) *(Deerns Italia S.pA.)*

5 Promotion of Greening in Urbanized Areas

The greening strategy implemented at the Jean Moulin schoolyard constitutes a deliberate recalibration of the site's vegetative composition to enhance microclimatic regulation and biodiversity potential.

A total of 934.3 m^2 of open-soil planting zones were introduced, enabling direct interaction between vegetation and subsoil hydrological systems. The vegetative assemblage was expanded through the addition of 37 deciduous tree specimens, specifically selected for their high canopy density (reaching up to 60% foliage cover), and their adaptability to urban stress conditions, including moderate drought tolerance. Dominant species include Celtis australis, Pyrus calleryana, and Liquidambar styraciflua, chosen for their combined shading capacity, seasonal evapotranspiration profiles, and minimal maintenance requirements.

These interventions were strategically distributed across thermally sensitive zones to maximize localized surface cooling through combined mechanisms of transpiration flux and solar occlusion. While the design intent aligns with the state-of-the-art bioclimatic principles, the lack of detailed vegetative metadata (such as leaf area index (LAI), phenological variability, and canopy stratification) precludes the full integration of volumetric cooling effects into advanced simulation models.

As a result, while qualitative performance outcomes are evident, the capacity for predictive modelling of dynamic vegetation–climate interactions remains constrained. Nonetheless, the expansion of vegetative infrastructure represents a critical vector for thermal mitigation, ecological connectivity, and pedagogical enrichment in urban school environments.

6 Carbon Evaluation of Greening Measures

The environmental performance of the Jean Moulin retrofit was evaluated through a comprehensive life cycle carbon analysis, structured in accordance with the French regulatory framework defined by the Loi AGEC (2020) and Décret 2021–950 (Gouvernement de la République Française, 2021). The assessment employed the QuantiGES methodology developed by ADEME (ADEME – Agence de la Transition Écologique, 2021), enabling full-spectrum quantification of greenhouse gas (GHG) emissions across materials, transport, and biogenic sinks over a 50-year reference horizon, illustrated in Fig. 9 and Table 2.

Material strategies prioritized the substitution of carbon-intensive components with low-emission alternatives. Most notably, the use of CEM III/A low-carbon cement in lieu of conventional Portland cement mixtures resulted in an estimated 72% reduction in embodied CO_2 per square meter. This material shift was augmented by dual-stream reuse protocols: in-situ reutilization of hardscape elements (e.g., pavers and site furniture) and ex-situ integration of recycled mineral aggregates, yielding net carbon savings of approximately -30.5 $kgCO_2e/m^3$ for granular substrates.

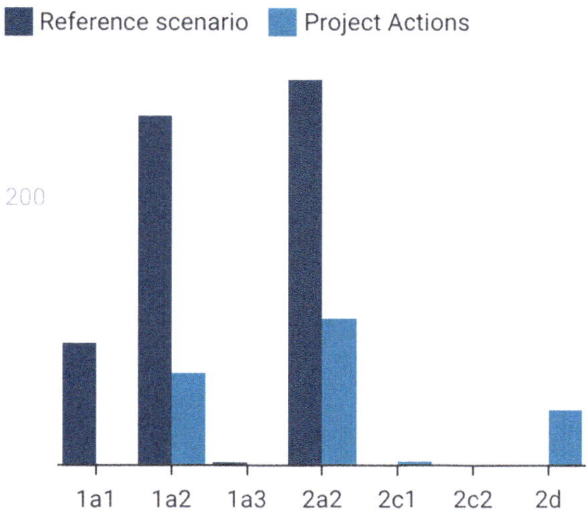

Fig. 9 Emissions bar chart of project and reference scenarios by category (tonCO2e) (Deerns Italia S.pA.)

Table 2 Emissions of project and reference scenarios by category (tonCO_2e). (Deerns Italia S.pA.)

Category	Reference scenario	Project Actions
1a1—Destruction of areas to be de-sealed	92	1
1a2—Transport of non-recyclable waste on site	262	69
1a3—Treatment of waste that is not recycled on site	2	0
2a2—Creation of new permeable surfaces and improvement of the albedo of permeable surfaces	289	110
2c1—Removal of old trees	0	3
2c2—Planting new trees	0	1
2d—Construction of new furniture	0	42

The aggregate GHG emissions for the project lifecycle were calculated at 225.4 tons CO_2e, representing a 65% reduction relative to a business-as-usual (BAU) scenario benchmarked against conventional public schoolyard retrofits (see Fig. 10).

Furthermore, as detailed in Table 3, vegetative carbon sequestration was modelled using standardized absorption coefficients: 296 tons CO_2e attributed to green surface treatments and 1885 tons CO_2e projected for the total arboreal biomass (existing and newly planted) over the analysis period.

An environmentally optimized procurement framework underpinned these outcomes, favouring biosourced materials such as untreated timber and organic mulch, alongside geosourced inputs including raw earth and crushed limestone. Transport emissions were systematically calculated using mass-distance factors (ton·km), calibrated to actual transportation routes to regional processing facilities.

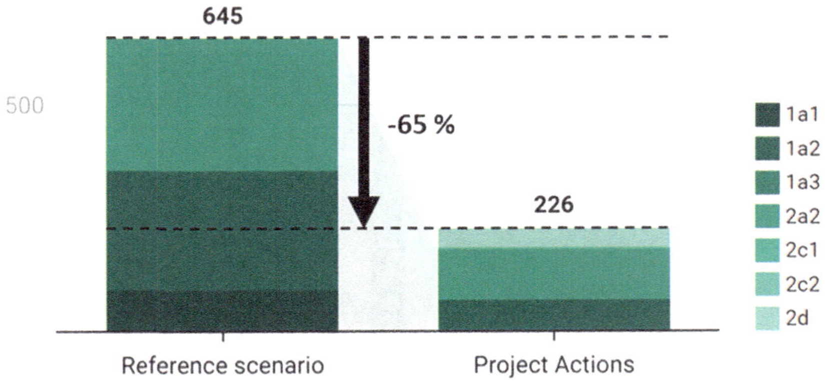

Fig. 10 Total emissions of project and reference scenarios (tonCO2e). *(Deerns Italia S.pA.)*

Table 3 Calculated quantitative impact of green spaces and trees. (Deerns Italia S.pA.)

Material Type	Description	Quantity	Unit	Carbon Impact ($kgCO_2eq./unit$)	Carbon Impact ($tonCO_2e$)
Green Spaces	CO_2 emissions reduction	5928	m^2	−150	−296
Trees	Planting and life cycle of existing trees	38	units	−1813	−68
Trees	Planting and life cycle of new trees	37	units	−49,132	−1817

The synthesis of low-carbon materials, circular resource strategies, and long-horizon biological sequestration mechanisms positions the intervention within best-practice benchmarks for low-impact urban greening. It also affirms the strategic relevance of integrating carbon performance criteria into early-phase design workflows for public-sector climate adaptation projects.

7 Discussion

7.1 *Capabilities of the Workflow*

The design-research integration demonstrates operational capability in combining drainage improvement, surface material optimization, and greening strategies for multi-scalar climate performance. The evidence supports that a strategic substitution of materials and spatial-functional reorganization can halve the surface impermeabilization index while nearly doubling the effective albedo. The ICE tool provides surface temperature simulations confirming measurable comfort gains up to 7.3 °C in the site. Additional simulations substantiate microclimatic performance through validated reductions in solar load and maintenance of effective ventilation.

The project illustrates a replicable process for applying quantitative environmental performance criteria in early-stage educational retrofitting. The method developed for the Jean Moulin project has already been applied to other schools within the Département des Hauts-de-Seine's climate adaptation program.

7.2 Limitations

No post-intervention empirical microclimatic monitoring was conducted, and simulation results remain unvalidated by field data. The absence of comprehensive vegetation descriptors (e.g., canopy height and seasonal dynamics) and the lack of advanced thermophysiological models (e.g., UTCI and PET) constrain the robustness of comfort projections. CFD analysis, although informative, is not dynamically coupled with thermal outputs.

The current implementation relies on simulations using present-day climate data. This approach does not yet account for projected climate variations. Future work will incorporate scenarios based on anticipated climate conditions to enhance the robustness of the analysis.

7.3 Applicability

Despite limitations, the intervention provides a methodological template for climate-responsive design in schoolyards. Its indicators (impermeability coefficient, albedo, surface temperature delta, biodiversity, and CO_2 balance) are standardized and reproducible. The approach aligns with European regulatory imperatives and is adaptable to similar institutional contexts.

8 Concluding Remarks

The Jean Moulin schoolyard retrofit exemplifies a robust, simulation-guided model for climate adaptation in urban educational environments. The intervention delivers demonstrable improvements across six critical environmental performance domains: thermal moderation, hydrological permeability, surface reflectance, user comfort, vegetative biodiversity, and lifecycle carbon impact. These outcomes are substantiated by high-resolution environmental simulations, quantified material metrics, and regulatory-aligned carbon accounting.

The design methodology, anchored in an iterative workflow of environmental modelling, material optimization, and lifecycle analysis, enables a data-driven calibration of spatial, morphological, and vegetative parameters to maximize microclimatic efficacy. The substantial reductions in surface temperature (up to $-$ 7.3° C), impermeability (-44%), and embodied carbon (-65% relative to BAU) position this case as a benchmark for evidence-based retrofit strategies within the European school infrastructure context.

The approach constitutes a transferable prototype for municipal authorities and design practitioners seeking to integrate environmental performance targets into the early design phase of public realm projects. By advancing a quantifiable, replicable framework for climate-sensitive design, the Jean Moulin project contributes materially to the discourse on urban microclimate governance and regenerative educational landscapes.

Acknowledgements The contribution of Atelier LD, the landscape architects, was essential to the project, with close interdisciplinary collaboration integrating landscape design and environmental analysis.

References

ADEME – Agence de la Transition Écologique. (2021). Méthode QuantiGES – Quantifier l'impact GES d'une action de réduction des émissions. *Guide méthodologique – 3e version*. ADEME.

Deerns Italia S.pA. (n.d.).

Départements des Yvelines et des Hauts-de-Seine. (2020, Ottobre). Annexe 03: Fiches Techniques - Version 03. *Référentiel de Qualité Environnementale des Bâtiments Départementaux – Livre 1*. Algoé consultants; TRIBU.

EPA (U.S. Environmental Protection Agency). (2008). Urban heat island basics. *Reducing urban heat islands: Compendium of strategies*. EPA.

Google. (2025, 11 05). https://www.google.com/maps. Retrieved from Google Maps.

Gouvernement de la République Française. (2021). *Loi AGEC – Décret n°2021–950 du 16 juillet 2021 relatif au tri des déchets de papier, métal, plastique, verre, textiles, bois, fraction minérale et plâtre*. Légifrance.

IWBI (International WELL Building Institute). (2025). *Optimization thermal comfort: T09 β outdoor thermal comfort*. Retrieved from Standard WELL v2, Q2 2025. https://v2.wellcertified.com/en/wellv2/overview

Li, D., & Sullivan, W. (2016). Impact of views to school landscapes on recovery from stress and mental fatigue. *Landscape and Urban Planning, 148*, 149–158.

Littlefair, P. J., King, S., Howlett, G., Ticleanu, C., & Longfield, A. (2011, Settembre 12). Site layout planning for daylight and sunlight: A guide to good practice. *BR209 2011 Edition*. IHS BRE Press.

Netherlands Standards. (2006). *NEN 8100: Wind comfort and wind danger in the built environment*. NEN.

Oke, T. R. (1982). The energetic basis of the urban heat island. *Quarterly Journal of the Royal Meteorological Society, 108*(455), 1–24.

Soga, M., Gaston, K., & Yamaura, Y. (2017). Gardening is beneficial for health: A meta-analysis. *Preventive Medicine Reports, 5*, 92–99.

A Site-Specific Methodology for the Sustainable Design of Green Roofs in Urban Climates

Cristina Passarelli

Abstract The progressive increase in urban temperatures, driven by soil sealing and urban densification, has intensified the need for effective and climate-responsive adaptation strategies. Among Nature-Based Solutions (NBS), green roofs are widely recognized for their potential to mitigate the Urban Heat Island (UHI) effect while providing multiple co-benefits, including thermal regulation, stormwater management, energy savings, and biodiversity enhancement. Despite this acknowledged potential, green roof design is still frequently based on generic prescriptions or aesthetic criteria, with limited consideration of site-specific climatic, ecological, and structural constraints. This often results in suboptimal performance, increased maintenance demands, and reduced long-term resilience. This paper proposes a holistic and site-specific framework for green roof design that integrates climatic classification, ecological bioindicators, solar radiation assessment, substrate characterization, and structural verification into a coherent decision-making process. Climatic conditions are first evaluated using the Köppen–Geiger classification to guide the selection of vegetation adapted to local temperature and precipitation regimes. Ecological compatibility is further refined through the use of Ellenberg indicator values, enabling plant species to be selected according to their preferences for light, temperature, and moisture. Spatial variability in rooftop microclimatic conditions is addressed through detailed solar exposure analysis, ensuring an appropriate allocation of species according to actual irradiance levels. Substrate selection is carried out in accordance with international technical standards, considering physical, hydrological, and ecological properties while ensuring compatibility with vegetation requirements. Structural feasibility is assessed through analytical verification and load simulations, accounting for both dry and saturated conditions of the green roof system. The framework is tested on a residential building in Portici (Naples), located in a Mediterranean Csa climate. The results demonstrate that the proposed methodology allows a progressive reduction of design uncertainty, leading to green roof solutions that are ecologically viable, structurally safe, and context-sensitive.

C. Passarelli (✉)
Department of Civil, Building and Environmental Engineering, University of Naples Federico II, Naples, Italy
e-mail: Cristina.passarelli@unina.it

L. Diana et al. (eds.), *Greening and Resilience*, Urban Sustainability,
https://doi.org/10.1007/978-981-92-0419-9_26

The approach provides a replicable and scientifically grounded tool for sustainable green roof implementation, supporting UHI mitigation and urban resilience.

Keywords Green roofs · Climate-adaptive green infrastructure · Framework for sustainable design · Urban heat island mitigation · Site-specific vegetative systems

1 Introduction

The rapid intensification of urbanization processes, combined with global climate change, has led to a significant increase in thermal stress within cities. One of the most evident manifestations of this phenomenon is the Urban Heat Island (UHI) effect, defined as the systematic increase in air and surface temperatures in urban areas compared to their rural surroundings (Oke, 1987; Rizwan et al., 2008). The UHI phenomenon is primarily driven by land sealing, high building density, reduced vegetation cover, and the widespread use of materials characterized by low albedo and high thermal inertia, such as asphalt and concrete (Berardi et al., 2014; Rizwan et al., 2008; Santamouris, 2014). As a result, urban environments experience elevated daytime and nighttime temperatures, increased cooling energy demand, deterioration of air quality, and heightened public health risks, particularly during heat waves (Bowler et al., 2010; Rizwan et al., 2008; Santamouris, 2014).

In recent decades, the UHI effect has been recognized not only as an environmental issue but also as a critical challenge for urban resilience and sustainability. Studies have demonstrated that UHI intensity can reach differences of 2–8 °C between urban and rural areas, with peak values exceeding 10 °C in densely built metropolitan regions under favorable meteorological conditions (Alexandri & Jones, 2008; Oke, 1987). These thermal anomalies significantly exacerbate energy consumption for space cooling, contributing to higher greenhouse gas emissions and creating a feedback loop that further intensifies urban overheating (Sailor, 2008; Santamouris, 2014). Consequently, mitigating and adapting to UHI effects has become a priority within international climate policies and urban planning strategies.

Within this framework, Nature-Based Solutions (NBS) and Green Infrastructure (GI) have emerged as effective and multifunctional approaches to address urban overheating while simultaneously delivering co-benefits in terms of biodiversity, stormwater management, air quality improvement, and human well-being (Bowler et al., 2010; De Montis, 2024; Ramyar & Rutherford, 2021). Among these solutions, green roofs represent one of the most promising interventions for the built environment, as they allow the transformation of otherwise impervious roof surfaces into thermally and hydrologically active systems (Oberndorfer et al., 2007). By combining vegetation, substrate, and technical layers, green roofs can reduce surface temperatures, enhance evapotranspiration, improve building energy performance, and contribute to the cooling of the surrounding urban microclimate (Alexandri & Jones, 2008; Perini et al., 2011; Sailor, 2008; Takebayashi & Moriyama, 2007).

Recent literature increasingly frames green roofs within the broader paradigm of climate-adaptive Nature-Based Solutions, emphasizing their role not only in heat mitigation but also in enhancing urban resilience to climate extremes. In this perspective, green roofs are recognized as adaptive systems whose effectiveness depends on their capacity to respond to local climatic stressors, such as heat waves, drought conditions, and seasonal variability, rather than on standardized or typology-based configurations (De Montis, 2024; Ramyar & Rutherford, 2021; Spyrou et al., 2024).

Extensive research has quantified the benefits of green roofs under different climatic conditions. Experimental and numerical studies report reductions in roof surface temperatures of up to 30–40 °C compared to conventional roofs, alongside decreases in ambient air temperature ranging from 1 to 3 °C at the local scale (Morakinyo et al., 2017; Santamouris, 2014; Shafique et al., 2018; Vijayaraghavan, 2016). Furthermore, green roofs have demonstrated the capacity to reduce peak stormwater runoff by up to 70–80%, attenuating pressure on urban drainage systems (Getter & Rowe, 2006; Speak et al., 2013), while also extending the lifespan of waterproofing membranes and improving acoustic and thermal insulation (Kolokotroni et al., 2006). These advantages make green roofs particularly attractive for the retrofitting of existing buildings, which constitute a substantial portion of the urban fabric in European cities.

Despite their well-documented potential, the design and implementation of green roofs are still frequently based on standardized typologies, prescriptive guidelines, or aesthetic considerations rather than on scientifically grounded, site-specific methodologies. International standards and guidelines—such as the FLL Guidelines (Forschungsgesellschaft Landschaftsentwicklung Landschaftsbau, 2018), ASTM standards (ASTM International, 2017), and UNI 11,235 (UNI, 2023)—provide essential technical requirements for materials, stratigraphy, and construction practices. However, they generally adopt a static and climate-neutral approach, offering limited support for adapting vegetation and substrate choices to local climatic conditions, microclimatic variability, and ecological compatibility. As a result, green roof systems are often designed using generic plant palettes and substrate configurations that may not be optimal for the specific thermal, radiative, and hydrological context of the site.

One of the most critical gaps in current practice concerns vegetation selection. In many projects, plant species are chosen based on availability, visual qualities, or generalized classifications (e.g., "extensive" or "intensive" systems), without a rigorous assessment of their ecological requirements in relation to local climate, solar exposure, and water availability (Shashua-Bar et al., 2009). This lack of ecological coherence can lead to plant stress, increased irrigation demand, higher maintenance costs, and, in some cases, system failure. Moreover, the interaction between climatic drivers—such as solar radiation, temperature extremes, wind exposure, and precipitation—and the physical properties of substrates is rarely evaluated in an integrated and quantitative manner. In response to these limitations, recent studies advocate data-driven and parametric design approaches capable of integrating climatic data, environmental simulations, and ecological indicators into decision-support workflows for green infrastructure design (Cascone, 2024; Wu et al., 2025). However, while advanced modelling tools are increasingly available, their application often remains fragmented, lacking a structured methodology that systematically links climate classification, ecological compatibility, microclimatic analysis, substrate characterization, and structural feasibility within a unified and replicable framework.

To overcome these limitations, recent research highlights the need for site-specific and climate-adaptive design approaches capable of linking environmental data, ecological indicators, and technical constraints within a unified decision-making framework (Cascone, 2024; Perini et al., 2011; Wu et al., 2025). However, despite growing interest in parametric modeling and environmental simulation tools, a structured methodology that systematically integrates climatic classification, ecological bioindicators, solar radiation analysis, substrate characterization, and structural verification is still largely absent from both scientific literature and professional practice.

In response to this gap, this paper proposes a site-specific and holistic methodology for the sustainable design of green roofs in urban climates. The framework integrates climatic classification through the Köppen–Geiger system (Peel et al., 2007), ecological bioindication using Ellenberg indicator values (Ellenberg et al., 1992), parametric solar radiation analysis, substrate selection based on technical standards, and structural verification according to national and European codes (European Committee for Standardization, 2002–2010; Ministero delle Infrastrutture e dei Trasporti, 2018). By combining these elements into a progressive and replicable workflow, the proposed methodology aims to support designers in making informed, scientifically grounded decisions that are coherent with local environmental conditions and building constraints. The effectiveness of the framework is demonstrated through its application to a real case study in a Mediterranean urban context, highlighting its potential for replication across different climates and urban settings.

2 Methodology

2.1 Conceptual Framework and Rationale of the Progressive Filtering Approach

Conventional green roof design approaches typically address climatic, ecological, and technical dimensions as separate or weakly connected layers, or rely on standardized typologies and prescriptive guidelines. In such frameworks, climate classification, vegetation selection, substrate definition, and structural verification are often treated as independent decisions, leading to solutions that are only nominally climate-adapted. This fragmentation frequently results in vegetation and substrate selections that are poorly suited to local climatic stressors or incompatible with the structural limits of existing buildings, requiring compensatory measures such as intensive irrigation or increased maintenance.

To overcome these limitations, the proposed methodology adopts a progressive reduction of design uncertainty, structured as a constraint-driven workflow. Each step introduces a site-specific constraint derived from measurable environmental or technical parameters, progressively narrowing the solution space. Rather than selecting vegetation directly from predefined typologies, the methodology reduces an initial vegetation database to a final set of site-compatible and usable species, ensuring coherence between climate, ecology, microclimate, substrate, and structure.

2.2 Step 1—Project Framing and Climate Classification

The first step is dedicated to the **preliminary framing of the project context**, which represents a fundamental reference for all subsequent analyses.

This phase defines the basic characteristics of the intervention, including geographic location, building typology, urban context, roof accessibility, and the nature of the project (new construction or retrofit).

Within this step, the climatic context of the site is identified through the Köppen–Geiger climate classification, allowing the project to be positioned within a broader climatic domain. This classification provides a first-order climatic filter, ensuring that the following ecological and design choices are coherent with long-term temperature and precipitation regimes.

The outcome of Step 1 is a clearly defined project and climatic framework, which anchors the methodology to the specific conditions of the case study and prevents generic or decontextualized design solutions.

2.3 Step 2—Climate-Driven Ecological Screening (Köppen–Geiger Classification and Ellenberg Indicators)

The second step defines the macro-climatic and ecological boundary conditions of the intervention site. Climatic descriptors—including temperature regime, precipitation patterns, humidity levels, and seasonal stress conditions—are collected and synthesized using the Köppen–Geiger climate classification system (Peel et al., 2007), which provides a consolidated representation of climatic regimes at the regional scale.

Within the proposed framework, Köppen–Geiger classification is not used to directly prescribe plant species. Instead, it defines the climatic envelope within which vegetation must operate. To translate this climatic information into species-level compatibility criteria, the methodology integrates Ellenberg ecological indicator values (Ellenberg et al., 1992) directly within this step.

Ellenberg indicators are employed as an interpretative interface between climate classification and vegetation selection. Indicators related to temperature (T), moisture (U), and, where relevant, light (L) are used to identify species whose ecological preferences are consistent with the climatic stresses characteristic of the identified Köppen–Geiger class. In this way, Ellenberg values operationalize climate classification, allowing the exclusion of species that are climatically incompatible despite generic classifications such as "extensive" or "intensive".

1. L(Light)–Preference for light intensity	
Indicates the plant's light requirement, from shade-loving to sun-loving species	
Value	**Preference**
1	Highly sciaphilous plant (deep shade)
2-3	Moderately sciaphilous (partial shade)
4-5	Semi-shade tolerant (light shade or half-shade)
6-7	Slightly heliophilous (prevalence of light)
8-9	Strongly heliophilous (full sun requirement)

2. T(Temperature)–Preference for thermal regime	
Indicates the adaptability of the plant to climate, from cold to warm environments	
Value	**Preference**
1	Species typical of very cold environments (high mountains, tundra)
2-3	Cold temperate climates
4-5	Areas with cold winters and mild summers
6-7	Warm temperate climates
8-9	Mediterranean and subtropical climates
10-12	Tropical climates

3. K(Continentality)–Adaptation to climatic variability	
Represents tolerance to seasonal climatic variations between summer and winter	
Value	**Preference**
1	Stable oceanic climates, without marked thermal excursions
5-6	Intermediate climates (moderately continental)
9	Extreme continental climates with strong thermal excursions

4. F(Soil moisture)–Tolerance to water availability	
Indicates the plant's water requirement in the soil.	
Value	**Preference**
1	Xerophilous plant (very dry, arid soils)
3-4	Adapted to moderately dry soils
5-6	Fresh soils (constant moisture)
7	Hygrophilous plant (very moist soils)
8-9	Hydrophyte (grows in aquatic environments)

5. R(Soil reaction, pH)–Preference for soil pH	
Indicates the plant's preference for acidic or alkaline soils.	
Value	**Preference**
1	Strongly acidophilous (pH < 4)
3-4	Acidophilous (pH 4–6.5)
5-6	Neutral (pH 6–7)
7-8	Basophilous (pH 7–8)
9	Alkaline (pH > 8.5)

6. N(Soil nutrients)–Preference for nutrient availability	
Indicates the amount of nutrients required by the plant.	
Value	**Preference**
1	Oligotrophic plant (very nutrient-poor soils)
3-4	Grows in moderately nutrient-rich soils
5-6	Fertile soils with good nutrient availability
7-8	Soils very rich in nutrients
9	Nitrophilous species (dependent on extremely rich, often nitrogen-loaded soils)

7. S(Salinity)–Tolerance to soil salinity	
Measures the ability to adapt to soils with high salt concentrations (e.g., coastal areas or brackish soils).	
Value	**Preference**
0	No tolerance to salinity
1-2	Low tolerance (slightly saline soils)
3-4	Moderate tolerance (environments with occasional salt inputs)
5-6	Halophytic species (highly saline soils, near coasts or lagoons)

The outcome of this integrated climatic–ecological screening is a first reduction of the initial vegetation database from the total number of candidate species (N) to a climate-consistent and ecologically plausible subset (N − X), where X represents species excluded due to incompatibility with the site's climatic envelope as described in Fig. 1.

2.4 *Step 3—Solar Radiation Analysis and Microclimatic Refinement*

While Step 2 ensures compatibility with the regional climate, it does not account for intra-site microclimatic variability, which is particularly pronounced on rooftops due to shading, orientation, and surrounding morphology. To address this limitation, Step 3 introduces a solar radiation-based refinement of the climate-consistent vegetation set.

Solar radiation is evaluated through parametric simulations, and the roof surface is subdivided into zones characterized by different levels of irradiance. Importantly, radiation analysis does not introduce an independent selection criterion; instead, it acts as a site-scale refinement of Ellenberg light (L) preferences already considered in Step 2.

By correlating simulated irradiance values with Ellenberg L ranges, the methodology ensures that species identified as climatically compatible can also be spatially

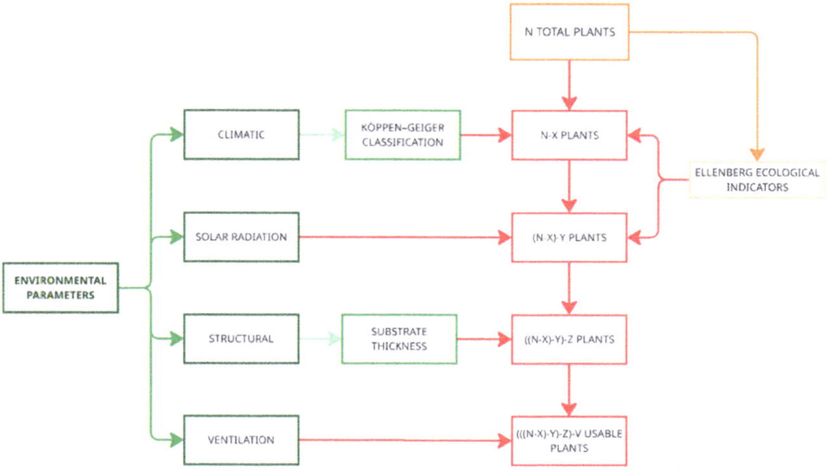

Fig. 1 Author's conceptual scheme illustrating the multi-criteria workflow developed for plant selection in green roof design

allocated to roof areas that meet their light requirements. Species that cannot be consistently matched to any roof microzone are excluded.

This step reduces the vegetation set from (N − X) to ((N − X) − Y), where Y represents species incompatible with the actual solar microclimate of the roof.

2.5 *Step 4—Substrate Thickness Definition and Structural Feasibility Constraint*

The fourth step integrates technical feasibility into the ecological screening process by linking substrate selection and thickness to the structural capacity of the existing roof. Reference values for substrate properties—such as dry and saturated bulk density and minimum thickness—are derived from established standards for green roof systems, including the FLL Guidelines (Forschungsgesellschaft Landschaftsentwicklung Landschaftsbau, 2018), ASTM standards (ASTM International, 2017), and UNI 11,235 (UNI, 2023).

Structural verification is performed according to the Italian Technical Standards for Construction (Ministero delle Infrastrutture e dei Trasporti, 2018; NTC, 2018) and the relevant Eurocodes (European Committee for Standardization, 2002–2010), ensuring that the combined weight of substrate, vegetation, and system layers remains within admissible limits under both dry and saturated conditions.

Species requiring substrate depths or system configurations incompatible with the roof's load-bearing capacity are excluded, leading to a further reduction from ((N − X) − Y) to (((N − X) − Y) − Z), where Z represents species excluded due to structural constraints.

2.6 Step 5—Ventilation Constraint and Definition of Usable Plant Species

The final step introduces ventilation and wind exposure as a site-specific constraint affecting plant survival and long-term performance. Rooftop environments often experience wind regimes that differ substantially from ground-level conditions, influencing mechanical stress, evapotranspiration rates, and microclimatic cooling effects.

In this methodology, ventilation acts as a final compatibility filter applied to the structurally feasible species set. Species that are not tolerant to the site's wind exposure conditions are excluded, resulting in the final subset of usable plant species, expressed as $(((N - X) - Y) - Z) - V$, where V represents species excluded due to incompatibility with the ventilation regime.

Methodological Significance

By integrating Ellenberg ecological indicators directly within climatic classification, the proposed framework establishes a continuous and logically coherent link between climate, ecology, and design. The progressive reduction approach transforms green roof design from a typology-based exercise into a site-specific, climate-adaptive decision-making process, explicitly addressing the limitations of conventional methods that treat climatic and ecological dimensions separately.

3 Case Study

The proposed methodology was applied to a real retrofit scenario involving a residential building located at Via Dalbono 18 in Portici (Naples, Italy), within a consolidated urban fabric characterized by medium building density, limited ground-level green spaces, and a predominance of sealed surfaces. This urban configuration makes rooftop greening a particularly relevant strategy for climate adaptation, as horizontal green infrastructure opportunities are scarce. The building consists of a reinforced concrete frame with a flat roof originally designed as a conventional non-accessible roof, representative of a large portion of the existing Mediterranean building stock constructed without provisions for additional permanent loads associated with green roof systems. This condition makes the case study especially suitable for testing a site-specific methodology capable of reconciling ecological ambition with structural feasibility (Fig. 2).

From a climatic perspective, the site falls within a Csa Mediterranean climate according to the Köppen–Geiger classification (Peel et al., 2007), characterized by hot and dry summers, mild winters, and a marked seasonal imbalance in water availability. These conditions represent one of the most critical climatic contexts for green roof performance, as prolonged summer drought and high solar radiation can severely affect vegetation survival and long-term functionality. In accordance with the methodological framework, climatic classification was not used as a prescriptive

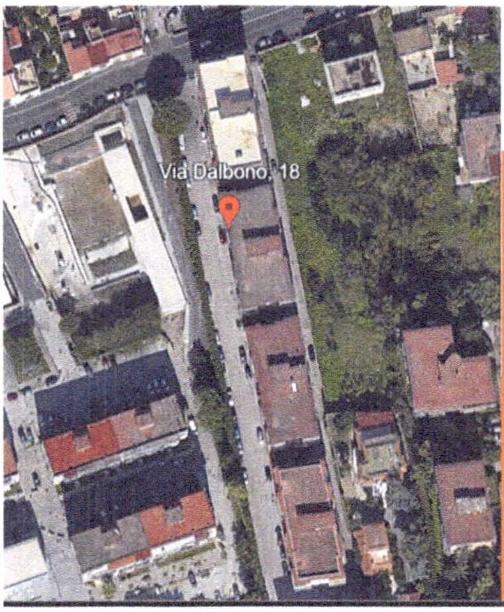

- PORTICI, VIA DALBONO 18
  BUILDING A – BARRA
  COORDINATES: 40°49'26'' N, 14°20'21'' E
  REINFORCED CONCRETE FRAME BUILDING
  CLIMATE CLASS: ZONE C – Csa

SB	Climate Zone	Regions / Extension	Average Temp. (°C)	Precip. (mm/year)	Average Humidity (%)	Notes
Csa	Hot dry Mediterranean	Sicily, Sardinia, Apulia, Basilicata, Calabria, Campania, Lazio, part of Abruzzo and Molise	16–20	400–1,000	65–75	Hot dry summers, mild winters

Parameter	Symbol	Average Value	Unit
Annual temperature	T_a	17.4	°C
Annual precipitation	P	870	mm
Average relative humidity	RH	67	%
Global radiation	R_s	1950	$kWh/m^2 \cdot year$
Average wind speed	u_2	14	km/h

Fig. 2 Location and climatic characterization of the case study building in Portici (Naples), including Köppen–Geiger climate classification and main environmental parameters

tool for vegetation selection, but rather as a means to define the climatic envelope within which plant compatibility had to be assessed. This envelope was translated into species-level criteria through the integrated use of Ellenberg ecological indicator values (Ellenberg et al., 1992), which allowed climatic stresses to be expressed in terms of plant ecological preferences, particularly with respect to temperature tolerance and moisture requirements.

Starting from an initial vegetation database comprising more than 100 herbaceous species and 46 moss species, a first climate-driven ecological screening excluded species whose Ellenberg indicators were incompatible with Mediterranean summer conditions, such as plants requiring persistently high soil moisture or characterized by low tolerance to thermal stress. This initial reduction transformed the broad database into a climate-consistent and ecologically plausible subset of species (N – X), ensuring that only vegetation inherently adapted to the regional climatic regime progressed to subsequent stages of the design process. However, climatic compatibility alone was considered insufficient to guarantee successful rooftop performance, as roof environments are characterized by strong microclimatic heterogeneity that cannot be captured by regional climate classification (Fig. 3).

For this reason, roof-level solar radiation was explicitly analyzed through parametric simulations, revealing pronounced spatial variability in irradiance driven by roof geometry, parapets, and the surrounding built environment. Rather than treating solar exposure as an independent selection criterion, simulated radiation patterns were interpreted as a site-specific refinement of Ellenberg light (L) preferences already embedded in the ecological screening. This approach enabled the subdivision of the roof into distinct microclimatic zones and the allocation of climatically compatible species to areas with corresponding solar exposure levels. Species whose

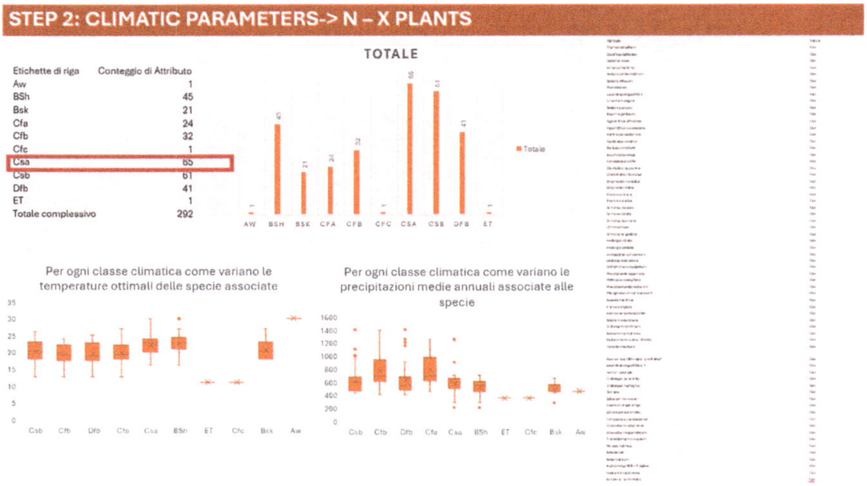

Fig. 3 Step 2 of the methodology: climate-based filtering of plant species using the Köppen–Geiger classification, leading to the selection of species compatible with the Mediterranean Csa climate

light requirements could not be coherently matched to any of the identified micro-zones were excluded, resulting in a further reduction of the vegetation set to (N – X – Y) (Fig. 4).

Following climatic and microclimatic refinement, the methodology integrated technical feasibility by addressing substrate thickness and structural capacity as

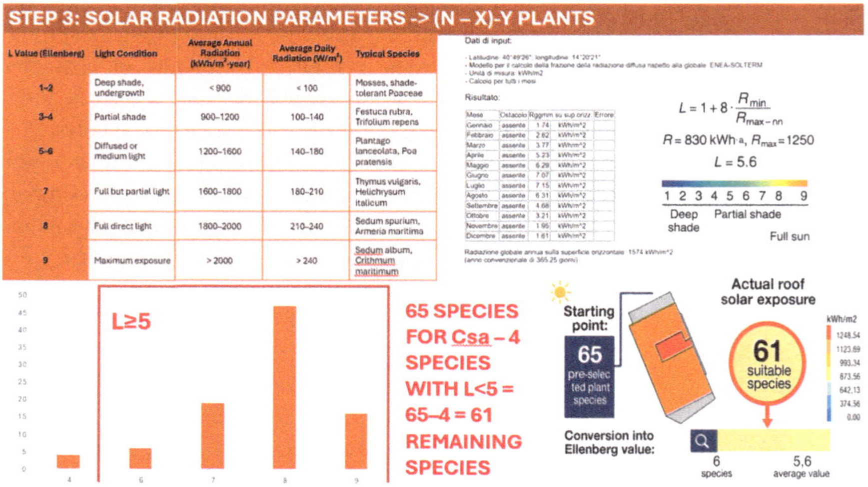

Fig. 4 Step 3 of the methodology: solar radiation–based filtering of plant species. Annual roof solar exposure is converted into Ellenberg light values, allowing the selection of species compatible with Mediterranean Csa conditions and the exclusion of light-incompatible plants

limiting factors for ecological implementation. Substrate properties and admissible ranges were derived from established technical standards for green roof systems, including the FLL Guidelines (Forschungsgesellschaft Landschaftsentwicklung Landschaftsbau, 2018), ASTM standards (ASTM International, 2017), and UNI 11,235 (UNI, 2023), which were critically interpreted rather than applied prescriptively. These standards provided benchmark values for dry and saturated bulk density and minimum substrate thickness, which were then evaluated in relation to the specific structural characteristics of the building.

Structural verification was carried out in accordance with the Italian Technical Standards for Construction (Ministero delle Infrastrutture e dei Trasporti, 2018; NTC, 2018) and the relevant Eurocodes (European Committee for Standardization, 2002–2010), leading to the identification of a maximum allowable permanent load of approximately 350 kg/m^2 under saturated conditions. Species requiring substrate depths incompatible with this structural threshold were excluded, further reducing the vegetation set to ((N − X − Y) − Z) and ensuring that all remaining options were not only ecologically suitable but also technically feasible within a retrofit context (Fig. 5).

Finally, ventilation and wind exposure were considered as an additional site-specific constraint, reflecting the aerodynamic conditions typical of rooftops in Mediterranean urban environments, where increased wind speeds can amplify evapotranspiration rates and mechanical stress on vegetation. Species not tolerant to these conditions were excluded, yielding the final subset of usable plant species (((N − X − Y) − Z) − V). Rather than producing a single prescriptive design solution, the application of the methodology resulted in a site-specific set of ecologically

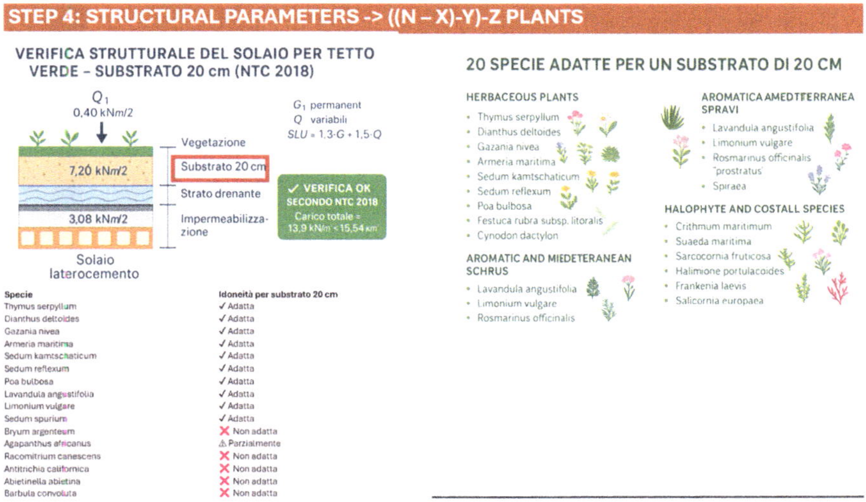

Fig. 5 Step 4 of the methodology: structural filtering of plant species based on substrate thickness and load-bearing verification according to NTC (2018), resulting in a final set of species compatible with a 20 cm green roof system

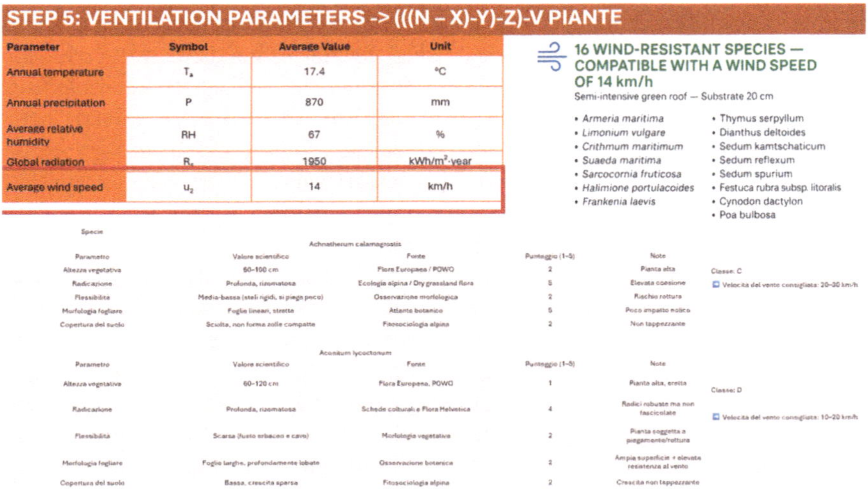

Fig. 6 Step 5 of the methodology: ventilation-based filtering of plant species, integrating site wind speed conditions to identify wind-resistant species suitable for semi-intensive green roofs

and technically compatible species that can be flexibly allocated across roof micro-zones. The case study thus demonstrates how the proposed framework operates as a decision-support tool that progressively reduces design uncertainty, translating a large and heterogeneous vegetation database into a coherent, climate-responsive, and structurally feasible green roof design strategy suitable for real-world retrofit applications in Mediterranean urban contexts (Fig. 6).

4 Discussion

The results of the case study highlight the potential of the proposed methodology to address several limitations that characterize both current regulatory frameworks and prevailing design approaches for green roofs. While the benefits of green roofs for urban heat island mitigation, stormwater management, and building energy performance are well documented in the literature, their practical implementation often remains constrained by generalized design prescriptions and a limited integration of site-specific environmental factors.

4.1 *Comparison with Normative Frameworks*

International and national standards for green roofs, such as the FLL Guidelines (Forschungsgesellschaft Landschaftsentwicklung Landschaftsbau, 2018), ASTM standards (ASTM International, 2017), and UNI 11,235 (UNI, 2023), play a fundamental role in ensuring minimum technical performance, construction quality, and durability of green roof systems. These documents provide detailed specifications regarding system stratigraphy, drainage layers, substrate properties, load assumptions, and maintenance requirements. However, as also highlighted in the comparative analysis conducted in the preliminary phases of this research, such standards adopt a predominantly prescriptive and climate-neutral approach.

In particular, normative documents generally do not provide structured guidance for vegetation selection based on local climatic conditions, nor do they include explicit links between climate classification, ecological requirements of plant species, and microclimatic variability at the roof level. Vegetation is often addressed through generic categories (e.g., "sedum-based systems", "extensive vegetation"), which may be insufficient to ensure long-term ecological performance under diverse climatic contexts, especially in water-limited or thermally stressed environments such as Mediterranean cities.

The proposed methodology does not aim to replace existing standards, but rather to complement them by introducing a decision-support layer that operates upstream of normative prescriptions. While standards such as FLL, ASTM, and UNI provide essential boundary conditions for substrate properties and structural safety, the framework integrates these requirements within a broader, climate-adaptive workflow. In this sense, normative constraints are treated as feasibility limits, rather than as drivers of vegetation selection, allowing ecological and climatic coherence to guide design choices before technical verification.

4.2 Advancement Beyond Conventional Design Approaches

Beyond normative limitations, conventional green roof design practices—both in professional applications and in a portion of the scientific literature—often address climatic, ecological, and structural dimensions in isolation. Climate is typically considered at a general level, vegetation is selected based on availability or aesthetic criteria, and structural verification is conducted as a final, often restrictive step. This sequential but disconnected approach increases the risk of incompatibility between vegetation requirements and site conditions, leading to higher maintenance demands or system underperformance.

The progressive reduction framework proposed in this study introduces a fundamentally different design logic, in which climate classification, ecological compatibility, microclimatic refinement, and structural feasibility are explicitly coupled. The integration of the Köppen–Geiger classification (Peel et al., 2007) with Ellenberg ecological indicator values (Ellenberg et al., 1992) represents a key methodological contribution, as it allows climatic information to be translated into quantitative, species-level selection criteria. This addresses a critical gap in both practice and literature, where climate-adapted vegetation is often discussed conceptually but rarely operationalized through measurable ecological parameters.

Furthermore, the explicit consideration of solar radiation as a refinement of ecological light requirements, rather than as an independent design variable, enables the methodology to account for roof-level heterogeneity without fragmenting the decision process. This approach aligns ecological theory with parametric design tools, bridging the gap between qualitative ecological knowledge and quantitative environmental simulation.

4.3 Implications for Retrofit Applications and Urban Resilience

The case study demonstrates that the proposed methodology is particularly well suited for retrofit scenarios, which represent the dominant condition in European cities. Existing buildings often impose strict structural constraints that limit substrate thickness and system weight, making traditional, typology-based green roof solutions impractical. By embedding structural verification within the progressive filtering process, the framework ensures that ecological ambition is continuously reconciled with technical feasibility, rather than constrained retrospectively.

From an urban resilience perspective, the methodology supports the deployment of green roofs as context-sensitive Nature-Based Solutions, capable of responding to local climatic stressors rather than applying uniform solutions across different regions. This is especially relevant in climates characterized by increasing thermal extremes and water scarcity, where inappropriate vegetation choices can undermine both environmental benefits and system durability.

4.4 Methodological Limitations and Future Developments

While the proposed framework enhances the coherence and transparency of green roof design, some limitations should be acknowledged. First, the methodology relies on the availability and quality of ecological indicator data, which may vary across regions and plant species. Second, although the framework supports site-specific adaptation, its application requires a minimum level of expertise in environmental modelling and parametric workflows, which may limit immediate adoption in all professional contexts.

Future developments will focus on addressing these limitations through the automation of the decision-making process, translating the methodology into an interactive digital abacus that integrates climatic data, ecological indicators, and technical constraints within a unified interface. Such an evolution would further enhance the accessibility and scalability of the framework, facilitating its application across different climatic zones and project typologies.

5 Conclusions

This study proposed a site-specific, climate-adaptive methodology for the sustainable design of green roofs in urban environments, addressing key limitations of both normative frameworks and conventional design practices. By structuring the design process as a progressive reduction of design uncertainty, the framework enables the coherent integration of climatic conditions, ecological compatibility, microclimatic variability, and structural feasibility within a single decision-making workflow.

A central contribution of the proposed approach lies in the explicit coupling of Köppen–Geiger climate classification with Ellenberg ecological indicator values, which allows climatic information to be translated into species-level selection criteria. This integration bridges a critical gap between climatological characterization and vegetation design, transforming abstract notions of "climate-adapted vegetation" into operational and measurable design inputs. The subsequent refinement through solar radiation analysis, substrate thickness definition, and ventilation constraints ensures that vegetation selection remains consistent with roof-level environmental heterogeneity and technical limitations, particularly in retrofit scenarios.

The application of the methodology to a representative Mediterranean case study demonstrated its capacity to transform a broad and heterogeneous vegetation database into a site-compatible and structurally feasible set of design options, without relying on predefined typologies or prescriptive vegetation lists. Rather than proposing a single optimal solution, the framework supports flexible and context-responsive design, enabling designers to adapt green roof configurations to local conditions while remaining compliant with existing standards.

From a practical perspective, the proposed framework provides a transferable decision-support tool for designers, planners, and public authorities involved in green roof implementation, particularly within dense and climatically sensitive urban contexts. By explicitly linking climate adaptation objectives with ecological performance and structural feasibility, the methodology enhances the reliability and effectiveness of green roofs as multifunctional Nature-Based Solutions, capable of contributing simultaneously to Urban Heat Island mitigation, microclimatic regulation, and building-scale environmental performance.

These aspects are especially relevant for Mediterranean cities, where rising temperatures, prolonged drought periods, and the predominance of existing building stock pose critical challenges to climate adaptation strategies. In this context, the proposed methodology supports retrofit-oriented green roof design that is both ecologically resilient and technically feasible, offering a scientifically grounded framework that can inform policy-oriented guidelines, incentive schemes, and urban resilience strategies. By enabling climate-sensitive and site-specific vegetation selection, the approach contributes to the broader transition toward adaptive, evidence-based green infrastructure planning in Mediterranean and comparable climatic regions.

References

Alexandri, E., & Jones, P. (2008). Temperature decreases in an urban canyon due to green walls and green roofs in diverse climates. *Building and Environment, 43*(4), 480–493. https://doi.org/10.1016/j.buildenv.2006.10.055

ASTM International. (2017). *ASTM E2397/E2397M: Standard practice for determination of dead loads and live loads associated with green roof systems.* ASTM International.

Berardi, U., GhaffarianHoseini, A., & GhaffarianHoseini, A. (2014). State-of-the-art analysis of the environmental benefits of green roofs. *Applied Energy, 115*, 411–428. https://doi.org/10.1016/j.apenergy.2013.10.047

Bowler, D. E., Buyung-Ali, L., Knight, T. M., & Pullin, A. S. (2010). Urban greening to cool towns and cities: A systematic review of the empirical evidence. *Landscape and Urban Planning, 97*(3), 147–155. https://doi.org/10.1016/j.landurbplan.2010.05.006

Cascone, S. (2024). Integrating green roof design within BIM-based decision-support frameworks for climate-adaptive urban environments. *Sustainability, 16*(8), 3271. https://doi.org/10.3390/su16083271

De Montis, A. (2024). Green infrastructure and adaptation to climate change: A planning perspective. *Sustainability, 16*(19), 8641. https://doi.org/10.3390/su16198641

Ellenberg, H., Weber, H. E., Düll, R., Wirth, V., Werner, W., & Paulissen, D. (1992). *Indicator values of plants in Central Europe.* Erich Goltze KG.

European Committee for Standardization. (2002–2010). *Eurocode: Basis of structural design and actions on structures (EN 1990–EN 1991).*

Forschungsgesellschaft Landschaftsentwicklung Landschaftsbau. (2018). *Guidelines for the planning, construction and maintenance of green roofing.*

Getter, K. L., & Rowe, D. B. (2006). The role of extensive green roofs in sustainable development. *HortScience, 41*(5), 1276–1285. https://doi.org/10.21273/HORTSCI.41.5.1276

Kolokotroni, M., Giannitsaris, I., & Watkins, R. (2006). The effect of the London urban heat island on building summer cooling demand and night ventilation strategies. *Solar Energy, 80*(4), 383–392. https://doi.org/10.1016/j.solener.2005.03.010

Ministero delle Infrastrutture e dei Trasporti. (2018). *Norme tecniche per le costruzioni (NTC 2018).*

Morakinyo, T. E., Ouyang, W., Lau, K. K. L., Ren, C., & Ng, E. (2017). Right tree, right place (urban canyon): Tree species selection approach for urban heat mitigation. *Urban Forestry & Urban Greening, 26*, 154–163. https://doi.org/10.1016/j.ufug.2017.06.010

Oberndorfer, E., Lundholm, J., Bass, B., Coffman, R. R., Doshi, H., Dunnett, N., Gaffin, S., Köhler, M., Liu, K. K. Y., & Rowe, B. (2007). Green roofs as urban ecosystems: Ecological structures, functions, and services. *BioScience, 57*(10), 823–833. https://doi.org/10.1641/B571005

Oke, T. R. (1987). *Boundary layer climates.* Routledge.

Peel, M. C., Finlayson, B. L., & McMahon, T. A. (2007). Updated world map of the Köppen-Geiger climate classification. *Hydrology and Earth System Sciences, 11*(5), 1633–1644. https://doi.org/10.5194/hess-11-1633-2007

Perini, K., Ottelé, M., Fraaij, A. L. A., Haas, E. M., & Raiteri, R. (2011). Vertical greening systems and the effect on air flow and temperature on the building envelope. *Building and Environment, 46*(11), 2287–2294. https://doi.org/10.1016/j.buildenv.2011.05.009

Ramyar, R., & Rutherford, P. (2021). Adapting cities for climate change through urban green infrastructure: A systematic review and framework for resilience-oriented planning. *Cities, 109*, Article 103016. https://doi.org/10.1016/j.cities.2020.103016

Rizwan, A. M., Dennis, L. Y. C., & Liu, C. (2008). A review on the generation, determination and mitigation of urban heat island. *Journal of Environmental Sciences, 20*(1), 120–128. https://doi.org/10.1016/S1001-0742(08)60019-4

Sailor, D. J. (2008). A green roof model for building energy simulation programs. *Energy and Buildings, 40*(8), 1466–1478. https://doi.org/10.1016/j.enbuild.2008.02.001

Santamouris, M. (2014). Cooling the cities—A review of reflective and green roof mitigation technologies to fight heat island and improve comfort in urban environments. *Solar Energy, 103*, 682–703. https://doi.org/10.1016/j.solener.2012.07.003

Shafique, M., Kim, R., & Rafiq, M. (2018). Green roof benefits, opportunities and challenges—A review. *Renewable and Sustainable Energy Reviews, 90*, 757–773. https://doi.org/10.1016/j.rser.2018.04.006

Shashua-Bar, L., Pearlmutter, D., & Erell, E. (2009). The cooling efficiency of urban landscape strategies in a hot dry climate. *Landscape and Urban Planning, 92*(3–4), 179–186. https://doi.org/10.1016/j.landurbplan.2009.04.005

Speak, A. F., Rothwell, J. J., Lindley, S. J., & Smith, C. L. (2013). Rainwater runoff retention on an aged intensive green roof. *Science of the Total Environment, 458*, 209–216. https://doi.org/10.1016/j.scitotenv.2013.04.001

Spyrou, C., Tsiros, I. X., Nastos, P. T., & Dimoudi, A. (2024). Green roofs as a Nature-Based Solution to mitigate urban heating during heatwaves: Microclimatic performance and adaptation potential. *Sustainability, 16*(22), 9729. https://doi.org/10.3390/su16229729

Takebayashi, H., & Moriyama, M. (2007). Surface heat budget on green roof and high reflection roof for mitigation of urban heat island. *Building and Environment, 42*(8), 2971–2979. https://doi.org/10.1016/j.buildenv.2006.06.017

UNI. (2023). *UNI 11235: Istruzioni per la progettazione, l'esecuzione, il controllo e la manutenzione delle coperture a verde*. Ente Italiano di Normazione.

Vijayaraghavan, K. (2016). Green roofs: A critical review on the role of components, benefits, limitations and trends. *Renewable and Sustainable Energy Reviews, 57*, 740–752. https://doi.org/10.1016/j.rser.2015.12.119

Wu, Y., Zhang, X., Li, J., & Chen, Y. (2025). Optimizing urban green roofs through a data-driven decision-support framework under climate stress conditions. *Land, 14*(9), 1742. https://doi.org/10.3390/land14091742

Assessing Green Walls as a Nature-Based Strategy to Improve Urban Microclimate and Reduce Energy Demand Under Current and Future Climate Conditions

Amir Dehghan Lotfabad and Seyed Morteza Hosseini

Abstract Urbanization and climate change are exacerbating the Urban Heat Island (UHI) effect, significantly increasing energy demand and challenging sustainable development. While Vertical Greenery Systems (VGSs) are recognized as a passive mitigation strategy, their large-scale impact on urban energy performance under future climate scenarios remains underexplored. This study evaluates the efficacy of Green Walls (GWs) as a nature-based solution at the urban scale in Turin, Italy, under both current and projected 2050 (RCP 8.5) climate conditions. Using a validated numerical simulation workflow with Ladybug Tools and EnergyPlus, the research quantifies the microclimatic and energy effects of large-scale GW implementation. The results indicate that while GWs induced only a modest reduction in ambient air temperature (up to 0.1 °C), they significantly reduced outdoor surface temperatures, with maximum decreases of 24 °C and 34 °C under current and future climates, respectively. Crucially, GWs enhanced building energy efficiency by reducing the Energy Use Intensity (EUI) by approximately 35% in both scenarios. This reduction effectively offset the projected energy burden of climate change, bringing the future EUI with GWs close to the current baseline. The findings demonstrate that the large-scale application of GWs is a scalable and effective strategy for mitigating UHI effects, reducing energy demand, and fostering climate-resilient urban environments.

Keywords Energy use intensity · Vertical Greenery · Urban resilience · Cooling demand · Climate adaptation

---

A. D. Lotfabad
Department of Environment, Land and Infrastructure Engineering, Politecnico di Torino, Turin, Italy

S. M. Hosseini (✉)
Department of Architecture, Design and Media Technology, Aalborg University, Copenhagen SV, Denmark
e-mail: smho@create.aau.dk

L. Diana et al. (eds.), *Greening and Resilience*, Urban Sustainability,
https://doi.org/10.1007/978-981-92-0419-9_27

Abbreviations

EUI	Energy Use Intensity
GCC	Global Climate Change
GHG	Green House Gas
GRs	Green Roofs
GWs	Green Walls
LWs	Living Walls
RH	Relative Humidity
T_a	Air Temperature
TMY	Typical Meteorological Year
T_s	Outdoor Surface Temperature
UHI	Urban Heat Island
UWG	Urban Weather Generator
VGSs	Vertical Greenery Systems

1 Introduction

Rapid industrialization and urbanization are driving economic growth but also causing significant environmental challenges, particularly increasing carbon emissions and global temperatures. A survey in EU countries showed that almost 78% of people feel environmental issues affect their daily life and health, with 68% of Italians considering climate change the most critical environmental issue (European Commission, 2024).

Green house gas (GHG) emissions from fossil fuels trap solar heat, contributing to warming. Despite growing policy efforts, Global Climate Change (GCC) mitigation has yet to yield significant results. Mitigation aims to limit environmental changes caused by GHG emissions. Projected emissions for 2030 indicate warming will exceed 1.5 °C, making it necessary to achieve global net-zero emissions within the next 50 years to keep warming well below 2 °C (Cambridge University Press, 2022). Considering the links between the Sustainable Development Goals and global climate, urgent action is needed to increase renewable energy use (Astiaso Garcia et al., 2023). Furthermore, the rising frequency of climate-related natural disasters highlights the need to shift from traditional to renewable energy sources to reduce GHG emissions (Attia et al., n.d.).

Urban areas are increasingly important, with two-thirds of the global population projected to live in cities by 2050 and cooling energy demand is anticipated to grow by a factor of 3.1 times by this time (Ma & Yu, 2024). Climate risks also threaten cities, which highlights the need for climate-resilient buildings (Bahadori et al., 2025). Buildings are major producers of GHG, consuming 34% of the world's total energy and contributing 37% of global CO_2 emissions to the accelerated GCC. However, building passive energy-saving strategies to mitigate the impacts of both GCC and the

UHI effect remains understudied, as the UHI effect intensifies energy consumption. The increase in the amount of impervious, hard-surfaced areas, dark-colored building materials, and unvegetated areas can lead to both increased urban temperatures in the form of UHI, as well as increased surface runoff of urban stormwater (Emilsson, 2021). Studies report that cooling load increases ranging from 4 to 65% and heating load reductions from 3% to over 100% (Hashemi & Mills, 2025).

Passive energy-saving strategies include enhancing the building envelope and architectural design by optimizing orientation, selecting appropriate materials, improving natural ventilation, and incorporating shading devices such as greenery and kinetic façades (Aghajanzadeh et al., 2023; Ganjimorad et al., 2024; Gholami et al., 2025; Heiranipout et al., 2025; Yazdi Bahri et al., 2025), along with thermal insulation measures like double- or triple-glazed windows (Fereidoni et al., 2023). Using green covers and vegetation on buildings is an effective passive strategy that restores urban green space, mitigates UHI effects, and improves building energy performance by lowering indoor temperatures, regulating humidity, and providing thermal insulation (Dehghan Lotfabad et al., 2025). VGSs are simple to construct, require fewer materials, have lower costs, are environmentally friendly, and provide significant economic and environmental benefits (Farrokhirad et al., 2024).

Although green façades have limited impact on air temperature (Susca et al., 2023), they are more effective than simply increasing the albedo of vertical surfaces. In this case, the shading effect of green façades reduces heat flux and building's heat load. In an experimental study of four small buildings (43 m^2, max height 3 m), an indirect green façade with an air gap lowered exterior wall temperatures by 23.1 °C and indoor temperatures by 1–5 °C, achieving energy savings of 45.75%, compared to only 6.43% for direct green façade coverage (Qi et al., 2025).

Increasing the coverage ratio would result in a better cooling effect (Dardir & Berardi, 2021). Combining Green Roofs (GRs) and façades on all elevations of a 10-story, 1,558 m^2 residential building achieved a maximum EUI reduction of 16.43% in a subtropical climate (Ribeiro et al., 2025). In a warm Mediterranean summer climate, full roof and façade coverage reduced thermal discomfort hours and energy use by 9.31% (Turhan et al., 2025). In arid climates, full wall and roof coverage reduced annual electricity consumption by 17–25% and indoor air temperature by 5 °C (Wahba et al., 2018). In a semi-arid climate, Living Walls (LWs) applied to 48 buildings (150 m × 120 m, max height 18 m) reduced air temperatures by up to 2.25 °C (average 1.35 °C), lowered cooling energy use by 15.85% in the east–west direction, and decreased exterior surface temperatures by up to 18 °C (Bakhtyari et al., 2024).

GWs would reduce exterior surface temperatures by up to 15 °C during peak hours. Applying GWs to the east and west façades of a three-story building also improved indoor thermal conditions, lowering cooling loads by approximately 64% (Ouldboukhitine et al., 2025). An experimental study in a tropical climate showed that 10 m^2 of LWs reduced east-facing exterior surface temperatures by up to 4.7 °C on average (Jayasooriya et al., 2025).

This study aims to evaluate the effect of GWs on building energy performance at the urban scale under both current and future climate conditions. While previous research has explored various VGSs and GRs, these studies are generally limited to single buildings or small areas. Importantly, no one has examined the impact of GWs on energy performance at the urban scale while also considering projected future climate conditions.

The objectives are to:

1. Assess the potential of GWs to improve building energy performance at the urban scale.
2. Investigate the effectiveness of GWs on EUI under future climate projections.
3. Bridge the gap between building-scale studies and urban-scale applications, providing a comprehensive assessment of GWs as a sustainable urban strategy.

By addressing these gaps, this study provides new insights into the urban-scale application of GWs for enhancing energy efficiency under both current and future climate conditions, an aspect not considered in previous literature.

2 Materials and Methods

2.1 Data Gathering

The background climate, which governs external gains in buildings, was acquired from the nearby Turin-Caselle Airport weather station as Typical Meteorological Year (TMY) data, excluding extreme events (Ladybug Tools, 2025). Meteorological data to assess future climate conditions for the year 2050, based on the RCP 8.5 scenario (Heiranipour et al., 2024), were obtained and morphed using the Urban Weather Generator (UWG) to generate localized projections. The urban model, including 2D and 3D building geometries, commercial buildings program (Regione Piemonte, 2024), and material properties, were derived from prior studies, ensuring a realistic representation of the built environment (Dehghan Lotfabad et al., 2025; Lotfabad, 2025).

2.2 Simulation Settings

The simulations workflow was conducted using Ladybug Tools and EnergyPlus workflow (see Fig. 1). The TMY files were modified, utilizing the UWG to account for local urban climate effects, including the UHI induced by urbanization (Hashemi & Mills, 2025; Mackey et al., 2017), enabling evaluation of its impact on cooling demand and site microclimate conditions. The simulation period focused on the hottest day of the year, 29th July, between 08:00 and 18:00, as most activities are conducted during this time (Naboni et al., 2020; Rezaei et al., 2024).

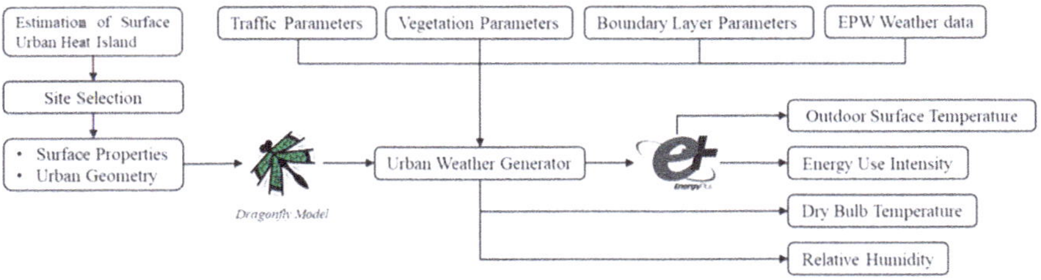

Fig. 1 Analysis flowchart of the GWs cooling impact on energy efficiency

The numerical simulations were validated against in-situ meteorological measurements, including Mean Radiant Temperature (Tmrt), to ensure model accuracy, achieving acceptable calibration metrics according to ASHRAE Guideline 14 (American Society of Heing Georgia, 2014) in Eqs. (1)–(3), with ±10% mean bias error (MBE) and ≤30% coefficient of variation of the root mean square error (CVRMSE) (see Fig. 2).

$$\mathrm{MBE} = \frac{\sum_{i-1}^{N_p}(M_i - S_i)}{\sum_{i=1}^{N_p} M_i} \tag{1}$$

$$\mathrm{CVRMSE} = \frac{\sqrt{\sum_{i=1}^{N_p}(M_i - S_i)^2/N_p}}{M_p}. \tag{2}$$

$$\left(\overline{M_p} = \frac{\sum_{i=1}^{N_p} M_i}{N_p}\right) \tag{3}$$

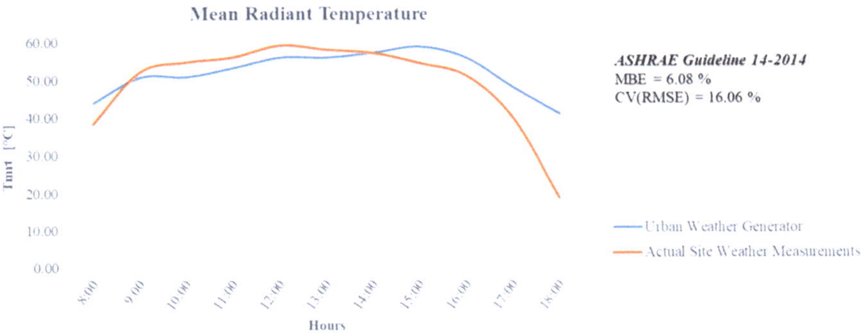

Fig. 2 Mean radiant temperature validation

2.3 Urban-Scale Analysis of GWs Effects

The urban-scale influence of GWs on energy performance was assessed in two steps, using consistent building and urban models across all scenarios (Table 1). The combined workflow (see Fig. 1), considering a cooling/heating system with a window AC, a baseboard gas boiler, and electric service hot water with an efficiency of 1, would enable the evaluation of the EUI of the model in all cases.

Step 1: Baseline and Urban Microclimate Conditions

To isolate the effect of urbanization, two climate datasets were prepared:

1. Baseline climate input (Scenario 1): TMY data under current climate conditions from the Turin-Caselle Airport weather station, representing regional conditions without urban modification.
2. Urban-morphed climate input (Scenarios 2 and 4): UWG-simulated TMY files, accounting for urban morphology and material thermal properties, boundary layer parameters, and anthropogenic heat release.

Both datasets were applied to the same building stock, and the EUI was calculated. The difference between these two simulations quantified the impact of the UHI effect on cooling demand. Projected climate change was incorporated using a morphed TMY file for 2050. The same UWG parameters as previously were applied to ensure comparability. This allowed for the assessment of how future climate conditions, combined with urban microclimates, will affect EUI on the urban scale. The study area (128 m × 136 m) was characterized by an average building height of 17.6 m, a footprint density of 0.37, and a facade-to-site ratio of 1.08, reflecting a moderately compact mid-rise urban fabric.

Step 2: Implementation of GWs

In the final step (Scenarios 3 and 5), GWs were introduced into the urban model across suitable façades being exposed to urban canyons. The vegetation layer was represented with its thermal and optical properties (Table 2). These properties were integrated into EnergyPlus simulations, enabling the analysis of both direct shading effects on outdoor surface temperatures and the indirect cooling of outdoor air

Table 1 Key parameters and assumptions for each simulated scenario

Scenario	Climate	UHI effect	Green Wall	Analyzed parameters			
				T_a	RH	T_s	EUI
1	Current	–	–	*	*	–	*
2	Current	*	–	*	*	*	*
3	Current	*	*	*	*	*	*
4	2050	*	–	*	*	*	*
5	2050	*	*	*	*	*	*

Table 2 GWs thermal and optical properties

Green Wall characteristics			
Air gap	0.05 [m]	Soil emission	0.9
Plant height	0.1 [m]	Stomata resistance	180 [s/m]
Leaf Area Index (LAI)	3[1]	Soil thickness	0.22 [m]
Substrate	0.05 [m]	Soil conductivity	0.35 [W/mk]
Leaf reflection	0.22	Soil density	1100 [kg/m3]
Leaf emission	0.95	Soil specific heat	1200 [J/kg k]
Soil reflection	0.3		

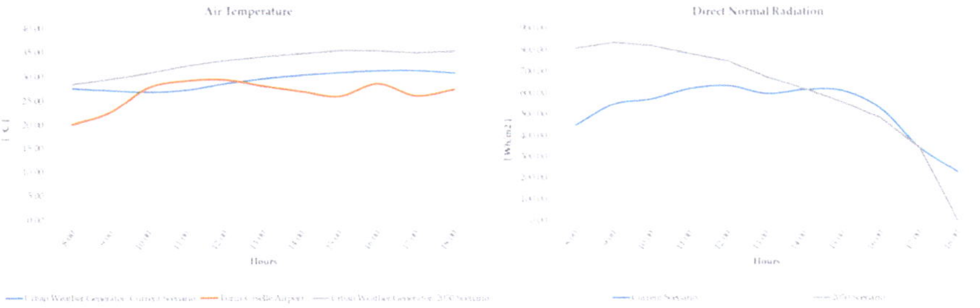

Fig. 3 Comparison of microclimatic conditions for the study area. Left: Diurnal air temperature profiles (08:00–18:00) for the base model without UHI effects (Scenario 1), the UWG under current conditions (Scenario 2), and the UWG projection for 2050 (Scenario 4). Right: Direct normal radiation under current and projected 2050 climate projection

resulting from evaporation. The simulations were repeated under both present-day and 2050 morphed climates to quantify the mitigation potential of GWs across time horizons.

3 Results

This section presents the simulation outcomes under current and future climate conditions, with and without the implementation of GWs. The results are reported in terms of outdoor air temperature (°C), direct normal radiation (Wh/m^2) (see Fig. 3), EUI at the end uses, and outdoor surface temperature (°C).

1 Susca et al. (2023).

3.1 Effects of GWs on Urban Microclimate Conditions and Outdoor Surface Temperature

The introduction of GWs resulted in only modest changes in average outdoor air temperature and relative humidity, with a maximum reduction in air temperature of 0.1 °C and a 1% increase in relative humidity under future climate projections. However, the shading effect of GWs led to notable reductions in outdoor surface temperatures (see Fig. 4). Under current climate conditions (Scenario 3), GWs would cause maximum reductions of 24 °C and an average of 1.8 °C. Evaluations of future climate projections (Scenario 5), showed that GWs would cause the maximum decrease up to 34 °C, with an average of 2.15 °C.

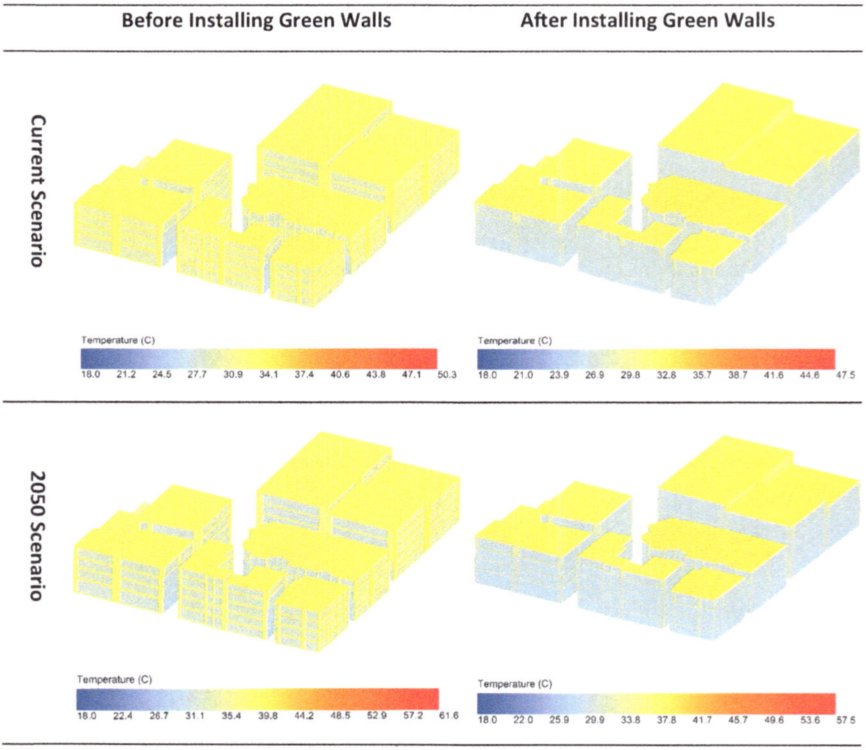

Fig. 4 Spatial distribution of surface temperature (°C) before and after GWs installation under the current and 2050 scenarios

Table 3 End-use Energy Use Intensity (kWh/m^2) for a 69,620 m^2 building under different UHI conditions: baseline without UHI, with UHI under the current climate, and with UHI under the 2050 climate scenario

End uses	Scenario 1	Scenario 2	Scenario 4
Cooling	0.251	0.301	0.481
Interior lighting	0.017	0.017	0.017
Electric equipment	0.056	0.056	0.056
Water systems	0.005	0.005	0.005
EUI	0.329	0.379	0.559
Total energy use (kWh)	9744.934	11,225.927	16,557.502

3.2 *UHI and GWs Impacts on Energy Performance Under Current and Future Climate*

Comparison between the baseline TMY data and the morphed TMY from the UWG shows that urbanization led to an increase in EUI of 15%, rising from 0.329 (kWh/m^2) to 0.379 (kWh/m^2) under current conditions. This highlights the additional energy burden caused by UHI effects at the urban scale (Table 3).

Under the 2050 scenario, EUI is projected to increase substantially to 0.56 (kWh/m^2), reflecting the combined effects of urbanization and climate change. Cooling demand alone accounted for 0.48 (kWh/m^2), confirming that climate projections significantly amplify cooling requirements at the urban scale.

Implementation of GWs significantly improved energy efficiency under both scenarios (Table 4): The Current scenario showed EUI decreased by 35%, from 0.379 (kWh/m^2) to 0.245 (kWh/m^2) because of a reduction in cooling demand by 44%, from 0.301 (kWh/m^2) to 0.167 (kWh/m^2). Meanwhile, the 2050 scenario showed that EUI would decrease by 35%, from 0.559 (kWh/m^2) to 0.363 (kWh/m^2) as cooling demand decreased by 41%, from 0.481 (kWh/m^2) to 0.285 (kWh/m^2).

Table 4 End-use Energy Use Intensity (kWh/m^2) for a 69,620 m^2 building after GWs installation under current and 2050 climate scenarios, including UHI effects

End uses	Scenario 3	Scenario 5
Cooling	0.167	0.285
Interior lighting	0.017	0.017
Electric equipment	0.056	0.056
Water systems	0.005	0.005
EUI	0.245	0.363
Total energy use (kWh)	7265.248	10,750.858

4 Discussion

This study assessed the influence of GWs on the urban microclimate and building energy performance under current and projected 2050 climate scenarios, accounting for the effects of the UHI. The results demonstrate consistent improvements in air temperature, relative humidity, surface temperature, and energy consumption following the implementation of GWs, findings that both align with and extend prior research. The simulations indicated a modest reduction in air temperature and a 1% increase in relative humidity. Although the reduction in air temperature was limited, this outcome is consistent with earlier study that reported only marginal cooling effects of green façades on ambient air conditions (Susca et al., 2023). The increase in relative humidity can be explained by evapotranspiration from the vegetation, which contributes to localized improvements in thermal comfort.

Substantial reductions in surface temperature were observed in spatial heat maps. Under current climate conditions, maximum reductions of 24 (°C) and an average reduction of 1.8 (°C) were recorded, aligning with experimental and numerical findings that reported reductions between 4.7 (°C) and 23 (°C) due to green façades and LWs (Bakhtyari et al., 2024; Jayasooriya et al., 2025; Ouldboukhitine et al., 2025; Qi et al., 2025). Under the 2050 scenario, maximum reductions reached 34 (°C), with an average of 2.1 (°C). The larger reductions in future projections may be attributed to higher baseline air temperatures and increased solar radiation, conditions under which vertical greenery demonstrates amplified cooling potential. These findings underscore that, among various vertical greenery systems, GWs emerge as the most effective strategy, particularly in terms of improving building energy efficiency (Arenghi et al., 2021). Moreover, the pronounced surface temperature reduction highlights the shading capacity of GWs, which diminishes heat flux into the building envelope, and supports the hypothesis that large-scale deployment with higher coverage ratios enhances cooling effectiveness (Bakhtyari et al., 2024; Dardir & Berardi, 2021; Herath et al., 2018).

The analysis also confirmed that UHI significantly intensifies building energy demand: under current conditions, incorporating UHI effects increased EUI by 15%, supporting evidence that UHI aggravate cooling loads (Hashemi & Mills, 2025). Projections for 2050 revealed a sharp increase in EUI to 0.56 (kWh/m^2), with cooling demand alone accounting for 0.48 (kWh/m^2). Reductions in heat transfer into buildings, which in turn lowered EUI. The introduction of GWs, however, resulted in a 41% reduction in EUI, surpassing reductions reported in earlier studies (Bakhtyari et al., 2024; Ribeiro et al., 2025). This stronger effect may be attributed to the large-scale application and higher coverage ratio modeled in this study, as well as the superior cooling potential of GWs compared with other strategies. Under future climate conditions, the total EUI with GWs decreased to 0.363 (kWh/m^2), which is nearly equivalent to the baseline model without GWs under the current climate. This demonstrates that large-scale implementation of GWs has the potential to offset the adverse impacts of climate change on building energy performance.

5 Concluding Remarks

This study demonstrated that large-scale implementation of GWs can play a critical role in mitigating the adverse impacts of UHI and climate change on building energy performance. While their influence on ambient air temperature was modest, GWs significantly reduced outdoor surface temperatures and lowered cooling demand, achieving up to 35% reductions in overall EUI under both current and projected 2050 climates. These reductions effectively offset much of the additional energy burden expected in a warming climate, bringing future EUI values close to present-day conditions.

Compared with previous studies, the stronger mitigation observed here is attributed to the urban-scale application and extensive façade coverage modeled. This highlights the importance of expanding green infrastructure research beyond building-level interventions to city-wide strategies that support sustainable urban development. By integrating GWs into urban design, cities can enhance energy efficiency, reduce environmental impacts, and foster climate-resilient urban environments that are better prepared for future challenges. For future studies, this research could be extended by analyzing the effect of different GWs coverage ratios and orientations in conjunction with urban morphology. This approach would provide a deeper understanding of how urban form and vegetation strategies together influence the cooling and energy-saving potential of vegetation-based systems at the district scale.

References

Aghajanzadeh, S. A., Hosseini, S. M., Lorzangeneh, M., & Tabana, M. (2023). The effects of orientation and width of space between buildings on ventilation of high-rise areas. *Journal of Daylighting, 10*(1), 99–116. https://doi.org/10.15627/jd.2023.8

American Society of Heating Georgia, R. & A. C. E. Atlanta. (2014). *Ashrae Guideline 14-2014: Measurement of energy, demand and water savings*. American Society of Heating, Refrigerating, and Air-Conditioning Engineers. Retrieved from https://books.google.it/books?id=zlJkAQAACAAJ

Arenghi, A., Perra, C., & Caffi, M. (2021). Simulating and comparing different vertical greenery systems grouped into categories using energy plus. *Applied Sciences, 11*(11), 4802. https://doi.org/10.3390/app11114802

Astiaso Garcia, D., Dionysis, G., Raskovic, P., Duić, N., & Al-Nimr, M. A. (2023). Climate change mitigation by means of sustainable development of energy, water and environment systems. *Energy Conversion and Management: X, 17*, 100335. https://doi.org/10.1016/j.ecmx.2022.100335

Attia, S., Petersen, S., Hoxha, E., Gobbo, É., Bertini, A., Dasse, M., Abu-Ghaida, H., Heiranipou, M., Al-Obaidy, M., Norouziasa, A., & Stephan, A. (n.d.). *Framework to model building carbon emissions*. Sustainable Building Design Lab. https://doi.org/10.13140/RG.2.2.15338.73925

Bahadori, E., Rezaei, F., He, B.-J., Heiranipour, M., & Attia, S. (2025). Evaluating urban heat island mitigation strategies through coupled UHI and building energy modeling. *Building and Environment, 280*, 113111. https://doi.org/10.1016/j.buildenv.2025.113111

Bakhtyari, V., Fattahi, K., Movahed, K., & Franz, A. (2024). Investigating the effect of living walls on cooling energy consumption in various urban microclimates, building heights, and greenery coverage areas. *Sustainability, 16*(2), 920. https://doi.org/10.3390/su16020920

Cambridge University Press. (2022). *Climate Change 2022: Mitigation of Climate Change. Contribution of working group I to the sixth assessment report of the Intergovernmental Panel on Climate Change.*

Dardir, M., & Berardi, U. (2021). Development of microclimate modeling for enhancing neighborhood thermal performance through urban greenery cover. *Energy and Buildings, 252*, 111428. https://doi.org/10.1016/j.enbuild.2021.111428

Dehghan Lotfabad, A., Hosseini, S. M., Dabove, P., Heiranipour, M., & Sommese, F. (2025). Impacts of vertical greenery on outdoor thermal comfort and carbon emission reduction at the urban scale in Turin, Italy. *Buildings, 15*(3), 450. https://doi.org/10.3390/buildings15030450

Emilsson, T. (2021). Chapter 18—Urban life and climate change. In T. M. Letcher (Ed.), *The impacts of climate change* (pp. 453–462). Elsevier. https://doi.org/10.1016/B978-0-12-822373-4.00009-4

European Commission. (2024, May). *Attitudes of Europeans towards the Environment.* Retrieved from https://europa.eu/eurobarometer/surveys/detail/3173

Farrokhirad, E., Rigillo, M., Köhler, M., & Perini, K. (2024). Optimising vertical greening systems for sustainability: An integrated design approach. *International Journal of Sustainable Energy, 43*(1). https://doi.org/10.1080/14786451.2024.2411831

Fereidoni, S., Nabisi, M., Fereidooni, L., Javidmehr, M., Zirak, N., & Kasaeian, A. (2023). An assessment of the impact of building envelope design on the tradeoff between embodied and operating energy. *Energy and Buildings, 298*, 113542. https://doi.org/10.1016/j.enbuild.2023.113542

Ganjimorad, M., Fernandez, J. D., & Heiranipour, M. (2024). Impact of wind in urban planning: A comparative study of cooling and natural ventilation systems in traditional Iranian architecture across three climatic zones. *Architecture Papers of the Faculty of Architecture and Design STU, 29*(4), 15–29. https://doi.org/10.2478/alfa-2024-0020

Gholami, H., Kamelnia, H., Mahdavinejad, M. J., & Sangin, H. (2025). Optimizing building configuration and orientation for social housing projects in Iran. *Iranica Journal of Energy and Environment, 16*(2), 289–308. https://doi.org/10.5829/ijee.2025.16.02.11

Hashemi, F., & Mills, G. (2025). On the impact of urban climate and heat islands on building energy performance: A critical review. *Energy and Buildings, 343*, 115946. https://doi.org/10.1016/j.enbuild.2025.115946

Heiranipour, M., Favoino, F., Juaristi Gutierrez, M., & Avesani, S. (2024). *Typical and extreme (heatwave) future weather files for building energy simulations: Case studies for Turin and Bolzano, Italy, and De Bilt, Netherlands.* Zenodo. https://doi.org/10.5281/zenodo.14274362

Heiranipout, M., Juaristi, M., Avesani, S., Favoino, F., & Serra, V. (2025). *Contrasting building performance and thermal resiliency: A simulation-based quantitative evaluation framework for evaluating the impact of building envelope technologies* (pp. 432–437). https://doi.org/10.1007/978-981-97-8313-7_60

Herath, H. M. P. I. K., Halwatura, R. U., & Jayasinghe, G. Y. (2018). Modeling a tropical urban context with green walls and green roofs as an urban heat island adaptation strategy. *Procedia Engineering, 212*, 691–698. https://doi.org/10.1016/j.proeng.2018.01.089

Jayasooriya, V. M., Liyanage, C. T., Muthukumaran, S., & Nilusha, R. T. (2025). Impact of green wall orientation on building energy performance in a tropical climate: An experimental assessment. *PLOS Sustainability and Transformation, 4*(1), e0000156. https://doi.org/10.1371/journal.pstr.0000156

Ladybug Tools. (2025). *EPW map.* Retrieved from https://www.ladybug.tools/epwmap/

Lotfabad, A. D. (2025). *Mitigating the urban heat Island effect by enhancing green structures and implementing double skin facades in urban areas* [Politecnico di Torino]. Retrieved from http://webthesis.biblio.polito.it/id/eprint/36282

Ma, Y. X., & Yu, A. C. (2024). Impact of urban heat island on high-rise residential building cooling energy demand in Hong Kong. *Energy and Buildings, 311*, 114127. https://doi.org/10.1016/j.enbuild.2024.114127

Mackey, C., Galanos, T., Norford, L., & Roudsari, M. S. (2017). Wind, sun, surface temperature, and heat Island: Critical variables for high-resolution outdoor thermal comfort. In *Proceedings of Building Simulation 2017: 15th Conference of IBPSA*, Vol. 15, pp. 985–993. https://doi.org/10.26868/25222708.2017.260

Naboni, E., Milella, A., Vadalà, R., & Fiorito, F. (2020). On the localised climate change mitigation potential of building facades. *Energy and Buildings, 224*, 110284. https://doi.org/10.1016/j.enbuild.2020.110284

Ouldboukhitine, S.-E., Bakkour, A., Khiati, S., & Belarbi, R. (2025). Modular green walls: A sustainable architectural solution for energy efficiency in oceanic climates. *Energy and Buildings, 342*, 115850. https://doi.org/10.1016/j.enbuild.2025.115850

Qi, S., Utaberta, N., Kiet, A. L. K., Yanfang, X., & Xiyao, H. (2025). Effects of green façade retrofitting on thermal performance and energy efficiency of existing buildings in northern China: An experimental study. *Energy and Buildings, 335*, 115550. https://doi.org/10.1016/j.enbuild.2025.115550

Regione Piemonte. (2024). *Geo Piemonte*. Retrieved from https://geoportale.igr.piemonte.it/cms/

Rezaei, F., Sangin, H., Heiranipour, M., & Attia, S. (2024). A multi-objective optimization of window and light shelf design in office buildings to improve occupants' thermal and visual comfort. *Journal of Daylighting, 11*(1), 55–68. https://doi.org/10.15627/jd.2024.4

Ribeiro, F. P., Oladimeji, O., de Mendonça, M. B., Boer, D., Maqbool, R., Haddad, A. N., & Najjar, M. K. (2025). BIM-based parametric energy analysis of green building components for the roofs and facades. *Next Sustainability, 5*, 100078. https://doi.org/10.1016/j.nxsust.2024.100078

Susca, T., Zanghirella, F., & Del Fatto, V. (2023). Building integrated vegetation effect on microclimate conditions for urban heat island adaptation. Lesson learned from Turin and Rome case studies. *Energy and Buildings, 295*, 113233. https://doi.org/10.1016/j.enbuild.2023.113233

Turhan, C., Carpino, C., Chen Austin, M., Özbey, M. F., & Akkurt, G. G. (2025). Impact of green wall and roof applications on energy consumption and thermal comfort for climate resilient buildings. *Urban Science, 9*(4), 105. https://doi.org/10.3390/urbansci9040105

Wahba, S. M., Kamel, B. A., Nassar, K. M., & Abdelsalam, A. S. (2018). Effectiveness of green roofs and green walls on energy consumption and indoor comfort in arid climates. *Civil Engineering Journal, 4*(10), 2284–2295. https://doi.org/10.28991/cej-03091158

Yazdi Bahri, S., Alier Forment, M., Sanchez Riera, A., Heiranipour, M., & Hosseini, S. N. (2025). Kinetic facades as a solution for educational buildings: A multi-objective optimization simulation-based study. *Energy Reports, 13*, 3915–3928. https://doi.org/10.1016/j.egyr.2025.03.021

Performance-Based Assessment of Extensive Green Roofs: Field Monitoring and Energy Simulation

Stefano Cascone, Salvatore Polverino, and Manfredi Saeli

Abstract This chapter evaluates the summer thermal behaviour and energy implications of three extensive green-roof technologies, focusing on how substrate composition and drainage/storage typology influence roof heat transfer. A comparative in-situ monitoring campaign was conducted on a university roof in Catania (Italy) from 11 to 25 August 2025. Three co-located systems were assessed: an "Experimental" configuration with a biochar-amended substrate and recycled-plastic drainage granules, a modular "Draining Modules" solution based on polypropylene drainage elements and a "Green Safe" system incorporating a felt drainage/storage layer for controlled water management. Temperatures and heat fluxes were recorded at 10-min intervals below canopy and at the substrate–drainage interface, with a conventional roof sector as reference. All green-roof assemblies strongly damped heat transfer at the substrate–drainage interface, maintaining stable conditions at depth under peak summer forcing, while near-surface behaviour differed among technologies. The Experimental and Draining Modules systems showed moderate surface heat-flux peaks, whereas the Green Safe system exhibited recurrent surface overheating and short-lived positive heat-flux spikes, consistent with transient moisture-mediated effects on the surface energy balance. Annual dynamic simulations of a simplified reference building across three Italian climates (Catania, Rome and Aosta) indicate only marginal differences in non-renewable primary energy demand among the solutions, with outcomes more consistent with assumed stratigraphic thermo-physical properties, particularly of the drainage/storage layer, than with short-term surface dynamics. Overall, the chapter provides comparative field evidence and a modelling-based appraisal to support technology selection in warm climates.

S. Cascone (✉)
Department of Structural, Geotechnical and Building Engineering, Politecnico di Torino, Turin, Italy
e-mail: stefano.cascone@polito.it

S. Polverino
Department of Architecture and Design, University of Genoa, Genoa, Italy

M. Saeli
Department of Architecture, University of Palermo, Palermo, Italy

L. Diana et al. (eds.), *Greening and Resilience*, Urban Sustainability,
https://doi.org/10.1007/978-981-92-0419-9_28

Keywords Biochar-amended substrates · Heat-flux attenuation · Summer overheating mitigation · Drainage-water retention stratigraphy · Envelope thermo-physical characterization

1 Introduction

Green roofs are adopted as nature-based envelope technologies to mitigate summer overheating, attenuate roof heat gains and help counter the urban heat island effect, particularly in Mediterranean climates where high solar loads and prolonged heat-waves can raise cooling demand. Their thermal response depends on coupled heat-moisture exchanges within the vegetation-substrate system and on the hydraulic and thermal role of the drainage and water-retention layer (Lin & Li, 2025). Extensive systems therefore implement different, complementary design strategies: modifying the growing medium through functional amendments (e.g. biochar) to influence moisture availability and thermo-physical properties; using modular drainage panels to provide standardized drainage/retention behaviour; and adopting felt-based drainage/storage layers that can promote capillary redistribution and controlled water management (Mihalakakou et al., 2023).

The objective of this study is to provide a comparative, performance-based assessment of three extensive green-roof configurations by examining how drainage-layer typology and substrate composition influence roof heat-transfer pathways and, consequently, the building energy balance. Specifically, the study quantifies transient temperature and heat-flux dynamics under peak summer forcing and evaluates whether the observed roof-level differences translate into measurable changes in annual building energy performance, expressed as non-renewable primary energy demand. In addition, the study discusses the potential practical relevance of substrate design choices (including the lower bulk density often associated with biochar-amended mixes) for retrofit contexts where dead-load constraints can limit green-roof feasibility.

To meet these aims, a mid-summer, high-resolution monitoring campaign was conducted on a university roof in Catania, where three technologies were installed side-by-side under comparable boundary conditions: an "Experimental" system with a biochar-amended substrate and recycled-plastic drainage granules, a modular "Draining Modules" system and a "Green Safe" system based on a felt drainage/storage element conceived for controlled sub-irrigation. Temperatures and heat fluxes were monitored at two representative planes (below the canopy and at the substrate–drainage interface), with a conventional roof sector used as reference.

Finally, the implications at building scale are assessed through annual dynamic simulations of a simplified reference building across Catania, Rome and Aosta.

2 Background Knowledge

The performance of extensive green roofs depends on coupled thermal-hydrological processes within the vegetation-substrate system and on the functional role of the drainage and water-retention layer. Commercial systems implement these mechanisms through different strategies, including substrate optimization via amendments (e.g. biochar), modular drainage elements integrating drainage and, in some cases, storage and felt-based drainage/storage layers that promote capillary redistribution and controlled water management. Because these approaches act on different physical pathways (moisture availability, latent versus sensible heat partitioning and heat transfer through the build-up), a balanced background is needed to frame comparative assessments among roof typologies.

Biochar is a porous, carbon-rich material produced through biomass pyrolysis. Owing to its high surface area, chemical stability and sorption capacity, it can improve nutrient retention and water-holding capacity in engineered growing media (Nguyen, Chau, Kumar et al., 2024) and has been explored in several green-infrastructure applications (Senadheera et al., 2024). Since around 2010, research on biochar in green roofs has addressed substrate optimization and plant performance, stormwater retention/runoff control and circular-economy strategies based on waste-derived materials (Lee & Kwon, 2024).

A key motivation in extensive roofs is the potential to increase water-holding capacity while reducing bulk density. Experimental evidence indicates that biochar additions can decrease bulk density and increase water retention, improving drought resilience and irrigation efficiency (Cao et al., 2014). Reported effects include changes in biological activity and plant growth (Chen et al., 2021; Olszewski & Eisenman, 2017) and improved leachate quality through reduced nutrient and organic carbon losses (Liao, Drake, et al., 2022; Liao, Sifton, et al., 2022b; Xiong et al., 2023). Biochar can also increase saturated water content, delay runoff onset and reduce peak outflow, with outcomes dependent on amendment level and rainfall intensity (Gan et al., 2021; Kuoppamäki et al., 2016; Nguyen, Chau, & Muttil, 2024). Increasingly, studies adopt a circular-economy framing, combining biochar with waste streams and recycled aggregates while noting the need for standardized selection protocols (Haeldermans et al., 2023; Kuoppamäki et al., 2023; Qiu et al., 2020; Vavrincová et al., 2024; Zhou et al., 2024). Biochar-amended roofs have also been linked to carbon stabilization and increased carbon storage (Chen et al., 2018; Picca et al., 2025; Zhang et al., 2023) and some studies suggest possible reductions in surface temperatures, although evidence on transient thermal effects under real outdoor forcing remains heterogeneous (Tan & Wang, 2023). Overall, despite extensive hydrological and environmental literature, direct field evidence linking biochar amendment to short-term heat-flux dynamics in peak summer conditions, especially in side-by-side comparisons against other roof typologies, remains limited.

A second common strategy uses modular drainage elements, typically polymeric panels, that provide repeatable drainage pathways, protect the waterproofing membrane and often include storage voids. Compared to granular drainage layers, these systems can improve installation repeatability and define geometric characteristics governing drainage and retention (Pirouz et al., 2023). By shaping storage volume and drainage routing, they can affect substrate moisture availability during dry periods and the timing of wetting/drying cycles after rainfall or irrigation, which in turn can influence near-surface temperature evolution and heat-flux dynamics. However, published evidence is often framed around hydraulic performance or constructability, while fewer studies provide high-resolution thermal and heat-flux measurements comparable across alternative drainage/storage solutions under identical boundary conditions (Juras, 2023).

A third typology employs felt or geotextile-based drainage/storage elements that combine drainage with distributed storage and capillary redistribution. Capillary transport can promote lateral spreading and upward moisture supply toward the substrate, supporting controlled water management and, in some configurations, sub-irrigation (Szlachetka et al., 2025). By sustaining substrate moisture, such designs aim to maintain evapotranspiration during dry conditions and buffer thermal loads via latent cooling. At the same time, performance can be sensitive to operational and boundary conditions (irrigation regime, rainfall frequency, plant cover development) and to local heterogeneities in wetting/drying, which may increase variability in near-surface thermal behaviour. Comparative evidence on transient heat-flux dynamics measured at multiple planes within the roof build-up remains limited (Chen et al., 2024).

Taken together, the three typologies act through different levers: biochar amendment targets substrate properties (water retention, bulk density and thermo-physical response); modular drainage panels target controlled drainage and storage through standardized geometry; and felt-based drainage/storage elements target capillary redistribution and controlled water management. In this chapter, these expected advantages are examined through high-resolution monitoring of temperature and heat flux at two planes (below canopy and at the substrate–drainage interface) under peak summer conditions and through annual simulations of non-renewable primary energy demand across three Italian climates, enabling verification of roof-level heat-transfer attenuation and its translation to building-scale energy outcomes.

3 Materials and Methods

3.1 Experimental Set-Up, Monitoring Campaign and Data Processing

The monitoring campaign was carried out on the flat roof (slope <2%) of a university building in Catania (Sicily, Italy), a coastal Mediterranean location. The experimental set-up comprises adjacent roof sectors implementing three extensive green roof technologies and a conventional reference sector. The construction stratigraphy of the green roof fields, roof segmentation and the overall monitoring architecture follow the same test site and configuration described by (Cascone & Vitaliano, 2025), ensuring consistency of build-ups, sensor locations and acquisition procedures across campaigns (Table 1). All vegetated sectors were planted with Gazania nivea to keep vegetation type uniform among technologies.

Each green roof sector includes waterproofing/root-resistant layers and an extensive build-up with a compacted substrate thickness of 15 cm. The three technologies differ in the drainage/water-storage element and substrate composition. The Experimental technology uses a commercial growing medium amended with biochar (20% by weight) placed above recycled-plastic drainage granules. The Draining Modules Technology uses a mineral-organic growing medium placed above modular polypropylene drainage panels with integrated storage/drainage functionality. The Green Safe Technology adopts a mineral-organic growing medium placed above a polyester felt-based drainage/storage element conceived for sub-irrigation. The reference sector corresponds to the original roof build-up without vegetated layers.

Four variables were monitored in each green roof sector using the same sensor typologies and acquisition chain adopted in the earlier campaign: surface temperature (°C), measured at the substrate surface below the vegetation canopy; external heat flux (W/m^2), measured at the same plane by a heat-flux plate installed on the substrate surface below the vegetation; internal heat flux (W/m^2), measured by a heat-flux plate placed at the interface between the substrate and the drainage layer; and internal temperature (°C), measured at the substrate–drainage interface, co-located with the internal heat-flux sensor. For the reference roof, internal temperature was also monitored at the corresponding internal plane used to represent the roof thermal response. Signals were recorded by a multi-channel data logger at a 10-min sampling interval, producing continuous time series suitable for diurnal-cycle analyses. Calibration, pre-deployment verification and periodic consistency checks followed the procedure reported by (Cascone & Vitaliano, 2025). A consistent heat-flux sign convention was adopted for all analyses: positive values indicate heat transfer directed downward into the roof assembly, whereas negative values indicate upward heat release.

The present study focuses on 14 consecutive days (11–25 August 2025), selected to represent peak-summer conditions and to provide a long sequence for robust day-to-day comparison. Data quality control included time alignment of channels, screening for missing or flat-lined segments and removal of isolated spikes inconsistent with the surrounding temporal pattern.

3.2 *Virtual Modelling and Energy Simulation*

In accordance with previous studies, the software TermiPlan 2021 (v. 10.0.1) was used to generate the building model and to quantify its energy performance in compliance with the European Directive 2012/27/EU, the UNI/TS 11300-5:2016 standard, and the EU/2024/1275 (EPBD) as transposed into national legislation by DPR 59/2009 and the D.M. 26 June 2009 and following integrations and modifications. In accordance with previous studies (Lombardo et al., 2024; Saeli et al., 2022), a simplified

Table 1 Investigated extensive green roof typologies and monitoring instrumentation

Item	Experimental	Draining modules	Green safe
Vegetation (common)	Gazania nivea	Gazania nivea	Gazania nivea
Substrate thickness (cm)	15	15	15
Substrate composition	80% standard substrate + 20% biochar (by weight)	Pumice/perlite/peat/compost	Commercial extensive mix
Drainage/storage thickness (cm)	6	8	6
Drainage/storage element	Bags of recycled PE granules + voids	PP panels + expanded clay + voids	Perlite cells + felt + voids
Base layers (common)	Waterproofing + root barrier (same for all test sectors)	Same	Same
Monitoring Plane A (below canopy)	Substrate surface: T_surf + q_surf	Substrate surface: T_surf + q_surf	Substrate surface: T_surf + q_surf
Monitoring Plane B (interface)	Substrate–drainage interface: T_int + q_int	Substrate–drainage interface: T_int + q_int	Substrate–drainage interface: T_int + q_int
Sampling interval	10 min	10 min	10 min

reference building model was developed, consisting of a single-storey structure with a conditioned floor area of 75 m^2 and an internal height of 3 m. The building envelope was modelled using conventional construction technologies representative of typical residential buildings. External and internal surfaces were finished with cement-based plaster, with a total thickness of 5 cm; each vertical opaque element included one transparent opening equipped with standard double-glazed clear glass windows; the load-bearing structure was composed of standardized masonry bricks with a thickness of 25 cm; the horizontal elements were modelled as a traditional composite floor system comprising reinforced concrete and hollow clay blocks. All materials and construction techniques were selected to reflect common residential practice. A detailed description of the geometrical and thermo-physical properties of the model is provided in the study by Saeli et al. (2023). Internal boundary conditions were defined to reproduce realistic residential operating profiles. Indoor air temperature was controlled within the range of 19–26 °C, in accordance with established comfort and operational assumptions for residential buildings (Saeli et al., 2010). Simulations were run over a full solar year (1 January–31 December) to ensure temporal completeness and consistency of climatic inputs. This set-up enables a robust assessment of the thermal and energy performance of the investigated systems under realistic operating conditions, while isolating their specific contribution to the overall building energy balance. Finally, the building model was oriented with its principal axes aligned with the cardinal directions. Simulations were performed for three representative Italian

Table 2 Representative average monthly temperatures in selected climatic zones (1981–2022)

Average monthly temperatures [°C]							
Month	Catania	Rome	Aosta	Month	Catania	Rome	Aosta
January	10.3	7.7	1.5	July	25.5	25.2	22.2
February	10.6	8.4	4.2	August	26.2	25.6	21.5
March	12.3	10.9	8.4	September	23.5	21.3	17.4
April	14.7	13.8	11.7	October	19.7	17.3	12.2
May	18.5	17.9	15.6	November	15.3	12.7	6.4
June	22.8	22.4	19.2	December	11.8	8.7	0.8

locations (Catania, Rome and Aosta) selected to capture the pronounced heterogeneity of Italian climatic conditions, spanning climatic zones from A to F. Climatic boundary conditions were defined in accordance with UNI 10349-1:2016 and official datasets provided by the Italian National Institute of Statistics. Table 2 summarizes the average climatic conditions for the selected sites.

4 Results

4.1 Thermal and Heat-Flux Dynamics During the August 2025 Monitoring Period

Figure 1 shows the surface temperature measured at the substrate surface below the vegetation canopy for the three green roof technologies over 11–25 August 2025. All series exhibit a pronounced diurnal cycle, but with different amplitudes. The Experimental and Draining Modules technologies remain within a narrow thermal band, with daily maximum below 28–29 °C and nighttime minimum around 22–24 °C, resulting in limited diurnal excursions. In contrast, the Green Safe technology displays larger daytime overheating, with repeated sharp midday peaks that reach 36–38 °C during the first part of the monitoring period and remain elevated (31–34 °C) thereafter, while nighttime values drop to 21–23 °C. Overall, Fig. 1 indicates that, despite comparable daily periodicity, the surface layer response of Green Safe is more sensitive to daytime forcing, whereas the other two technologies provide stronger surface-temperature moderation.

Figure 2 reports the external heat flux at the same surface plane. All technologies show daytime positive flux and nighttime negative flux, confirming a daily reversal of the heat-transfer direction. However, the magnitude and waveform differ among systems. Experimental technology shows the most attenuated response, with daytime peaks 10–18 W/m^2. Draining Modules presents moderate and broader positive peaks, 20–30 W/m^2. Green Safe exhibits narrow, high-intensity positive spikes, with daytime peaks in the range of 60–85 W/m^2, indicating intense short-duration

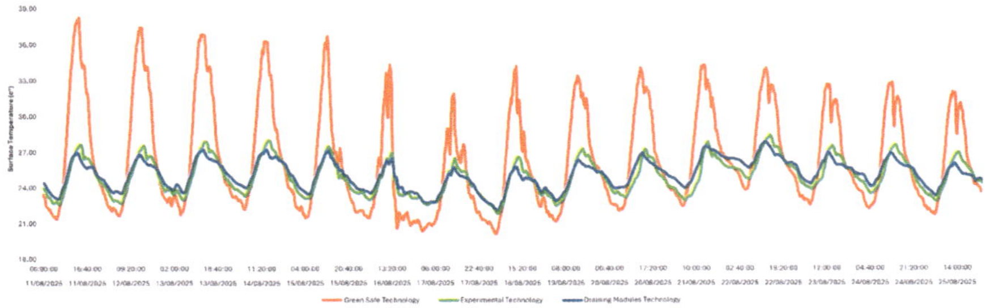

Fig. 1 Time series of surface temperature measured at the substrate surface below the vegetation canopy for the three green roof technologies during 11–25 August 2025

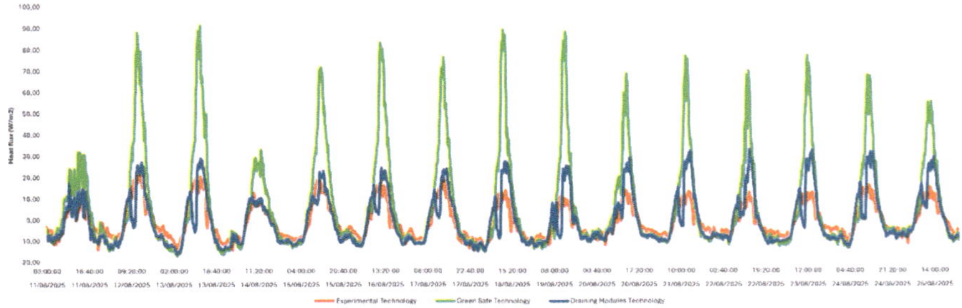

Fig. 2 Time series of external heat flux measured at the substrate surface below the vegetation canopy for the three green roof technologies during 11–25 August 2025

heat transfer at the surface level. Nighttime negative flux values are comparable across technologies (around −10 W/m^2 with minima approaching −15 to −17 W/m^2), indicating that the main inter-technology differences are concentrated in the daytime loading phase.

Figure 3 shows the internal heat flux at the substrate–drainage interface. Compared with Fig. 2, internal flux magnitudes are reduced for all technologies, demonstrating attenuation of the surface forcing through the substrate layer. The Experimental technology shows smaller positive excursions (near 0 to +1 W/m^2) but deep nighttime minima (~ −4 to −5 W/m^2), indicating a negative internal balance with limited daytime reversal. The Draining Modules technology exhibits the largest internal oscillation, with daytime maxima reaching +3 to +4 W/m^2 and nighttime minima down to −5 to −6 W/m^2. Green Safe displays the smallest internal amplitudes, confined between −2.3 and +1.2 W/m^2, with smoother transitions and reduced trough depth relative to the other technologies. Overall, Fig. 3 highlights that the substrate–drainage interface experiences much weaker heat-flux variability than the surface plane, with technology-specific differences emerging in the amplitude of the diurnal wave and the frequency/magnitude of positive reversals.

Figure 4 presents internal temperature at the substrate–drainage interface and includes the reference roof for comparison. The reference roof undergoes extreme

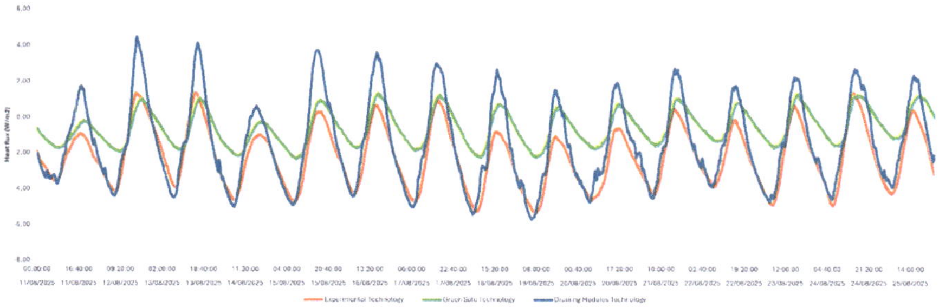

Fig. 3 Time series of internal heat flux measured at the substrate–drainage interface for the three green roof technologies during 11–25 August 2025

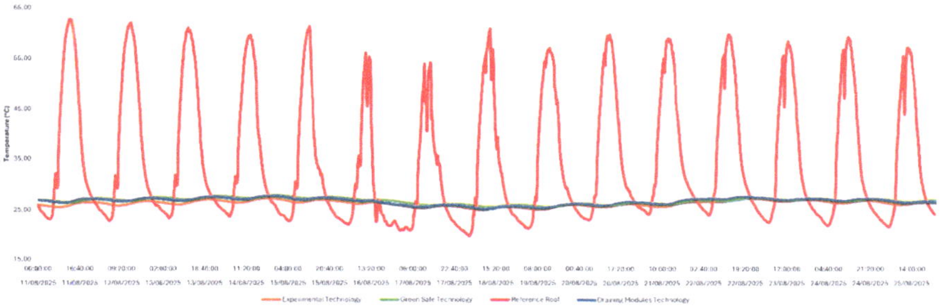

Fig. 4 Time series of internal temperature measured at the substrate–drainage interface for the three green roof technologies and the reference roof during 11–25 August 2025

daily overheating, with daytime peaks reaching 58–62 °C and nighttime minimum near 20–24 °C, producing large diurnal swings. In contrast, all green roof technologies maintain stable internal temperatures clustered around 26–27 °C, with small daily oscillations within 1 °C and overall ranges of only 2–3 °C over the full period. Within the green roofs, differences are subtle: the Experimental technology tends to be cooler on average, while Draining Modules and Green Safe remain aligned, all maintaining a damped internal thermal regime relative to the reference (Fig. 4).

4.2 Energy Simulations Analysis

The simulation software TermiPlan provides a comprehensive set of outputs describing the energy performance of the analysed building and identifies the requirements necessary to ensure compliance with current regulatory frameworks. In the present study, the Global Energy Performance index (EPG) is considered, an index representing the annual non-renewable primary energy demand per unit of conditioned floor area, expressed in kWh/m^2. In accordance with the ISO 52016-1:2017, the EPG index is calculated as the sum of the non-renewable primary energy contributions associated with all relevant end uses, including space heating, space cooling, domestic hot water production, ventilation, and auxiliary systems, as applicable. Each energy demand is first evaluated at the final energy level and subsequently converted into primary energy through the application of conversion factors. As prescribed by UNI/TS 11300-5, system efficiencies, distribution losses, emission subsystems, and control strategies are explicitly modelled, ensuring consistency with the standardized national calculation methodology. The resulting EPG value therefore constitutes a synthetic and normalized performance indicator, enabling direct comparison between different building configurations and ensuring alignment with Italian and EU energy performance regulations. A reference scenario, corresponding to the building model without any green roof system implementation, was also simulated as benchmark energy performance. Modelled technological configurations are shown in Table 3, and the results of the energy simulations are presented in Fig. 5.

From the energy simulations, at first instance, it resulted that—for all the considered technical combinations of layers—the building model performed much better in warmer climates (e.g. Catania) rather than in colder ones (e.g. Aosta). This behaviour is primarily attributable to the higher heating energy demand required in colder climates, which dominates the annual energy balance for heating and significantly increases the non-renewable primary energy consumption. In contrast, in warmer climates, the reduced space-heating load partially compensates for the increased cooling demand, resulting in a lower overall EPG value. Moreover, the designed thermo-physical characteristics of the building envelope, which are representative of conventional residential construction in Italy, are more effective in mitigating cooling loads than heating losses, particularly under Mediterranean climatic conditions. As a

Table 3 Target characteristics of the layers adopted into the modelled system

Technology	Layer	Composition	Thickness cm	λ W/m·K	ρ kg/m³	Cp J/kg·K
Experimental	Drainage	Bags of recycled PE granules + voids	6	0.09	160	1900
	Substrate	80% standard substrate + 20% biochar, by weight	15	0.17	650	1500
Draining Modules	Drainage	PP panels + expanded clay + voids	8	0.10	350	1100
	Substrate	Pumice/perlite/peat/compost	15	0.20	700	1500
Green Safe	Drainage	Perlite cells + felt + voids	6	0.05	140	900
	Substrate	Commercial extensive mix	15	0.22	680	1500

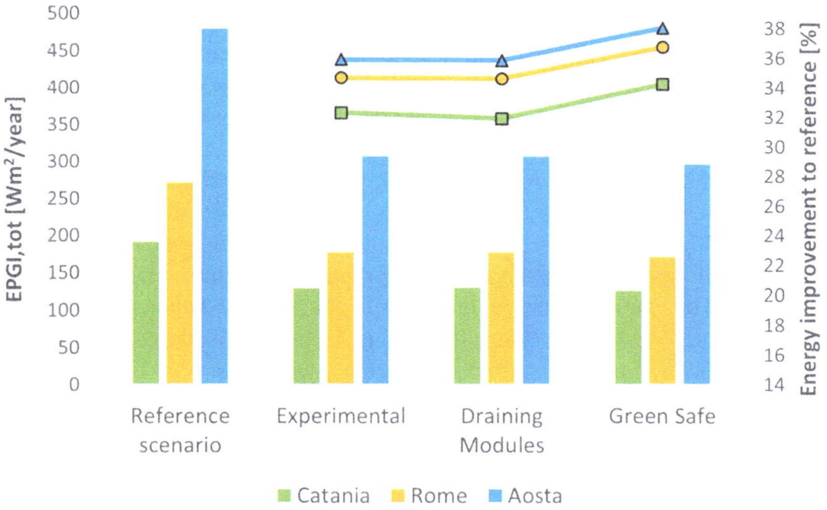

Fig. 5 Global energy performances ($EP_{G,tot}$ – kWh/m^2) (bars) and energy improvement (%) (lines) of the considered building model per location and technological configuration, including the reference scenario

consequence, passive strategies such as increased thermal inertia and moderate insulation levels could provide greater relative benefits in warm climates, whereas in cold climates their impact is less pronounced due to the predominance of transmission losses and prolonged heating periods. Moreover, it is noted that, as expected, the reference scenario showed the highest EPGs (worst building energy performance) due to the lack of insulating layers. This is obviously unlikely to happen in real-world scenarios but it is here extremely useful to isolate the energy improvement of the model due to the considered green roofs technologies.

It is also observed that, for a given geographical location (Catania, Rome or Aosta), EPG shows only marginal differences among the three analysed technologies (Experimental, Draining Modules and Green Safe). Nevertheless, a consistent performance hierarchy emerged across all climatic scenarios, as previously discussed. Specifically, the Green Safe technology systematically exhibited the lowest EPG values, followed by the Experimental solution, while the Draining Modules configuration showed the highest overall building energy demand. From a general standpoint, the biochar-based substrate (characterized by reduced thermal conductivity and lower bulk density compared to conventional substrate mixes) represents an effective compromise between commercially available solutions. This configuration combines improved thermal insulation performance with acceptable structural and constructive characteristics, resulting in a measurable, albeit limited, reduction in non-renewable primary energy demand. Moreover, the lower density of the biochar-based mix (among the three substrate options) contributes to a reduced additional load on the supporting structure, potentially facilitating its integration in retrofit applications without requiring significant structural reinforcement. Finally, the simulation results indicated that the relative energy performance improvements (%) are most pronounced under the coldest climatic conditions. In particular, the building model exhibits the highest energy performance increase in Aosta, with reductions in overall energy demand of up to approximately ~38%, followed by Rome with ~37%, and Catania with ~34%. For the considered experimental layers, the resulting reductions were ~36%, ~34.5%, and ~32%, respectively. The observed trend suggests that the implemented energy strategy (green roof) is particularly effective in climates characterized by higher heating requirements (cold climates), in other words, where the building envelope improvement yields greater relative benefits due to the dominance of space (m^2) over heating loads in the overall annual energy balance. Conversely, in warmer climates, the lower baseline for heating demand limits the achievable relative improvements, despite comparable absolute energy savings.

5 Discussion and Implications

The monitoring campaign indicates that, under peak Mediterranean summer forcing, the three extensive technologies provide comparable thermal conditions at the deeper roof interface while exhibiting different near-surface dynamics. Temperatures at the substrate–drainage interface remained within a narrow band with limited diurnal oscillations and heat flux at the same interface was attenuated for all systems. This supports the ability of extensive green roofs to damp roof heat transfer through thermal mass and moisture-mediated processes, even with limited substrate thickness.

At the below-canopy plane, technology-specific responses emerged. The Experimental and Draining Modules solutions showed lower daily maxima and moderate surface heat-flux peaks, whereas the Green Safe system exhibited recurrent near-surface overheating and narrow, high-intensity positive heat-flux spikes. A plausible explanation is that Green Safe's drainage/storage element and moisture distribution altered the timing of conductive and latent exchanges near the surface. Short-lived spikes may reflect transient surface energy-balance changes (e.g. rapid wetting/drying, localized moisture gradients, or intermittent solar loading due to canopy heterogeneity). These effects can coexist with strong damping at depth, explaining why Green Safe shows larger surface excursions but small internal heat-flux amplitudes. This also indicates that surface temperature alone is not a sufficient proxy for roof heat gain.

Linking these short-term results to annual energy simulations requires accounting for differences in time scale and represented physics. The monitoring captures a limited mid-summer window, when behaviour is sensitive to moisture availability and evapotranspiration. The simulations report annual performance (EPG) integrated over all seasons and driven by standard weather files and assigned layer properties; therefore, summer surface-level differences are not expected to translate into annual indicators, particularly because the roof is only one envelope component.

A more consistent correlation is obtained by comparing simulations with monitored variables closer to the building energy balance. Heat flux and temperature at the substrate–drainage interface (q_int and T_int) are more representative of heat transfer toward the roof structure than surface-plane variables (T_surf and q_surf). The similar interface damping observed for all systems helps explain the modest spread in annual EPG among technologies.

The simulated ranking should be interpreted in light of modelling assumptions. The roof is represented through layer thicknesses and thermo-physical properties, without explicit coupled hydrothermal processes (moisture-dependent properties, capillary redistribution, evapotranspiration dynamics, irrigation/rain-driven wetting cycles). Under this scope, EPG is more sensitive to effective conductive resistance (including the drainage/storage layer) than to short-lived surface spikes.

Overall, biochar-related expectations are partially confirmed: the Experimental system exhibits the most attenuated daytime surface heat-flux peaks, while all three systems show stable interface behaviour. Annual energy differences remain modest. Limitations include the short monitoring window and the lack of concurrent substrate moisture and vegetation measurements.

6 Concluding Remarks

This chapter combined high-resolution field monitoring and dynamic energy simulations to assess the summer thermal behaviour and energy implications of three extensive green-roof technologies installed under comparable conditions in Catania (Italy). Monitoring confirms that all systems damp roof heat transfer under peak Mediterranean forcing: at the substrate–drainage interface, temperature oscillations and heat-flux amplitudes were reduced and internal-facing temperatures remained stable compared with the conventional roof reference, which exhibited severe overheating.

The Experimental (biochar-amended) and Draining Modules solutions presented lower daily maxima and more moderate heat-flux peaks, whereas the Green Safe system displayed recurrent short-lived surface overheating and high-intensity positive heat-flux spikes. This suggests that drainage/storage configuration and moisture management can influence the surface energy balance without compromising deeper-layer attenuation and it highlights the value of combined temperature-heat flux indicators rather than surface temperature alone.

Annual simulations across three Italian climatic scenarios (Catania, Rome and Aosta) indicate that changes in non-renewable primary energy demand are modest and technology-dependent, consistent with differences in effective thermo-physical behaviour of the roof build-ups.

Future work should extend monitoring across seasons and include irrigation/rain logs, substrate moisture measurements and vegetation coverage metrics to support stronger causal interpretation and calibrated modelling. Also, further more detailed energy simulation will be implemented to better assess thermos-hygrometric features of the proposed solution vs the commercial products.

References

Cao, C. T. N., Farrell, C., Kristiansen, P. E., & Rayner, J. P. (2014). Biochar makes green roof substrates lighter and improves water supply to plants. *Ecological Engineering, 71*, 368–374.

Cascone, S., & Vitaliano, S. (2025). Innovative green roof technologies in Mediterranean climate: Implications for sustainable design of the built environment. *Building and Environment, 273*.

Chen, H., Du, X., Lai, M., Nazhafati, M., Li, C., & Qi, W. (2021). Biochar improves sustainability of green roofs via regulate of soil microbial communities. *Agriculture, 11*(7), 620.

Chen, H., Ma, J., Wang, X., Xu, P., Zheng, S., & Zhao, Y. (2018). Effects of biochar and sludge on carbon storage of urban green roofs. *Forests, 9*(7), 413.

Chen, N., Deng, Q., Chen, Q., & Wang, Z. (2024). Green roof heat transfer coefficient measurement and impact of plant species and moisture. *Energy and Buildings, 303*, 113805.

Gan, L., Garg, A., Wang, H., Mei, G., & Liu, J. (2021). Influence of biochar amendment on stormwater management in green roofs: Experiment with numerical investigation. *Acta Geophysica, 69*(6), 2417–2426.

Haeldermans, T., Puente Torres, J., Vercruysse, W., Carleer, R., Samyn, P., Vandamme, D., Yperman, J., Cuypers, A., Vanreppelen, K., & Schreurs, S. (2023). An experimentally validated selection protocol for biochar as a sustainable component in green roofs. *Waste, 1*(1), 176–194.

Juras, P. (2023). Rainwater retention test of various green roofs: Influence on membrane temperatures and evapotranspiration. *Buildings, 13*(8), 2058.

Kuoppamäki, K., Hagner, M., Lehvävirta, S., & Setälä, H. (2016). Biochar amendment in the green roof substrate affects runoff quality and quantity. *Ecological Engineering, 88*, 1–9.

Kuoppamäki, K., Prass, M., & Hagner, M. (2023). Crushed concrete and biochar: A sustainable solution for vegetated roofs. *Urban Forestry & Urban Greening, 88*, 128082.

Lee, J., & Kwon, E. E. (2024). Biochar in green roofs. *Journal of Building Engineering, 89*, 109272.

Liao, W., Drake, J., & Thomas, S. C. (2022). Biochar granulation, particle size, and vegetation effects on leachate water quality from a green roof substrate. *Journal of Environmental Management, 318*, 115506.

Liao, W., Sifton, M. A., & Thomas, S. C. (2022). Biochar granulation reduces substrate erosion on green roofs. *Biochar, 4*(1), 61.

Lin, H., & Li, X. (2025). The role of urban green spaces in mitigating the urban heat Island effect: A systematic review from the perspective of types and mechanisms. *Sustainability, 17*(13), 6132.

Lombardo, L., Campisi, T., & Saeli, M. (2024). Spent coffee grounds-based thermoplaster system to improve heritage building energy efficiency: A case study in Madonie Park in Sicily. *Sustainability, 16*(15), 6625.

Mihalakakou, G., Souliotis, M., Papadaki, M., Menounou, P., Dimopoulos, P., Kolokotsa, D., Paravantis, J. A., Tsangrassoulis, A., Panaras, G., Giannakopoulos, E., & Papaefthimiou, S. (2023). Green roofs as a nature-based solution for improving urban sustainability: Progress and perspectives. *Renewable and Sustainable Energy Reviews, 180*, 113306.

Nguyen, C. N., Chau, H.-W., Kumar, A., Chakraborty, A., & Muttil, N. (2024). Biochar amendment in green roof substrate: A comprehensive review of the benefits, performance, and challenges. *Applied Sciences, 14*(16), 7421.

Nguyen, C. N., Chau, H.-W., & Muttil, N. (2024). A field study to investigate the hydrological characteristics of newly established biochar-amended green roofs. *Water, 16*(3), 482.

Olszewski, M. W., & Eisenman, S. W. (2017). Influence of biochar amendment on herb growth in a green roof substrate. *Horticulture, Environment, and Biotechnology, 58*(4), 406–413.

Picca, G., Goñi-Urtiaga, A., Lozano de Sosa, L., Colombo Rodriguez, M. V., Fernández Navarro, I., Plaza, C., & Panettieri, M. (2025). Sustainability in urban agriculture: The role of biochar in enhancing the productive capacity of compost-based growing media for rooftop farming. *Urban Forestry & Urban Greening, 107*, 128774.

Pirouz, B., Palermo, S. A., Becciu, G., Sanfilippo, U., Nejad, H. J., Piro, P., & Turco, M. (2023). A novel multipurpose self-irrigated green roof with innovative drainage layer. *Hydrology, 10*(3), 57.

Qiu, D., Peng, H., Li, T., & Qi, Y. (2020). Application of stabilized sludge to extensive green roofs in Shanghai: Feasibility and nitrogen leaching control. *Science of the Total Environment, 732*, 138898.

Saeli, M., Batra, V. S., Singh, R. K., Tobaldi, D. M., & Labrincha, J. A. (2023). The coffee-house: Upcycling spent coffee grounds for the production of green geopolymeric architectural energy-saving products. *Energy and Buildings, 286*, 112956.

Saeli, M., Capela, M. N., Campisi, T., Paula Seabra, M., Tobaldi, D. M., & La Fata, C. M. (2022). Architectural technologies for life environment: Spent coffee ground reuse in lime-based mortars. A preliminary assessment for innovative green thermo-plasters. *Construction and Building Materials, 319*, 126079.

Saeli, M., Piccirillo, C., Parkin, I. P., Binions, R., & Ridley, I. (2010). Energy modelling studies of thermochromic glazing. *Energy and Buildings, 42*(10), 1666–1673.

Senadheera, S. S., Withana, P. A., Lim, J. Y., You, S., Chang, S. X., Wang, F., Rhee, J. H., & Ok, Y. S. (2024). Carbon negative biochar systems contribute to sustainable urban green infrastructure: A critical review. *Green Chemistry, 26*(21), 10634–10660.

Szlachetka, O., Witkowska-Dobrev, J., Baryła, A., & Dohojda, M. (2025). Experimental evaluation of the sustainable performance of filtering geotextiles in green roof systems: Tensile properties and surface morphology after long-term use. *Sustainability, 17*(14), 6242.

Tan, K., & Wang, J. (2023). Substrate modified with biochar improves the hydrothermal properties of green roofs. *Environmental Research, 216*, 114405.

Vavrincová, L., Pipíška, M., Urbanová, J., Frišták, V., Horník, M., Machalová, L., & Soja, G. (2024). Sewage sludge biochar as a sustainable and water-safe substrate additive for extensive green roofs. *Sustainable Chemistry and Pharmacy, 39*, 101604.

Xiong, W., Li, J., Wang, H., Wu, Y., Li, D., & Xue, J. (2023). Biochar addition and the runoff quality of newly constructed green roofs: A field study. *Sustainability, 15*(5), 4081.

Zhang, G., Tang, B., Bian, H., Huang, Y., Jin, H., & He, B.-J. (2023). Investigating the impacts of biochar addition to substrate on mitigating carbon dioxide emissions in green roofs (pp. 131–140).

Zhou, J., Deng, Q., Chen, Q., Chu, B., Li, Y., & Wang, Z. (2024). Waste-green infrastructure nexus: Green roof promotion by digestate and digestate biochar from food waste. *Bioresource Technology, 402*, 130845.

Vertical Greenery Systems: Some Remarks on the Management Phase

Alessandro Lo Faro, Grazia Lombardo, Angela Moschella, and Gaetano Sciuto

Abstract In order for Vertical Greenery Systems (VGS) to be truly sustainable greening strategies, it is the task of technicians and researchers to plan operations such as management and maintenance already in the design phase. This chapter highlights the main maintenance-related actions. They are influenced by the classification of VGS (green facade or living wall). The large size of the installations requires maintenance work to be carried out at height, ensuring the safety of operators. Some indications are also given on the costs, sometimes apparently high, of installing VGS. Future prospects point to the search for low-cost greening solutions to be applied to existing building stocks and not only to new factories.

Keywords Green façade · Maintenance · Costs · Safety · Sustainability

1 Introduction

The accelerated urbanization of cities has resulted in the increasing fragmentation of nature in towns and peri-urban areas. Nature-based strategies for climate change resilience have the potential to enhance the regulation of bioclimatic parameters and the quality of life experienced by people (Lafortezza & Sanesi, 2019). Nature can be regarded as a tool that provides ecosystem services, which can be categorized as cultural, regulatory and provisioning. It is evident that these services have the capacity to contribute towards a reduction in carbon emissions, the resolution of social inequality issues through job creation or food production, and the enhancement of human well-being and health. This is achieved by incorporating the services into urban planning at an early stage (Oquendo Di Cosola et al., 2022). In particular, the Green Infrastructure (GI) approach is predicated on the implementation of nature-based solutions, which are intended to transcend the scope of mere climate mitigation. Rather, the approach is intended to provide a theoretical and practical reflection on

A. L. Faro (✉) · G. Lombardo · A. Moschella · G. Sciuto
Department of Civil Engineering and Architecture, University of Catania, Catania, Italy
e-mail: alessandro.lofaro@unict.it

L. Diana et al. (eds.), *Greening and Resilience*, Urban Sustainability,
https://doi.org/10.1007/978-981-92-0419-9_29

how to proactively obtain ecosystem services, thereby strengthening shared advantages deriving from positive socio-ecological interactions. For this reason, GI has been acknowledged as an effective tool to increase territorial and urban resilience. Green infrastructure helps ensure the sustainable provision of ecosystem goods and services while increasing the resilience of ecosystems. The concept is central to the overall objective of ecosystem restoration, which is now part of the 2020 biodiversity target (European Environment Agency, 2011). Vertical Greenery Systems (VGS) are a fundamental component of GI, particularly in high-density urban environments. They exploit the vertical surfaces of buildings, addressing the shortage of horizontal green spaces without requiring the occupation of new land. VGS can be divided into green facades (GF) and living wall (LW) based on the different plants and support structures used (Lo Faro et al., 2022). The key distinction lies in the medium used: green façades employ a ground-based medium for creeper plants, whereas living walls feature a vertical medium. In temperate climates, living walls often use pocket, geotextile felt, and modular systems, while green façades typically utilize cable wire, mesh, and grid systems. The extensive use of VGS, especially when we adopt them as greenery strategy for existing buildings, requires preliminary authentication of multiple indicators to ascertain the most sustainable solution. The positive results of VGS may be influenced by various macro criteria, including the type and classification of GF, façade design, cost and benefit (Moschella et al., 2025). The macro cost/benefit criterion also concerns the high initial installation and maintenance cost of VGS, which represents some of the most tenacious barriers to their application (Huang et al., 2021). The literature quantifies several economic benefits of VGS, which, when considered over the lifetime of these systems, can compensate for their high initial costs and assist users in adopting VGS in buildings.

2 Background Knowledge

In scientific literature, many papers deal with maintenance of VGS. Rather than proceeding with a structured analysis of the state of the art, it was deemed more useful to refer to the most recent review papers, doing a critical review. It was possible to find in them some summary overview that has been particularly useful for our assessments. For example, Khan (2024) evaluated 17 studies in accordance with the PRISMA guidelines to determine various costs, benefits, and economic indicators associated with VGS. The VGS type and context affect economic viability and sustainability. Also, according with PRISMA, Alkadri et al. (2025) included 98 studies in the final review addressing key gaps in VGS by focusing on three main areas: (1) identifying trends and future research directions from academic databases regarding vegetation use; (2) exploring how vegetation can enhance VGS performance through plant maintenance technologies such as mediums, fertilizers, lighting, and irrigation; (3) identifying key plant species most suited for VGS, with a focus on benefits such as air quality, ecology, energy efficiency, acoustics, thermal performance, and visual appeal. This research highlights the need for region specific, evidence-based guidelines that integrate plant traits, maintenance requirements, and multifunctional performance. Farrokhirad et al. (2024) analyzed the VGS design according to an integrated and interdisciplinary approach, addressing key gaps in current practice. An import aspect that they underline is the user input that is primarily limited to selecting VGS type, with minimal involvement in maintenance. They suggest a more holistic approach that engages users throughout the design and maintenance process. Several standards, principally German, were also analyzed in order to provide a comprehensive overview of the design and installation of VGS. The analysis also offered guidance on maintenance measures.

3 Maintenance Remarks for VGS

The main cost components of implementing VGS in buildings include initial costs and maintenance costs (Rosasco & Perini, 2018). Initial costs encompass the acquisition and installation of plants, growing media, support structures, irrigation systems, waterproofing, drainage, as well as labor and transportation expenses. Maintenance costs involve regular upkeep activities such as watering, fertilizing, pruning, pest control, repair or replacement of components, ongoing monitoring, and labor. Actual costs and payback periods are influenced by external variables including climate conditions, selected plant species, building morphology, and local practices of working. Direct green facades are low-cost and low-maintenance, especially using ground-rooted climbers, making them suitable for mild climates (see Fig. 1).

VGS type	Sub type	Description	Initial cost	Maintenance cost	Climate
Green façade (GF)	Direct a)	Climbing plants rooted at base, attach directly to wall	Low	Low	Temperate, mild
	Indirect b)	Climbing plants supported by cables/mesh	Low/moderate	Low	Temperate, mild
Living wall (LW)	Modular system	Pre-vegetated panels, hydroponic or substrate based	High	High	All (with irrigation)
	Container system c)	Modular container with soil	High/moderate	Moderate	All (with irrigation)
	Hydroponic d)	Felt/nutrient soil	High	Very high	All (with irrigation)

Fig. 1 Comparison of different types of VGS and their cost analysis (image of the authors)

Indirect facades add support structures, slightly increasing cost but maintaining similar climate limitations. Living walls, instead, use modular or hydroponic systems with irrigation, allowing diverse plants to use in various climates but have higher cost and maintenance due to system complexity. So, our attention is more oriented to green façades, because they are a more economically viable solution. Typical maintenance tasks for a green structure include:

- Pruning plants;
- Removal of dead plant material;
- Replacing plants that have died;
- Ensuring windows, downpipes, air vents, etc., are kept clear;
- Maintenance of irrigation and fertilization equipment;
- Protecting the irrigation system from frost;
- Maintenance of climbing aids (if they have been planned);
- Tidying and litter pick.

The maintenance activities for a VGS in a temperate country are summarized in Table 1. A maintenance plan for a green structure needs to be allowed at the design stage. From the very beginning of the design process, factors relating to maintainability must be considered. Careful integration of façade access equipment and strategic planning is essential. In the case of multi-story buildings, the maintenance system should be designed to be as easy as possible, so as not to inconvenience building occupants or management.

Maintenance operation times vary depending on the type of plants, species and associated maintenance activities. Not all the activities are carried out during specific months or weeks of a typical year. Therefore, the use of façade access equipment for vertical greenery maintenance can increase the time and costs required for façade cleaning and maintenance. There are several options that could be considered in order to increase efficiency, such as:

(a) Multiple façade access systems allow for multitasking, thereby reducing overall equipment usage time and minimizing wear and tear during the system's service life;
(b) The greenery layout is to be simplified in order to reduce the time and effort required to maneuver the façade access equipment around hard-to-reach areas;
(c) The repetitive pattern of greenery arrangement is intended to ensure a consistent level of maintenance and cleaning across the entire façade;
(d) Selecting an access strategy that requires the least amount of time to set up;
(e) Inclement weather can have a detrimental effect on the utilization of façade access equipment, which may result in a delay to cleaning activities and an increase in the overall time required for the task to be completed;
(f) The deployment of experienced personnel who are familiar with the sequence of maintenance operations can lead to greater efficiency and effectiveness. This has the potential to reduce the necessity for repetitive training of new staff, thereby saving time and resources;

Table 1 A typical VGS maintenance schedule in a temperate country

Maintenance operations	Category	Typical schedule
Weed control	Trees	Monthly from April to October
	Shrubs	
	Perennials, ornamental grasses and ferns	
	Grass	Twice annually in March/April and October
Mulching	Trees	Top Up March/April after fertilizer application, check and top up if required in October
	Shrubs, perennials, ornamental grasses and ferns	
Fertilizer/feeding	Trees	In March/April
	Shrubs & perennials	
	Grass	Spring application
Watering	Trees	Between April and October
	Shrubs	
	Perennials, ornamental grasses and ferns	
	Grass	
Pruning	Trees	October to end of February
	Shrubs	
	Perennials	December–Spring
	Ornamental grasses	January–February
	Ferns	March
Dead and dying stock	Trees	End of November—end of March
	Shrubs	
	Perennials, ornamental grasses and ferns	April–May
	Grass	April
Inspect/repair/tighten/replace damaged parts	All category	Annually
Tidying and litter pick	All category	Bi-weekly
Leaf litter	All category	Bi-weekly September to end of November
Plant inspections	All category	October and after strong winds
Re-firm	All category	
Fixings/tree ties/guys	All category	
Inspect and keep clear	All category	Monthly

(g) Streamlined procedures and workflows can be standardized and optimized for efficiency. The team can learn and identify improvements based on their collective experience.

According to DIN 18916:2016-06 standard (Vegetation technology in landscaping—Plants and plant care) maintenance measures can be divided into three different phases: completion, development, and upkeep maintenance. To ensure lush growth of VGS, completion maintenance must be performed immediately after installation on the ground. The goal of completion maintenance is to achieve a state suitable for further development. Development maintenance aims to achieve the functional state of vegetation and lasts approximately 2–3 years. Once this functional state or green vegetation target has been achieved, maintenance services begin to ensure the functional state of the vegetation (upkeep maintenance). To this end, it is necessary to draw up a maintenance plan to be followed on an ongoing basis. The importance of maintenance measures is underlined, also, in another German standard, FLL:2018 "Facade greening guidelines – Guidelines for the planning, construction and maintenance of greening façade". The guideline reiterates that facades can only be used for greening if it is possible to guarantee that the necessary maintenance measures will be carried out.

The maintenance required may vary and depends on the outcomes the customer wants. The main challenges of VGS maintenance are those associated with working at a height. Qualified personnel with great knowledge of the system used are required. This is the reason why the maintenance services are often hired along with the installation company. For routine maintenance activities, such as watering and cleaning, a maintenance reach that allows easy access to the entire green façade is important. Equipment such as roof mounted building maintenance units (BMUs, used especially in UK or for very high buildings) and scissor lifts may be required. The minimum reach of such equipment will depend on the specific height and design of the VGS. In other cases, we have to provide for suspended cradles attached to brackets or using work positioning systems. In both cases, a specific project is required for the installation of these devices. The most common solution involves the use of Mobile Elevating Working Platform (MEWPs). In these cases, it is necessary to check in advance the access to the site, the maneuvering space for the vehicles, and the height to be reached (see Fig. 2).

The effectiveness of VGS depends on the health of the plant species involved. Maintenance work must be planned over time. The dimensions of VGS may require work that could expose workers to the risk of falling from height (as already mentioned), as well as other risks associated with the use of electrical and mechanical equipment and chemicals. It is essential to ensure that workers can access and work in higher sections of the greenery securely. This can be achieved through proper training, the use of appropriate equipment, or the implementation of fall protection measures. The use of pruning and trimming tools while on abseil ropes can present certain challenges, particularly when abseilers rely on their feet and knees to position themselves. It is essential that the impact of platform buffers or abseiling foot loads is taken into account during the initial design phase.

Fig. 2 Equipment for VGS maintenance: Work positioning systems on the left and mobile elevating working platform on the right (image of the authors)

Different types of LW and GF may require different approaches. For example, some installations may be designed with modular planters that can be easily accessed and maintained at ground level, while others may involve climbing or reaching higher sections. Arborists and horticulturists must be trained in using suspended platforms, MEWPs, and abseiling systems to access various parts of the living wall safely. Suspended platforms on BMUs are equipped with buffers to soften the impact. The plant species selected and their arrangement must be able to withstand these minimal contact forces. It is essential to ensure that the LW can withstand the tools and equipment required for maintenance in order to guarantee both safety and the efficiency of the process. It is important to note that there is no single, universal solution to the question of the minimum maintenance reach, as each installation may possess unique characteristics. The minimum reach should be determined based on a comprehensive assessment of the specific VGS design and maintenance requirements.

4 Discussion and Implications

The selection of plant species is a pivotal factor influencing the overall performance and sustainability of VGS. It is essential that this process is conducted taking into account the specific characteristics of both the site and the installation, such as exposure to sunlight, wind intensity, wall orientation, and availability of water for irrigation. In addition, it is crucial to consider the influence of local climatic variables to ensure the accuracy and relevance of the results, such as temperature, humidity, and pest control.

It is common practice for a significant portion of activities to be managed manually, which requires the physical presence of the facility manager.

At present, remote management tools are used solely for the purpose of controlling the flow of irrigation; several experts use mobile applications to monitor conditions through integrated sensors and complementary valves.

Another important aspect of VGS management is the assessment of maintenance costs (see Table 2). The scientific literature generally reports costs in dollars, as the experiments are carried out all over the world. The costs have been converted from dollars to euros, here. Actual costs and payback periods are influenced by external variables including climate conditions, selected plant species, building morphology, and local maintenance practices. Maintenance tasks such as pruning and replacing plants annually can represent up to 50% of the total expenses. Maintenance costs may include pipe deterioration and replacement, which can make up 15% of the expenses, while transportation costs can range from 6 to 10% of the total. Additional maintenance factors, including the degradation and replacement of irrigation pipes, can contribute up to 15% of the maintenance budget. Transport and shipping costs further add to the financial problem, constituting approximately from 6 to 10% of the overall budget. Because maintenance expenses are often similar of installation costs, especially when we refer to LW, there is a clear need for more economically sustainable solutions. The distinction between maintenance costs and initial and disposal costs is predicated on the premise that maintenance costs are recurring expenses incurred throughout the lifetime of the VGS, whereas initial and disposal costs are one-time expenditures (Khan, 2024).

Installation and maintenance costs obviously affect the payback period, which, when calculated, is always very long, even up to 25 years. Shorter PBPs are achieved for particularly low-cost solutions and in contexts where labor costs are also affordable, as in developing countries.

5 Conclusion

Stakeholders should carefully evaluate the specific elements of each case when assessing the feasibility of implementing VGSs in urban environments. The assessment of economic sustainability can be improved if all the possible collateral benefits of VGSs are quantified and analyzed to provide a comprehensive understanding of these systems. In this context, the management of VGSs cannot be overlooked, as it represents, as specified in this chapter, a significant cost item. The beneficial effects of VGS are closely linked to the well-being of the vegetation they host. This vegetation has its own management and maintenance requirements, which are strongly influenced by the climatic characteristics of the site, the type of VGS and, therefore, the choice of plant species. A well-designed VGS can also contribute to cost

Table 2 Economic viability of different types of VGS. The cost in euros has been updated based on the monetary revaluation proposed by the Italian National Institute of Statistics (ISTAT), referring to 2025

Type	Method	Country	Year	Plant type	Energy consumption reduction (%)	Initial cost (€/m^2)	Maintenance cost (€/m^2)	Payback period (years)
Modular living wall	Experiment	Nigeria	2016	Vines Herbs Succulents Shrubs	NA	34.66	124.30	3
Indirect green façade	Simulation	Italy	2016	Climbing plant Shrubs Evergreens	NA	576.79	245.44	NA
Direct green façade					NA	239.01	230.58	NA
Indirect green façade	Cost Benefit Analysis	Italy	2018	Climbing plant Shrubs Evergreens	43	318.13	11.80	25
Direct green façade	Experiment	China	2015	Grass	NA	28.71–38.28	2.87	NA
Indirect green façade					NA	38.28–72.21	2.87	NA
Modular living wall					NA	96.57–772.56	9.65	NA
Modular living wall	Sampling	Malaysia	2016	Climbing plant	37	42.89	NA	NA

reduction, given the impact of design choices on maintenance costs over the system's lifecycle.

A simpler technological system and plant species requiring little maintenance can reduce the overall cost (initial + maintenance) of VGS (Rosasco & Perini, 2018). In terms of cost-effectiveness, GF is a more viable option when compared to LW, making it a more suitable choice for the implementation of VGS in building applications (Khan, 2024).

It is recommended that future research concentrate on context-specific studies, particularly in developing regions, in order to enhance our understanding of the potential for mitigating climate change and improving urban sustainability (Khan, 2024). It is evident that this would result in a significant increase in the adoption of VGS, thereby leading to the creation of urban environments that are characterized by enhanced sustainability, improved environmental quality, and greater healthiness for citizens.

The maintenance of VGS represents a significant expense (almost 50% of the total cost) which, therefore, cannot be overlooked in order to meet the sustainability requirements of the project. In the case of medium–high installations (height >8 m), safety measures for gardeners must be provided for at the design stage.

To serve broader environmental and social benefits across diverse economic settings is crucial to make VGS more cost-effective for adoption them in many countries. Without such innovations, the application of VGS is likely to remain restricted to aesthetic improvements on buildings and only wealthy nations will be able to afford the widespread use of these greening strategies.

Acknowledgements This research was funded by European Union-EU Next Generation, Italian PRIN 2022 PNRR Programme under "Greenwork: An interdisciplinary framework for urban health and urban resilience enhancement based on greening strategies on buildings and open spaces" project, CUP E53D23019120001. The present study has been conducted within the frame, also, of the project "SIA3 - Soluzioni InnovAtive e strategie di sostenibilità per città, territori e società - Innovative solutions and strategies for sustainability of Cities, Land and Society," financed by University of Catania.

References

Alkadri, M. F., Yolanda, A., Alifa, R. P., Purbaya, R., Susanto, D., Istiani, N. F. F., & Suryanegara, M. (2025). Systematic review of plant selection in vertical greenery systems for urban sustainability: Current research, knowledge gaps, and future directions. *Energy & Buildings, 346*, 116136.

DIN 18916:2016-06, Vegetation Technology in Landscaping—Plants and Planting Work (Vegetationstechnik im Landschaftsbau - Pflanzen und Pflanzarbeiten).

European Environment Agency. (2011). *Green infrastructure and territorial cohesion* (EEA Technical report, No 18/2011).

Farrokhirad, E., Rigillo, M., Köhler, M., & Perini, K. (2024). Optimising vertical greening systems for sustainability: An integrated design approach. *International Journal of Sustainable Energy, 43*(1), 2411831.

FLL:2018, Facade greening guidelines—Guidelines for the planning, construction and maintenance of greening façade (Fassadenbegrünungsrichtlinien - Richtlinien für die Planung, Bau und Instandhaltung von Fassadenbegrünungen).

Huang, Z., Tan, C. L., Lu, Y., & Wong, N. H. (2021). Holistic analysis and prediction of life cycle cost for vertical greenery systems in Singapore. *Building and Environment, 196*(November 2020), 107735.

Khan, I. H., & Munawer, T. (2024). A systematic review of economic sustainability of vertical greenery systems, buildings. *Construction Economics and Building, 24*(1–2), 119–143. https://doi.org/10.5130/AJCEB.v24i1/2.8710

Lafortezza, R., & Sanesi, G. (2019). Nature-based solutions: Settling the issue of sustainable urbanization. *Environment Research, 172*, 394–398.

Lo Faro, A., Moschella, A., Sciuto, G., Lombardo, G., Nocera, F., & Salemi, A. (2022). The building stock rehabilitation: The contribution of Vertical Greenery Systems (VGS). In *Rehabend 2022, 9th Euro-American Congress on Construction Pathology, Rehabilitation technology and heritage management*, Circulo Rojo, Granada (pp. 1476–1484).

Moschella, A., Lo Faro, A., Lombardo, G., Sciuto, G., Sturiale, L., & Trovato, M. R. (2025). Vertical greenery systems for building stocks: Application of type solutions and evaluation models. In R. Albatici et al. (Eds.), *Envisioning the futures designing and building for people and the environment* (Vol. 2, pp. 547–569). Springer Cham.

Oquendo Di Cosola, V., Olivieri, F., & Ruiz García, L. (2022). A systematic review of the impact of green walls on urban comfort: Temperature reduction and noise attenuation. *Renewable and Sustainable Energy Reviews, 162*, 112463.

Rosasco, P., & Perini, K. (2018). Evaluating the economic sustainability of a vertical greening system: A cost-benefit analysis of a pilot project in Mediterranean area. *Building and Environment, 142*, 524–533.

GPSR Compliance

The European Union's (EU) General Product Safety Regulation (GPSR) is a set of rules that requires consumer products to be safe and our obligations to ensure this.

If you have any concerns about our products, you can contact us on ProductSafety@springernature.com

In case Publisher is established outside the EU, the EU authorized representative is:

Springer Nature Customer Service Center GmbH
Europaplatz 3
69115 Heidelberg, Germany

Batch number: 10920814

Printed by Printforce, the Netherlands